Nanotechnology Cookbook

Nanotechnology Cookbook

Practical, Reliable and Jargon-free Experimental Procedures

Andrew M. Collins
School of Chemistry, University of Bristol

AMSTERDAM • BOSTON • HEIDELBERG • LONDON • NEW YORK • OXFORD
PARIS • SAN DIEGO • SAN FRANCISCO • SINGAPORE • SYDNEY • TOKYO

Elsevier
The Boulevard, Langford Lane, Kidlington, Oxford OX5 1GB, UK
Radarweg 29, PO Box 211, 1000 AE Amsterdam, The Netherlands

First edition 2012

Notice
No responsibility is assumed by the publisher for any injury and/or damage to persons or property as a matter of products liability, negligence or otherwise, or from any use or operation of any methods, products, instructions or ideas contained in the material herein. Because of rapid advances in the medical sciences, in particular, independent verification of diagnoses and drug dosages should be made

British Library Cataloguing in Publication Data
A catalogue record for this book is available from the British Library

Library of Congress Cataloging-in-Publication Data
A catalog record for this book is available from the Library of Congress

ISBN: 978-0-08-097172-8

For information on all Elsevier Publications visit
our Web site: www.store.elsevier.com

Working together to grow
libraries in developing countries

www.elsevier.com | www.bookaid.org | www.sabre.org

ELSEVIER   BOOK AID International   Sabre Foundation

Contents

A companion site for the book is available at: http://www.elsevierdirect.com/companion.jsp?ISBN=9780080971728

Acknowledgements

A great many have contributed time and expertise to this project. Since I get a single page to thank everybody involved, I have instead opted to scatter the names of those patient and wonderful contributors throughout this book. I could not have done this without them. In particular the Bristol Centre for Functional Nanomaterials within the Nanoscience and Quantum Information Centre have been an invaluable source of knowledge and have helped me at every step.

I would like to thank the Elsevier team who have worked hard to make this book a physical reality.

Thanks to Stephen Mann for getting me into materials chemistry and nanotechnology and to his group who helped me at every opportunity.

Thanks to Simon Hall without whom this book would simply not have happened.

Thanks to Jon Jones and Sean Davis for their continued patience during the period I was writing.

Finally a big thanks to my Mum, Dad, brother Steve and my wife Gait who got me through a very long and busy year with their loving support.

Chapter 1

Introduction

I am going to avoid beginning this book with a dry definition of nanotechnology in favour of conveying my excitement about the potential advances this field of study can provide. Nanoscale science and technology draws from a multitude of research interests and is not confined to any individual scientific discipline. The multi-platform approach to engineering at length that scales somewhere between the thickness of a hair down to a few atoms across requires a harmonious and collaborative approach from the 'pure' sciences (physics, biology and chemistry) and often this results in new ideas and avenues of investigation for the enquiring mind. The capacity to manipulate and control matter in ultra-small dimensions is more than just a technical challenge to be overcome for its own sake. Materials can begin to exhibit new properties when fashioned on the nanoscale which are unseen in the larger everyday bulk. Gaining new functionality from existing materials simply by virtue of the way in which they are structured is highly attractive and nanoscience opens up the opportunity to tailor-make a physical material through the hierarchy of atomic, chemical, crystal, micro, meso and macro composition. The terms 'top down' (starting with something large and making it smaller) and 'bottom up' (starting with the raw chemical, element or atom and working up from there) are often used as catch all terms for describing the manufacturing process of nanomaterials, but these phrases do not do justice to the plethora of cross-disciplinary techniques required to produce various nanostructured materials. It would not be correct to say that nanotechnology is a recent development. Products that utilise nanomaterials such as microchips, solar cells, sporting goods, clothes and medicines have already been available for decades in various guises. It would be more fitting to say that the *concept* of nanotechnology has recently risen to prominence in terms of academic development and public interest. Industry is investing heavily in this area with particular emphasis on electronic goods, information technology, materials, power storage, catalysis and energy generation. It is important to remember that nanotechnology is not limited to extremely small items, and many everyday scale items incorporate nanomaterials as part of a greater functional whole. Nanotechnology holds promise for major advances in drug delivery and the diagnosis and treatment of disease. In parallel with the purely scientific and commercial progress, the impact that nano-enabled products and medicines have on the environment and on our own bodies is being monitored so that the transition from the laboratory to the world at large can be made safely. To develop and study nanomaterials requires an interdisciplinary approach and many universities all over the world now run dedicated courses to train the next generation of nanoscale scientists.

Starting out in my own research on the nanoscale, I often found that I had to jump out of my role as a chemist and into the role of a biologist or physicist to get things done on a practical level. After speaking to other colleagues and students, I quickly discovered that I was not alone in having to get to grips with an alien technique or experimental procedure to compliment my practical work. Although learning new experimental approaches is fun, it can take precious time to come up to speed. Having someone by your side who knows what they are doing is the best way to learn something for the first time (for me at least), but this option is not always available. Similarly, a clear and well-written experimental section can be followed, but these often assume prior knowledge and it can be easy to omit a simple detail that could make all the difference to a given procedure working out as you hope it will. Personally, I always wanted a book at hand that would spell everything out for me step by step without having to dig through papers or knock on doors. Something like the teaching manual books that I followed as an undergraduate which broke each stage of a preparation down and explained why it was done. When I met Graham, the Elsevier acquisitions editor, in Texas two years ago, we discussed the possibility of a 'Cookbook' for nanomaterials and from then on I began to collect procedures and began to put this book together. Funnily enough, putting the book together involved large amounts of digging through papers and knocking on doors but then it is my hope that you, the reader, won't have to do as much of this yourself.

In this book I aim to provide you with some practical techniques drawn from the realms of physics, biology and chemistry as starting points for incorporating various aspects of nanotechnology, that you may otherwise be unfamiliar with, into your own research. I have tried to maintain a teaching manual approach to every synthesis in order to maintain clarity. In contrast to many textbooks, I have used a minimal approach to references to encourage a fresh direction of thought in how you move forward with your work. Each of the synthetic methods listed as recipes within this cookbook have been sorted into sections based on the classical discipline from which they are derived. This is done only to give you a frame of reference for the kind of scientific setting in which you will most likely be attempting the experiment and does not infer that one type of procedure cannot be integrated with another. The hybridisation of scientific approaches can lead to discovery and I hope that you find this to be the case when undertaking any of the recipes outlined herein.

Andrew M. Collins
January 2012

A Quick Definition

A nanometre is 1×10^{-9} m. To put this in perspective, the distance of a carbon–hydrogen bond is around 1 Å (or ~1×10^{-10} m or ~100 pm). In broad terms, nanotechnology deals with the manipulation, fabrication and application of materials having at least one spatial dimension less than 100 nm (or 0.1 μm) in size.

In all recipes the term 'water' is used to indicate deionised and distilled pure water unless otherwise stated.

Chapter 2

Safety

This is the most important section of the book.

Many of the experiments in this book deal with chemical and biological hazards and hazardous reactions. If performed incorrectly or without adequate safety procedures and equipment, there is a real danger that the experimentalist or others around them will be injured or killed. The products of these experiments may be, in and of themselves, toxic and/or hazardous and should be contained or disposed of correctly after use.

The above statement may read as sensationalist scaremongering but is absolutely true. Accidents in the laboratory happen all the time (there are weekly bulletins posted by the American Chemical Society's Chemical and Engineering News). Even when everything appears to be correct with a reaction, there can still be factors beyond your control that result in dangerous problems. The reader should make their safety and the safety of others the main priority when undertaking any work. In this section we will explore some of the considerations surrounding the safety aspect of working with nanomaterials and some of the new protocols to help minimise the potential risks of working with them. Most of this section covers general best practice procedures and the reader should always default to local laboratory practices and regulations. Nanoscopic materials can behave in radically different ways from the bulk counterpart. This means that even materials considered Generally Recognised as Safe (GRAS status) in bulk cannot be considered so after formulation on the nanometre scale. Currently, no separate risk classification exists for nanomaterials and they are treated with the same risk management and disposal procedures that are in place for chemical and biological hazards. The field of nanotoxicology continues to grow and preliminary research shows that nanomaterials do behave in unique ways biologically. This is of some concern as products containing nanomaterials have already been in circulation for some years and the amount produced grows year on year. Some discussion of this is provided at the end of the section. Overall, working with nanomaterials means that you will be responsible for safely producing, handling and disposing of them with control of the associated immediate and long-term risks. This ethos of responsibility is encompassed by the phrase 'from the cradle, to the grave'.

GENERAL LABORATORY PROCEDURE

All laboratories should have a health and safety framework in place. This includes a hierarchy of people responsible for managing safe laboratory practice and its implementation. People trained as emergency first aiders should be on-site with a contact number on visible display in the laboratory. For every reaction you do, you should perform a risk assessment on it before starting. This is not only a legal requirement in most countries but will also allow you to think out the experiment and spot any potential accidents before they happen. A procedure for this is provided further on in this section. You should be aware of the following in any chemical or biological laboratory:

- Fire plan and fire extinguishers. In chemical laboratories, foam or carbon dioxide extinguishers are used. Water extinguishers are normally avoided as some reagents will catch fire on contact with water.
- The contact point for first aid and the location of the first aid kit.
- The emergency shower and emergency eyewash.
- Chemical spill kits. These kits are usually a large bucket of inert adsorbent granules. In the event of a spill, the granules can be used to soak up the liquid and collected to be disposed of.
- Reagent lists with the location of the chemicals or general reagent class listed on a map. This makes things easier to find day to day and also helps to keep track of dangerous and hazardous materials.
- A secure area for solvent storage (if solvents are used) that is adequately vented. Solvents are normally stored in cabinets built into fume hoods for this purpose.
- Solid and solvent waste disposal receptacles or procedures. Private companies will deal with the accumulated waste in the correct way. With the possible exception of washing glassware with water almost everything produced in a laboratory cannot and should not be poured down the sink.

Nanotechnology Cookbook. DOI: 10.1016/B978-0-08-097172-8.00002-3

- Dedicated glass bins and sharps bins. Both glass and needle waste cannot be mixed with regular waste for any reason.
- Never work alone in a laboratory. Some places have out of hours procedures, but these will always rely on having someone in the same room as you to summon aid in case of an accident.
- Hands must be washed before leaving the laboratory. Even if you have been wearing gloves.

Overall you should never tolerate unsafe practice in a laboratory. This may seem obvious, but it is easy for people to become complacent when they get comfortable with their work. One of the most effective methods of making the laboratory environment less accident-prone is to engender a culture of intolerance to lazy or unsafe practice. From a social point of view, people can sometimes be uncomfortable confronting others about this, but remember, the life you save may be your own. I have heard of a student who was too afraid of confronting a senior researcher about their use of hydrofluoric acid open on the bench instead of a fume hood. The matter was settled by an anonymous note delivered to an even more senior member of staff who then arranged to catch the offender red-handed in a surprise laboratory visit.

Even when every safety procedure is followed as carefully as possible, the unexpected can still happen. Below is a photo of a fume cupboard where a reaction under reflux suddenly 'ran away'. The uncontrolled reaction overpressured the glassware and released that pressure with enough force to shatter the front of the fume hood. In this case no one was hurt but it demonstrates that even with safeguards in place and an approved assessment of the risk, it can all go wrong. After a reaction has been safely conducted, you must still control where your products eventually end up. The photograph in fig. 2.1b shows a bin into which 'made safe' chemical waste had been deposited but unexpectedly caught fire overnight. Although independently the waste was benign, a chance combination must have spontaneously ignited. Once more, this type of accident is almost impossible to predict but does help to refine the procedures that will prevent incidents like it from happening again (Fig. 2.1).

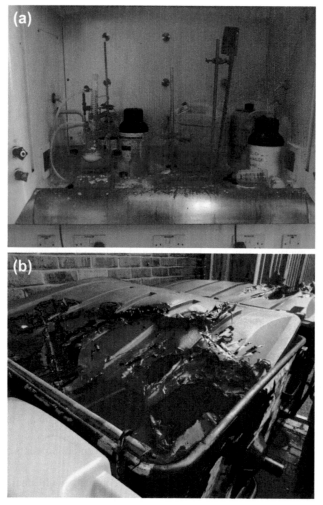

FIGURE 2.1   (a) A fume hood that has suffered catastrophic damage after a reaction has gone out of control and blown up. (b) The aftermath of a bin fire ignited by incompatible chemical waste.

PERSONAL SAFETY EQUIPMENT

There are some items that should be with you at all times when you are in a laboratory. A barrier in some form should be between you and whatever you are working with. Gloves and laboratory coats should remain in laboratory areas and never transferred into offices or group spaces:

- Safety spectacles. *This is the number one item of equipment to protect you from losing your eyesight.* Whenever you are in a laboratory, even if you are not working directly at a bench you should keep safety spectacles on at all times. Explosions in other areas of the laboratory can propel glass and other chemicals at high velocity and spectacles are the only thing that will save your sight. Something nasty could get in your eye at any time. Someone could drop a beaker of acid right next to you and it could splash in your face. You should feel naked without a pair on your face and resist the urge to place them on your head when working close up. Speaking as a former laboratory demonstrator, people will often take them off without thinking. You must notify the owner to immediately put them back on. Confront anyone not wearing safety spectacles in a laboratory and never let someone say 'it's OK, I'm not working with anything dangerous'.
- A laboratory coat. This is the first defence you have against spills on your clothes and can be quickly removed if it catches fire. Laboratory coats can be washed by specialist companies and should not be washed with normal clothes under any circumstances.
- Gloves. There are many different types of disposable gloves and, depending on the chemicals they will be exposed to, not all are suitable for all purposes. *You must check that the gloves you are using are compatible with your experiment before starting!* Nitrile gloves are commonly used around both chemical and biochemistry laboratories, but it is important to remember that these can be degraded by some chemicals and may offer only a partial barrier to the skin. Latex gloves used to be common but are no longer used as people can become sensitised to using them and they can cause a serious allergic response. A chart grading the protection offered from four common types of gloves against various solvents is provided below (Sourced from Ansell).

Chemical	Nitrile	Neoprene	Poly(vinyl chloride)	Natural rubber
Acetic acid (glacial)	G	E	F	E
Acetone	Nr	G	Nr	E
Ammonium hydroxide	E	E	E	E
Aqua regia	F	G	G	G
Benzene	P	Nr	Nr	Nr
Butanol	E	E	F	E
Chloroform	Nr	Nr	Nr	Nr
Dichloroethane	F	P	P	Nr
Dimethyl formamide	Nr	G	Nr	E
Dimethyl sulfoxide	E	E	Nr	E
Ethyl acetate	Nr	F	Nr	G
Ethyl ether	E	E	Nr	Nr
Formaldehyde	E	E	E	E
Hexane	E	E	Nr	Nr
Hydrochloric acid (10%)	E	E	E	E
Hydrofluoric acid (48%)	E	E	G	G
Hydrogen peroxide (30%)	E	E	E	E
Isopropanol	E	E	G	E
Methanol	E	E	G	E
Methyl methacrylate	P	N	Nr	P
Methylene chloride	Nr	Nr	Nr	Nr
Nitric acid (70%)	Nr	G	F	Nr
Octanol	E	E	F	E
Phosphoric acid	E	E	G	G
Sodium hydroxide (50%)	E	E	G	E
Sulphuric acid (10%)	G	E	G	E
Sulphuric acid (95%)	Nr	F	G	Nr
Tetrahydrofuran	Nr	Nr	Nr	Nr
Toluene	F	Nr	Nr	Nr
Trichloroethylene	Nr	Nr	Nr	Nr
Triethanolamine	E	E	E	G

Risk Assessment

Most reagents and solvents used in a chemistry laboratory are in some way harmful. In a biological laboratory, there is the added risk of being exposed to an infection. A risk assessment must be performed before starting any new experimental procedure and checked with the person responsible for laboratory safety. Every chemical is labelled in some way to denote any potential hazards associated with it. All chemicals will also come with a Materials Safety Data Sheet (an MSDS) which will detail the associated hazards more explicitly. If you do not have the MSDS sheet, then these can be downloaded or obtained directly from the supplier. As of 2010, the European legislation regarding the labelling of chemicals has changed so that a globally standardised system of warning labels and hazard phrases is used (Health and Safety Executive, UK). This system is called Classification, Labels and Packaging (CLP) and replaces the older system of Chemicals Hazard Information and Packaging (CHIP). The CHIP system of classification is still more widely used and consists of Risk phrases and Prevention phrases which will be listed as R- and P-, respectively, followed by a number. The number relates to a particular hazard which can be read from a chart as given below. The new CLP system is similar but used the terms Hazard phrase and Safety phrase listed as H- and S-, respectively, followed by a number. The newer global hazard symbols under CLP classification are also included in the table.

These hazard/risk phrases will help you assess how dangerous it is to use a particular reagent and how to go about dealing with the risk it presents. When performing a risk assessment, write out the reaction you intend to do and list all of the chemicals involved. Check their risk phrases as given on the warning label (if any) and MSDS. List any potential hazards involved and how you will deal with them on a practical level. Also check the literature to see if any of the chemical combinations being used could fume, explode or do anything other than the expected reaction. A good place to start is to check in regularly updated handbooks like Brethericks 'Hazards in the Chemistry Laboratory' (Urben, 2006) and 'Hazardous Chemicals Handbook' (Carson and Mumford, 2002). Many of these risk phrases will denote a substance dangerous enough that you will need to perform a special COSHH (Control of Substances Hazardous to Health) assessment and get clearance from your safety officer before proceeding. *Include a procedure for what you will do if it becomes a worst case scenario (e.g., the reaction catches fire).* A copy of the risk assessment should be kept with your safety officer and another copy should be placed in your laboratory book as a reference though it is also a legal requirement in most laboratories.

Symbol	Abbreviation	Hazard	Description
Explosive	E	Explosive	Chemicals that explode.
Oxidising	O	Oxidising	Chemicals that react exothermically with other chemicals.
Flammable	F	Highly flammable	Chemicals that may catch fire in contact with air, only need brief contact with an ignition source, have a very low flash point or evolve highly flammable gases in contact with water.
	F+	Extremely flammable	Chemicals that have an extremely low flash point and boiling point, and gases that catch fire in contact with air.
	T+	Very toxic	Chemicals that at very low levels cause damage to health.

Symbol	Abbreviation	Hazard	Description
Toxic	T	Toxic	Chemicals that at low levels cause damage to health.
	Carc Cat 1 & 2	Category 1 or 2 carcinogen	Chemicals that may cause cancer or increase its incidence.
	Muta Cat 1 &2	Category 1 or 2 mutagen	Chemicals that induce heritable genetic defects or increase their incidence.
	Repr Cat 1 & 2	Category 1 or 2 reproductive toxin	Chemicals that produce or increase the incidence of non-heritable effects in progeny and/or an impairment in reproductive functions or capacity.
Corrosive	C	Corrosive	Chemicals that may destroy living tissue on contact.
Harmful	Carc Cat 3	Category 3 carcinogen	Chemicals that may cause cancer or increase its incidence.
	Muta Cat 3	Category 3 mutagen	Chemicals that induce heritable genetic defects or increase their incidence.
	Repr Cat 3	Category 3 reproductive toxin	Chemicals that produce or increase the incidence of non-heritable effects in progeny and/or an impairment in reproductive functions or capacity.
	Xn	Harmful	Chemicals that may cause damage to health.
	Xi	Irritant	Chemicals that may cause inflammation to the skin or other mucous membranes.
Dangerous for environment	N	Dangerous for the environment	Chemicals that may present an immediate or delayed danger to one or more components of the environment.
Compressed gas		Compressed gas	Contains gas under pressure.
Irritant		Less serious harmful substances, irritants	Applies to many harmful substances normally marked with an X symbol.
Potential long term health effects		Long-term damage	Includes respiratory sensitisers and carcinogens.

R1	Explosive when dry
R2	Risk of explosion by shock, friction, fire or other sources of ignition
R3	Extreme risk of explosion by shock, friction, fire or other sources of ignition
R4	Forms very sensitive, explosive metallic compounds
R5	Heating may cause an explosion
R6	Explosive with or without contact with air
R7	May cause fire
R8	Contact with combustible material may cause fire
R9	Explosive when mixed with combustible material
R10	Flammable
R11	Highly flammable
R12	Extremely flammable
R14	Reacts violently with water
R14/15	Reacts violently with water, liberating extremely flammable gases
R15	Contact with water liberates extremely flammable gases
R15/29	Contact with water liberates toxic, extremely flammable gases
R16	Explosive when mixed with oxidising substances
R17	Spontaneously flammable in air
R18	In use, may form flammable/explosive vapour—air mixture
R19	May form explosive peroxides
R20	Harmful by inhalation
R20/21	Harmful by inhalation and in contact with skin
R20/21/22	Harmful by inhalation, in contact with skin and if swallowed
R20/22	Harmful by inhalation and if swallowed
R21	Harmful in contact with skin
R21/22	Harmful in contact with skin and if swallowed
R22	Harmful if swallowed
R23	Toxic by inhalation
R23/24	Toxic by inhalation and in contact with skin
R23/24/25	Toxic by inhalation, in contact with skin and if swallowed
R23/25	Toxic by inhalation and if swallowed
R24	Toxic in contact with skin
R24/25	Toxic in contact with skin and if swallowed
R25	Toxic if swallowed
R26	Very toxic by inhalation
R26/27	Very toxic by inhalation and in contact with skin
R26/27/28	Very toxic by inhalation, in contact with skin and if swallowed
R26/28	Very toxic by inhalation and if swallowed
R27	Very toxic in contact with skin
R27/28	Very toxic in contact with skin and if swallowed
R28	Very toxic if swallowed
R29	Contact with water liberates toxic gas
R30	Can become highly flammable in use
R31	Contact with acids liberates toxic gas
R32	Contact with acids liberates very toxic gas
R33	Danger of cumulative effects
R34	Causes burns
R35	Causes severe burns
R36	Irritating to eyes
R36/37	Irritating to eyes and respiratory system
R36/37/38	Irritating to eyes, respiratory system and skin
R36/38	Irritating to eyes and skin
R37	Irritating to respiratory system
R37/38	Irritating to respiratory system and skin
R38	Irritating to skin

R39	Danger of very serious irreversible effects
R39/23	Toxic: danger of very serious irreversible effects through inhalation
R39/23/24	Toxic: danger of very serious irreversible effects through inhalation and in contact with skin
R39/23/24/25	Toxic: danger of very serious irreversible effects through inhalation, in contact with skin and if swallowed
R39/23/25	Toxic: danger of very serious irreversible effects through inhalation and if swallowed
R39/24	Toxic: danger of very serious irreversible effects in contact with skin
R39/24/25	Toxic: danger of very serious irreversible effects in contact with skin and if swallowed
R39/25	Toxic: danger of very serious irreversible effects if swallowed
R39/26	Very Toxic: danger of very serious irreversible effects through inhalation
R39/26/27	Very Toxic: danger of very serious irreversible effects through inhalation and in contact with skin
R39/26/27/28	Very Toxic: danger of very serious irreversible effects through inhalation, in contact with skin and if swallowed
R39/26/28	Very Toxic: danger of very serious irreversible effects through inhalation and if swallowed
R39/27	Very Toxic: danger of very serious irreversible effects in contact with skin
R39/27/28	Very Toxic: danger of very serious irreversible effects in contact with skin and if swallowed
R39/28	Very Toxic: danger of very serious irreversible effects if swallowed
R40	Limited evidence of a carcinogenic effect
R41	Risk of serious damage to eyes
R42	May cause sensitisation by inhalation
R43	May cause sensitisation by skin contact
R42/43	May cause sensitisation by inhalation and skin contact
R44	Risk of explosion if heated under confinement
R45	May cause cancer
R46	May cause heritable genetic damage
R48	Danger of serious damage to health by prolonged exposure
R48/20	Harmful: danger of serious damage to health by prolonged exposure through inhalation
R48/20/21	Harmful: danger of serious damage to health by prolonged exposure through inhalation and in contact with skin
R48/20/21/22	Harmful: danger of serious damage to health by prolonged exposure through inhalation, in contact with skin and if swallowed
R48/20/22	Harmful: danger of serious damage to health by prolonged exposure through inhalation and if swallowed
R48/21	Harmful: danger of serious damage to health by prolonged exposure in contact with skin
R48/21/22	Harmful: danger of serious damage to health by prolonged exposure in contact with skin and if swallowed
R48/22	Harmful: danger of serious damage to health by prolonged exposure if swallowed
R48/23	Toxic: danger of serious damage to health by prolonged exposure through inhalation
R48/23/24	Toxic: danger of serious damage to health by prolonged exposure through inhalation and in contact with skin
R48/23/24/25	Toxic: danger of serious damage to health by prolonged exposure through inhalation, in contact with skin and if swallowed
R48/23/25	Toxic: danger of serious damage to health by prolonged exposure through inhalation and if swallowed
R48/24	Toxic: danger of serious damage to health by prolonged exposure in contact with skin
R48/24/25	Toxic: danger of serious damage to health by prolonged exposure in contact with skin and if swallowed
R48/25	Toxic: danger of serious damage to health by prolonged exposure if swallowed
R49	May cause cancer by inhalation
R50	Very toxic to aquatic organisms
R50/53	Very toxic to aquatic organisms, may cause long-term adverse effects in the aquatic environment
R51	Toxic to aquatic organisms
R51/53	Toxic to aquatic organisms, may cause long-term adverse effects in the aquatic environment
R52	Harmful to aquatic organisms
R52/53	Harmful to aquatic organisms, may cause long-term adverse effects in the aquatic environment
R53	May cause long-term adverse effects in the aquatic environment
R54	Toxic to flora
R55	Toxic to fauna
R56	Toxic to soil organisms
R57	Toxic to bees
R58	May cause long-term adverse effects in the environment
R59	Dangerous for the ozone layer

(*Continued*)

R60	May impair fertility
R61	May cause harm to the unborn child
R62	Possible risk of impaired fertility
R63	Possible risk of harm to the unborn child
R64	May cause harm to breast-fed babies
R65	Harmful: may cause lung damage if swallowed
R66	Repeated exposure may cause skin dryness or cracking
R67	Vapours may cause drowsiness and dizziness
R68	Possible risk of irreversible effects
R68/20	Harmful: possible risk of irreversible effects through inhalation
R68/20/21	Harmful: possible risk of irreversible effects through inhalation and in contact with skin
R68/20/21/22	Harmful: possible risk of irreversible effects through inhalation, in contact with skin and if swallowed
R68/20/22	Harmful: possible risk of irreversible effects through inhalation and if swallowed
R68/21	Harmful: possible risk of irreversible effects in contact with skin
R68/21/22	Harmful: possible risk of irreversible effects in contact with skin and if swallowed
R68/22	Harmful: possible risk of irreversible effects if swallowed

Special Considerations for Chemical and Physical Laboratories

- When working in a chemistry laboratory assume that nothing is safe to touch without gloves on.
- Most experiments will be conducted inside a fume hood so that the risk of breathing in fumes or particulates is minimised. Ensure the fume hood you are working at is clearly labelled with your experiment and check with others who may use it that their experiments do not introduce risk or vice versa. It is not unknown for people to unknowingly use open flames next to beakers full of solvents. Many chemicals require special storage conditions under an inert gas or else require storage in a desiccator. Make sure chemicals are returned to their proper places after use.
- Sometimes reactions may need to be left running overnight. In this circumstance you must check that everything has been made failsafe as best as it can be before leaving it. A safety assessment must be performed before hand.
- Solvent waste is collected in marked containers and as much as possible similar solvents or type of waste should be kept together and not mixed. Solvent wash bottles of distilled water, methanol, ethanol, isopropanol and acetone are normally on hand. It is important that the solvents contained within the bottles do not get mixed up and are clearly marked. Acetone will dissolve many types of plastic and must not be poured down the sink. Have a dedicated solvent waste container for acetone.
- Lasers should be screened and shielded. Watches, computer screens and any similar reflective surface should be removed from the vicinity of the experiment involving the laser. Laser blocking eyewear should also be used as appropriate. Laser laboratories will have controls in place to prevent people walking in at the wrong time including lockable doors and 'laser in use' signs.

A Note on Acids and Bases
Chemical laboratories are the most likely environments to encounter strong acids and bases. Concentrated acids are highly corrosive and have a tendency to fume. They should be worked with exclusively in a fume cupboard. When diluting a strong acid solution, particularly H_2SO_4, the *acid should be added into the water* and not the other way around. The disassociation of the acid in the water is exothermic and if water is added into the acid then the solution can spit or splash. Some of the recipes in this book involve the use of hydrofluoric acid (HF) and this acid is extremely unpleasant. It rapidly destroys tissues and penetrates deep within the skin where it can also decalcify bone. Special measures are required for its use and these include the presence of a calcium gluconate gel when working with HF. This can be applied to areas of skin that have come into contact with HF to form an insoluble calcium fluoride. However, any casualty exposed to HF must be taken immediately to hospital. Wherever possible, avoid working with it.

Strong basic solutions are also corrosive to exposed tissue but you will not necessarily be able to feel it happening. Basic solutions attack the nerves first so that you will not be able to feel any pain. Areas of skin exposed to strongly basic solutions will feel slippery like a soap solution has been applied. This is in fact exactly what is happening, as the strongly basic solution will convert fatty tissues into surfactant molecules. Areas exposed to base should be rinsed with copious amounts of water immediately.

Special Considerations for Biological Laboratories

The risk an infectious agent presents falls into one of four biosafety levels with level one being relatively harmless up to level four which is the most dangerous (Health and Safety Executive, 2005). The recipes in this book do not go beyond biosafety level two and will involve genetic modifications of non-infectious organisms of minimal toxicity. Should you find yourself working with

FIGURE 2.2   The symbol for biohazard.

infectious agents ensure that appropriate vaccinations and medical consultation have been provided to you before starting work by the person responsible for laboratory safety. All biosafety laboratories and equipment must be labelled with a biohazard warning symbol (Fig. 2.2).

Level 1: Biological or biologically derived materials or agents that are not known to cause harm in humans. Examples of level 1 biohazards include commercially produced bacteriophage DNA, *Nitzschia* (diatom) cultures from a natural source which are non-toxic and non-infectious, a fibroblast cell line which is non-infectious, yeast cells, some strains of *Esherichia coli*, and *Bacillus subtilis.*

Level 2: Biological or biologically derived materials or infectious agents that can cause mild, non−life threatening illness or symptoms in humans. Level 2 biohazards are not transmissible by aerosol. The scope of this book does not cover working with infectious agents such as lyme disease, salmonella and some types of hepatitis though these are worked within level 2 biosafety laboratories. However, any genetically modified organism or organism that is to be modified in an experiment comes under level 2 classification. This includes the sections of this book concerned with the genetic modification of yeast and *E. coli* cells. Any experiment involving genetic modification will require a special risk assessment and any modified organism must be contained and/or destroyed within the laboratory facilities.

Maintaining the sterility of your experiment is the key to getting good results. When cultures have to be worked with outside an incubator or storage there is a risk they may become contaminated by foreign dust, spores, virus or bacteria. To prevent this, most biological work involving cultures is performed in a laminar flow cabinet. This type of cabinet draws in air through a high efficiency particle air (HEPA) filter and directs it in a steady stream out of the front of the hood towards the user. HEPA filters will remove any particles in the air over 300 nm. Inside the cabinet there are no sharp corners or joints so that surfaces can be kept clean. A laminar flow hood is pictured below. In this case a burner has been placed in the cabinet to prevent contamination of open cultures. The volumetric pipettes are dedicated to this particular hood as is the waste bin located in the corner (Fig. 2.3).

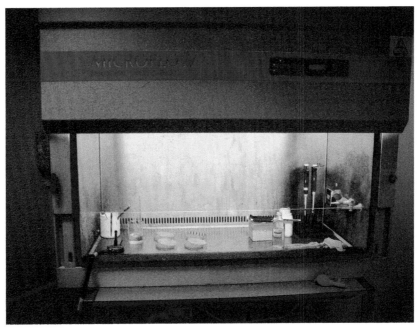

FIGURE 2.3   A laminar flow cabinet used for yeast transformation.

Biosafety Level 2 Procedures

- Hand washing before and after leaving the laboratory is doubly important when working in a biolaboratory. Infections can come in with you as well as make it out back with you.
- Laboratories should have controlled access so that only qualified people can gain access.
- All glassware and equipment such as spatulas need to be autoclaved before use. Autoclave tape is normally affixed to the outside of glassware or a paper container with metal items within. This tape has stripes which appear black after autoclaving so you know that the item it is attached to has been sterilised.
- All surfaces must be wiped down with fresh tissue and a 70% ethanol solution before each use. Surfaces in a biolaboratory should be impermeable to liquids and non-porous. Never trust anyone else to have wiped a surface down. Many laboratories have spray bottles of this solution on each bench. Many biolaboratories use open flames for sterility control. *Ensure ethanol is never used near an active flame as this solution is flammable.*
- A 10% sodium hypochlorite solution should be on hand for decontaminating used tips and glassware with. This is normally maintained in a large vessel into which items are placed and left for a few hours or overnight. Take care to ensure that items are placed slowly into the bleach as glass will sometimes break in them and you could cut yourself. Bleach waste can be diluted and disposed of in a sink.
- Swabs and dry biohazard waste (pipette tips and small centrifuge tubes) should be disposed of in hard-walled plastic containers ('sweetie jars') or disposable Biobins and disposed of by incineration.
- Bacterial solid waste (used agar gel plates or similar) should be disposed of in a yellow biohazard marked bag lining dedicated yellow plastic biohazard bins. This will also be incinerated.
- All work with liquid should be conducted in trays to catch spillage with. This includes filling tubes for use in a centrifuge or any other operation involving pouring solutions. Bleach or ethanol solution can be added to sterilise trays after use. Minor spills can be mopped up with tissue and bleach or ethanol.
- Waste liquids (used growth medium or similar) that may still contain live organisms need to be autoclaved before they can be disposed of in a sink. Alternatively, bleach can be added to 10% of the total liquid volume and left overnight (Fig. 2.4).

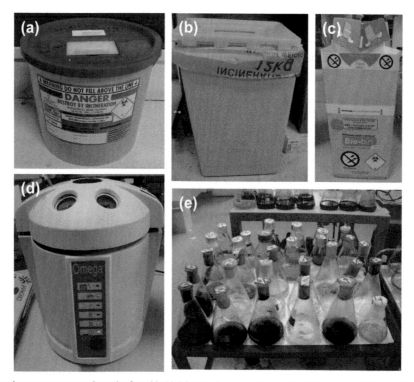

FIGURE 2.4   Photographs of various waste receptacles to be found in biolaboratories. (a) A sharps bin with a special lid for pulling off needles with. (b) A yellow solid biowaste bin. (c) A benchtop Biobin carton for pipette ends and other solid biowaste. (d) An autoclave used for sterilisation. (e) A tray of liquid biowaste ready for disposal. The tape with black lines on it denotes that the waste has been autoclaved and is ready to be poured away.

Special Considerations for Nanomaterials

The field of nanotoxicology has grown from a need to address understandable concerns over how nanotechnology may present a risk to both personal health and the environment (Donaldson et al., 2004; Nel et al., 2006; Weiss and Diabaté, 2011). The unique way in which nanomaterials interact chemically and physically means that working with them entails fresh approaches to handling and disposal (Aitken et al., 2004; DEFRA, 2007; Teeguarden et al., 2007). Their high surface area can mean a higher chemical reactivity and their shape can radically alter how they interact with cells, membranes and barriers. The breakdown products of nanoparticles in the body can also raise concerns. For example, fluorescent cadmium selenide quantum dots are used for biolabelling but have the potential to breakdown into toxic cadmium salts. When quantum dots are used in vivo, the nanoparticles are encapsulated with a bioinert shell which prevents degradation but there is still the risk that, as a particulate material, the quantum dots might accumulate somewhere in tissues. The particles might not be flushed from the system by existing clearance mechanisms only to degrade at a later date. Conversely, apparently benign compounds and elements can become toxic when synthesised as a nanostructure. This type of interaction has been extensively investigated in the case of carbon nanotubes which have been found to present an inhalation risk similar to asbestos. Asbestos is a naturally occurring silicate mineral that crystallises into long, thin fibres. It was used widely as an insulation material in buildings until it was found that people who worked with it were prone to a multitude of serious lung disorders. Ultra-fine asbestos fibres can be inhaled into the lungs where they become lodged. Because silicates are biologically inert, the body is incapable of breaking these fibres down chemically. The anisotropic morphology is also too large for macrophages in the blood to engulf the fibres and transport them away, so they remain in the lungs where they cause damage. Similar results have been observed in animal models exposed to carbon nanotubes and it is highly likely that the direct inhalation of carbon nanotubes can lead to lung complications (Hurt et al., 2006). Carbon nanotubes are already present in many commercially available products and this raises the question of how, or if, these might be released into the environment and what damage they might cause. Nanotechnology is rapidly making the transition from the laboratory and into the marketplace. Antibiotic silver in clothes, light scattering particulates in makeup, titania nanoparticles used in invisible sunscreen, and stronger, lighter construction materials are only a small example of the many products now available with nanotechnology as a key selling point. Currently, the production of an existing chemical compound or element as a new nanomaterial does not require any additional testing in order to be commercialised and released (The Royal Society, 2004). The need for commercial exploitation must be balanced with a thorough investigation of the long-term health risks of nanomaterial-based products. Globally, this will change as the research in the field points towards separate testing and controlling of new nanomaterials produced in the laboratory. Already, intensive research is underway to standardise methods to assess and quantify nanoparticle hazards in line with biological and chemical safety standards (Hallock et al., 2009; Bhabra et al., 2009). For now, a set of best practice procedures should be used to minimise the, possibly unknown, risks that may arise from any nanomaterial you work with. The following list has been adapted from references and the best practice policy used within the Nanoscience and Quantum Information Centre (NSQI) in Bristol.

- Have the material's safety data sheets for the chemicals/materials from which the nanomaterial is composed but be aware that *the nanoscopic product may have a radically different toxicology.*
- Dry nanoparticulate powders present the greatest risk from exposure by inhalation or by skin contact. Avoid contact of powders with exposed skin. Full personal protection equipment should be worn when handling them. The level of coverage should be proportionate to the level of hazard the material may present. It is recommended that eyewear also has side protection. Gloves should be long sleeved and avoid wearing a watch in which particles could accumulate. Used gloves should be disposed of in a sealable bag. A P100 or APF40 rated filter mask should be used wherever there is a risk of inhalation.
- *Do not work with dry nanoscopic powders in a fume hood or a laminar flow cabinet.* Either of these options will blow a fine particulate into the environment externally or into your laboratory area. A glove box free from air movement and static is suitable for handling dry nanopowders and containing them if they become dispersed. Some glove boxes may be fitted with a HEPA filter unit and it is advisable to switch this off. There is a risk that particles smaller than ~300 nm may make it through the filter (Fig. 2.5).
- *There must be no source of ignition within the glove box or near dry nanoparticles.* The increased surface area and reactivity of some nanomaterials means that, if dispersed in the air, they will burn rapidly and explosively.
- Surfaces should be wet cleaned or wiped and the wipes disposed of in a sealable bag should be incinerated. All nanomaterial waste should be treated as potentially hazardous.
- Nanomaterials dispersed in liquids or contained within solid matrices present a lower personal risk though contact with skin or the inhalation of aerosols must be avoided.
- Liquid nanoparticle waste should be centrifuged to separate out solids where possible. The solid fraction should be treated as hazardous solid waste. In cases where the particulate is too small to centrifuge from solution, it can be salted out or chemically converted to a soluble species provided the reclamation methods are deemed safe and appropriate.

FIGURE 2.5   A glove box used for handling dry nanoparticles.

FIGURE 2.6   The nanomaterials hazard sign used in the Nanoscience and Quantum Information Centre at the University of Bristol.

- Nanoparticle solid waste should be sealed in a plastic container, such as a falcon tube, and sealed further with film before being disposed of in a separate nanomaterial specific hazardous waste container. Label containers and keep a record of what has been disposed of and in what volumes.
- Biological waste mixed with nanomaterials should be treated with a 10% sodium hypochlorite solution before disposal. Ensure the bleach will not have an unexpected reaction with the nanomaterial.
- Keep a record of personal exposure. There are no definite guidelines on nanoparticle exposure in the work place and no metric by which nanoparticle exposure is measured. Follow existing chemical and biological standard practice. Principally, keep a good record of what you have been working on within a laboratory book.
- A warning sign, similar to other hazard labels, to let people know they are entering an area in which nanomaterials are made and used. No official sign yet exists but here is a copy of the sign used at the NSQI. It might be slightly unfair to fullerenes to use them as the warning label design. C_{60} is arguably considerably less harmful than carbon nanotubes to human health. However, the fullerene shape is recognisable to many people unfamiliar with the field and set against the standard warning label background it can visually delineate specific nanomaterial risks from warning labels used for chemical or biological hazards (Fig. 2.6).

REFERENCES

Aitken, R.J., Creely, K.S., Tran, C.L., 2004. Nanoparticles: An Occupational Hygiene Review Prepared by the Institute of Occupational Medicine Nanoparticles: An Occupational Hygiene Review.

Bhabra, G., et al., 2009. Nanoparticles can cause DNA damage across a cellular barrier. Nature Nanotechnology 4 (12), 876–883.

Carson, P., Mumford, C., 2002. Hazardous Chemicals Handbook, Elsevier.

DEFRA, 2007. Characterising the Potential Risks Posed by Engineered Nanoparticles: Second UK Government Research Report. DEFRA Government Research Report.

Donaldson, K., et al., 2004. Nanotoxicology. Occupational and Environmental Medicine 61 (9), 727–728.

Hallock, M.F., et al., 2009. Potential risks of nanomaterials and how to safely handle materials of uncertain toxicity. Journal of Chemical Health and Safety 16 (1), 16–23.

Health and Safety Executive, 2005. Biological Agents: Managing the Risks in Laboratories and Healthcare Premises.

Hurt, R., Monthioux, M., Kane, A., 2006. Toxicology of carbon nanomaterials: status, trends, and perspectives on the special issue. Carbon 44 (6), 1028–1033.

Nel, A., et al., 2006. Toxic potential of materials at the nanolevel. Science 311 (5761), 622–627 (New York, N.Y.).

Teeguarden, J.G., et al., 2007. Particokinetics in vitro: dosimetry considerations for in vitro nanoparticle toxicity assessments. Toxicological Sciences: An Official Journal of the Society of Toxicology 95 (2), 300–312.

The Royal Society, 2004. Nanoscience and Nanotechnologies: Opportunities and Uncertainties (July).

Urben, P. (Ed.), 2006. Bretherick's Handbook of Reactive Chemical Hazards, Elsevier.

Weiss, C., Diabaté, S., 2011. A special issue on nanotoxicology. Archives of Toxicology 85 (7), 705–706.

Chapter 3

Common Analytical Techniques for Nanoscale Materials

In this section we will explore some routinely used techniques for the analysis and characterisation of nanomaterials. This is by no means an extensive list and highlights only the most commonly employed and accessible methods of characterisation.

PRINCIPLES OF ELECTRON MICROSCOPY

Adapted from Chescoe and Goodhew (1992)

Electron microscopy is based on the fundamentals of quantum mechanics, that particles can have wave-like properties, and in particular the de Broglie equation that describes a moving particle as having a wavelength:

$$\lambda = \frac{h}{mv}$$

where λ = wavelength in metres, h = Planck's constant and mv = momentum (mass in kilograms times the velocity in metres per second). When a beam of electrons is accelerated through a potential difference, the electrons have a de Broglie wavelength which is much shorter than that of light. Hence the resolution of an image generated from an electron beam can potentially be far higher than that attainable using visible photons. An analogous formula for the wavelength of electrons may be derived by substituting the mass and velocity term for a momentum value derived from electronic charge:

$$\lambda = \frac{h}{(2m_e eV)^{1/2}}$$

where m_e is the rest mass of an electron, e is the electron charge and V is the voltage through which the electron is accelerated. This formula is accurate at low voltages but, for electron microscopy, electrons used for imaging are first accelerated to near relativistic velocities through high voltages. Particles moving at sizable fractions of light speed are better described by equations that incorporate a Lorentz contraction term. This theory is a component of special relativity and its effects become more pronounced as matter approaches the speed of light. If a particle has a length at rest of l_0, then the observed length l' for an observer in a separate frame of reference is given by:

$$l' = \frac{l_0}{\left(\dfrac{1}{\sqrt{1 - \left(\dfrac{v}{C}\right)^2}}\right)}$$

where v is the velocity of the particle and C is the speed of light in metres per second. The electrons used for microscopy experience a Lorentz contraction and so a more exact formula for the wavelength of an electron moving at speed is:

$$\lambda = \frac{h}{\left(2m_e eV\left(1 + \dfrac{eV}{2m_e C^2}\right)\right)}$$

For an electron beam accelerated through a voltage of 120 keV the non-relativistic formula gives an electron wavelength of 0.035 Å. With the Lorentz contraction applied this becomes 0.033 Å. This may seem marginal given that a carbon−carbon bond length is around 1 Å but the difference can alter the outcome of electron diffraction measurements used in electron microscopy for determining crystal structure.

In addition to imaging, an electron beam passing through a sample will cause it to eject X-ray photons of an energy characteristic to the element from which they were emitted. The X-rays are the product of the inelastic scattering of electrons by a material

Nanotechnology Cookbook. DOI: 10.1016/B978-0-08-097172-8.00003-5

whereby core electrons of the surface atoms present can be 'knocked out' generating a vacancy, which is subsequently filled by an electron from an outer shell with energy lost in the form of an X-ray. The X-rays emitted from a certain element have characteristic energies reflecting electronic transitions within the atom and are identified based on the initial vacancy (K is the most inner orbital vacancy followed by L and M) and from where the electron came that filled it ($\alpha = 1$ shell out, $\beta = 2$ shells out and so on). In both transmission and scanning electron microscopy (SEM) this is used to perform energy dispersive X-ray analysis (EDX) which can build up an elemental map of the sample you are looking at. This means that an electron microscope can be used to determine the elemental composition of the sample being imaged.

TRANSMISSION ELECTRON MICROSCOPY

A transmission electron microscope (TEM) accelerates an electron beam through a potential difference of 20−200 KeV. This beam is aligned through a specimen, and where samples are electron dense, less electrons will penetrate. This forms an image of light areas where the electron beam has remained intact and dark areas where electron scattering has occurred. The beam is focussed and aligned by electromagnetic lenses through high vacuum (10^{-5} mm Hg) and although all lenses contribute to the formed image, during operation only the projector and intermediate lens currents are altered. The image focus is maintained by alteration of the focal length of the objective lens (Fig. 3.1).

The theoretical resolving power of the microscope is given by:

$$\text{resolving power} \approx 0.61 \frac{\lambda}{\alpha}$$

where α is the angular aperture of the objective lens and λ is the wavelength.

The resolution is increased at smaller wavelengths (larger accelerating voltages) and as the objective lens is made larger. However, the maximum resolution is not achieved due to lens aberrations which increase as the objective lens is increased. The contrast is increased as the objective lens aperture is decreased due to the improved contribution of non-scattered electrons. This also lowers the resolution, so a compromise between resolution and contrast requirements needs to be made during analysis.

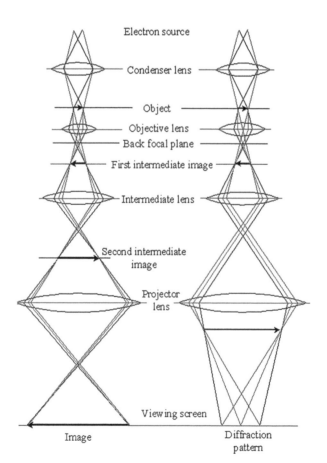

FIGURE 3.1   Diagrammatic representations of a transmission electron microscope (TEM) in both imaging mode (left) and diffraction mode (right). *Adapted with permission from Chescoe and Goodhew (1992). Copyright 1992, Oxford University Press.*

A variety of electron sources are available but the most commonly used is a thermionic emission gun. A current is passed through a tungsten filament and electrons are ejected and accelerated by a potential difference (120 keV) towards an anode. The centre of the anode is hollow and electrons pass through a Wehnelt cap which acts to draw electrons from a specific region of the filament. The electron beam then enters the condenser system of the TEM where the beam is demagnified and the diameter and convergence angle are controlled. In this stage a spot size is selected and an aperture physically restricting the beam may be selected. A smaller spot size equates to higher resolution, and physical restriction of the beam leads to less X-rays being generated by the sample when the beam strikes.

At the objective lens the first image is formed and, similar to traditional light optics, lens aberrations like astigmatism must be corrected. This is accomplished by altering the magnetic field of the objective lens. Further contrast in the image is generated by the insertion of the objective aperture which filters scattered and unscattered electrons. At this point the image is magnified 50−100 times and further magnification is performed using an array of projector lenses before the beam strikes a phosphorous screen to produce an image. The depth of focus is very large, extending over many metres. Due to this, the positioning of photographic equipment in the column is not a critical factor in obtaining a sharp image.

The TEM may also be used as a diffraction chamber in order to elucidate the crystal structure of the sample you are looking at. The intermediate lens is focussed onto the back focal plane of the objective lens. The diffraction pattern which is generated is then magnified by the projector array. Using diffraction mode imaging produces a spot pattern or rings for crystalline materials. In amorphous materials no pattern will be observed. Single crystals generate a spot pattern and polycrystalline materials produce ring patterns. The lattice planes and d-spacings can be worked out from the diffraction pattern by measuring the distance of the dots from the central spot made by the electron beam in the pattern. The projected pattern can be recorded and is related to the height of the diffraction beam, or camera length l by the relationship:

$$rd = \lambda l$$

where r is the distance from the central spot to the diffraction spot, d is the lattice spacing of the crystal, λ is the electron wavelength (0.033 Å at 120 keV) and l is the camera length (normally around 100 cm in most commercially available microscopes). This technique is particularly useful in deriving the crystal structure of nanoparticles or where larger crystal growth is difficult. The area of diffraction may be selected by the insertion of a selected area aperture at the image point of the intermediate lens.

SAMPLE PREPARATION FOR TEM

Most TEMs use small copper grids coated with a thin film of polymer or carbon for imaging. The grids can be purchased with a carbon film on them ready for use or you can try to make your own. Grids are made by using an electric arc to sputter a very thin film of carbon onto a freshly cleaved piece of mica. In the meantime a Büchner funnel is fitted with filter paper and filled with water. The outlet is sealed so that water does not drain out of the funnel and into the flask. Uncoated mesh grids are arranged onto the filter paper underwater ready to be coated. The carbon film can be floated off onto the surface of the water which is then drained slowly so that the carbon film settles on top of the uncoated copper or nickel mesh grids. Making your own can be a time consuming affair and for the novice it is recommended that you initially purchase preformed grids. This way you can embark upon a project without fear of seeing nothing under the microscope due to a broken carbon film. As an example of basic grid preparation we will look at making a TEM grid using a gold colloid. Metal colloids are perfect for imaging under an electron microscope as heavier elements are electron dense and will give a good contrast under the beam. Hold a grid using tweezers then add a droplet of the colloid to the surface of the grid. Leave it for around 30 s then drain the excess fluid away by dabbing the edge of the grid very gently against a piece of filter paper. Then allow the grid to dry in air or resting on a filter paper. If you suspect your sample is too dilute then you could allow the whole droplet to dry onto the surface. Often with a colloid the sample has been stabilised in an excess of surfactant. This can be washed away by applying a droplet of water (or other suitable solvent) to the grid then wicking this away with filter paper (Fig. 3.2).

Biological samples are composed mostly of carbon, nitrogen and oxygen which will have a low contrast under an electron beam. This limitation can be overcome by staining biological samples either positively or negatively. Positive stains infiltrate the biological structure with a heavy metal and will make the sample appear dark against a light background. A negative stain will make the background appear dark so that the sample stands out. Uranyl acetate is a commonly used stain for this. A virus for example, is dispersed into an aqueous solution containing an organic buffer such as HEPES or PIPES. A phosphate buffer would leave salt residues after drying, so it is best to avoid using it for grid preparation. The virus solution is allowed to dry on the grid. A droplet of 1−3% uranyl acetate in water (pH ~4) is then allowed to sit on the grid for 20 min before being wicked off. The uranyl acetate will negatively stain the grid so that the virus appears bright under the microscope. *This is radioactive and must be handled using specialist procedures in a dedicated area.* Alternatively, if the biological sample must be maintained at a neutral pH, a solution of 1% ammonium molybdate in water can be used. The pH is adjusted to 7 before using ammonium hydroxide solution. *Heavy metal stains are toxic and correct safety procedures must be followed when handling them.* Another problem with biological imaging is that often the samples are very delicate. They can deform upon drying and are also sometimes damaged by the intensity of the electron beam during analysis. These problems can be overcome by embedding the sample in a resin and taking thin sections using

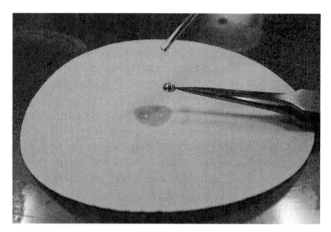

FIGURE 3.2   Photographs showing basic grid preparation of a gold colloid for TEM analysis.

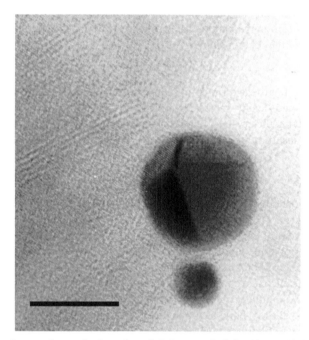

FIGURE 3.3   An example of a transmission electron micrograph taken using a digital camera. Scale bar 10 nm. This is an image of gold nanoparticles formed on the surface of a hydrogen titanate nanorod. The difference in the atomic mass of the elements in the sample gives rise to the contrast so that the heavier metal appears darker. The particle is multiply twinned and it is possible to see the lattice places of the [111] plane in the gold.

a microtome. The thin sections are floated onto a grid for analysis. Sample stability under the beam can be achieved by using cryogenic techniques which freeze the sample during analysis. These techniques are beyond the scope of this overview but the facilities and know how for using them should be on hand with whoever looks after your electron microscopy facilities (Fig. 3.3).

SCANNING ELECTRON MICROSCOPY

This technique gives topographical information about a sample. The electron beam is less powerful than in a TEM and the accelerating potentials are normally operated at 2–50 keV. Higher accelerating voltages penetrate deeper into the surface of a sample and so are used for energy dispersive X-ray analysis (EDX) and high resolution work, some samples may experience charging or damage under these conditions and generated subsurface artefacts may interfere with resolution. An electron beam is rastered over the surface of a specimen. When the electron beam strikes the sample, inelastic collisions may cause electrons to be detached from the surface with energies around <50 eV. These are called secondary electrons and are collected by a detector with a positive potential. This information is then translated into an image on a screen where the contrast is related to the topological detail of the sample. Characteristic X-rays are also generated in this process and these may be employed in compositional analysis. Elastic collisions give rise to backscattered electrons possessing a high energy. These may also be imaged by the application of

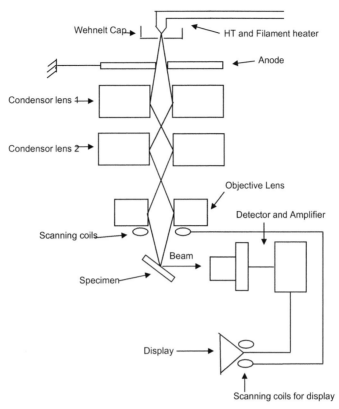

FIGURE 3.4 Diagrammatic representation of a scanning electron microscope (SEM).

a weak negative bias to the detector. The contrast in a backscattered image arises mainly from the atomic number of the specimen. Most imaging work with an SEM is conducted using secondary electron imaging. The magnification achieved is the ratio of the scanned area to the derived image. For example if a 1 mm^2 raster area on the sample is used to form a 100 mm^2 image, then the magnification is 100 times. The image is focussed, not by electromagnetic lenses, but by focussing the electron beam into the smallest possible diameter on the sample surface. However, astigmatism corrections are performed by the induction of a compensating asymmetric field around the electron beam (Fig. 3.4).

SAMPLE PREPARATION IN SEM

SEM is capable of looking at significantly larger samples than TEM and objects of around 1 cm^3 can be routinely imaged depending on the make of the microscope. Samples must be conductive or else coated in a conductive medium. Carbon coating deposited by electrical arcing is used for low resolution work. More commonly gold or a platinum/palladium alloy is deposited by sputter or plasma coating over the sample surface. Electrically insulating samples will appear overly bright or will seem to flash as the rastering electron beam builds up charge on the surface. It is possible to operate an SEM at low vacuum pressures in a mode called 'environmental SEM' which allows damp or vacuum unstable samples to be imaged. Image resolution is reduced in this mode but there is no need to coat the sample with a conductive medium as low levels of moisture prevent charge from building up (Fig. 3.5).

SCANNING TUNNELLING MICROSCOPY

The scanning tunnelling microscope (STM) is an imaging technique capable of atomic scale resolution. The fundamental operating principle is radically different to that of an optical or electron microscope in that the resolution is not dependent on the wavelength of the particle used for imaging (Binnig et al., 1982). Instead, the image is built up by employing the phenomenon of electron tunnelling to get a current to pass between an electrically conductive sample and a sharp metal tip held only a few Ångstroms from the sample surface. Three-dimensional (3D) images at near sub-atomic resolution can be reached using this technique under the right conditions. The tip (or sometimes the sample) is mounted on a piezoelectric scanner which can move it back and forth and up and down with absolute control over the spatial position. The tunnelling current is dependent on the separation distance between the sample and the tip. As the tip is moved relative to the surface, changes in the tunnelling current

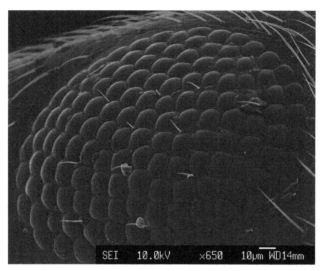

FIGURE 3.5 An example of a scanning electron micrograph of the eye of an ant. The ant was mounted on a conductive carbon sticky pad on an aluminium stub and then sputter coated in a 15-nm-thick layer of gold.

represent a change in the separation distance. The piezo stage is adjusted via a feedback loop to compensate for the change and so maintain a constant current. This process keeps the tip at a constant height above the surface and the current information related to the X and Y position of the scanner build up a topographical map of the area being scanned. This is known as constant current imaging and provides an image of the height the tip has had to move to maintain the current. On atomically flat samples it is also possible to scan the tip at a constant height over the surface and build up an image from the fluctuations in the current; this is known as constant height imaging.

It is important to use the term tunnelling as current does not flow in a classical sense between the sample and the tip. At no point does an electron travel physically from one surface to another and instead the electron appears to instantaneously transfer itself from the sample and into the tip. This behaviour is explained by a quantum mechanical effect which becomes more dominant at the nanoscale. The Heisenberg uncertainty principle states that the exact position and momentum of a particle cannot be known precisely. More specifically, the more you know about the particle position (x) then the less you will know about its momentum (p) and vice versa. This imprecision can be quantified as:

$$\Delta x \cdot \Delta p = \frac{\hbar}{2}$$

$$\hbar = \frac{h}{2\pi}$$

The implication of this formula is that the position of an electron is never known for certain, or if it is then its energy is uncertain. The act of measuring a particle will reveal its position but will alter its energy or else you might measure its energy but this will displace its position. For example, the shapes of atomic orbitals can be thought of as probability maps of where the most likely places to find an electron of a given energy state around an atom are. Atomic orbitals are not strict boundaries, however, and there is always a non-zero probability that an electron might be somewhere else entirely. The electron in that orbital will only be revealed when the measurement to find it is made. In the context of an STM measurement, an orbital presented on the surface of the sample represents the most probable position for an electron to be. The probability of finding an electron above the surface of the sample, beyond the boundary of an orbital, decreases exponentially with distance. When the STM tip is placed only a few Ångstroms from the surface there, is a probability of spontaneously finding an electron from the sample 'inside' the tip. Conversely, there will be an equal probability of finding an electron from the tip 'inside' the surface of the sample. The net exchange of electrons tunnelling between the tip and the sample remains equal until a potential difference is applied between them. This skews the tunnelling probability so that more electrons will tunnel into one component than the other depending on the direction of the applied electrical bias. As long as a voltage is maintained then a current of tunnelling electrons is produced and this can be used for imaging. The tunnelling current (I) is given by the formula:

$$I \propto V e^{(-2kW)}$$

where w is the separation distance between the tip and the surface. The term e is Euler's number and not electronic charge. K describes the energy potential across the vacuum:

$$k = \frac{\sqrt{2m_e(U - E_e)}}{\hbar}$$

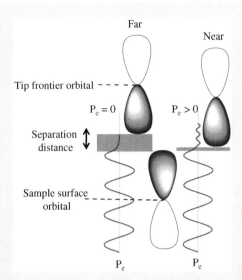

FIGURE 3.6 A schematic depicting electron tunnelling between a sample orbital and the orbital in an STM tip. P_e denotes the probability of finding an electron at a given position. For large separation distances (far), the chance of finding an electron in the tip orbital is practically, but never quite, zero. When the tip orbital approaches the surface (near), the probability that the electron will tunnel from the sample to the tip increases exponentially as the distance closes.

where m_e is the rest mass of an electron, U is the potential energy between surfaces, E_e is the energy of the electron and $\hbar$ is Planck's constant divided by 2π (Hornyak et al., 2009) (Fig. 3.6).

A wide range of STMs are available commercially and these fall broadly into the categories of large ultra-high vacuum setups or smaller devices which may be operated under ambient conditions. Aside from the scale and complexity of the STM being used, the most important factor in obtaining a good STM image is isolating the equipment from ambient electrical, thermal and mechanical noise. Housing the microscope in an earthed metal enclosure fitted with sound proofing can cut out much of the ambient electrical and sonic noise that will blur images. The microscope should be suspended in some way and this can be done by mounting the STM on an air table or by resting it on a base suspended by stiff springs. If you mount the microscope on springs, then take care to ensure it is balanced and that the resonance of the springs does not interfere with the microscope operation. If your assembly rocks around on the springs with a frequency faster than 10 Hz, you will see noise in your imaging. When it comes to isolating the STM microscope you really cannot do enough and a combination of multiple noise cancelling strategies is a good approach. Samples for STM must be conductive or semiconductive and should be mounted on a conductive base. Highly oriented pyrolytic graphite (HOPG) is often used as a substrate. The hexagonally ordered carbon is used as a calibration standard for calibrating scanners with atomic resolution and makes a good starting point for something to look at. A clean HOPG layer can be prepared by cleaving sheets from the graphite block using tape (Fig. 3.7). Try peeling the tape from different directions until the substrate appears mostly flat. Sometimes the graphite layers can fray giving the surface a torn appearance. Aromatic organic molecules make a good candidate for imaging with STM. In high vacuum systems these are allowed to condense onto the substrate surface under vacuum before imaging. Under ambient conditions it can be difficult to dry down or spin coat a monolayer of your aromatic molecular sample onto HOPG. Instead it is better to dissolve the molecule in a non-volatile oil and image the sample under the solvent. This can be as simple as getting the tip to engage with the HOPG surface, then disengaging it before adding a droplet of the sample on top and giving the system time to equilibrate. Aromatic interactions between your molecule and the HOPG substrate will ideally form a pattern on the surface which can be imaged. Getting a good STM image is highly dependent on the quality of sample preparation and the tip used. Getting these two parameters correct will take patience. Instructions on making good-quality tips are provided elsewhere.

ATOMIC FORCE MICROSCOPY

With thanks to John Mitchels

One limitation of STM is that the sample has to be conductive in order for tunnelling to occur. To image non-conductive samples, the atomic force, or scanning probe, microscope was developed (Binnig et al., 1986). These names are often abbreviated to AFM or SPM respectively but essentially both describe a setup where a nanometre sharp tip, mounted on the end of a cantilever, is drawn across a sample surface and a change in the tip is monitored to build up an image. A deflection in the tip is induced by the attractive or repulsive forces at work when the tip is in close proximity to the sample. These forces can be described by the Lennard-Jones

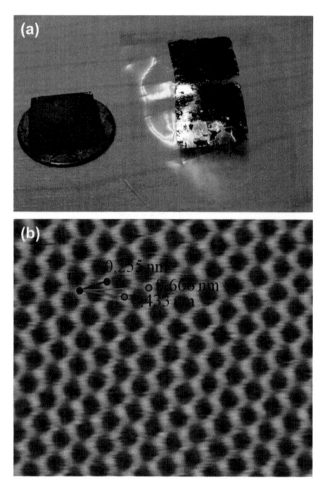

FIGURE 3.7 (a) An HOPG substrate is cleaved using tape and in preparation for imaging with an STM. (b) An STM height image of HOPG. The lattice parameters of the HOPG can be measured and used for calibration as noted.

potential which shows that, as two surfaces approach one another, there is an initial attraction due to van der Waals forces (Atkins and De Paula, 2010). This attraction encompasses molecular dipole interactions, induced dipole interactions and hydrogen bonding. At a closer range, this force is dependent on the elemental composition of the material, but not much larger than a few Ångstroms, the force between surfaces becomes strongly repulsive. This is because the atoms of the respective surfaces are close enough that their electrons repel one another as described by the Pauli exclusion principle (Fig. 3.8).

Since it was first developed, a wide range of experimental variations have become available which can monitor magnetic, electrical, semiconductor and force interactions along with topographical detail at the near atomic scale on a samples surface. The

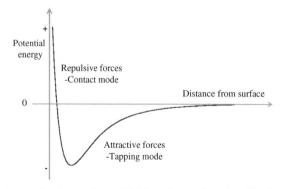

FIGURE 3.8 A plot of potential energy against separation distance for an AFM tip and a sample surface. The shape of the plot is described by the Lennard-Jones potential. The repulsive and attractive forces experienced at various heights above the surface can be used for imaging with contact and tapping mode imaging, respectively.

most basic configuration for an AFM is contact mode imaging where a cantilever tip is scanned across a sample surface and is maintained with a constant height or force above it. The deflection in the tip is measured by reflecting a laser beam off the back of the cantilever into a dual axis photodetector which can monitor the deflection in horizontal and vertical planes. A sample is mounted on top of a piezoelectric stage and the tip is bought into contact (or near contact) with the sample. Contact mode is performed within a few Ångstroms of the sample surface where repulsive forces are dominant. Either the tip or the sample is moved relative to the other and the deflection in the cantilever causes a change in the current detected in various parts of the photodetector. This signal is relayed back to a computer-driven controller which can adjust the height of the tip to maintain it at a constant deflection over the surface. The cantilever is maintained at a given 'setpoint' above the sample surface. The setpoint for contact mode AFM describes the level of force between the cantilever and sample that must be maintained for imaging. It does not necessarily mean that the cantilever is physically touching the surface as there are short-range attractive and repulsive forces which also serve to tell the computer when the tip is near the surface. This is analogous to holding a magnet near a metal surface where you do not touch the metal with the magnet before you start to feel them drawn together. For imaging, the setpoint is adjusted so that the minimum force is used whilst still being able to scan over the surface. This prolongs the lifetime of the tip and prevents damage to the sample surface. The speed at which the height of the cantilever is raised or lowered during scanning is determined by the integral and proportional gains used in the feedback loop. Proportional gain determines the amount of response the feedback loop will give based on the immediate cantilever deflection. For example, if the tip moves over a raised area, then it will begin to bend upwards. The tip will be raised slightly if the deflection in the cantilever is slight but should be raised higher and more quickly if the cantilever deflection is large. The proportional gain determines the level of proportional response to the immediate cantilever deflection. This can often produce a margin of error in the scan as the proportional gain can over- or under-shoot the target setpoint. To minimise overshoot in either direction integral gain is employed which looks at the cumulative error between the measured setpoint and the target setpoint. By adjusting the integral gain, the rapid fluctuation errors from the proportional gain are smoothed out. The integral gain generally tends to have a larger effect on the imaging process. If it is set too high then sine wave-like or 'buzzing' artefacts can be seen in the generated topological image.

During an AFM scan, the tip will be drawn in adjacent rows across the surface in a forward and back direction for each row moving along the sample. This is known as 'trace' and 'retrace' and when setting up for imaging with AFM, a plot of the trace and retrace from the same row should overlay one another. In general the gains should be as high as they can go before you begin to see artefacts in the trace and retrace plots being used to build up the image. Changes in the scan parameters such as the scan size alter the speed of the tip relative to the surface being imaged and so the gains may need to be altered each time to give the best image. Many commercially available systems can automatically adjust scan parameters to produce an optimal image. In contact mode the cantilever experiences lateral deflection as well as vertical deflection so it is possible to examine the frictional forces experienced between the tip and the sample while imaging the height at the same time. A contact mode AFM can be used for sensitive measurement of the force interactions of a stationary point on the sample. This technique is called force spectroscopy and on most scanning probe microscopes it is accurate enough to measure forces in the pico-Newton range. This is the order of force magnitude at which proteins fold themselves together. By anchoring one end of a coiled up protein to a fixed surface, the other terminus can be caught on an AFM tip which is retracted to pull the protein apart (Rief, 1997). The deflection of the cantilever as this occurs can be used to calculate the force with which the helices constituting the protein are bound. AFM tips with known force constants can be purchased or else, the force constant can be determined by experimental means. An example force plot (sometimes called force curves) on a solid substrate like mica is provided (Fig. 3.9b). This is a plot of the force, or deflection, against the vertical Z height with respect to the sample. The approach (the red portion of the plot) will show a flat line as the tip approaches the surface which may then snap down with a small negative force due to attractive interactions. As the tip continues to push into the solid mica surface a linear 'constant compliance' deflection is observed as the tip bends. If the sample under the tip were soft, then this region would appear curved which denotes the deformation of the sample under the tip. When the tip is retracted from the surface (the black portion of the plot) the measured deflection decreases linearly and will overlay the approach plot. If the material under the tip was deformed plastically then the red trace would be curved once more but constantly at a lower value as there is nothing pushing against the tip to deflect it as it pulls out. When the measured force reaches zero on retraction it can be seen that the force begins to become negative until a point where it suddenly snaps back to zero. This region is where the tip is attracted to the surface so that as the cantilever is raised relative to the surface, the tip is deflected opposite to the direction it was when it was pushed in. A good demonstration of the shape of a force curve is to tie a magnet on the end of a flexible ruler and use it as a 'tip' in proximity to a metal sheet. At a given height above the metal sheet the magnet will be pulled down quickly onto the metal. As you depress the ruler into the surface it will begin to bend until you pull it back and it straightens out. When you try to lift the magnet from the surface the ruler bends towards the metal until the force is great enough to snap it from the surface. There are a number of forces at work which cause this type of attraction at the nanoscale. Electrostatic interactions, van der Waals interactions and capillary forces from surface bound water are the most common explanation.

The second most common imaging technique using AFM is known as intermittent contact or 'tapping' mode where the tip is rapidly cycled up and down a few tens of Ångstroms over the surface of the sample in the attractive force domain. This overcomes some of the limitations that contact mode can experience when trying to image soft or delicate samples due to

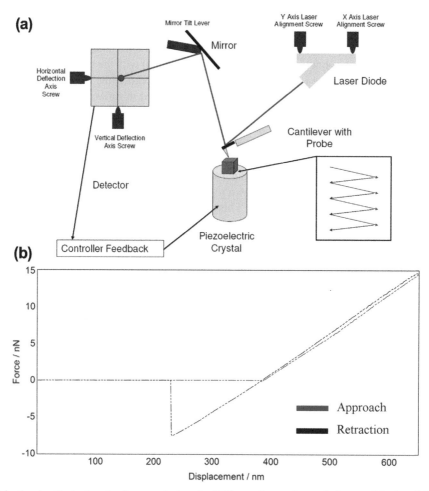

FIGURE 3.9 (a) A schematic showing the basic setup for a contact mode AFM experiment. A sample is moved back and forth on a piezoelectric stage underneath a tip maintained at a constant height above the surface. The deflection in the cantilever is measured by bouncing a laser of the back of a cantilever and aiming it into a photodetector with a mirror. (b) An example force plot of force against movement in the Z-direction for a tip approaching a mica surface. The red plot is the force experienced by the tip as it moves towards the sample surface and then pushes into it. The red plot is the force experienced by the tip as it is pulled back out.

friction. Tapping mode imaging is ideal for use with samples such as biological materials, self assembled polymer structures and lipid or surfactant assemblies as the tip is in contact with the sample only briefly during scanning. Instead of maintaining the tip with a constant force on the sample surface, the cantilever is driven to oscillate at a resonant frequency by a piezoelectric actuator so that the tip end is moving vertically up and down. The laser bouncing off the back of the cantilever also moves up and down in time with the cantilever and produces a sine wave signal in the vertical portion of the photodiode. The computer controller uses the amplitude of this sine wave signal in the feedback loop and attempts to maintain a constant root mean square (rms) value for the tip oscillation by raising or lowering the sample during imaging to build up an image. A good mental image to see the difference between contact and tapping mode is to think of trying to image a bowl of thick custard and marshmallows with your finger by touch. In contact mode you would drag your finger across the custard gently, so as not to disturb it, but may knock a marshmallow out of place when you encounter one. If you tap the custard gently with your finger, at a rate far faster than you move your finger across, then you can detect the marshmallows without knocking them out of place. Because the technique is designed to minimise the tip—sample friction there should be no lateral force detected and this means you cannot image torsional forces in the same way as contact mode imaging. In lieu of this, tapping mode imaging can provide information about relative hardness/softness of the sample being imaged by monitoring the phase of the sine wave signal while imaging and how this changes from the established target setpoint amplitude the feedback loop is aiming for. If the tip encounters something hard, then this will shift the cantilever oscillation out of phase with the sine wave signal the computer is expecting. Conversely, if the tip encounters something soft or sticky, the oscillation of the cantilever may be momentarily slowed and so, again, moves out of the expected phase. The phase differences can be measured and used to build up a phase image at the same time as a height image is processed (Fig. 3.10).

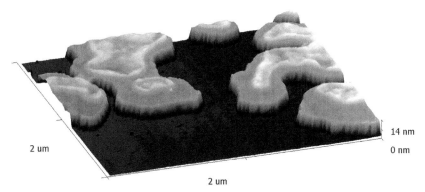

FIGURE 3.10 A 3D tapping mode AFM image of the triblock polymer F127 assembled in molecular layers on a mica substrate.

POWDER X-RAY DIFFRACTION

With thanks to Matteo Lusi and Mairi Haddow

A crystal structure comprises a collection of atoms or molecules arranged spatially with a repeating pattern. The smallest repeating motif of the pattern is called the 'unit cell' and no matter what the crystal is composed of, it will have a symmetry belonging to one of seven unique crystal systems. The arrangement of the molecules and atoms inside the unit cell can also have a characteristic symmetry and in total there are 230 unique symmetry combinations for unit cells. These are called space groups and the space group of a particular crystal structure will describe all of a crystals translational and symmetry operations. In practice this means that the shape of a large crystal composed of any number of unit cells can be predicted along with the positions of the atoms and molecules it is composed of. A basic unit cell is described by the length, width and height of the cell along with the three angles of each of the axes relative to one another. The lengths of the unit cell are labelled a, b and c. The angles are referred to as α, β and γ. The symmetry of the unit cell defines the crystal system of which there are seven as outlined in the table. The Miller indices describe repeating lattice planes in the crystal which are imaginary planes that cut the unit cell into $1/h$ pieces along the a axis, $1/k$ parts along the b axis and $1/l$ parts along the c axis. Lattice planes that satisfy Bragg's law (given further on) give rise to reflections when a crystal sample is observed by an X-ray beam of a particular wavelength (Fig. 3.11).

Crystal system	Defining symmetry	Unit cell dimensions and angles	
Triclinic	None		
Monoclinic	1 two-fold axis		$\alpha = \gamma = 90°$
Orthorhombic	3 perpendicular two-fold axes		$\alpha = \beta = \gamma = 90°$
Rhombohedral	1 three-fold axis	$a = b = c$	$\alpha = \beta = \gamma \neq 90°$
Hexagonal	1 six-fold axis	$a = b$	$\alpha = \beta = 90°\ \gamma = 120°$
Tetragonal	1 four-fold axis	$a = b$	$\alpha = \beta = \gamma = 90°$
Cubic	4 three-fold axes	$a = b = c$	$\alpha = \beta = \gamma = 90°$

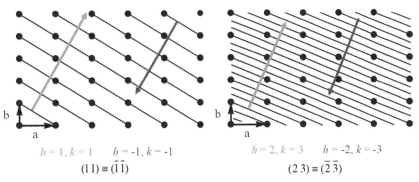

$$h = 1, k = 1 \quad h = -1, k = -1$$
$$(1\,1) \equiv (\bar{1}\,\bar{1})$$

$$h = 2, k = 3 \quad h = -2, k = -3$$
$$(2\,3) \equiv (\bar{2}\,\bar{3})$$

FIGURE 3.11 A schematic representation of various lattice planes in a two-dimensional crystal with an h and k axes.

The crystal structure of a solid material can be determined by passing X-rays of a wavelength comparable to interatomic distances through a sample and watching how that beam is diffracted. Ideally, an X-ray beam is passed directly through a sample and the pattern is recorded on the opposite side. This is known as single crystal X-ray diffraction and it is not easy to perform on nanomaterials since they are too small to position accurately in an X-ray beam. An easier option is to dry down a large amount of the nanomaterial to be tested and use powder X-ray diffraction (PXRD) to determine the structure. In a PXRD experiment a dried powder of the sample is collected on a flat silicon wafer or packed into a plastic holder or capillary tube. X-rays from a metal anode are projected onto the sample over a range of angles with respect to a detector mounted on the other side of the sample. The X-rays will be scattered as they pass through the sample by electrons around the constituent atoms. For particular angles of incident radiation scattered x-ray photons interfere coherently with one another to give a 'reflection' in an X-ray detector. The angle at which constructive diffraction occurs is related to the interatomic distances of the lattice planes by the Bragg equation:

$$\lambda = 2d_{hkl}\sin\theta$$

where d_{hkl} is the interatomic spacing for a particular crystallographic plane [hkl], θ is the angle of incident radiation and λ is the wavelength of the X-rays. Most commercial PXRD machines use a copper source which produces X-ray photons with a wavelength of 1.542 Å. Because a powder is used, there is an assumption that no one particular crystal axis is presented more than another. This is not always the case and, for example, a surfactant can dry into a lamellar film on a silicon wafer which will produce some dominant reflections and some diminished. This is known as preferred orientation (Fig. 3.12).

The pattern produced from a PXRD experiment is a plot of the reflected radiation intensity against the incident angle. The angle at which the reflection is observed relates to a repeating interatomic distance within the crystal under observation. To give rise to a reflection, there must be at least several unit cells having a coherent order for the constructive interference to be strong enough to be detected. For a given material there are characteristic d-spacings between the lattice planes and these can be assigned to Miller indices for a specific crystal structure. Some reflections will appear more strongly than others and the relative intensities of

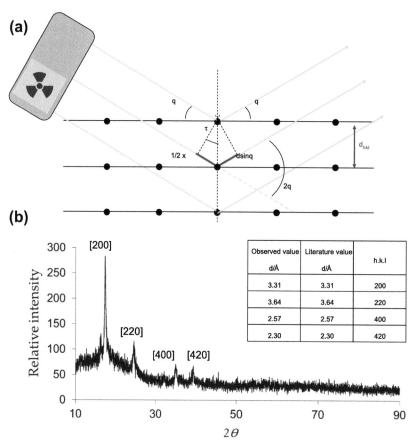

FIGURE 3.12   (a) A schematic of a powder X-ray diffraction experiment. The condition of diffraction is that interfering waves are in phase. This happens when the path difference of the interfering waves 'reflected' from successive planes satisfies the Bragg equation. The beam and detector are rotated through a range of angles and some will satisfy the Bragg equation to produce a detectable reflection. (b) A typical X-ray powder diffraction pattern with the Miller indices marked over the major reflections. The d-spacings were indexed to JCPDS card no. 46−907 for Prussian blue. Inset—the corresponding d-spaces for the Miller indices.

reflections can also be used in crystal structure identification. In this way each material has a structural 'fingerprint' which can be compared to an existing database to get a pattern match to a known crystal structure. There may be more than one crystal structure present in your sample as many materials have multiple polymorphs. Most powder diffractometer manufacturers produce pattern matching software which will let you compare the pattern you obtain to potential structures. The International Centre for Diffraction Data (ICDD), also formerly known as the Joint Committee of Powder Diffraction Standards (JCPDS), issues cards for thousands of existing crystal structures that list the d-spaces, expected relative intensities and the Miller indices to which they correspond. A powder pattern can be indexed by inspection using one of these cards. Alternatively, the Cambridge structural database holds thousands of digital structure files and provides free software for viewing them with. Not only will these programs generate images of a crystal structure but can also provide a calculated PXRD pattern which you can use for indexing your sample structure.

The broadness of the peak can be used to estimate the spherical size of the crystallite using the Scherrer equation as the peak width is inversely proportional to the crystallite size. This can be used to provide a rough estimate of nanomaterial dimensions by selecting a peak and measuring its width at half the peak height. This gives a value in degrees θ called B to avoid confusion in the formula. The diameter, l, of the crystal is given by:

$$l = \frac{(0.9\lambda)}{(B\cos\theta)}$$

where θ is the incident angle for the peak centre and λ is the X-ray wavelength. The value 0.9 represents the Scherrer constant for a spherical particle with cubic symmetry. This value can vary between 0.6 and 2 depending on the symmetry or anisotropy of the crystal or nanoparticle (Langford and Wilson, 1978).

UV-VISIBLE SPECTROSCOPY

With thanks to David Williams

The purpose of ultraviolet-visible spectroscopy (UV-vis) is to measure the wavelength-dependent absorbance of light from various chemical systems. By controlling the wavelength of emitted light (from the ultraviolet to the infra-red regions if necessary) the optical properties of a system can be investigated. At each wavelength of light that is transmitted through a liquid sample, a value for the absorbance can be calculated which is related to the concentration of the optically active compound in solution. The formula relating chemical concentration to optical absorbance is known as the Beer-Lambert law:

$$A(\lambda) = \log_{10}\left(\frac{I_0}{I}\right) = \varepsilon.c.l$$

where A is the absorbance value at a given wavelength (λ), I_0 and I are the intensities of the light passing through the sample, c is the concentration of the solution in moles per litre, l is the path length of the light through the sample (normally given as centimetres) and ε is the molar extinction coefficient in $L\,mol^{-1}\,cm^{-1}$. Using this formula it is possible to derive the concentration of a solution by measuring its absorbance in a UV-vis spectrometer. For chemicals homogenously dispersed in a solution this equation is reliable but this does not necessarily hold true for solutions containing nanomaterials as these will scatter light passing through them in high concentrations. In addition, they may not have a colour absorbance derived from classical concentration-dependent chemical chromophores. For a molecular species in solution, the absorbance is determined by the electronic configuration of the chemical. This can be described by a molecular orbital diagram where there exists discrete energy gaps between the highest occupied molecular orbital (HOMO) and molecular orbitals that are unoccupied and of higher relative energies; the lowest of these being the lowest unoccupied molecular orbital (LUMO). If the energy that it takes to excite an electron from the HOMO to the LUMO (or other unoccupied orbital) corresponds to light within the range of 190−800 nm, then the transition can be observed using UV-vis. Many nanomaterials exhibit colour interactions in this range but these arise from a property that may be independent of electronic transitions in the material. One example of this is colloidal gold which has a ruby red colour due to plasmon resonance of the collected electrons on the particle surface. For special cases like this, the Beer-Lambert law cannot be used to deduce the concentration directly from absorbance as the optical absorbance of the particles depends on the size of the gold particle in addition to the concentration. Despite the breakdown of the concentration formula, the UV-vis spectrum of a gold colloid can be used to determine the size of the gold particles and, for nanomaterials, this type of spectroscopy can be used to relate particle size to optical behaviour. Another example using this approach is in the quick estimation of the particle size of semiconductor quantum dots related to their colour. The analysis of both semiconductor and gold colloids using UV-vis spectroscopy is explored elsewhere.

Running a solution-based UV-vis spectrum can be as easy as filling a cuvette with a sample. An ultraviolet lamp and a visible lamp are both used as light sources which are run through a series of optics through a prism which splits the light into component wavelengths for transmission through the sample. It is better to use a quartz cell as these have superior optical transparency in the UV region and are resistant to solvents. Make sure you know if you are working with quartz or plastic disposable cells as often the difference is not immediate. Some organic solvents will dissolve disposable cuvettes and it is not unknown for a full cuvette to empty its contents into a spectrometer and ruin the machine. Samples should be dilute enough that the peak absorbance is somewhere around a value of 1. The absorbance value in a spectrometer is an arbitrary value calibrated to produce a concentration value from a known standard. Modern spectrometers will remain accurate even if you are measuring absorbance values higher than 1. The exact tolerances vary between machines but should be listed in the manual. Always run a background control with the continuous medium for the sake of comparison. Some applications may require you to measure the absorbance spectrum of a dry film or powder that is too thick for light to penetrate through. For example, you might wish to know the absorbance spectrum of a blade of grass or the peak absorbance of a semiconductor thin film coated on glass. For applications like this solid state diffuse reflectance UV-visible (SS−UV-vis) spectroscopy can be employed. SS−UV-vis splits a light beam into two channels. One of these reflects from the surface of a mirror and provides the value I_0. The other beam is directed onto the surface of the sample being tested and is either reflected or absorbed with an intensity value I. So that there is not direct reflection of the incident light into the detector from a rough or reflective surface, the ambient 'diffuse' light from the sample surface is measured inside a detector sphere coated with a powdery reflector. The solid state spectrum generated is a percentage comparison of the ratio of reflected light from the sample against the reflected light of the control at any given wavelength. As there is no transmission of light through the sample there is no concentration-dependent absorbance and so the Beer-Lambert law cannot be used for SS−UV-vis spectroscopy. A solid state spectrum will provide you with the position of peak absorbance within a sample.

For both liquid and solid state spectrums, any peak should have a Gaussian distribution though this can be wide or sharp depending on the sample. If you observe sharp kinks or abrupt drops in intensity or absorbance, then this is an indication of misaligned optics. This aberration shows up particularly strongly where the visible lamp changes over to the ultraviolet lamp. Machine manuals should have procedures to remedy this problem.

DYNAMIC LIGHT SCATTERING AND ZETA POTENTIAL MEASUREMENT

With thanks to Gait Collins and David Williams

Dynamic light scattering (DLS) is a rapid benchtop technique for characterising the size of a dilute solution of nanoparticles. Light can be scattered whenever it passes through a medium that is polarisable or has a dielectric constant different from unity. Photons interact with electrons bound in the material, which re-radiates the light in a scattered direction. When a coherent beam of light (such as a laser) having a known intensity, I_0, interacts with colloidal spherical particles undergoing Brownian motion, the intensity of the scattered light fluctuates as a result of constructive and destructive interference. This is called Rayleigh scattering and the intensity of the scattered light, I, is modelled by the Rayleigh approximation formula:

$$I = I_0 \cdot \left(\frac{1 + \cos^2\theta}{2R^2} \right) \cdot \left(\frac{2\pi}{\lambda} \right)^4 \cdot \left(\frac{n^2 - 1}{n^2 + 2} \right) \cdot r^6$$

where λ is the wavelength of light, n is the refractive index of the material, r is the particle size and R is the distance of the detector from the particle. These fluctuations can be analysed automatically through the use of autocorrelation to derive particle diffusivity in the form of the diffusion coefficient. For a monosized particle the diffusion coefficient, D, can be related to particle diameter using Stokes' diffusion law and the Einstein equation for Brownian motion giving the Stokes−Einstein equation:

$$d_h = \frac{kT}{3\eta\Pi D}$$

where d_h is the hydrodynamic diameter in metres, k is the Boltzmann constant ($1.3806503 \times 10^{-23}$ m^2 kg s^{-2} K^{-1}), T is the temperature in Kelvin and η is the viscosity of the solvent medium in Pascal seconds. The calculated hydrodynamic radius is usually slightly larger than the actual solid particle core. This can be due to the presence of a diffuse solvent, ion or surfactant layer on the particle surface. These surface interactions affect the rate at which particles move through the solvent and so the calculated diameter is slightly larger than the core. In a DLS experiment a colloidal sample is placed into a cuvette which is then placed into the path of a laser beam. The time between fluctuations in light intensity are detected, correlated and analysed by a photon correlator.

The result of a DLS experiment is a correlation function, G, which represents the relation between the average intensity at time t and that at time $(t + \tau)$, where τ is the time difference (or sample time):

$$G(\tau) = <I(t) \cdot I(t + \tau)>$$

when τ is large, the intensities $I(t)$ and $I(t + \tau)$ are independent of one another but at very short delay times, they are closely related and can thus be correlated. This function is, in effect, a measure of the probability of a particle moving a given distance in a time τ. For monodispersed particles, the correlation function is:

$$G(\tau) = A\left[1 + B\exp\left(\frac{-2\tau}{T_c}\right)\right]$$

where A is the baseline of the correlation function, B is the intercept of the correlation function and T_c is the relaxation time for diffusion. This quantifies how far from some arbitrary position the particle might be expected to be after a certain time:

$$T_c = \frac{1}{Dq^2}$$

where q is the scattering wave vector given by:

$$q = \frac{4\pi n}{\lambda}\sin\left(\frac{\theta}{2}\right)$$

where n is the refractive index of the solution, λ is the wavelength of the laser and θ is the scattering angle. As previously mentioned, the mean hydrodynamic radius of particles in the solution can then be calculated from the diffusion coefficient using the Stokes–Einstein relation (Fig. 3.13).

Analysis of the autocorrelation function is done automatically using software. The output from a light scattering experiment will be a series of charts that plot the size distributions of the particles. Normally, the software will be able to tell you automatically about the quality of the sample and if it is monodisperse or not. The size distributions will be provided as intensity, number and volume distributions and each of these is derived from the signal intensity. A number plot will show you the relative numbers of particles of different sizes present in the solution. An intensity plot will show you the relative intensity of the light signal from particles of different sizes. A volume plot will show you the relative volume of the particles observed within the solution and is derived from the intensity signal. Light is scattered with an increasing intensity as the particle diameter becomes larger. The increase in signal intensity scales up relative to the diameter to the sixth power. Therefore even a few large particles present in the sample being tested will produce large peaks. For example, if you are viewing a gold colloid which you expect to be 10 nm in diameter and the sample is contaminated with around 1% 300-nm particles, you might expect to see a small peak centred at 10 nm and a relatively huge peak centred at 300 nm. Depending on the sample quality, as determined by the software, you can ignore the larger particles as outliers. On a more practical level it is important to run samples through a 200-nm filter using a syringe before performing a DLS experiment.

This type of equipment is often coupled with a zeta potential analyser. The zeta potential of a particle in solution is a measure of its surface charge. Many colloids are stabilised by electrostatic interactions where the like charges of a surface coating repel one another and prevent the particles from flocculating. About a charge stabilised particle there exists an electrical double layer which has an interior composed of strongly bound charged ions called the Stern layer and a diffuse outer layer of counter ions. In solution, the Stern layer coats the particle surface and as the particle moves through solution it can drag other ions along with it in the diffuse layer. Depending on the charge of the Stern layer, the diffuse layer may be larger or smaller. The depth of the ionic cloud at which the ions are no longer drawn along with the bulk of the particle as it moves is known as the slipping plane. If viewed with DLS, then the machine will view the diameter of the core particle and the extra width of the slipping plane over the surface which together makes the total hydrodynamic radius. The charge of the particle at the edge of the hydrodynamic radius is the zeta potential. This figure will usually be in tens of millivolts and can be positive or negative. For stable colloids of electrostatically stabilised particles, the zeta potential will be greater than plus or minus 25 millivolts. The zeta potential of a colloid is sensitive to changes in pH and for every colloid there will be a pH where the surface charge approaches zero. This is known as the isoelectric point and is normally the most unstable condition for a colloid. Similarly, the electrostatic repulsion of charged particles can be screened by increasing the ionic strength of a solution. Screened particles will flocculate and this is called 'salting out'. To measure the zeta potential of colloidal particles, samples are placed into an electric field or field gradient and their relative motion is monitored. The velocity at which the particle moves is called the electrophoretic mobility (U_e). This depends on the field strength, solution viscosity (η), zeta potential of the particles (z) and the dielectric constant (ε) of the

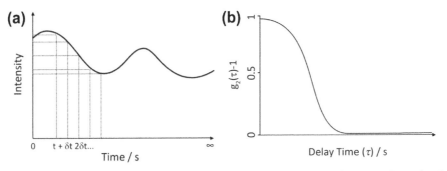

FIGURE 3.13   (a) The fluctuation of scattered light intensity with time and (b) the autocorrelation function ($g_2(\tau)-1$) as a function of the delay time (τ).

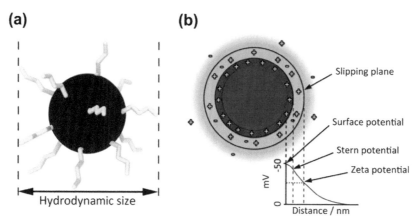

FIGURE 3.14 Diagrams of (a) the hydrodynamic structure of a simple surface-modified particle in solution and (b) a representation of the double layer at the surface of a negatively charged particle highlighting the different components.

medium through which they move. The electrophoretic mobility is measured directly and is related to the zeta potential by the Henry equation:

$$U_e = \frac{2\varepsilon z f(ka)}{3\eta}$$

where the term *f(ka)* is Henry's function and describes one of two approximation equations used for modelling the electrophoretic mobility. The Smoluchowski equation is used for spherical colloids in aqueous conditions and has a value of *f(ka)* = 1.5. The Huckel equation is used for non-aqueous mediums and has a value of *f(ka)* = 1.0. Most of these parameters can be automatically selected using the analysis software. As with DLS measurement, samples should be passed through a filter to prevent unwanted particulates skewing the results (Fig. 3.14).

BET SURFACE AREA MEASUREMENT

The total internal and external surface area of a sample can be determined by the isothermal adsorption of nitrogen (Rouquerol et al., 1999). This method is called the Brunauer, Emmett and Teller (BET) method where the uptake of nitrogen at various partial pressures is measured and used to form a plot called an isotherm. From the shape of the isotherm a variety of information can be determined and further analysis can give pore size and distribution information.

Langmuir initially proposed a model for surface area measurement using the adsorption of a monolayer of inert gas at the surface:

$$\left(P + \frac{a}{V^2}\right)(V - b) = RT$$

If the adsorption of the gas is measured at a temperature well above the condensation temperature, then only a monolayer is formed.

$$\text{Total surface area} = (\text{Number of adsorbed gas molecules}) \times (\text{Area per molecule})$$

The BET method accounts for the formation of multiple layers adsorbed to a substrate surface.

$$\frac{1}{W[(Po/P) - 1]} = \frac{1}{WmC} + \frac{(C-1)}{WmC}\frac{P}{Po}$$

where *W* is the weight of the nitrogen adsorbed at a given *P/Po* and *Wm* is the weight of the gas to give monolayer coverage and *C* is a constant that is related to the heat of adsorption. A linear relationship between $1/W[(Po/P)-1]$ and *P/Po* is required to give the quantity of the nitrogen adsorbed. This region of the isotherm is normally between *P/Po* = 0.05−0.3 which represents the multi-layer adsorption on the substrate. The intercept is taken to be zero and the slope is used to calculate the surface area (Fig. 3.15).

I = Microporous materials (<2 nm) beyond the range of BET measurement
II = Non-porous materials
III = Weak adsorption interaction

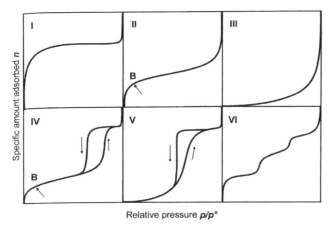

FIGURE 3.15 The six main types of gas physisorption isotherms according to IUPAC classification. *Adapted with permission from Rouquerol et al. (1999). Copyright 1999, Academic Press.*

IV = Mesoporous materials
V = A relatively rare isotherm indicating weak adsorbent—adsorbate interactions. The hysteresis loop is derived from pore filling and emptying.
VI = Also rarely observed, this isotherm indicates the layer by layer adsorption of gas on to a uniform surface

Calculating the surface area requires knowledge of the average surface area (molecular cross sectional area) occupied by the adsorbate molecule in a complete monolayer, a_m such that:

$$A_s(BET) = Wm \cdot L \cdot a_m$$

and

$$a_s(BET) = \frac{A_s(BET)}{m}$$

where $A_s(BET)$ and $a_s(BET)$ are the total and specific surface area of the adsorbent, respectively (of mass, m), and L is the Avagadro constant. If it is assumed that the monolayer is close-packed, then for nitrogen adsorbents, $a_m (N_2) = 0.162$ nm^2 at 77 K.

In addition to surface area calculation, it is possible to derive the pore sizes and pore area of a sample. The Lippens and deBoer method of t-plot analysis is used to derive the matrix surface area. This states that the multi-layer adsorption curve for nitrogen at different pressures at a constant temperature is identical for a variety of adsorbents. Hence the thickness, t, of a gas layer as a function of partial pressure:

$$t = \left[\frac{13.99}{\log(Po/P) + 0.34} \right]^{1/2}$$

A plot of the volume of nitrogen adsorbed at different partial pressure values as a function of t should give a straight line passing through the origin for a non-porous material. The gradient of the slope (V/t) can be used to plot the specific surface area S from the relationship $S = 15.47$ (V/t). A positive intercept indicates micropores and a deviation from a straight line indicates mesoporosity.

REFERENCES

Atkins, P., De Paula, J., 2010. Atkins' Physical Chemistry, ninth ed. Oxford University Press.

Binnig, G., Quate, C., Gerber, C., 1986. Atomic force microscope. Physical Review Letters 56 (9), 930—933.

Binnig, G., Rohrer, H., Gerber, C., 1982. Surface studies by scanning tunneling microscopy. Physical Review Letters 49 (1), 57—61.

Chescoe, D., Goodhew, P., 1992. The Operation of Transmission and Scanning Electron Microscopes Royal Microscopical Society. Oxford University Press, Oxford.

Hornyak, G., et al., 2009. Introduction to Nanoscience and Nanotechnology. C. Press, Boca Raton.

Langford, J.I., Wilson, A.J.C., 1978. Scherrer after sixty years: a survey and some new results in the determination of crystallite size. Journal of Applied Crystallography 11 (2), 102—113.

Rief, M., 1997. Reversible unfolding of individual titin immunoglobulin domains by AFM. Science 276 (5315), 1109—1112.

Rouquerol, F., Rouquerol, K., Sing, K., 1999. Adsorption by Powders and Porous Solids, Principles, Methodology and Applications. Academic Press, London.

Chapter 4

Chemical Techniques

THE SOL−GEL PROCESS

Alkoxides

Sol−gel chemistry is a catch all term for the reaction of an organometallic compound, usually an alkoxide or chloride, under aqueous conditions to form a solid product. This product can be anything from a dense glass monolith, a high surface area molecular filter, an aerogel to a metal oxide or nitride coating or nanoparticle. An advantage of the sol−gel method is that it is normally performed under ambient temperatures and pressures and can produce a wide range of different functional materials. Ebelmen demonstrated the formation of the first metal alkoxide from $SiCl_4$ in 1844 (Ebelmen, 1846). It was discovered shortly thereafter that exposure to atmospheric moisture formed a gel. Almost a century later Geffcken adapted the chemistry towards the preparation of oxide films (Geffcken, 1939) and in the 1930s the Schott glass company utilised the gelation of alkoxides in making glass (Hurd, 1938). Since then, industry has made use of developments in sol−gel chemistry and precursor design in the preparation of ceramics and glasses at room temperatures (Levene and Thomas, 1979). The sol−gel procedure has been adopted in many fields and the technique is of great importance to nanotechnology. The attraction is that a solid can be precipitated with great control from a solution and this means that colloids can be coated or thin films can be formed. Sol−gel techniques are used extensively in biotemplating and bioencapsulation procedures as the conditions of formation are generally very mild.

The chemistry of the sol−gel process consists of an initial hydrolysis and condensation of an alkoxide precursor to form a gel followed by ageing, solvent extraction and finally drying (Brinker and Scherer, 1990). The process may be catalysed by the presence of an acid or a base both of which will produce dense or diffuse networks respectively by alteration of the alkoxide hydrolysis kinetics. The selection of the precursor and catalyst depends ultimately on what you would like to make. The recipes outlined in this section will be concerned mostly with silica and transition metal based nanoparticle and gel formulations. Silica is ubiquitous throughout the field and has some special uses for those looking to work with biological nanomaterials so this will form a natural focus for this chapter. The d-block elements are of great importance from an electronics point of view as they form the bulk of semiconductor materials; classical electronics uses components produced by physical top down techniques like photolithography or chemical etching and so are explored elsewhere in this book. By using the sol−gel process it becomes possible to use bottom up techniques to construct new and exciting nanomaterials with the high compositional and geometric tolerances demanded at the fore front of emerging technologies.

The general chemistry of alkoxide hydrolysis and condensation to form a metal alkoxide is outlined below. A metal oxide formed using acidic conditions will form a dense network and is good for making strong films or monoliths for optical applications. An oxide formed under basic conditions will form a less dense network structure which will find useful applications in catalysis. Mechanistically, the sol−gel process starts with a water molecule coordinating through its oxygen lone pair to an empty *d*-orbital on the transition metal cation. Hydrogen will dissociate from the now partially acidic coordinated water molecule forming a proton free in solution and leaving behind a hydroxyl group linked to the metal centre. The binding of the water to the metal centre to form the hydroxyl will displace an alkoxide group which forms an alcohol with the left over hydrogen. The mechanism by which the substitution occurs is thought to be a bimolecular nuclear displacement reaction (S_N2) where the rate of the substitution is dependent on steric factors influenced by the length of the alkoxide chain. That is to say an ethoxide chain (two carbons long) will be displaced by water faster than a butoxide chain (four carbons long). This basically means that the bulkier the alkoxide chains are on a metal alkoxide then the slower it will be to react with water (Fig. 4.1).

During acid catalysis an alkoxide group coordinated to the metal centre is protonated at the oxygen which draws electron density from the inorganic (Ti, Al, Si) centre leaving it more susceptible to nucleophilic attack by a water molecule. This in turn decreases the positive charge on the alkoxide which promotes its separation as an alcohol. Base-catalysed hydrolysis is similar mechanistically. Under basic conditions the hydroxyl group directly attacks the inorganic centre and drives the inversion of the lattice tetrahedron as an alkoxide anion is ejected. During the condensation step the metallic hydroxyl ligands undergo a nucleophilic

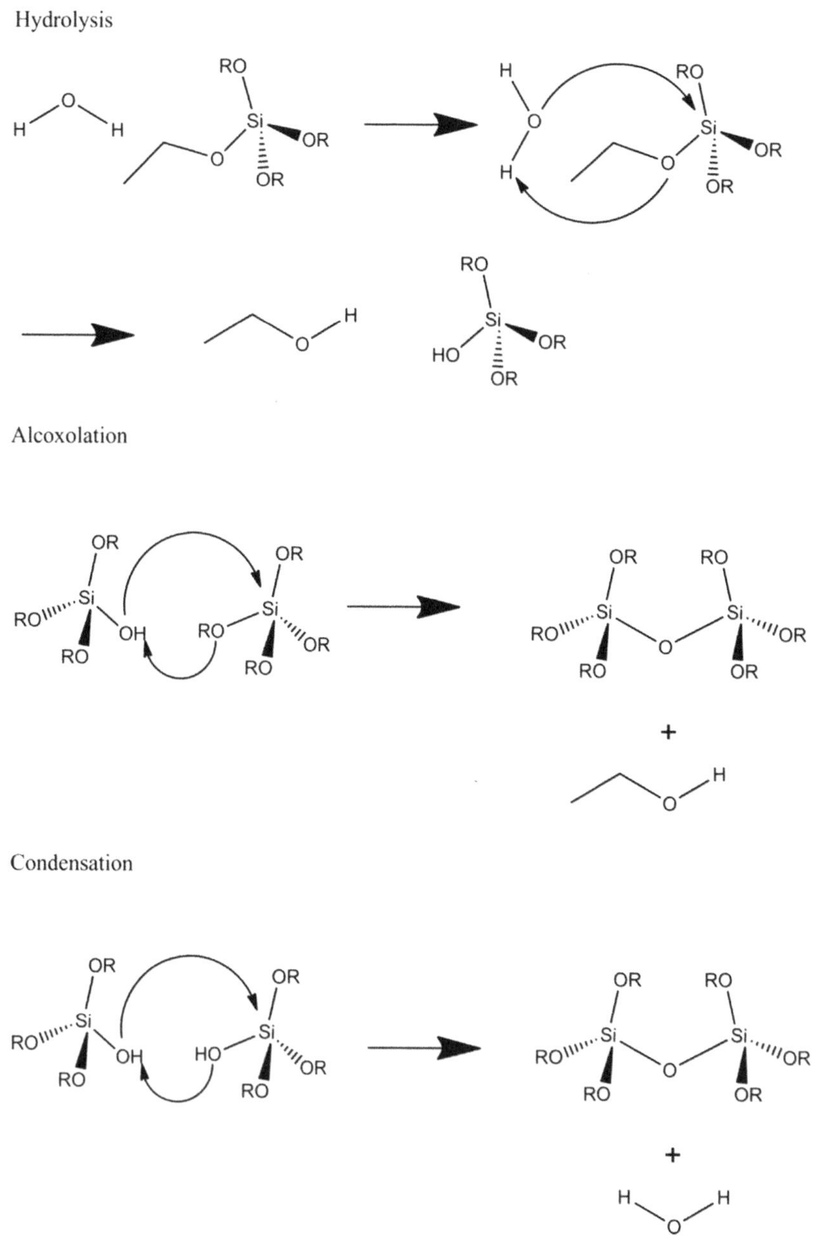

Hydrolysis

Alcoxolation

Condensation

FIGURE 4.1   A schematic of the sol–gel reaction of a silicon alkoxide with water. The three main stages of the reaction are hydrolysis, alcoxolation and finally condensation to form a solid oxide network.

substitution to evolve water and a metal–oxygen–metal bond. At this point the metal oxide formed precipitates as a solid. The solution contains a mixture of alcohol and water along with the catalyst (Fig. 4.2).

For the hydrolysis of a titanium alkoxide the stoichiometric relationship may be written as:

$$nTi(OR)_4 + (4n + x - y)H_2O \rightarrow nTiO_2 + xH_2O + yR(OH)$$

The evolved solid will still contain alkyl chains trapped in the lattice structure and at this point ageing begins, where the remaining alkyl chains trapped in the solid network continue to react and form bridging oxygen bonds between adjacent metals. This induces shrinkage of the precipitated gel known as syneresis. The pH of the reaction solution is most influential on the rate of contraction at this point. By ageing the precipitate in the catalytic solution, further oxygen metal bonds will be formed causing the structure to increase in density and contract further. This contraction drives the liquid phase products out of the structure. The capillary forces generated in the pores during drying can be relatively huge and cause damage in the form of cracking in the gel. It is for these reasons that if products have to be dried then it must be done slowly under well controlled humidity conditions. Catastrophic cracking may be overcome by gentle drying or hypercritical drying, where the liquid phase is gradually replaced by supercritical CO_2. Both these types of drying method will be explored further on.

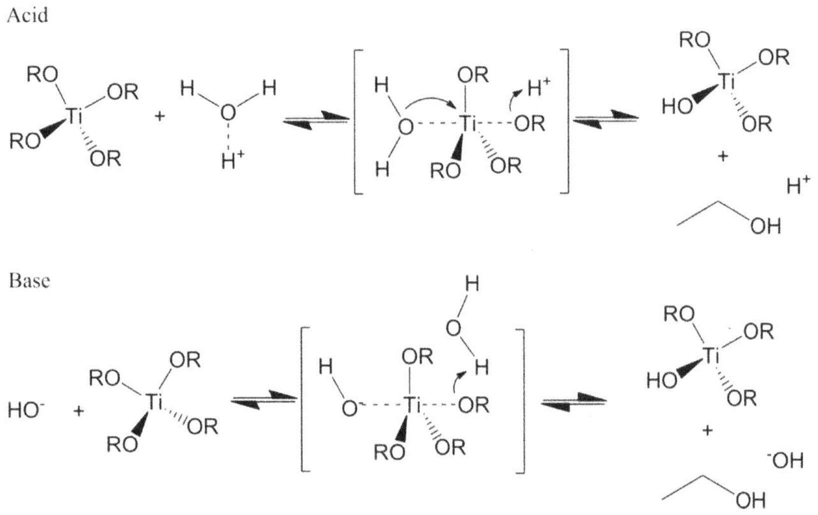

Acid

Base

FIGURE 4.2 Schematic representation of an acid catalysed alkoxide and a base-catalysed alkoxide.

SILICA

To begin with we will explore the use of sol gel chemistry to produce various morphologies of silica. SiO_2 permeates the sol gel field and much of what may be done with the formation of glass can be applied to the formation of metal oxides and more exotic materials such as nitrides. The recipes outlined will move from particles to particle coating and then on to films and monoliths of these oxides and chemically modified or doped forms of the oxides. One major attraction of glass is that it is highly inert biologically and resistant to chemical attack. Silicone rubber materials are used routinely for implants and even at the nanoscale silica is relatively safe to employ as a drug delivery or biomarking agent. It is also possible to entrap proteins, enzymes and even whole cells to produce biologically functional glasses. This entails some level of consideration over the design of the reaction as precursors which generate alcohol will denature delicate biological structures as the glass condenses. Of course, alkoxides are not the only way to produce glasses and sodium silicate is one way to bypass the use of alcohols in the formation of a silica gel.

SODIUM SILICATE

Adapted from Conradt (2008)

As shown previously, an important step in the formation of a glass is the hydrolysis of the silicon centre to form a hydroxide species. $Si(OH)_4$ is short lived, even at low concentrations, in solution below pH 7 as it condenses to form a polymeric silicon oxide network. However, a metastable sodium form of silicate can be produced that, under basic conditions, is quite stable. Silica is dissolved in a strong NaOH solution to produce the sodium silicate. The chemical formula of sodium silicate can be viewed as Na_4SiO_4 though in water this will vary in elemental ratio. In a hydrated form the formula will be $Na_2SiO_3 \cdot xH_2O$ though the silica is more close to the form of $SiO_2(OH)_2^{2-}$ in solution. Treatment with acid replaces the choride with a proton to form a silicic acid which can then undergo condensation. Acid exchange in sodium silicate is done by passing the solution through and acidified ion exchange column. Sodium silicate may therefore be thought of as a salt based sol gel precursor as opposed to an organometallic one like an alkoxide. This is especially useful where the synthesis requires an absence of organic groups or alcohols being evolved such as cell encapsulation. Sodium silicate is also used in the preparation of mesoporous molecular sieves such as MCM-41 and the recipe for this will be explored later.

SILICA NANOPARTICLES

Adapted from Costa et al. (2003)

Silica nanoparticles are generally very useful in nanomaterials chemistry. They are easily chemically modified due to reactive hydroxyl groups at the surface and can be impregnated relatively easily with dyes and fluorescent labels. Opalescent materials get their iridescent sheen as a result of periodically arranged matter at the surface inducing destructive and constructive interference in reflected light. If silica particles are made with a narrow size distribution they can be dried down from a colloidal dispersion to produce a synthetic opal. This can be used as a template to form an inverse opal displaying photonic band gap properties. Silica nanoparticles can even be used to control the stability of an emulsion in the place of a surfactant. These are called Pickering emulsions and by control of the hydrophobic and hydrophilic interactions a 'dry water' powder can be formed. These themes are explored elsewhere so for this recipe we will focus on a classic method for the production of silica nanoparticles having a controllable diameter. Before the refinement of the sol gel method, silica nanoparticles were produced mainly by the pyrolysis of silicon tetrachloride in a hydrogen torch. This was not an ideal synthesis and the particles produced had size distributions around 100 nm with a tendency towards crashing out. A solution-based route towards the production of monodisperse particles was hit upon and this is known as The Stöber Process (Stöber, 1968).

The Stöber Process has become a classic method for producing silica nanoparticles in a controlled manner having a narrow (less than 10% or lower) polydispersity. A tetraethylorthosilicate (TEOS) precursor in ethanol solution is hydrolysed with a small amount of water to form a silicic acid species. The addition of ammonia solution then catalytically drives the condensation reaction of the silicic acid to form silica particles. These particles are also self stabilising as the surface hydroxyl groups will remain electrostatically charged depending on the pH of the colloid. The isoelectric point of silica is ~pH 2−3 which means that above this pH, the surface charge will shift to being predominantly negatively charged $Si-O^-$ groups. As this preparation method uses ammonia to catalyse the condensation reaction the conditions are very basic and the formed colloid is stable. There are limits though and if the particles are too large then the sedimentation force of gravity will overcome the electrostatic repulsion. The amounts in this recipe may be adjusted as it is the concentrations of reagents that are important to the size of particle produced. For a fixed amount of TEOS precursor the amount of ammonia added to the solution as a condensation catalyst has a large effect on the size. The more ammonia added to the solution the larger the particles will be. This may strike some as counterintuitive considering that we are adding a catalyst and this should make the condensation mechanism happen faster. For most nanoparticle production methods the faster the nucleation is then the smaller the particle produced will be. If we were not to add the ammonia but only water then the TEOS would still eventually hydrolyse and form a glass. The formation of silicic acid (if there is water present) means that the solution is already nucleating before the addition of the catalyst albeit slowly. As the nucleation is slow at this point small oligomeric units of glass that make contact with one another may not necessarily condense when in contact with one another. In this way many particles can form in solution from small oligomers. When the ammonia is added the tendency to collect more glass at the surface by condensation reactions increases and the particles become larger. In this recipe particles will be formed between ~15 and 115 nm in diameter, though you should experiment with how large you can get them before they crash out. If these sizes are unsuitable for your end application then you can redisperse them in ethanol and repeat the addition of TEOS and ammonia until you achieve the desired size. Some details of this process are given later.

To make silica nanoparticles by the Stöber process you will need:

1. A 10% HCl in water solution, water and bench ethanol for cleaning glass ware with. Clean glass ware is very important as any particulate rubbish matter can provide nucleation points for the silica and you will end up with uncontrolled growth or wide polydispersity.
2. 4 mL (3.732 g, 0.0179 mol) of tetraethylorthosilicate (TEOS) or $Si(OC_2H_5)_4$. This should be kept under an inert atmosphere and stored in a dessicator when not in use. Exposure to atmosphere will cause it to hydrolyse slowly.
3. 50 mL of absolute ethanol, Bench ethanol is not pure enough as it contains benzene.
4. A saturated ammonium hydroxide solution. This will be 28−30 wt% ammonia in water. The amount you use depends upon the size of the particle required but will be no more than 4 mL in this recipe.
5. A 100 mL screw top bottle.
6. A volumetric pipette capable of dispensing 4 mL and 10 mL and pipette tips.
7. An ultrasonic bath set to 37 °C. If one is unavailable then a rocker or magnetic stirrer in a water bath could also be used.
8. A drying oven set to 80 °C.

To make silica particles by the Stöber process:

1. Clean your glassware using the 10% HCl solution then rinsing with water and then rinsing with ethanol. Allow it to dry thoroughly before use.
2. Using a volumetric pipette add 4 mL of the TEOS to 50 mL of absolute ethanol in the 100 mL screw top bottle. Stir by swirling the bottle.
3. Add a given amount of ammonia solution depending on what size of particle you would like. Again swirl the solution after addition. You should not see any precipitation or cloudiness immediately. It will take around ten minutes for the alkoxide to be converted into a silicic acid species after which point the ammonia will catalyse the condensation step. The table below shows the amount of ammonia solution to add and the produced particle size as observed by TEM:

Volume of ammonia solution/mL	Particle size/nm
2	15
2.5	33
3	47
3.5	76
4	115

4. Screw the bottle shut and place into the sonic bath at 37 °C. Turn the ultrasonics on and leave for 2 h. The bath will become hot during this time so add or exchange water to keep the temperature correct. Turn the sonic bath off as you do this.
5. Turn off the heat and ultrasonic bath and remove the bottle containing the now formed silica nanoparticles. Open the bottle and allow the ethanol to evaporate in the drying oven. This may take some time. Alternatively you might try centrifuging portions of the solution to isolate the silica. You can also keep them in the ethanol solution. There should not be any further reaction in the system so they should not grow any larger unless you add more TEOS and ammonia.

The produced colloids can be analysed by dynamic light scattering, transmission electron microscopy and atomic force microscopy. Interestingly, the observed diameters of the particles will change depending on the analysis technique used. If the particles are suspended in ethanol they will appear larger than those suspended by water when viewed by DLS by up to 10 nm. During formation it is possible that some of the alkoxide remains trapped within the silica framework of the particle. This means that the particle is, to some extent, alcophillic and in the presence of ethanol the low density network can swell. Swelling is also observed in water due in part to the presence of hydroxyl groups but this effect is less pronounced than with ethanol. Similarly, a difference is observed between AFM analysis and TEM with particles appearing 20–40% smaller by the latter technique. Under the TEM particles are exposed to high vacuum in order to allow the electron bean to pass through the sample unhindered. This also removes any water from the system so that the particle observed is unswollen silica. In the AFM, samples are normally observed under atmospheric pressure and humidity, meaning the silica particles can draw in moisture from the air. It may seem strange that a glass can adsorb water when on the macroscopic scale we do not see glass soaking up rain water! However we must also think of those little packets of silica gel that are packed in with electronics and shoes as dessicants. The glass in this case is not a dense SiO_2 melt but rather a network of Si−O−Si, Si−OH and Si−OC_2H_5 which imparts some flexibility and allows for swelling. The presence of unsubstituted alcohol groups can be detected by performing infra red spectroscopy on the dried powder. If alcohol is present then a small peak for C−H stretching will be seen somewhere around 2900 cm^{-1}. There will be Si−OH stretches between 3200 and 3600 cm^{-1} and Si−O−Si deformation between 1100 and 1000 cm^{-1}. Typically the alcohol stretching will be less strong in the larger particles as will the Si−OH stretching as the particles are more dehydrated after formation. The different size ranges observed by a range of techniques are outlined in the table below (Fig. 4.3):

FIGURE 4.3 An atomic force microscope height image of a collection of silica spheres made by the Stöber process dried down onto mica. (*Image taken by James Fothergill, University of Bristol*)

You are by no means limited to TEOS as the starting precursor and many other synthesis methods use alkoxides with increasing chain lengths such as propoxides or alkoxides. If employing a silicon based alkoxide it is important to select a co-solvent of suitable compatibility with the alkoxide chain length. Generally longer alcohol chain lengths will hydrolyse more slowly than shorter ones so if you wish to take advantage of a slower rate of hydrolysis you might use tetrabutylorthosilicate (TBOS) dissolved in butanol. If the TBOS is dissolved in ethanol then, by the very nature of the sol gel reaction, the butoxide ligands will quickly be substituted to form TEOS once more and a few butanol molecules in the excess of ethanol solvent. This can be useful for fine tuning a sol gel reaction by creating blends using specific solvent mixtures. Methoxide coordinated silica precursors will form smaller particles than butoxide coordinated ones though this, as mentioned before, must be weighted against the effect that the strength of ammonia catalyst has on the produced size.

TITANIA NANOPARTICLES

Titania is another material that has found a multitude of uses realised by sol gel techniques. TiO_2 ceramics have numerous applications and are used in paint pigments, abrasives, electronics coatings, membranes and filters (Airoldi and Farias, 2004). Titania has three common crystal forms, anatase, rutile and brookite. Of these three, anatase is the most utilised as a semiconductor material with photocatalytic and photoelectric properties. These properties arise from the electronic band structure which allows promotion of electrons from the materials ground state up in to the conductance band when stimulated by ultraviolet and visible light. For example, anatase TiO_2 can decompose organic materials when dispersed in a polluted aqueous medium under irradiation with sunlight (Dobosz and Sobczyński, 2003; Chatterjee and Mahata, 2001). Hydroxyl groups bound to the surface of the titania can receive the electron from the conduction band to form a reactive hydroxyl radical. This degrades surface bound organics on the particle via the following scheme:

$$TiO_2 + hv \rightarrow (e^-/h^+)$$

$$O_2 + e^- \rightarrow O^{2-}$$

$$2O^{2-} + 2H_2O \rightarrow H_2O_2 + 2OH^- + O_2$$

$$H_2O_2 + e^- \rightarrow OH^. + OH^-$$

The efficiency as a catalyst has been shown to improve by the addition of metallic nanoparticles, notably silver and gold, to the titania surface (Subramanian et al., 2001). These prolong the duration of the charge separation after photoexcitation and so greatly increase the radical formation rate (Fig. 4.4).

Titania is also bioactive (Hench, 1997) and this has led to its use in the adhesion of titanium implants to bone in medical procedures. Hydroxyl groups present on the surface lead to a negative surface charge, allowing for the easy adsorption of ions and biomolecules. This in turn promotes the nucleation of hydroxyapatite, the major mineral component of bone. The porosity and

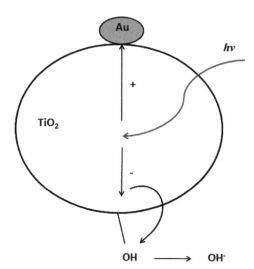

FIGURE 4.4   Schematic representing the photoexcitation of titania to produce an electron-hole pair. The charge separation is stabilised by the surface bound gold forming an Au^+ cation.

surface roughness are critical factors in determining the strength of the interface between the implant and the new bone layer. TiO_2 is used as a surface coating before the implantation of titanium plates, pins and prosthetic bones into patients because it readily bonds to the natural bone and surrounding tissue. Until recently this was facilitated by coating the plate or pin with a titania powder and fusing it by heat treatment. For most commercial applications titania powder is produced in bulk by a number of means. Most commonly employed is the addition of water to a titanium alkoxide to produce bulk precipitates or monoliths. Finer powders can be generated by aerosol preparation methods which spray titanium alkoxides or chlorides in to a combustion chamber. There pyrolysis is induced by heat or high energy light in the presence of an oxidant. Some control over the polymorph formed has been demonstrated in these systems although further heat treatment can be employed to give anatase and rutile phases. The grain and quality of the powder produced is controllable and alteration of molar fraction injected, the temperature and the reaction duration within the flow chamber have been shown to alter size and distribution. In addition, alkoxides have been employed more than chlorides due to the more facile reaction kinetics. In this way powders as fine as 12–20 nm have been produced with no trace of organic components as might be seen in sol–gel systems (Alexandrescu et al., 2004).

High surface area titania can be prepared as a colloidal dispersion by controlling the hydrolysis and condensation of a titanium precursor in the presence of a catalyst. For example, anatase phase titania can be formed as 20 nm crystals by hydrothermal treatment of titanium (IV) isopropoxide and addition of pH 0.5 nitric acid. Initially, a precipitate of aggregated amorphous nanoparticles is formed. The added acid acts to peptise the particles, which redisperses and stabilises them. Continued heating results in the transformation to the crystalline anatase phase of titania. Subsequently these particles can be dip coated onto surfaces, producing high surface area (80 m^2/g–160 m^2/g) thin films (Al-Salim et al., 2000). This method produces porous films which are of particular use in sensing, photocatalytic and chromatography applications. A selection of different sol gel methods of producing titania nanoparticles are outlined here.

MAKING TITANIA NANOPARTICLES USING TITANIUM TETRACHLORIDE AS A PRECURSOR

Adapted from Serpone et al. (1995)

$TiCl_4$ is an air and water sensitive precursor and is possibly the most difficult to work with out of the methods given here. It is quite volatile and has a tendency to fume if exposed to atmospheric moisture. The particles produced will be in the range 2–4 nm and are not made in the presence of a stabilising agent. Control of the dispersion pH is therefore important in preventing flocculation from happening. It is similar in route to the Stöber process except instead of base catalysis the titanium precursors are acid catalysed to hydrolyse rapidly. The beauty of using the chloride salt of titania is that, as the chloride ion is displaced by an OH group, it can react with the left over proton to form a molecule of HCl. The acidic conditions peptise the titania produced and so the particle formed is around 20–30% of the anatase crystal form without having to heat treat the dried powder. One of the tricky technical issues is controlling the temperature which is vital to controlling the size of the particles produced. If the $TiCl_4$ is equilibrated to −20 °C before addition to water maintained at 0 °C then the diameter of particles formed will be in the 2–4 nm range. If the precursor is at room temperature and is added to the water at 0 °C then the particle diameter will be around 13 nm. With the precursor at room temperature and the water kept at 10 °C then the particle diameter will be nearer 30 nm. As the water temperature is increased above this the particle diameter also increases but the formed colloids become unstable and the size distributions become wider. In this recipe we will use a syringe to deliver the precursor from a chilled reservoir to the water as this is a quick and easy way of doing it. You may find you wish to place a cooling jacket around a dropping funnel and allow the titanium chloride to drop directly into the water so that the temperature is more rigorously controlled. This recipe will outline the preparation of the smaller 2–4 nm titania particles but can be changed as mentioned earlier to produce the larger particles (Fig. 4.5).

To make 2–4 nm titania particles from $TiCl_4$ you will need:

1. A fume hood. This experiment should be carried out in a fume hood with a nitrogen or argon gas line fitted so that moisture sensitive precursors can be worked with. You will be pouring the precursor into water at one stage but this should happen under your control.

2. Titanium tetrachloride. This will come as a liquid. *It is corrosive and air and moisture sensitive! Work with under nitrogen! If it is exposed to water it can react exothermically and will fume in damp atmospheres.*

3. A peltier cooling system and bath full of antifreeze. The bath should be large enough to hold a 100 mL beaker but will only be used to cool less than 5 mL of the titanium precursor. Check that it can reach as low as −20 °C. If it cannot try adding more ethylene glycol to the antifreeze solution as this sometimes helps. The recipe for antifreeze is equal volumes of ethanol, water and ethylene glycol. If you do not have a peltier system you can try using a freezer to keep the precursor in. If you do this make sure it is sealed in a round bottomed flask such that no moisture can get in. Don't forget that it's in there once you are finished.

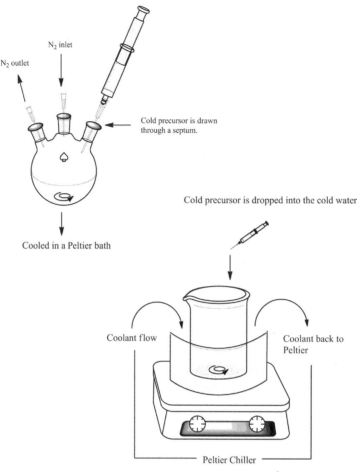

N₂ inlet

N₂ outlet

Cold precursor is drawn through a septum.

Cold precursor is dropped into the cold water

Cooled in a Peltier bath

Coolant flow

Coolant back to Peltier

Peltier Chiller

FIGURE 4.5 Experimental setup for the hydrolysis of TiCl₄.

4. A 10 mL triple necked round bottomed flask with a clamp to keep it stable. This will hold the precursor in the cold bath and allow you to remove small aliquots with a syringe. The necks should have rubber septums fitted. You will need two needles that can penetrate one septum each so you can flush the round bottomed flask with nitrogen.

5. A large chiller bath maintained as close to 0 °C as you can get it without the water freezing though ~1−2 °C will still result in smaller particles. Depending on the size of titania particle you would like being able to turn the temperature up to 10 °C is useful. Ideally this will need to be on top of a magnetic stirrer though overhead mechanical stirring can also be used.

6. A large 2 L beaker with a magnetic stir bar.

7. A 5 L beaker for dialysis and dialysis tubing and clips. Ensure the pore size of the dialysis membrane is smaller than 2 nm so you do not lose the colloid. You will need enough tubing to handle close to 1 L of solution in 100−200 mL aliquots allowing for half the length of each tube for folding over and sealing the bag. You will need 4 clips per bag—two for each end.

8. A rotary evaporator fitted with a cold trap. The cold trap should be filled with dry ice and ethanol. This will be used to dry the colloid into a powder and remove the HCl evolved from the reaction.

9. A pH meter for checking solution pH and a thermometer for checking the solution temperature.

10. Depending on the pH of the colloid you wish to produce you will need phosphate buffer salts for stabilising the colloid at pH 7.

11. A 5 mL graduated glass syringe fitted with a long needle that can reach the bottom of the 10 mL round bottomed flask. This will be used to transfer the precursor to the water.

To make a colloid of 2−4 nm titania particles from TiCl₄:

1. Flush the 10 mL round bottomed flask with nitrogen or argon by passing the gas through a pipe attached to one of the needles in the septum. Make sure there is a second needle in another neck to let the gas back out or something will pop off. After 1−2 min all the oxygen and moisture should be removed. Once this is done remove the needles so that the flask is sealed.

2. Depending on how your stock of TiCl₄ is stored it may be in a screw top bottle or a bottle fitted with a rubber septum or crown cap. This recipe is based on having a septum sealed bottle. If it is a screw top then you will need to extract an aliquot of the

precursor for use in a glove box under nitrogen or argon. Using the syringe extract 3.5 mL of $TiCl_4$ through the septum and transfer this to the nitrogen flushed 10 mL round bottomed flask by injecting it through a septum. If possible keep the syringe chilled in a freezer or on a cold plate protruding from the coolant bath so that it is cold before use.

3. Turn on the Peltier cooled bath and adjust the temperature to $-20\ °C$. Lower the flask into the bath so that the coolant is above the precursor and clamp it in place. Leave it ten minutes to equilibrate to temperature.

4. While you wait add 900 mL of water to the 2 L beaker and set it in the cold bath at $0\ °C$. Check that the water is at the correct temperature before proceeding. Add the stir bar and stir as vigorously as possible. Overhead stirring is an acceptable alternative if the use of a cooled water bath prevents the use of magnetic stirring.

5. Using the cold syringe, remove $TiCl_4$ from the flask and add the precursor dropwise to the water over the course of two minutes. It is likely that the precursor may warm up in this time so avoid handling the syringe in a way that will transfer heat into the barrel. You may find it easier to extract small amounts at a time with the syringe though the decrease in volume may allow the precursor to heat up faster.

   The titanium tetrachloride is hydrolysed within 2 min of addition to the water and then begins to condense into titania. The pH in this period will drop to pH 1.8. By 12 min the particle diameter will be somewhere around 2 nm and grows to around 2.4 nm by an hour. The reaction should be complete after 3 h.

6. If the formed colloid is maintained in the chiller bath below $5\ °C$ it should remain stable for many hours. If the colloid is raised to room temperature then it will begin to precipitate out of solution due to ionic pressure and the concentration of titania. This can be overcome by dialysis.

   Remove 100−200 mL aliquots of the colloid and place into dialysis bags. It is never a good idea to put all your solution into one dialysis bag as clips are notorious for coming undone and you might loose the lot. Bags can also split from internal osmotic pressure as water permeates in to dilute out a concentrated salt. Dialysis tubing should be soaked in water to loosen it up before use then one end should be folded over and clipped shut. Add the colloid to be dialysed either by pouring it into the bag with a funnel or by carefully pipetting the colloid in. The pipetting method is slow but allows you to control how quickly you add the sample so that it doesn't over flow from the tubing. The bags should be sealed by folding over the end of the tube and using two clips in case one fails.

7. Dialyse the aliquots, together if you are doing the whole lot, in 4 L of water in the 5 L beaker for an hour. Repeat once more. The pH of the titania colloids inside the bag should be somewhere around pH 2.5.

8. Transfer the aliquots from the dialysis bags into a rotary evaporator fitted with a cold trap and dry. This will remove the water to form a dry powder and will also remove the excess HCl where it will be captured in the cold trap. *Remember to clean out the cold trap after use.*

9. Collect the white formed powder. At this point it will be easiest to run solid state analysis such as powder X-ray diffraction of BET surface area measurement. If a higher portion of the anatase crystal phase is required then the powder should be heated between 350 and $500\ °C$ for 2 h in air to convert the amorphous material to the anatase phase. The smaller the particle formed is then the lower the sintering temperature needs to be to induce a phase change. After heating, check that the correct phase has been produced by powder x-ray diffraction or electron diffraction in a transmission electron microscope. The titania can now be redispersed in water to the desired concentration.

10. The concentration of titania colloid you make up in solution is up to you. Remember that if it is too concentrated then the colloid will most likely precipitate out of solution unless a stabilising agent is added.

A good place to start is to make up a solution equivalent to 0.5 g in 1 L of water. The as prepared dispersion will have a pH of ~3.1 as the surface of the titania particles will be acidic. This dispersion is stable for about a week. If a base, like sodium hydroxide, is added rapidly to raise the pH then the colloid will also be stable from pH 9.5−12. For applications at a neutral pH this will be near the zero point charge of the titania and it will be unable to stay separated by electrostatic repulsion. In this case make up the titania in a 2×10^{-3} M buffer of sodium hydrogen phosphate. Depending on the preparation methods you might find there is some variance in the zero point charge related to pH. The zero point charge can be determined by titration of the colloidal dispersion from acidic to basic conditions and monitoring the pH. The point of inflection in the acid to basic transition of the plot will be the location of the zero point charge. Additionally, by waiting for a few seconds between each addition of base you will be able to see if the colloid is flocculating or not. The pH value where precipitation occurs is the point of zero charge on the particle surface. This method can be time consuming and, if you have access to one, it is quicker to use a zeta potential analyser to determine the surface charge at various different pH. For titania this will be somewhere between pH 3.5 and 6.7. The relationship for pH at the zero point charge is defined as:

$$pK_1 = MOH_2^+ \leftrightarrow MOH + H^+$$

$$pK_2 = MOH \leftrightarrow MO^- + H^+$$

$$pH_{zpc} = 1/2\ (pK_1 + pK_2)$$

Transmission electron microscopy of the particles will allow you to determine the average diameter and particle distribution. If the temperature control has been successful then there should be a lower than 10% size distributions. Aside from the particle characterisation it is interesting to look at the photocatalytic properties of the titania powder. A powder consisting of nanoparticles would be expected to have a high surface area. For comparison, commercially available Degaussa P-25 titania nanopowder has a surface area around 50 m^2 g^{-1}. A larger surface area means that there is a lot of interfacial activity between the particle and the liquid phase so that there is a correspondingly higher number of surface bound hydroxyl groups that can be turned into free radicals when the titania is photoexcited.

The efficacy of the titania as a photocatalyst can be determined by adding a coloured organic dye to a colloidal dispersion of a known concentration or surface area and monitoring the concentration by UV-Visible spectroscopy (Hall et al., 2006). Methylene blue is the most commonly used organic dye for this and has a peak absorbance of 660 nm. A dispersion of titania can be added to the methylene blue solution and irradiated with a known power output light source. Roughly 0.1 g of titania in 25 mL of 6 mg/L methylene blue in water is a good place to start to get an idea of the catalytic rate. This will need adjustment depending on how efficient (or poor) the produced particles are as catalysts. As the titania is activated by UV light you will need to use a light source that has an output in the UV range. A sun lamp held at a known distance can be used. Be careful when using a UV lamp that the power output is not likely to cause burns or retinal damage. This is unlikely with a sun lamp but it is still advisable to partition off the light so that there is no light leakage into the laboratory area. It is also important, for long running periods, that the heat of the lamp will not set off any heat detectors or fire alarms or more importantly start any fires. Most laboratory fume hoods are fitted with these detectors and it is very embarrassing to have a whole department evacuated because you left a light on in the wrong place. The reaction will also need to be performed in a quartz cell as glass will filter out UV light which will defeat the purpose of the experiment. Include a magnetic stir bar to keep everything homogenised.

To run the photocatalytic degradation measurement: Calculate the extinction coefficient of your methylene blue solution by running absorbance spectra for three different known concentrations. Take 3 aliquots from your titania/methylene blue mixture at regular time intervals without illumination to check the system is equilibrated. For each aliquot run it in a centrifuge until the titania has sedimented out so that you have a top solution containing only methylene blue suitable for running on the UV-visible spectrometer. Then turn on the light source and take aliquots at regular intervals and centrifuge them so you can track the methylene blue concentration with the spectrometer. Removing the aliquots might arguably change the concentration of reactants but hopefully you will be removing the same amount of catalyst and dye each time so that the concentrations within the quartz reaction vessel remain constant. By relating concentration to peak absorbance you can work out the rate at which the dye is being photocatalytically degraded. This is normally run as a comparison against Degaussa P-25 titania as a standard which is essentially made commercially exactly the same way as this particular nanoparticle recipe.

The peak absorbance of the titania colloid in solution phase UV-visible spectroscopy can be used to determine the band gap structure (Precker, 2007). The band gap energy(E_g) is usually quoted in electron volts and this can be calculated from the wavelength of light that is required to promote an electron into the conduction band. The UV-visible spectrum of titania will show a maximum absorbance in the UV range and the position of the peak in nanometres can be converted to electron volts using the following formula.

First the energy of the photon E is calculated in Joules from the wavelength:

$$E = \frac{hC}{\lambda}$$

where h is Planck's constant, C is the speed of light and λ is the wavelength (in metres). To convert the value of E in joules to a value in electron volts E_g:

$$E_g = (6.21450974 \times 10^{18}) \times E$$

In bulk titania the peak absorbance is around 385 nm which correlates to a band gap of ~3.2 eV (5.1269×10^{-19} J). In the colloidal titania the peak absorbance is blue shifted to a lower wavelength (higher energy) and will display a band gap somewhere around 3.35 eV. It has been observed previously that the absorbance peak shifting in nanoscopic titania is not directly related to the particle size as is observed in other semiconductors such as cadmium selenide (Serpone et al., 1995). From 2 up to 30 nm no significant shift in the peak absorbance in the UV-visible spectrum is noted. Because of this the effective mass model formula for relating a band gap shift ΔE_g to the diameter of the particle cannot be directly used as it is with other semiconductors. This formula is explored more in depth in the cadmium selenide recipes section but essentially it relates the quantum size effect of band gap broadening to the diameter of an electron/hole pair known as an exciton and the size of the particle. The blue shifting in titania may be an effect of other transition states within the crystal lattice. Perhaps this is just as well if we think about the end application of titania as a photocatalyst. Blue shifting in titania means that we would need higher energies to promote an electron into the conduction band where it can then form a free radical hydroxyl at the particle surface. Much work has been done on bringing the active range of titania into the visible wavelength and this is normally achieved by doping the system with a metal to introduce defects (Ma et al., 2005). Another recipe will demonstrate a simple method to bring the activity of titania well into the visible spectrum but first we must look at a few other preparation methods to produce titania over some wider size ranges and with less water sensitive reagents.

PREPARING SUB 4 NM NANOPARTICLES USING AN ALKOXIDE TITANIUM PRECURSOR

Adapted from Choi et al. (1994)

In the place of $TiCl_4$, titanium tetraisopropoxide can be used. This variation uses smaller amounts and is considerably less technically challenging than the first recipe. Because there is a lack of choride ion formation *in-situ* to reduce the pH in the water an acid must also be added.

To make a titania colloid from a titanium alkoxide you will need:

1. 1.25 mL (0.00412 mol) titanium tetraisopropoxide ($Ti(OCH(CH_3)_2)_4$). This is still water sensitive but not as reactive as the $TiCl_4$. It should be stored in a dry box when not in use. Brief exposure of the alkoxide to atmosphere will not result in fuming and it is alright to work with open on the bench for short periods. It is a liquid and can be pipetted direct from the bottle using a volumetric pipette.
2. 25 mL absolute ethanol that is water free in a glass vial.
3. A 400 mL beaker containing ~200 mL water.
4. A pH meter and a nitric acid solution. The strength can be variable as you will be using to adjust the pH of the water to 1.5. A 1 M strength solution should be suitable. You will need a teat pipette to add the acid to the water with.
5. A magnetic stirrer and stir bar. On top of this should be a chiller bath maintained between 0 and 4 °C. The colder it is then the smaller the particles should be. The bath should be large enough to sit the 400 mL beaker in.
6. A rotary evaporator.
7. A centrifuge and suitable centrifuge tubes. You will need an excess of water and bench ethanol for rinsing.

To make a titania colloid from a titanium alkoxide:

1. Using a volumetric pipette add 1.25 mL of the $Ti(OCH(CH_3)_2)_4$ to 25 mL of absolute ethanol in a glass vial. The precursor can be transferred directly from the bottle but take care to not leave it open for long periods. Wherever possible use the precursor under an inert atmosphere. Mix thoroughly.
2. Place the 400 mL beaker containing water and a magnetic stir bar into the chiller bath and turn on the magnetic stirrer. While waiting for the temperature of the water to equilibrate to below 4 °C you can adjust the pH to 1.5 using nitric acid dispensed by teat pipette. Go slowly and drop by drop as it will change very quickly. Once the pH has been adjusted make up the volume of the water to 250 mL. Let the temperature equilibrate again under vigorous stirring.
3. Add the precursor in ethanol solution to the cold water dropwise under stirring. Once it has all been added leave the solution stirring overnight. This will form a transparent colloidal solution of anatase phase titania which will remain stable indefinitely if kept below 5 °C. Further steps need to be taken if you wish to get rid of the acid and isolate the dry powder.
4. Use a rotary evaporator to dry the colloid to form a white powder. This will be ~30 wt% HNO_3 so it is important to resuspend the powder in water and centrifuge it. Repeat this process 2—3 times exchanging the water and then centrifuging it. The acid can also be neutralised by the addition of sodium hydroxide but excessive concentrations of the formed $NaNO_3$ can 'salt out' the colloid.
5. The dried isolated titania can now be resuspended in solution and at a loading of 0.5 g/L the pH should be somewhere around 3.

These colloidal particles are uncoated and will behave similar to those in the $TiCl_4$ derived recipe. It is possible to use many different alkoxide chain lengths to obtain different sized particles as the rate of hydrolysis will vary. Larger particles can be made by using a glycolated titania precursor and the method for this type of preparation is outlined below.

MAKING LARGER TITANIA SPHERES FROM TITANIA GLYCOLATES

Adapted from Pal et al. (2007), Jiang et al. (2003)

An extra level of control can be imparted by exchanging a monodentate ligand such as the alkoxide for a bidentate ligand such as ethylene glycol. This means that all four ligands coordinated to the metal centre are exchanged for just two which can bind on two sites each. In this recipe larger spherical particles of titania hundreds of nanometres across can be made. Due to the high refractive index of titania (2.4 for the anatase and 2.9 for the rutile polymorphs, respectively) it can be of use in the making of optically active materials such as synthetic opals. A titania glycolate is formed by adding a titanium alkoxide like titanium tetrabutoxide to an excess of ethylene glycol according to the formula:

$$Ti(OC_4H_9)_{4(l)} + 2OHCH_2CH_2OH_{(l)} \rightarrow Ti(OCH_2CHO)_{2(l)} + 4C_4H_8OH_{(l)}$$

It is important to remember that the titania is still in the 4+ oxidation state (d^0 electrons) and that a nucleophillic substitution has occurred not a redox reaction. The chain length of the alkoxide makes no difference to the production of the glycolate as it will be substituted to form an alcohol which will play no further role in the titania formation. The produced glycolate precursor is highly

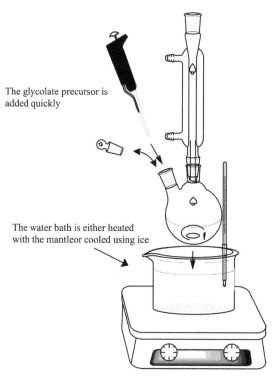

The glycolate precursor is added quickly

The water bath is either heated with the mantle or cooled using ice

FIGURE 4.6   The experimental setup for making titania spheres from titania glycolate.

stable to atmospheric moisture and can be worked with open on the bench once prepared. It is in fact quite difficult to get it to hydrolyse in many solvents save for acetone with a small amount of water in. Spherical particles can be formed over the range 100–500 nm by controlling the concentration of titania glycolate and the reaction temperature (Fig. 4.6).

To make titania spheres from titania glycolate you will need:

1. A glove box filled with an inert atmosphere such as argon or nitrogen. This may be slightly over the top as titanium tetrabutoxide is not very quick to react but it is always worth working to the highest standards you can reasonably achieve. If something goes wrong with the recipe you will not be left wondering if the titanium butoxide was prehydrolysed before you even began.
2. 0.5 mL (1.46×10^{-3} moles) of titanium tetrabutoxide ($Ti(OC_4H_9)_4$). The density of the butoxide is 1 g/mL meaning it can be transferred directly using a volumetric pipette to give 0.5 g. This will need to be inside the glove box.
3. 5 mL of ethylene glycol in a 20 mL beaker with a magnetic stir bar. This will be inside the glove box.
4. A magnetic stirrer within the glove box.
5. 100 mL of high purity acetone in a two necked 200 mL round bottomed flask with a quick fit neck for a reflux condenser and one with a glass stopper in. This should be made to 0.3% water by the addition of 0.3 mL of water. It is important that the water content does not exceed 0.5% or else the formed spheres will have a high polydispersity. Add a magnetic stir bar. *Acetone is flammable.*
6. A heating mantle with magnetic stirring and a water bath on top. This water bath should be large enough to accommodate the round bottomed flask and also to hold ice if required. In order to control the size of the formed titania spheres the temperature of the acetone will have to be controlled between ~2 and 50 °C. The mantle should be in a fume hood with a water line for a reflux condenser. You will need ice for cooling depending on the size of particle required.
7. A reflux condenser that fits to the top of the round bottomed flask. You will need a clamp and *make sure the pipes are clipped to the glassware and taps!*
8. A centrifuge and suitable centrifuge tubes for isolating the product. You will need an excess of water and ethanol for rinsing.
9. A furnace capable of reaching as high as 850 °C for calcining the amorphous spheres and converting them into the crystalline polymorphs. A crucible for holding the sample in is also needed.

To make titania spheres from titania glycolate:

1. In the glove box add the 0.5 mL of titanium tetrabutoxide to the 5 mL of ethylene glycol under magnetic stirring. Leave this stirring at room temperature for 8 h to ensure that all the butoxide is substituted by the glycol. After this time remove the formed titanium glycolate in ethylene glycol solution from the gas box. It should be stable to air with no visible precipitate forming over time.

2. Setup the reflux condenser for the 0.3 wt% water in acetone solution in the water bath. Depending on the end size of particle (post calcination in a furnace) you would like to get select a temperature from the chart below for the water bath:

Water bath temperature °C	Diameter/nm	Standard deviation
2	83	8
15	95	14
26	122	14
35	125	11
50	166	19

To get below room temperature then you can use ice to cool the water bath. Keep in mind this can be tricky to balance exactly. For temperatures above ambient conditions it is important to remember that acetone has boiling point of 56 °C and will evaporate quickly if left to air. Using a reflux condenser should enable you to keep the loss of acetone to atmosphere to a minimum. *Heating acetone can be dangerous so make sure there are no sources of ignition nearby.* Turn on the magnetic stirrer.

3. After the acetone solution has equilibrated to the desired temperature add all of the titanium glycolate in ethylene glycol solution at once through the side neck in the round bottomed flask. Replace the stopper and turn the stirrer up as high as possible. This is a 0.014 mol/L solution of titania precursor.

4. After 15 min turn off the magnetic stirrer and allow the solution to age for 1 h. The temperature can be allowed to return to ambient conditions at this point. After the initial hydrolysis and condensation the densification during the ageing process is not temperature dependant. The solution should turn white as the spheres of amorphous titania form in the first 10 min.

5. Turn off the heating mantle and *make sure the reflux condenser is turned off.* There should be no sediment in the solution but if there is, give the flask a swirl before transferring the solution to centrifuge tubes for washing.

6. Making sure the centrifuge tubes are balanced (usually to within 0.02 g of each other) and centrifuge at 13,200 rpm for 5 min. Check that the titania particles have sedimented before pouring away the top solution and adding a 50% water in ethanol mix and redispersing. Repeat this 5 times to ensure that any remaining ethylene glycol has been rinsed away. You should be left with roughly 1 g of structurally amorphous titanium dioxide spheres. At this point they can be converted directly in an oven or you may wish to arrange them into an opal structure. The procedure for this is given in the section on opal preparation.

7. In order to convert the as-formed spheres to anatase or rutile you will need to calcine them in an oven. For anatase the spheres should be heated to 500 °C for 2 h. For rutile the spheres should be heated to 850 °C for 2 h. A heating rate of between 1 and 10 degrees/minute is acceptable. The spheres will shrink by 30% in diameter after conversion to anatase and by 40% after conversion to rutile. One of the tricky factors is that the particles will tend to fuse together if they are small. Spheres below 200 nm will convert to the crystal polymorphs at lower temperatures so one idea to prevent fusing is to heat the spheres for longer at a lower temperature such as 350−400 °C. If you are attempting to make opal then the fusing is actually beneficial as it will hold the large crystal together and prevent fracturing.

Much of the decrease in size during heating is due to the presence of ethylene glycol groups still bound to the central titanium. IR analysis of the solid powder before calcination will show C−H stretching modes characteristic of the ethylene glycol at $3350 \, cm^{-1}$. This band will disappear if the spheres are heated above 350 °C due to the oxidation of the glycol units. As the titania is heated the Ti−O stretching mode will shift from $640 \, cm^{-1}$ to $465 \, cm^{-1}$ which represents a change from the amorphous form to the crystalline anatase form.

It is possible to try this experiment in other solvents such as alcohols, dimethylsulfoxide, dimethylformamide and acetonitrile but the size distribution of the formed spheres will become wider. The size of the spheres in the acetone based synthesis can also be altered by adjusting the concentration of the titanium glycol precursor added but this will influence the size distribution. A concentration of the precursor in acetone of 0.003 mol/L will give smaller particles (50−100 nm) but they will have a size distribution of less than 10%. Up to 0.077 mol/L will produce particles around 600 nm but these will have a size distribution above 10%.

MAKING BLACK TITANIA NANOPARTICLES

Adapted from Chen et al. (2011)

Titania is certainly a useful material for its light harvesting and photocatalytic properties but is limited in that it can only absorb a small portion of the solar spectrum. Shrinking the band gap between the highest occupied molecular orbital (HOMO) and the lowest unoccupied molecular orbital (LUMO) has been the focus of much research. By shrinking the band gap the wavelength of light required to promote an electron into the conduction band can be bought into the visible region. As mentioned previously, a metal or non metal dopant can be added to the titania which may redshift the absorbance spectra but this approach has been limited in its success. Currently, the best achievable doping has been through the use of nitrogen to introduce defects in the anatase structure. This is done by heating a powder of anatase phase TiO_2 at 500 °C under a nitrogen stream in a tube furnace for 2 h. After this process the titania takes on a yellow colour which is indicative that it has some absorbance in the visible range. Despite this the efficiency in light energy conversion is still only around 8%. The doping works by introducing crystal defects in the surface of the anatase crystal. In a phase pure particle the band gap will be well defined with a sharp cut off point between the bottom of the LUMO and the top of the HOMO. When the crystal lattice is distorted it begins to blur these edges because the defects offer ways in which an induced charge can 'short cut' into the conduction band through a lower energy transition. For example, in a metal doped material the guest metal d-orbitals can accept or donate electrons into the conduction bands depending on their oxidation state. Nitrogen doping works by forming a small amount of TiN in the crystal lattice. This introduces lattice defects of the type that accept lower energy transitions and allows for some overlap into the lattice of the lower energy nitrogen $2p$ orbital. Sulphur and carbon have also been shown to work as doping materials by the same mechanism of introducing defects. Although still higher than anatase alone in photocatalytic activity the improvement of absorbance in the visible region is still meagre when compared to the activity under UV illumination. This problem has been addressed by using hydrogenation to induce defects in the anatase lattice structure. The resulting powder is called 'black titania' and consists of a pure anatase nanocrystal surrounded by a shell of titania and titania hydride defects. Importantly, the defects are localised to the surface of the nanoparticle and this is advantageous over previous doping strategies. The surface defect layer acts not only to lower the absorbance into the visible and infra red region but acts as charge separation traps. Stabilised charge separation not only gives an electron more time to form a radical but, as the trap is at the particle surface, radical formation with surface hydroxyls is much more likely. It can truly be considered a nanoengineered material, as the structure at the particulate scale is just as important as the crystal structure in deriving the properties.

In this recipe you will make black titania which can be used as a photocatalyst or even to generate hydrogen from water and sunlight! The formation of the nanoparticles is rather standard compared to the previous recipes in that you will be performing an acid catalysed hydrolysis. The key difference to the hydrolysis step is that the nanoparticles are formed in the presence of an organic template. F127 is a triblock polymer that self assembles into micelles in solution. Each molecule of the polymer has a hydrophobic poly(propylene oxide) middle and two hydrophilic poly(ethylene oxide) tails. The hydrophobic portions of the polymer tend to aggregate together to minimise contact with the water while the hydrophilic portions point out from the formed micelle. Alkoxide precursors tend to be slightly hydrophobic (even though they react with water) so that titanium isopropoxide in the reaction solution will be sequestered into the core of the polymer micelle. The alkoxide in the core is not really prevented from hydrolysis with water free in solution but it does act to template the formed particle during condensation and the polymer will coat the surface preventing aggregation. During the heating step the organic polymer will be burned away leaving only pure titania nanoparticles ready for the hydrogenation step. The most important part of this preparation is the hydrogenation step where the titania powder will be placed in pure hydrogen and heated at a high pressure. If you have the setup you may wish to try this with other forms of anatase phase titania such as thin films or porous monoliths.

To make black titania you will need:

1. 3.7 mL (~0.0125 mol) of titanium tetraisopropoxide. You will need a volumetric pipette and tips for dispensing the alkoxide.
2. 30 mL of absolute ethanol in a 100 mL round bottomed flask with a glass stopper. You will also need a cork ring to stand the flask in.
3. 3.2 mL of a 2 M solution of hydrochloric acid in water.
4. 0.8 g of F127 triblock polymer. This is also known as Poloxamer 407.
5. A heating mantle and oil bath with a magnetic stirrer. Depending on the depth of the oil bath you may need to use a clamp to hold the round bottomed flask in place during heating.
6. A rotary evaporator that can fit with the round bottomed flask and is fitted with a splash guard. You will need a cold trap to collect the acidic ethanol and water that will come off. Dispose of the waste in the appropriate solvent waste container.
7. A drying oven that can reach 110 °C and a furnace that can reach at least 500 °C. Ideally the furnace will have a programmable ramp and cooling rate. You will need a crucible for the furnace.
8. A vacuum line or pump attached to a vacuum chamber.
9. A gas sorbtion analyser setup for hydrogen. This is a fairly specialised piece of equipment designed for looking at how hydrogen is adsorbed and released from various materials. The system you use will need to contain ~1 g of formed titania powder in a sealed vessel under hydrogen at a temperature of 200 °C and pressure of 2 mega pascals for 5 days. *Hydrogen gas is explosive. Observe all applicable safety protocols when working with it!*

To make black titania:

1. Dissolve the 0.8 g of F127 in the 30 mL of ethanol and add the 0.32 mL of 2 M HCl using a volumetric pipette. Add a magnetic stirrer.
2. Setup the round bottomed flask in the heated oil bath under magnetic stirring and set the temperature to 40 °C. You will probably need a clamp around the neck of the flask to hold it in place. Let the temperature equilibrate.
3. Using a volumetric pipette add the 3.7 mL of titanium isopropoxide all at once to the stirring solution. Place the glass stopper in the neck and leave this solution stirring at 40 °C for 24 h. You may feel more comfortable setting up a reflux condenser on top of the round bottomed flask. As the round bottomed flask is much larger in volume than the reaction solution there should not be enough of a pressure build up to pop off the cork. The boiling point of ethanol is 56 °C but this will be higher in the reaction due to the presence of water. As with any reaction keep an eye on it for the first hour to make sure the stopper does not come out. Titania nanoparticles will be formed in this step.
4. Turn off the magnetic stirrer and heating mantle and remove the flask from the oil bath. The solution should be transparent without opalescence which would indicate large particles have formed. Connect the round bottomed flask to the rotary evaporator making sure the splash guard is in place and the glassware is clipped together. Set the water/oil bath on the rotary evaporator to ~40 °C and remove the ethanol and water. HCl should also be removed by this step. The drainage bulb of the evaporator should be kept in a cold trap so that any acid is not pulled through the vacuum pumps. Because there is F127 in the ethanol solution it is possible that the solution will begin to foam under the vacuum. If this happens remove the flask from the evaporator and transfer the solution into a large beaker. Place this in a drying oven and let the bulk of the ethanol evaporate to air. Take care that the oven is vented into a fume hood and that there are no ignition sources nearby.
5. Once the liquid has been removed from the flask place into an oven and dry at 110 °C for a further 24 h to drive off the last of the water.
6. At this stage the formed nanoparticles are still mostly amorphous. To convert them to the anatase form they need heat treatment. This will also get rid of the organic template by converting it to CO_2. Transfer the dried powder into a crucible and place into a furnace. Heat to 500 °C for 6 h at a ramp rate of 0.3 °C/min in air. A white powder should remain but this will be extremely fine so take care that it does not get blown away when you move it about.
7. The nanocrystals of titania should now be in the photoactive anatase phase. To make them active in the visible region they need to be hydrogenated to introduce defects and give the powder a black colour. Place the powder into the gas chamber for treatment. Expose it to vacuum to degas the surface for 1 h. After this flood the chamber with hydrogen to a pressure of 2 mega-pascals and heat to 200 °C. Leave the sample under these conditions for 5 days. The titania will absorb 0.3 wt% of hydrogen into its surface layer to form defects shifting its absorbance into the visible range. *Hydrogen is explosive, ensure all of the required laboratory safety procedures are followed when using it.*
8. Remove the hydrogen from the chamber and bring back to ambient conditions. You should now have a black powder ready for analysis.

UV/visible solid state diffuse reflectance should show an absorbance onset in the near infra red with an absorbance across the visible spectrum. Titania normally reflects light in these regions which is why it appears white. The band gap will now be closer in energy to ~1 eV meaning that the black titania should have a far higher catalytic activity when illuminated with visible light. It can be interesting to try this material in the place of normal titania in solar cells and looking at the difference in efficiency. Another exciting potential application is in sunlight driven water splitting reactions. The original source paper describes a system where the black titania is used to generate hydrogen from a methanol and water mixture. The titania is loaded with 0.6 wt% platinum as this acts as a catalyst for the water splitting reacting by stabilising the charge separation generated in the titania. A rate 10×10^{-3} mol/h of hydrogen was produced under a sunlight simulator which corresponds to an efficiency of 24 % in converting sunlight energy to hydrogen energy. Just as with other anatase phase systems the black titania will degrade organic pollutants in water under sunlight and this can be measured by a methylene blue degradation experiment.

COATING NANOMATERIALS USING THE SOL GEL METHOD

One of the exciting aspects of nanomaterials is that they can be engineered structurally as well as chemically. Coating is a vitally important aspect of nanomaterials processing. The capacity to coat a colloid with a second layer of heterogenous material can impart an extra level of functionality. Coating is used to make nanoparticles biocompatible, chemically resistant and thermally resistant. Some fluorescent biomarkers can be toxic but will pass through the body if coated in glass. In contrast a sol−gel coating can be applied to a template with an interesting morphology such as pollen before the template is burned away to give a replica. Both protective and replication recipes and methods will be explored. These techniques can be applied in the manufacture of many core shell nanoparticles.

SILICA COATING A GOLD COLLOID

Adapted from Liz-Marzán et al. (1996), Kobayashi et al. (2011)

This recipe uses two classics of sol–gel templating, sodium silicate solution and the Stöber process, to coat colloidal gold particles with a layer of silica. Many colloids require a stabilising agent and often these will have been selected on their steric or electrostatic repulsion and how well they coat the particle. Depending on factors like charge and pH, the surfactant layer may repel the application of a precursor to the particle surface in solution. It could be tempting to try the approach of adding a precursor like an alkoxide and condensing it by the Stöber method in the hope that it will nucleate upon the surface. This can work if there is an oxide layer on the particle upon which the silica can grow but for many materials, like noble metals, there is not. For noble metal colloids a silica coating is helpful in forming a colloidal powder that can be dispersed in both oil and water phases. It can also protect the metal from extremes of pH that might degrade the surface though this may be rare if a noble metal is used. If you were to dry down a solution containing gold nanorods you would see the plasmon bands disappear as the particles came into contact with one another and it would look like bulk gold. An optically transparent layer of glass upon the surface can prevent the rods from coming into contact with one another and so the plasmon band is preserved even in the dried bulk. In this recipe a citrate stabilised gold colloid will be coated with silica. The most important step is to anchor a silane centred precursor to the surface of the gold so that silica can be grown upon it. Gold itself is 'vitreophobic' and provides no anchor point as it has no oxide layer to which the silica could condense. To get around this a modified silicon alkoxide is used that has replaced an alkoxide group with an amine terminated propyl group. This organosilane is anchored to the gold surface by the amine leaving the three remaining alkoxide ligands free to undergo hydrolysis and condensation with a glass precursor. An initial layer of glass is applied to this monolayer by using a sodium silicate solution to build up a 2–4 nm thick layer. The glass shell is thickened by using the Stöber method and further organosilane precursors can be added to functionalise the glass depending on whether it is going to be dispersed in a polar or non polar solvent (Fig. 4.7).

To make a glass coated gold colloid you will need:

1. 500 mL of citrate stabilised gold colloid containing the equivalent of 5×10^{-3} mol/L. For the best results they should be around 15 nm in diameter. You can use the recipe for this found in the gold nanoparticle section. This should be in a 1 L beaker.

2. 3-aminopropyltriethoxysilane (3-APTES) which will have a molecular weight of 221.37 g/mol. Shortly before beginning the recipe a solution of the 1×10^{-3} mol/L will need to be prepared. This can be done by using a volumetric pipette to add 0.0234 mL (0.0221 grams) of 3-APTES to 50 mL water under stirring. Make this up to 100 mL in a beaker and use immediately.

3. Roughly 100 mL of a 0.54 wt% solution of sodium silicate in water. Sodium silicate ($Na_2O(SiO_2)_{3-5}$) is normally sold as a 27–30 wt% solution in water. The correct concentration can be prepared by adding 2 mL of a stock 27 wt% sodium silicate solution and adding it to 98 mL of water. Only 20 mL is actually used but some will be lost during the ion exchange column step depending on how large the column is.

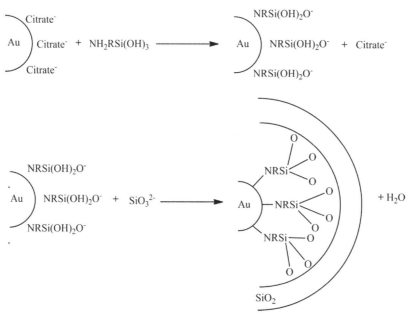

FIGURE 4.7 A schematic of the surface reactions involved in the formation of a thin silica shell on citrate-stabilized gold particles.

4. A cationic ion exchange column or resin. This will come in the form of beads of a zeolite which can be packed into a glass column supported by a clamp. It is used to adjust the pH of the sodium silicate so that it will form into silicic acid and from there condense to form silica. Cationic ion exchange resin normally comes loaded with sodium so in order to activate the resin you will need to rinse it through with acid first. This will replace the sodium ions with hydrogen ions. As the sodium silicate passes through the column the free sodium is swapped again with the hydrogen to form silicic acid. The solution is allowed to elute through the column and will need to be collected in a beaker at the bottom. The column should be wide enough to accept 20−30 mL of solution in one go. Keep in mind that the taller the column is the more pressure will be forcing the solution through so having a longer column can be marginally quicker.

5. A pH meter or pH paper for checking the pH of the sodium silicate solution.

6. An excess of 1 mol/L solution of HCl in water for activating the ion exchange column. A litre should be more than enough.

7. A magnetic stir bar and stirrer.

8. A rotary evaporator capable of coping with at least 500 mL of water. If one is unavailable the colloid can be concentrated by evaporation though this will be significantly slower.

The items listed from this point are for the growth of the silica shell on the gold using the Stöber method.

9. Tetraethoxysilane (TEOS) which will form more silica on the initial glass layer. The amount to be used will be determined by a formula given in the recipe.

10. 400 mL of pure ethanol. This will be added to a reduced volume of the colloid so that the conditions are correct for shell growth.

11. 2 mL of 30% ammonium hydroxide in water to act as a catalyst for condensation.

The items listed from this point are for the modification of the glass shell so that the particles can be dispersed in a non polar solvent.

12. 3-(trimethoxysilyl)propyl methacrylate (TPM). This is an organic functionalised silica precursor.

13. A centrifuge and suitable centrifuge tubes for isolating the glass coated colloid from ethanol.

To make a silica coated gold colloid:

1. Setup the ion exchange column and use a clamp to stand it over a receiving beaker. Flush the column through with the 1 M HCL. The first time the acid passes through the column you should see a drop in the pH as the water will now contain salt instead of acid. After two or three flushes the sodium in the column should have been completely exchanged. The pH of the eluted solvent should be the same as when it went in to the column if the exchange resin has been fully charged.

2. Place the magnetic stir bar in the 500 mL of gold colloid and set it to stir as fast as possible. Using a volumetric pipette add 2.5 mL of the 1×10^{-3} M 3-APTES in water solution. Allow this to sir with the colloid for 15 min. During this time ligand exchange will occur at the gold surface so that the organosilane precursor coordinates to the surface of the gold. The alkoxide groups are exchanged within minutes so that the aminopropylsilica becomes a triol. The isoelectric point of these groups is pH 2−3 which means that at pH 7 they remain negatively charged and repel each other through electrostatic interactions. This will keep the colloid stable.

   Do not be tempted to add excess 3-APTES in the hope that it will form a thick organosilica layer via hydrolysis. Too much and the electrostatic charge from the triol groups will not be enough to prevent the colloid from crashing out. The ligand exchange step is purely to provide an anchor point for glass formation on the surface using a sodium silicate solution.

3. Add 20−30 mL of the 0.54 wt% sodium silicate solution to the cationic ion exchange column (depending on the volume it can take) and allow it to pass through. Collect the eluant in a clean beaker and measure the pH. You want the pH of the sodium silicate to drop to somewhere between pH 10 and 11. At this pH the solubility of the sodium silicate is reduced so that it will begin to condense into glass. Because the solution is still basic the progress of the reaction is slowed enough that it will not self nucleate before there is a chance to add it to the colloid. If the pH is not low enough then pour the aliquot through the column again and it should drop further. If it does not then the column may need another charging with HCl.

4. Add 20 mL of the pH adjusted sodium silicate solution and add it to the 500 mL of gold colloid under stirring. The pH of the solution should drop to ~pH 8.5. After 10 min of stirring turn off the stirrer and leave the solution to stand for 24 h. During this time the sodium silicate will form small glass oligomers that will react with the silica alkoxide groups anchored to the particles. A shell of glass 2−4 nm thick will condense upon the surface of the gold colloid. At this point you may wish to use the particles as they are. The next step concerns the thickening of the glass shell using the Stöber process.

   If no further modification is to be made then it is a good idea to dialyse the colloid or centrifuge it and rinse with water to get rid of any excess silica. Over time it will either cause the particles to condense together or it will coprecipitate to form solid silica particles.

For the next step it is important to know the dimensions of both the gold particle and the glass shell. TEM analysis will help you determine this.

5. Transfer the colloid into a large round bottomed flask suitable for use with a rotary evaporator. You will need to reduce the volume from 500 mL to 100 mL. Take care that the colloid does not dry out completely. Any unreacted silicate solution

will condense upon drying and may cross link the separate particles so that a bulk solid is formed. Simple evaporation to air will also work but this will take much longer.

Alternatively use a centrifuge to draw the coated gold particles from solution and then redisperse them in a water:ethanol ratio of 1:4. This can also take a long time as the colloid will need to be centrifuged at over 6000 rpm for at least 6 h to isolate the particles. This can be worth it as it will result in a colloid free of excess silicates

6. Take the 100 mL of colloid and add pure ethanol until the ratio of water to ethanol is 1:4. Using a volumetric pipette ad 2 mL of ammonia solution to the colloid. The thickness of glass shell required depends on how large the starting gold particles were. This recipe is for gold particles ~15 nm in diameter. If the gold particles you are using are smaller (for the same molarity of gold in solution) then you will need more TEOS as the surface area will be relatively higher. If they are larger then you will need relatively less. For particles ~15 nm in diameter the shell thickness will increase from 6—60 nm with increasing TEOS concentration from 0.3×10^{-3} to 10×10^{-3} mol/L (Fig. 4.8).

It is a good idea to start off slowly. At higher concentrations of TEOS more self nucleation will occur. That is to say glass particles not containing gold cores will form. To produce a roughly 10 nm thick glass shell add 0.3 mL of TEOS to the solution using a volumetric pipette to the colloid under stirring. Leave this under mild stirring for 12 h.

7. After this time the thickness of the glass shell can be checked by TEM. Repeat the addition of TEOS until the desired shell thickness has been achieved. The colloid is ready for analysis at this point. If you wish to use the glass coated colloid in a non polar solvent proceed to the next step.

8. At this stage the surface of the particles is still coated with hydroxyl groups which stabilise the particles in aqueous conditions electrostaticaly. In a polar solvent these groups do not posses charge and will not prevent particle aggregation. For the particles to be stabilised in an oil they will need to be given yet another coating so that bulky organic groups protrude from the surface and prevent aggregation by steric repulsion. This is accomplished by adding 3-(trimethoxysilyl)propyl methacrylate (TPM). The surface bound hydroxyl groups on the glass coating nucleophillically substitute the alkoxide moieties leaving a surface bound methacrylate group in the place of three hydroxyls. The glass will have an organic layer covalently bound in place which will stabilise the particle in oil.

To the 500 mL of alcosol add 0.5 mL of TPM and stir magnetically for 45 min at room temperature.

9. Split the colloid into aliquots and place into centrifuge tubes. Centrifuge the colloid at 3000 rpm for an hour. There should be a sediment of the glass coated particles in the bottom. Redisperse the colloid in pure ethanol and repeat the centrifugation to remove all water from the sample.

10. The colloid can be dried and redispersed in ethanol or toluene for analysis.

The primary method of analysis of the silica coated gold nanoprticles is by transmission electron microscopy. The electron dense gold cores can easily be seen surrounded by a less electron dense glass corona. Depending on how well defined the size range of the produced particles is you may observe the spherical particles packing together to give ordered arrays. UV/visible spectroscopy of the coated particles should show that the plasmon band intensity is increased in comparison to uncoated gold colloid. This is due to the refractive index of the glass allowing the particles to absorb more light at specific wavelengths. Though this is useful for spectroscopic studies the glass shell can become a hindrance above 60 nm in thickness. At this point the glass causes large amounts of scattering which will decrease the observed plasmon absorbance along with increased absorbance at the blue end of the spectrum.

Thiolated organosilanes can be used instead of amines as an anchor point to the gold but be warned—they smell terrible. It is also important to use less harmful functionalities if the end application is in the biological field. Gold coated with glass can make an excellent contrast agent for X-ray imaging as the gold is X-ray opaque and the glass passivates the particles biologically.

This procedure can also be applied to coat CdS nanoparticles stabilised in the presence of citric acid (Correa-Duarte et al., 1998). The only modification to the recipe is to exchange the 3-aminopropyltriethoxysilane for 3-mercaptotriethoxysilane. The mercapto

FIGURE 4.8 A plot of shell thickness as a function of TEOS concentration. *(Reprinted with Permission from Kobayashi et al. (2011). Copyright 2011 Elsevier)*

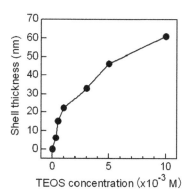

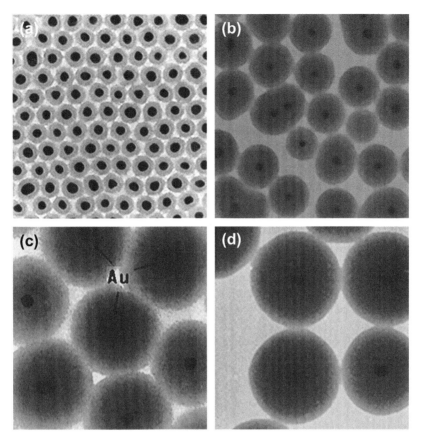

FIGURE 4.9 Transmission electron micrographs of silica coated gold particles produced during the extensive growth of the silica shell around 15 nm Au particles with TES in 4:1 ethanol/water mixtures. The shell thicknesses are (a, top left) 10 nm, (b, top right) 23 nm, (c, bottom left) 58 nm, and (d, bottom right) 83 nm. *(Reprinted with Permission from Liz-Marzán et al., 1996).*

group will bind preferentially to cadmium ions on the particle surface. A more in depth look at making cadmium sulphide and other semiconductor particles is given elsewhere (Fig. 4.9).

GLASS COATING AN ORGANIC CRYSTAL TEMPLATE

Adapted from Miyaji et al. (1999)

This recipe is a template directed sol−gel synthesis that will produce large hollow silica tubules 50−300 μ in length and 100 nm to 7 μ wide depending on the crystal template used. These are formed by coating a crystal substrate with a desirable morphology using a silica precursor in the presence of a basic catalyst as in the Stöber process. The template is then removed to leave the silica shell behind. In this case it is an organic crystal with a needle like morphology composed of ammonium *dl*-tartrate or ammonium oxalate. The first report of this templating mechanism described the silica as forming around an assembly of tartaric acid held together by hydrogen bonding (Nakamura and Matsui, 1995). This was later determined to be incorrect and that the tartaric acid was forming crystals with the ammonia being used as a catalyst for the glass condensation. The crystals are insoluble in ethanol but can easily be dissolved in water. As the sol gel reaction occurs under ethanol with only a small amount of water present it is possible to nucleate glass over the surface of the crystal before dissolving it by exposure to water. This recipe will work with a variety of crystals that behave in the same way so that it is possible to select a crystal for its size or shape and then make a replica of it. Hollow tubes made from the ammonia tartrate crystals will be 0.1−1 μ in width and 200−300 μ long. Hollow tubes made using the ammonium oxalate template will be 1−7 μ wide and 50−100 μ long. The thickness of the glass wall formed will be around 300 nm though this can vary depending on the rate of hydrolysis.

To make hollow silica tubules using an ammonia tartrate template you will need:

1. 0.77 mL (0.73 g, 0.0035 mol) of tetraethylorthosilicate (TEOS) which will be used as the silica precursor.
2. 5 mL of pure ethanol as the reaction medium in a 20 mL glass vial.
3. 0.02 g (1.33×10^{-4} mol) of *dl*-tartaric acid. The *dl*- prefix denotes that a mixture of both the left and right handed forms of tartaric acid is used. This is also called a racemic mixture and plane polarised light passing through a solution in water will not be rotated

one way or another compared to the pure *dextro* or *laevo* forms. The chirality of a chemical can have a great influence on how itself assembles but in this case both forms will turn into large needles having an identical morphology upon reaction with ammonia. The important difference is that the solubility of the racemic mixture is much lower than that of the pure enantiomeric forms. This means that the crystals are more stable and do not dissolve in the little amount of water present used to initiate hydrolysis. Additionally, the crystals composed of the individual enantiomers have a tendency to aggregate together which is no good for templating. If a single enantiomer must be used then reduce the exposure time to the TEOS during reaction to 10 min.

4. 2 mL of a 27% ammonium hydroxide solution which will act as both a catalyst and form the crystal template.
5. At least 50 mL of water and a volumetric pipette capable of measuring 0.06 mL of water.
6. A fine sieve or mesh. This will be used to separate and rinse the tubes. If a fine sieve is unavailable then a nitrocellulose 200 nm filter paper can be used.

To make hollow silica tubules using an ammonia tartrate template:

1. Dissolve the 0.02 g of tartaric acid in the 5 mL of ethanol.
2. Using a volumetric pipette add the TEOS and mix.
3. Using a volumetric pipette add 0.06 mL of water and mix. This amount of water will initiate hydrolysis but is not enough to dissolve the template crystals once formed.
4. Using a volumetric pipette add 2 mL of the ammonium hydroxide solution. A precipitate consisting of long white needles should form immediately. This is the ammonium tartrate and will have the formula $(NH_4)_2C_4H_4O_6$. The molar ratio of TEOS:H_2O:C_2H_5OH:*dl*-tartaric acid is 1:1:24.3:0.038. Give the vial a gentle swirl by hand and leave to stand for 30 min. Stirring will break up the crystals.
5. After this time swirl the vial gently and decant the sample through the sieve to collect the white precipitate. By this point the organic crystals should have a coating of silica. It is important that the reaction is not left to occur longer than 30 min or else a gel will form in the vial.
6. Rinse the collected tubes with repeated washings of water. This will dissolve away the organic crystal template leaving behind the hollow silica tubes.

The silica tubes are now ready for analysis. Below is the experimental variation to use an ammonium oxalate precursor.
To make hollow silica tubules using an ammonium oxalate template you will need:

1. 0.38 mL (0.36 g) of TEOS.
2. 5 mL of water in a 20 mL vial.
3. 0.4 g of oxalic acid.
4. 2 mL of 27% ammonium hydroxide solution.
5. A magnetic stirrer and stir bar. In contrast to the ethanol based procedure with the tartrate a balance between the rate of hydrolysis and the need to keep the crystals separate is required. The solution will be stirred for the hydrolysis and condensation step.

To make hollow silica tubules using an ammonium oxalate template:

1. Dissolve the oxalic acid in the 5 mL of water.
2. Add 2 mL of the ammonium hydroxide solution. Do not stir at this point. A white precipitate of crystals will form.

FIGURE 4.10   Scanning electron micrograph of hollow silica fibres prepared in the presence of incipient ammonium oxalate crystallization. Scale bar = 5 μm. *(Reprinted with Permission from Miyaji et al. (1999). Copyright 1999 American Chemical Society)*

3. Immediately after the precipitation add 0.38 mL of TEOS and a magnetic stir bar. Stir rapidly for one hour. Without the stirring at this stage the solution will form a gel. The TEOS:H$_2$O:Oxalic acid ratio in this reaction is 1:199.3:2.5.

4. Turn off the stirrer and decant the contents of the vial onto a sieve. You will need to rinse the isolated tubes repeatedly with water to remove the ammonium oxalate crystal as they are less soluble than the ammonium tartrate crystals. Alternatively the organic template can be removed by calcination above 400 °C though this may induce distortion and cracking in the tubes. The formed silica tubes are now ready for analysis.

The hollow silica replicas are most easily imaged by scanning electron microscopy. A cross section of the silica tubes will reveal walls 200−300 nm thick and either a triangular or rectangular shape denoting the morphology of the underlying crystal. X-ray diffraction patterns of the tubes should show a broad peak at $2\theta = 22°$ which is characteristic of amorphous silica. Any other peaks observed will correspond to organic crystals that have not been removed (Fig. 4.10).

HOW TO COAT A VIRUS TEMPLATE WITH SILICA

Adapted from Shenton et al. (1999), Rong et al. (2009)

Biological templates are incredibly useful in nanotechnology. Evolution is powerfully economical in design and function which is reflected in the nanoscale domains at which the machinery of life operates. A multitude of biological components found in nature serve as starting points for engineering at the nanoscale and much of this book is concerned with the use of these components from chemical, functional and morphological standpoints. In this recipe we will examine the simple aspect of using a virus with a uniform morphology as a template for coating with silica. One thing to keep in mind when working with biological templates is that they are not homogenously charged over their surface. The amino acid building blocks that make up a protein structure all have different behaviours at a given pH. This property is vital in the folding of a string of proteins together to form the familiar coiled and flat ribbon structures that come to mind when thinking of pictures of protein structures. The implications for this when using a virus or protein as a template is that the pH must be controlled and that the surface charges of the template and the electrostatic charge of the precursor must be complimentary.

The tobacco mosaic virus is a model template for the inorganic replication of a protein. It is comprised of a hollow protein tube 300 nm long by 18 nm wide. One major attraction of using a virus as a template is the consistency of the virion dimensions and a biological template of this sort will almost always exhibit a high fidelity if cultivated in large quantities. The virus is also stable in a wide range of pH and thermal environments and retains structural integrity between pH 2 and 10 and temperatures as high as 60 °C. TMV is therefore well suited to exploitation as a template due to its utility over a broad spectrum of synthetic conditions. It is relatively safe to work within the laboratory environment and does not require extensive protocols beyond standard biohazard level one. This means it is not infectious to humans and that keeping the working materials contained within an area cleaned with ethanol is normally sufficient for most laboratory work. In this recipe TEOS will be used to coat the TMV in silica. The coating reaction is performed at pH 2.5−3 which is below the isoelectric point of TMV (~pH = 3.5−4) (Oster, 1951). Amino acid groups within the protein which are presented to the structural exterior will be positively charged and provide a nucleation site for the anionic silicic acid molecules generated from the alkoxide precursor by acid catalysed hydrolysis. Glass will condense over the surface of the protein to form a shell roughly 3 nm thick when the reaction is complete. Interestingly, the lowering of the pH results in a neutralisation of the glutamic and aspartic acid groups at the tail ends of the proteins. Under neutral pH these groups tend to repel each other but with the electrostatic repulsion negated under the synthesis conditions, the TMV lines up end to end resulting in long glass coated wires of the virus. If the end application requires the formed rods to be of a uniform length then this can be difficult to overcome in the final product. If uniform lengths are required then it might be an idea to use another virus as a template. For every virus there are a multitude of strain variants though normally only a handful will have laboratory relevance. Different strains will have differences in the amino acid sequence. In turn this alters the functional groups that present themselves to the exterior surface so select specific viral strains that are better suited to act at templates under different pH, temperature and salt conditions depending on what you are going to do with them.

Though TMV is well characterised and understood there is another virus that can be employed to form one-dimensional anisotropic particles that do not link end to end during templating. The filamentus *fd* virus is slightly thinner and longer than an individual TMV virion having a diameter of 6.6 nm and a length of 880 nm. Like TMV the isoelectric point is around pH 4 so that the surface is positively charged below this. The preparation route is similar to the one outlined below though there can be some variation in the produced morphology and often a wheat sheaf like structure is obtained (Zhang, 2007)

To make glass coated tobacco mosaic virus you will need:

You will be working with a biological agent. Though it is of minimal harm to humans you should observe correct biohazard procedures for use and follow your laboratory *guidelines for handling biological waste.*

1. A strain of tobacco mosaic virus *vulgare* (U1). You will need a stock concentration of 40 mg/mL kept in a 10×10^{-3} mol/L solution of sodium phosphate buffer at pH 7. This can be stored in a fridge for use in other recipes.

2. 10 µL of tetraethylorthosilicate (TEOS) in 100 µL of ethanol. This will give a TEOS concentration of 0.05 mol/L.
3. A volumetric pipette that can measure out 100 µL.
4. A pH meter with a small probe that can be used with volumes of liquid. If you do not have access to one then pH paper can be used and the pH measured by dipping a glass pipette into the solution and dabbing it onto the paper. This will not be as accurate as a meter but the aim is to get the pH below the isoelectric point of the virus. As long as the solution does not go below pH 2 the virus will retain its shape.
5. Weak (0.01 mol/L) solutions of HCl and NaOH for adjusting the sample pH with. Use a teat pipette to deliver small volumes when adjusting the pH.
6. A centrifuge and centrifuge tubes that can handle ~5 mL of liquid.

To make glass coated tobacco mosaic virus:

1. Add 0.1 mL of the stock TMV solution to 3.9 mL water in a small vial.
2. Add the ethanol containing the TEOS to the TMV in water solution and adjust the pH to between pH 2.5 and 3.5 using the weak HCl or NaOH solutions. The preparation will still work if you adjust the pH to 4 but there will be minimal charge to attract the anionic silica species to the surface as this is close to the isoelectric point. This will obviously change the working concentrations but the ratio of silica precursor to TMV is the important factor. You should aim to not alter the reaction volume by more than 10%.
3. Leave the solution to stand for 24 h. Depending on results you may find that a standing period of 12 h works best.
4. Centrifuge the solution at 10,000 rpm for 15 min to form a pellet at the bottom of the centrifuge tube. Decant the top solution and resuspend the pellet in water. Repeat this twice more to get rid of unreacted precursor.
5. The silica coated TMV is now ready for analysis.

Grids for electron microscopy can be prepared by adding 20 µL of the suspension onto a carbon coated copper TEM grid. There should be no need to stain the grid for imaging as the silica coating should provide contrast under the beam.

Transmission electron microscopy should reveal that the produced inorganic wires can be up to 3 µ in length as a result of virion alignment at acidic pH. The dispersion can also be dried on mica or graphite for scanning probe microscopy. One of the interesting things about performing SPM on the formed fibres is that you can perform compressive force measurements. A force plot of a squeezed silica coated TMV tubule will show a dip in the positive deflection which is where the tube collapses on itself, a bit like compressing a cardboard tube (Fig. 4.11).

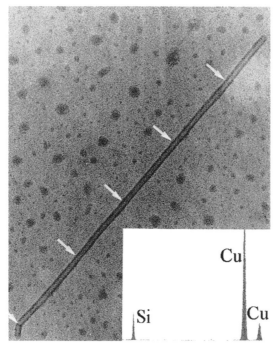

FIGURE 4.11   Self assembled silica/TMV nanotubular super structure. The white arrows mark the ends of five individual TMV particles, each 300 nm in length that have aggregated end to end along with a thin external shell of amorphous silica. Inset: Corresponding EDX spectrum showing Si peak (Copper peaks are from TEM grid). *(Adapted with Permission from Shenton et al. (1999). Copyright 1999 Wiley VCH)*

HOW TO COAT CARBON NANOTUBES IN SILICA AND OTHER OXIDES USING THE SOL GEL PROCESS

Adapted from Ling et al. (2011)

A method for producing carbon nanotubes is given elsewhere in this book and a more in depth look at the properties and uses of this staple of nanotechnology can be found there. In this recipe we will explore the use of carbon nanotubes as templates for glass coating and extend the range of materials to other metal oxide materials. The chemical stability of carbon nanorods means that you can be comparatively rougher with them chemically than you might dare with a biological template. Carbon nanotubes have a tendency to flocculate under aqueous conditions so for use in water it is necessary to modify them by either covalent surface modification or by associative coating with a surfactant. Because it is easier, this recipe will use a surfactant layer to keep the nanotubes stable in water. A cationic or anionic surface charge in the surfactant layer will have little influence over the absorbance of the silica as a small amount of 3-aminopropyl silane is added to the precursor solution to act as an anchor to the carbon surface. This is similar to the technique employed in the glass coating of gold recipe to overcome the vitreophobic citric acid layer on the gold particles.

To coat carbon nanotubes in silica you will need:

1. Just over 5 mg of multiwalled or single walled carbon nanotubes (MW- or SW-CNT's respectively). This can be weighed out using a four figure balance and keeping them in a small centrifuge tube to make sure they do not blow away. *Take care that the dry CNT's are contained so that there is no danger of inhalation. They are generally safe to work with when wet but can cause problems in the lungs if inhaled. They are too large for the lungs natural clearance mechanisms and will remain stuck there.*
2. A surfactant to be used as a dispersant for the CNT's. This could be cationic such as cetyltrimethylammonium bromide (CTAB) or anionic such as sodium dodecyl sulfonate (anionic). For convention in the recipe we will use CTAB. 5 mL of a 1% by weight solution of the surfactant in water should be prepared in a screw top vial suitable for use in a sonic bath. It is a good idea to have a piece of foam to use as a float for the vial in the bath.
3. 25 mL of a 1.85×10^{-3} mol/L solution of citric acid adjusted to pH 3 in a 50 mL beaker. The solution may not need adjustment. This can be prepared by adding 0.0355 g of anhydrous citric acid to 100 mL of water. The citric acid will act as a catalyst for the hydrolysis of the silica precursors.
4. A solution containing 5 µL of 3-aminopropyltriethoxysilane (3-APTES) and 45 µL of tetraethoxysilane (TEOS) in 100 mL of pure ethanol in a 150−200 mL beaker. This will give a 1:9 molar ratio of 3-APTES:TEOS.
5. A 1 mol/L solution of ammonium hydroxide in water. This can be made by diluting ~70 mL of a 27% ammonium hydroxide stock solution to 1 L with water. You do not need to be absolutely accurate as this will be used to adjust the solution pH to make it basic and catalyse the condensation of the silica.
6. A pH meter and a small Pasteur pipette for adjusting the reaction solution using the ammonia.
7. A sonic bath and magnetic stirrer with a magnetic stir bar. Both of these will be used to make sure the CNT's stay dispersed during the reaction.
8. A centrifuge with suitable centrifuge tubes. You will also need water for washing excess surfactant from the CNT's and ethanol for washing the product.
9. A vacuum oven capable of reaching 60 °C. If a vacuum oven is unavailable then a normal drying oven will do but take care to add another ethanol rinsing step.

To coat carbon nanotubes in silica:

1. Disperse the dry CNT's in the 5 mL of surfactant solution using the sonic bath. Screw the lid on the vial, sit the vial in a piece of foam and leave for an hour. The solution should be homogenously black with no sediment in the bottom.
2. Centrifuge the CNT dispersion so that the CNT's form a pellet in the bottom of the tube and decant away the top solution. Redisperse in fresh water and repeat 3−4 times to remove excess surfactant.
3. Redisperse the washed CNT's in a small aliquot of the citric acid solution using the sonic bath and then transfer the aliquot back into the 25 mL citric acid solution. Sonicate again to ensure the CNT's are dispersed evenly throughout.
4. Pour all of the citric acid solution containing the CNT's into the 100 mL of ethanol containing the 1:9 molar ratio of 3-APTES:-TEOS under stirring at room temperature. Sonicate the beaker every 20 min for 1 min in the sonic bath before putting the solution back onto stir. At this stage the 3-APTES will coordinate to the surface of the CNT's and hydrolysis step of the sol gel reaction will occur.
5. After 1 h adjust the pH to between pH 8 and 9 by adding the ammonium hydroxide solution dropwise to the solution under stirring and monitoring with the pH meter. Don't overshoot pH 9 or the silica will condense as spheres and not coat the CNT's.
6. Leave the solution to stir for 12 h after which the CNT's should be coated in silica.
7. To extract the silica coated CNT's centrifuge the solution. Use large capacity tubes as you will be trying to extract very small amounts of material. You may find that once the stirring is stopped the silica coated CNT's will quickly sediment to the bottom

of the beaker especially if the pH is neutralised. Once sedimented, the product can be drawn from the bottom using a Pasteur pipette and transferred into a centrifuge tube. Centrifuge the product 3 times using ethanol to rinse and redisperse the pellets. Wherever possible try to transfer separate aliquots into the same tube by using half as much ethanol to redisperse each and then adding the dispersions together.

8. The final centrifuged pellet of silica coated carbon nanotubes can now be dried in a vacuum oven at 60 °C to drive off the ethanol and unreacted precursor. This step should also condense the silica coating. The scale of the coating is small enough that we do not need to worry about the cracking from capillary forces that might be observed in larger scale gels or coatings. The silica coated CNT's are now ready for analysis.

High resolution transmission electron microscopy of the silica coated CNT's should reveal a 15 nm thick layer of amorphous silica surrounding a CNT core. It is possible to lower the thickness of the shell by adding less precursor to the ethanol prior to coating. Adding only 20, 30 or 40 μL will result in a shell thickness of ~3, 5 and 10 nm, respectively. If you are using a MW-CNT template you should be able to see carbon layers atop one another with the amorphous silica layer on the exterior. If necessary the carbon core can be removed by calcination at 550 °C in air. The core will burn away as carbon dioxide leaving only the silica behind.

COATING MULTIWALLED CARBON NANOTUBES IN ZIRCONIA

Adapted from Rao et al. (1997), Satishkumar et al. (2000), Sheu et al. (1992)

Much of the coating procedures seem to be a simple case of adding a template and reagent and letting the reaction happen. Of course, it is more complex in practice and so coating a CNT with other metal oxides is not as easy as substituting the silica precursor for another metal centred alkoxide. In the next recipe we will coat the surface of the carbon nanotubes in ZrO_2. Zirconium is below titanium in the same group and therefore has similar chemical behaviour. Alkoxides from this group tend to hydrolyse far quicker than their silicon based counterparts so keeping the reaction free from moisture at stages where the precursor used is important in getting a homogenous coating and preventing heterogeneous zirconia nucleation. The cubic phase of zirconia has a fluorite structure and is perhaps the most useful crystalline form of this material. All zirconia phases have a high dielectric constant which means a coating on a nanoscale component will behave as an electrical insulator. However, the cubic polymorph also demonstrates high chemical resistance and toughness and is even used as a replacement for diamond in a number of applications. The material is regularly used as a dental implant for crowning teeth and because of its ivory colour matches the look of natural teeth surrounding it. In contrast to diamond it exhibits a low thermal conductivity meaning it is useful for heat shielding up to ~2500 °C and high temperature furnaces use a chemically doped form of cubic zirconia as an insulating material. Thin films of modified zirconia are also employed in jet engines to protect parts that are routinely exposed to hot abrasive gases. Cubic zirconia is only stable at high temperatures and upon cooling it converts into a tetragonal phase and then a monoclinic phase. Both of these transitions are accompanied by a shearing and expansion in the structure which will cause a solid lump of zirconia to turn to powder. In order to take advantage of cubic zirconia's properties at room temperature the molecular structure must be stabilised by adding a dopant. Yttrium oxide is most commonly employed in this respect.

In this sol gel based preparation a dispersion of multiwalled carbon nanotubes is treated with a strong acid so that the surface becomes functionalised with acidic groups. These can then act as nucleation sites when the rods are treated with a zirconium alkoxide. A coating of amorphous ZrO_2 will form which can be converted into a crystal polymorph by heating. Under the conditions required to burn out the core CNT, the zirconia will be a mixture of the cubic and tetragonal crystal structures. If the pure tetragonal phase is required then the tubes must be heated further which can often lead to structural collapse from the stresses induced by the phase transition. One solution to this problem is to dope the zirconium by adding a dissolved yttrium oxide prior to gelation at the carbon nanotube surface. This way the formed gel will be doped with the right amount of yttrium oxide prior to hydrolysis and condensation and the tubular morphology will be retained after heating and cooling. A variation of the zirconia templating mechanism that incorporates yttrium oxide will be included at the end of the preparation (Fig. 4.12).

To coat multiwalled carbon nanotubes (MW-CNT) in zirconia you will need:

1. 200 mg of MW-CNT. This can be weighed out on a four figure balance into a screw top vial. You will need a few vials that can accommodate 3–4 mL of liquid with screw top lids stable in strong acids and that can be safely sonicated for the recipe.

2. A high temperature furnace capable of reaching at least 600 °C and a high temperature crucible. Ideally it will be the ceramic sort that has a glaze on the surface. This will make it easier to collect products after calcination by scraping them out with a spatula.

3. 3 mL of a 68 %wt (~16 mol/L) concentrated HNO_3 in water solution. This is the concentration at which nitric acid is commercially supplied. This will be used to activate the surface of the CNT's by forming carboxylic acid groups on the surface which

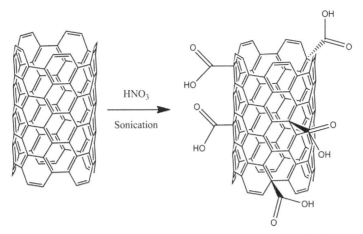

FIGURE 4.12   A schematic of the reaction of a strong acid with carbon nanotubes to produce surface bound hydroxyl and carboxyl groups suitable for zirconia nucleation.

will make them receptive to zirconia nucleation. *This is highly corrosive! Wear appropriate safety gloves and face mask when handling. Work within a fume hood—even during the sonication step.* Do not use a volumetric pipette for measuring out acids at this strength. It is almost a certainty that the acid will fume and corrode the springs inside the pipette. The best practice is to use a small glass measuring cylinder and use a Pasteur pipette to measure out the correct amount.

4. An 800 W sonic bath. This is used to disperse the CNT's but also provides energy to drive the reaction during the functionalisation step.

5. A centrifuge with centrifuge tubes large enough to handle ~4 mL. Samples can be split into smaller aliquots. This will be used to separate the CNT's from solutions.

6. 2 mL of zirconium (IV) isopropoxide 70 %wt in isopropanol solution. This type of solution in alcohol is how it is normally provided. Depending on results you may wish to try getting rid of some or all of the solvent by distillation. The resulting alkoxide will be highly sensitive to moisture and very viscous so it should only really be done if you need the pure reagent.

7. A gas tight dessicator through which an inert dry gas such as argon or nitrogen can flow. At many points in this reaction it is important to keep out moisture.

8. Excess methanol for rinsing that has been dried. Bench methanol can be dried by adding dried calcium sulphate to a stock bottle. This will draw out the moisture but you must ensure the methanol is filtered before use or you could contaminate your sample with foreign particles.

To coat multiwalled carbon nanotubes (MW-CNT) in zirconia:

1. Place the 200 mg of MW-CNT's into the crucible and heat to just below 600 °C for 20 min. Most carbon nanotubes should be thermally stable in air to this temperature. This step will remove other allotropes of carbon from the sample which will burn off as CO_2. Use a heating rate of between 1 and 5 °C/min to ensure the oven does not overshoot the set point.

2. Transfer the MW-CNT's from the crucible and into a screw top vial. This will require some patience scraping the CNT's out with a spatula. Alternatively a few drops of water delivered by a Pasteur pipette can be used to rinse the CNT's into the bottom of the crucible and then transferred into a screw top vial. It might be tempting to use the acid to rinse the crucible but this might attach the ceramic and contaminate your sample!

3. *In a fume hood and with all necessary precautions* add 3 mL of concentrated nitric acid solution to the screw top vial containing the CNT's. Seal the vial and ensure there is no concentrated acid on the exterior of the vial. There should be an airspace above the solution so that the vial will not pop if it overheats in the sonic bath.

4. Place the sealed vial in a float, place into the sonic bath at room temperature and sonicate. Leave the CNT's in the nitric acid to sonicate for an hour during which time carboxylic acid groups will form on the surface of the tubes. Some recipes advocate refluxing the CNT's in the acid and if a sonic bath is unavailable then setup a reflux condenser and boil the CNT's in an excess of nitric acid for 12 h. Should you choose this option remember *you are boiling nitric acid—make sure the reflux set up is correct, made safe and in a fume hood.*

5. Transfer the solution into a centrifuge tube and centrifuge at 13,200 rpm for 10 min. A pellet should form in the bottom of the tube. It may appear colourless or white. Pour away the acid into ~ 50 mL of a 1 mol/L solution of NaOH to neutralise it before disposal.

6. Redisperse the pellet of acidified CNT's in water and centrifuge. Repeat this three times. Allow the CNT's to air dry then drive off the water by drying the CNT's at 110 °C. If you have a vacuum oven this will also help to speed up the drying.

You do not want water to be present for the addition of the alkoxide. At this stage you may check that the CNT's have been acidified. An IR spectrum of the dried sample should reveal a sharp OH stretch at 3300 cm^{-1} and a C=O stretch at 1600 cm^{-1}.

7. Place the dried CNT's inside a screw top vial and stand it in the dessicator. Flush the dessicator with dry argon or nitrogen for 10 min. You must make sure that the gas flow does not disturb the CNT's within the vial or there is a chance they will get blown out! Open the dessicator and replace the lid quickly.

8. Using a volumetric pipette draw up 2 mL of the isopropoxide solution. This is preferably done in a dry atmosphere. Open the lid of the vial containing the carbon nanotubes and add the zirconium isoporpoxide. Place back into the dessicator and flush through with the inert gas for another 10 min. After this time seal the vial shut and remove from the dessicator.

9. Sonicate the sealed vial for 2 min. Dry it thoroughly with tissue to remove moisture from the exterior.

10. Leave the vial to stand in the dessicator under argon or nitrogen for 48 h. This will give the zirconium isopropoxide time to complex to the surface through the carboxylic acid groups.

11. After this time transfer the solution to a centrifuge tube and centrifuge to form a pellet of nanotubes in the bottom. Pour off the zirconium isopropoxide and then redisperse the CNT pellet in 2 mL of the strong nitric acid. *Take care!* The strong acid should hydrolyse the surface bound alkoxide. You do not want to leave this step for too long so the moment the pellet is redispersed then centrifuge it once more and rinse the sample with methanol. The source reference for this recipe uses dilute HF but this can be very hazardous unless your laboratory is setup for handling it.

12. After redispersing and centrifuging in methanol three times remove the pellet of zirconia coated CNT's from the centrifuge tube and dry in a crucible at 100 °C for an hour in the furnace.

13. After an hour at 100 °C set the furnace to 450 °C and leave the sample at this temperature for 48 h. This will oxidise any remaining precursor at the surface and will convert the amorphous zirconia to a mixture of cubic and tetragonal crystal phases.

14. If you wish to remove the carbon interior then further heat the sample at 700 °C for 48 h. The interior will be converted to carbon dioxide leaving a hollow zirconia structure. The sample is now ready for analysis.

Transmission electron microscopy of the sample should reveal hollow zirconia tubes about 40 nm in diameter with a central hollow core of 6 nm in diameter. The Tube length may vary greatly depending on the length of the template CNT's. If the CNT core was not removed then the layered structure of the multiwalled tube should still be visible. The crystallography of the zirconia tubes will show that the material is a mixed phase of monoclinic and tetragonal crystal structures. If the sample is heated further to obtain a pure phase then a mismatch of thermal expansion in the cubic and tetragonal phases will result in a catastrophic fracturing of the sample. This can be overcome by introducing 8 mole % or higher of yttria (Y_2O_3) as a stabilising agent for the cubic phase. In this next recipe we will make a gel by dissolving yttrium oxide in nitric acid and adding it to a solution containing $ZrOCl_2$. Upon the addition of ammonia the precursors will condense to form a loose gel network of yttrium doped zirconia which can then be subsequently dissolved in acid again to form a sol. Acid modified carbon nanotubes are then added to the sol and stirred so that they become coated. The solution is then allowed to dry out which drives the further condensation and formation of a zirconia coating upon the nanotubes.

To make yttrium stabilised zirconia coated nanotubes you will need:

1. About 300 mg of acid treated MW-CNT's. These can be produced in the same manner as the previous recipe by sonication in strong nitric acid.

2. 1.5 g (4.6×10^{-3} mols) of $ZrOCl_2 \cdot 8H_2O$ weighed into a ~50 mL beaker. The beaker should be wide enough that you can remove the gel that is going to form in it with a spatula.

3. 0.1 g (4.4×10^{-4}) of Y_2O_3 for use as a dopant this will give just below a 10% molar doping of yttrium oxide into the formed ZrO_2. The yttrium oxide should be kept in a 10 mL glass vial stable to strong nitric acid.

4. 3 mL of 16 mol/L concentrated nitric acid. *Work with this in a fume hood!* As before measure out strong acids with a measuring cylinder and Pasteur pipette.

5. At least 10 mL of a 27% solution of ammonium hydroxide in water. This is going to be used to induce the formation of the gel by adjusting the pH. *Work with this in a fume hood!*

6. An ice bath to stand the beaker in with a thermometer. The ice bath should be small enough that it can sit on top of a magnetic stir plate. During the pH adjustment the solution will heat up if the temperature rises more than one or two degrees above zero add more ice. It is a good idea to have a clamp holding the beaker in place. Even if the beaker is stable when you start the water can rise as the ice melts causing the beaker to float and you to lose your sample.

7. A pH meter and a magnetic stirrer with a stir bar.

8. A drying oven set to between 60 and 80 °C. You will also need a furnace capable of reaching 700 °C and a crucible.

To make yttrium stabilised zirconia coated nanotubes:
All handling of the concentrated acid solution should be done in a fume hood.

1. Dissolve the $ZrOCl_2 \cdot 8H_2O$ with 10 mL of water in the beaker.

2. *Carefully* add 3 mL of the 16 mol/L nitric acid to the 0.1 g of Y_2O_3. The yttrium oxide should dissolve.

3. Pour the dissolved yttrium oxide into the beaker containing the $ZrOCl_2 \cdot 8H_2O$ solution and swirl gently. Make sure the solution is homogenously mixed.

4. Place the beaker in the ice bath and monitor the pH with the metre. It will be strongly acidic so raise the pH to 9 using the concentrated ammonia solution. This step must be done dropwise with a Pasteur pipette or else it will heat up vigorously. Keep the solution just above 0 °C. The beaker should be swirled by hand to mix. As the pH reaches 9 the solution will form a gel network which will make swirling the solution impossible.

5. Once the gel has formed pour away excess solution into a proper waste solvent container. Add an excess of water to the beaker and swirl by hand for five minutes. This step will wash out excess salts in the gel formed from the neutralising step with ammonia. Repeat this three to four times.

6. The gel must now be converted into a sol which will coat the acidified carbon nanotubes. This is accomplished by adding 10 mL of the 1 mol/L nitric acid solution and stirring the gel magnetically. This should be enough for the gel to redissolve but if it is not add more dropwise until solid lumps are no longer visible. Stirring quickly will help but make sure the acid is not splashing out the sides.

7. Add the 300 mg of acidified CNT's to the zirconia/yttria sol and leave to stir for 24 h.

8. After 24 h, transfer the beaker containing the sample into a drying oven. Allow the water to evaporate away. The time this takes will depend on the temperature of the drying oven and the amount of ventilation it has. Keep the sample in the oven until it is visibly dry.

9. Scrape the powdery product out of the beaker and transfer to a crucible. Heat the crucible at 100 °C in air for 12 h. This step will drive the condensation and solidification of the stabilised zirconia coating and eliminate any remaining water or nitric acid.

10. Further heat the sample to 450 °C for 12 h to obtain yttrium stabilised zirconia coated carbon nanotubes. If you wish to remove the CNT's from the sample to yield hollow tubes of stabilised zirconia then heat at 700 °C for a further 12 h. A slow heating rate is preferable to a fast one as it will minimise stress from thermally induced phase changes. The product is now ready for analysis.

Transmission electron microscopy of the hollow yttrium stabilised zirconia should reveal a tubular structure of 4 to 50 nm in diameter having a wall thickness of around 6 nm. If the hollow tubes are extensively fractured or demonstrate a poor morphology then add a few drops of ethylene glycol to the precipitate prior to high temperature treatment. Electron diffraction or powder X-ray diffraction patterns should be indexed to JCPDS card number 30−1468 for a yttrium stabilised zirconia. Raman spectroscopy should show bands at 260, 322, 463,467 and 640 cm^{-1} for the distorted zirconia and weaker bands at 302, 348 and 367 cm^{-1} represent the yttria phase (Azad, 2006).

USING SOL GEL DIP COATING TO FORM THIN FILMS, THIN POROUS FILMS AND REPLICAS

So far we have examined methods to produce particles from sol−gel precursors and how to coat those particles or others at the nanoscale. In this section we will look at using sol gel chemistry to form thin coatings on structures at the meso and macroscale. This can be used to make homogenous thin films to protect sensors or to replicate more complex morphological structures such as pollen. This is often as simple as dipping the template structure into a sol−gel solution and then drawing it out.

DIP COATING

Dip coating is an easy concept although it can actually be technically challenging to get it right. In practice there are a number of subtle nuances that must be considered to ensure the formed film is homogenous, crack free and adhered to the substrate properly (Brinker et al., 1991). A substrate is lowered into a viscous sol and then drawn vertically out slowly. The meniscus of the sol is adsorbed to the surface of the substrate and condenses in air as the substrate is drawn out of the sol. This creates a homogeneous solid thin film on the surface of the substrate. The surface tension of the sol at the air−liquid interface and the viscosity of the solution determine how thick the liquid layer on the substrate will be and how rapidly this will drain back into solution under gravity. If this is counterbalanced with the substrate drawing rate from the solution then a gel will have time to condense into a smooth oxide layer on the surface. It really is that simple in theory and remains a mainstay of coating technology because it requires almost no specialist equipment when compared to other (Fig. 4.13).

The more viscous a solution is then the thicker the layer on the substrate is when pulled out of the bath. This means that by ageing a gel prior to the dipping procedure the thickness of the deposited film can be modified. The moment this layer is exposed to the atmosphere it will begin to dry and the sol will condense into an oxide framework. If the rate of chemical condensation is faster than the rate at which the film dries out then a solid coating is produced. If the solvent evaporates away too fast then the film will not have had a chance to densify properly. This results in a low density inhomogeneous network structure. For this reason the humidity and ageing of the film after it has been cast are vitally important to getting it right. A condensing gel will most often undergo some form

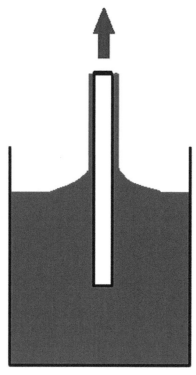

FIGURE 4.13   A schematic representation of the dip coating process.

of shrinkage and in a thin film there is a mismatch in the scale of contraction in the dimension of thickness versus the films area. If dried too rapidly then this can cause cracking due to tensile contraction stresses leaving the film flaky or hazy looking (Strawbridge and James, 1986). The following recipes should provide a decent practical overview on getting transparent and uniform films on a variety of biological and inorganic substrates.

It is important to remember that the precursor baths in all these recipes are ageing from the moment water and acid are added to the alkoxide. The speed at which this occurs depends on the recipe and how much the reaction is catalysed by the presence of an acid. Mostly, the formation of a gel will occur on the order of days. The older the sol solution is then the more oligomers will form into larger and larger networks and eventually particles. Very thick gels can sometimes be broken up under magnetic or manual stirring for use but films produced from old solutions will consist of agglomerated particles and be highly porous. This is sometimes desirable and the methods to produce both regular porous films and smooth homogenous films are provided here. Impermeable films are used for protective coatings or coatings where optical clarity is a requirement. Micro and meso porous films can be used for catalytic activity where an increased surface area is required. Of course, porosity in a glass can be produced in a more controlled manner by using a surfactant as a self assembling template. The technique is known as *evaporation induced self assembly* and is very useful in making ordered porous arrays of controlled and well spaced dimensions in a dip coated film. A surfactant is added to the dip coating solution in an amount just below its critical micelle concentration (CMC). During dip coating the surfactant is drawn up along with the precursor. As the liquid evaporates from the substrate surface the surfactant is concentrated below the CMC in the gel and self assembles into micelles or other structures. The condensation of the oxide then occurs around the micellular assembly to produce patterns upon drying. The configuration of the self assembled surfactants, or phase, depends on the surfactant used and the final concentration. By selection and control of these parameters a range of regular and oriented pores can be engineered into a dip coated film on the meso and macroscale.

DIP COATING A GLASS FILM

Adapted from Lu et al. (1997), Tate et al. (2010)

In this recipe we will cover most of the basics of dip coating by constructing a dip coating setup and coating a substrate with a silica film. The produced film can be an impermeable solid but we will also explore how to make porous films as these tend to be more widely used for nanomaterials chemistry. The porosity of the silica film will be controlled by the addition of CTAB or the non-ionic surfactant commercially known as P84. P84 is a triblock polymer consisting of a fourteen carbon alkyl chain tail, a twelve unit poly(propylene oxide) centre and a seventeen unit poly(ethylene oxide) end. Like all surfactants, it will self assemble into a range of

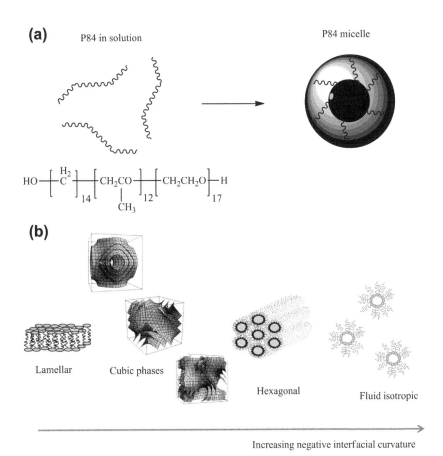

(a) P84 in solution   P84 micelle

(b)

Lamellar   Cubic phases   Hexagonal   Fluid isotropic

Increasing negative interfacial curvature

FIGURE 4.14 (a) A schematic of the self assembly of P84 to form a micelle in water. (b) a variety of liquid crystal structures can be formed by a surfactant at different concentrations or pressures. *(Images Thanks to Dora Tang, University of Bristol)*

phases depending on its concentration in solution. In an ionic surfactant like CTAB, the cationic head group will actively dissolve in water while the hydrophobic alkyl tail group will cluster together. In a non-ionic surfactant the driving force behind the self assembly into a particular phase is determined by the relative hydrophobicity/hydrophillicity of the component polymer units. The poly(propylene oxide) and the alkyl tail are hydrophobic while the poly(ethylene oxide) unit is relatively happy to disperse in water.

On self assembly of surfactants a number of different phases may form depending on the structure and concentration of surfactant in solvent. These phases can be tuned by pH, temperature, pressure and concentration and it is advisable to refer to a surfactant's phase diagram for details. In this recipe P84 will be used as it is relatively easy to get a lamellar, hexagonal or double gyroid phase when the surfactant assembles in the drying film. These phases are then templated by condensing silica into well ordered patterns on the surface. The surfactant is removed by calcination to leave a highly ordered glass 'negative' of the surfactant phase. Selection of the resulting phase and packing of the porosity in the glass is achieved by altering the concentration of the P84 in the gel and the age of the gel prior to coating.

For all the dip coating recipes it is very important to have a perfectly clean substrate. A variety of in depth cleaning procedures for substrates are given elsewhere but a brief review is provided here for ease. In this recipe the substrate will be a silicon wafer but as this is an expensive commodity it is also possible to use glass (Fig. 4.14).

MAKING A DIP COATING CHAMBER

Controlling the humidity under which the deposited films are formed is absolutely vital in ensuring the solvent evaporation rate is optimal for templating the surfactant and also for ensuring capillary stresses do not destroy the gel as it condenses. You will need two chambers for dip coating, one is for ageing your films in and the other is for performing the dip coating itself.

To make the dip coating chamber:

Schematics are outlined below for the simple dip coating setup. The chamber itself must be made of something impermeable to water so a large plastic box with a resalable lid is perfect. It does not need to be hermetically sealed so a cheap storage box will suffice. It does need to have a flat bottom or at least a bottom you can make flat. The height required depends on the size of the

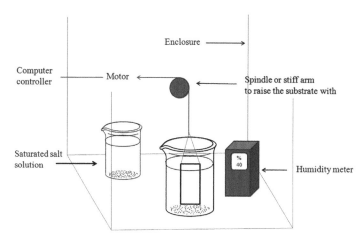

FIGURE 4.15 The basic design for a dip coating machine and chamber.

substrate you are going to coat. For coating glass microscope slides a vertical clearance of 40–50 cm will allow you enough room for lowering the sample in to a beaker stood upright containing the sol gel solution and then pull it clear again (Fig. 4.15).

You will need to monitor and control the humidity inside the box. Humidity sensors are relatively cheap and easy to obtain from an electrical components supplier or hardware shop. Opening the box up to make regular checks will alter the humidity so you will need to have the sensor placed through a hole in the side of the box and then use film or putty to seal up any gaps. The humidity can be controlled using a large bottle with a wide neck filled three quarters full with a saturated sodium chloride solution. The salt solution will lose water to the air if the environment is too dry and draw water in if the atmosphere is saturated. You may find that the box will behave like a greenhouse when sealed and naturally equilibrate to a specific humidity. Above 40% relative humidity is desirable for the stable ageing of films. If the humidity is not above 40% then you will need to pass air through the bottle to force moisture into the chamber. If you do this then make sure the flow is low enough that salt solution is not spraying out of the bottle. Also put a splash guard between the bottle lip and the dip coating apparatus. Any stray salt solution getting into the dip coating bath or on the film will wreck it. When transferring samples in and out of the box, remove the lid by sliding it slowly sideways. This will help the heavier moist air remain in the box. It is inevitable that some of the humidity will be lost during the transfer but it is good practice to try and keep the film exposed to the correct conditions every chance you get.

A glass or PTFE beaker can be used as the receptacle for the dip coating solution. Using a beaker only slightly wider than the sample to be coated will help you minimise how much solution is needed. Solutions can be reused or left to age separately to change the coating properties. If using glass then remember that the beaker should be thoroughly cleaned and dried after each use so that nucleation on the beaker surface is kept to a minimum. In this recipe a 25 mL beaker will be deep enough to completely coat a small wafer a few centimetre square.

For dipping and removing the substrates at a precisely controlled rate you will need an electrical motor with a spindle and a means of holding the substrate in place. Commercial systems are available that can be controlled with a high degree of accuracy and are programmable to pull at different speeds during a run. These will use a fixed arm instead of a pulley arrangement for lowering and raising the sample. If possible you should take this approach, as the substrate motion must be as smooth as possible for the best results. For the sol gel scientist on a budget you can assemble this yourself using an electric motor controlled by a computer. Most universities have a license for Labview which is a visually based programming language and can be used to control a USB driven motor. Similarly, a prototyping robotics kit will provide motors and components to let you build your own equipment. The motors must be capable of moving the sample at a rate of between 0.1 and 10 mm/s. In this recipe the draw speed will be 1 mm/s but you may find it necessary to alter this depending on the viscosity of the solution and your results. If you have to use a pulley arrangement then use two strings over guide wheels. Both strings will be attached to a cross beam to which a clip for the substrate is affixed. In either arrangement make sure the sample is directly vertical and does not twist as it comes out of the solution.

To make the humidity chamber:

This will be exactly the same setup as the dip coating chamber but without the dip coating equipment in it. Fit a plastic box with a humidity meter and a litre bottle three quarters full of saturated salt solution. Pass a low rate air flow through the salt solution to bring the humidity up if required.

To make a dip coated glass film on a substrate you will need:

1. A substrate for coating. This could be a flat silicon wafer, glass or a tin oxide electrode depending on the end application. In this recipe we will be working with a silicon wafer $2 \times 5 \times 0.2$ cm in dimension. The substrate you choose can vary widely as long

as the dip coating bath is deep and wide enough to accommodate it and you have enough solution. Whatever you choose it must be cleaned thoroughly before use.

2. Alconox or similar detergent for cleaning glassware. This is formulated from the anionic surfactant sodium dodecyl sulphate and a mixture of carbonates and phosphates. Make ~ 100 mL of this to a 2 wt% solution in water in a deep beaker.

3. A hot plate in a fume hood *away from anything flammable!* This will be used to heat the detergent solution for cleaning glassware. You will need Teflon tweezers for getting the substrate in and out of the alconox and the aqua regia.

4. If required, a setup for cleaning with aqua regia. This can be made by adding 5 mL of 70 wt% HNO_3 dropwise to 15 mL of 37 wt % HCl. *This must be done in a fume hood with full protective equipment. The solution will fume upon mixing and proper procedures for handling and disposal must be followed. More extensive instructions on the use of aqua regia can be found in the gold nanoparticles section of this book.* Have copious amounts of water ready for rinsing.

5. A magnetic stirrer and stir bar.

6. Two large (~100 mL) high density poly(ethylene) (HDPE) screw top bottles. These will be used for storing and mixing the dip coating solution.

7. A solution of P84 in 35.53 mL (28 g, 0.6 mol) pure ethanol. The concentration of P84 required is determined by which packing phase you are hoping to obtain. Specifically the silicon to surfactant ratio is critical. Given below is a table of the phases that can be obtained with the molar ratios of Si:P84 and the weight of P84 required to produce them.

Phase structure	Molar ratio of Si:P84	Weight of P84 in solution in grams
Lamellar	1:1	12.92
Double Gyroid	1.2:1	10.78
Hexagonal	1.3:1	9.23

To prepare the surfactant solution, add the desired weight of P84 into ethanol in one of the screw top HDPE containers. Shake gently to mix the contents and then leave at room temperature for 24 h before use. Give the contents a shake every few hours to ensure the P84 is homogenously mixed.

It is a good idea to purify the P84 before use. This can be accomplished by dissolving 50 wt% in pure ethanol. Any rubbish in the surfactant should then settle out of the solution. Decant and collect the top solution or centrifuge and allow the ethanol to evaporate to harvest the purified P84.

8. 32.64 mL (25.72 g, 0.558 mol) of pure ethanol and 12.70 mL of a 0.0177 mol/L solution of HCl in water. The HCl solution can be prepared by serial dilution from a 37 wt% stock solution. The concentration of a 37 wt% solution is 11.85 mol/L. Add 1 mL of concentrated HCl to 9 mL of water to make a 1.185 mol/L solution. Now take 1 mL of this solution and add to 65.94 mL of water to give 0.0177 mol/L solution. This may seem fairly roundabout but it is important to balance both the molarity of both the acid and water in relation to the TEOS. Store this in the other HDPE bottle.

9. 26.14 mL (24.39 g or 0.117 mol) of tetraethylorthosilicate (TEOS). This will be added to the ethanol water mix but must be done only when you are ready to proceed with the dip coating.

10. A furnace capable of reaching at least 400 °C with a programmable ramp rate accurate to less than 0.5 °C/min.

11. The dip coating chamber and ageing chamber equilibrated to a relative humidity of 40%. Also double check the program you wish to run is ready on whatever your setup is for dipping. There will be time to do this after mixing the gel solution as you will have to wait for the gel to age until use.

12. If you put together your own program for dipping then set the arm or pulley to lower the substrate into the gel at a rate of 1 mm/s and stopping a few millimetres before reaching the clip. If the gel gets onto the clip it can drip onto the film making it uneven. The drawing rate required to produce all the structural variants in the silica film is 1 mm/s.

To make a dip coated glass film on a substrate:

1. Heat the alconox solution until it is boiling and allow it to simmer. Place the substrate into the boiling solution using tweezers and rest it on the side of the beaker so that both surfaces are exposed as much as possible. Leave for 5 min then remove and turn off the hot plate. Dry in air and if possible stand the substrate on its side leaning against a petri dish wall or similar.

2. *In a fume hood* rinse the substrate thoroughly with water and place into the aqua regia using the tweezers. Once again expose both sides. Leave for 20 min then remove and rinse using excess water. This step is optional depending on how your preliminary results turn out and the alconox cleaning step should clean the surface well enough for deposition. Dry in air before use.

3. Add the TEOS to the ethanol/HCl solution and stir magnetically for 20 min precisely. Afterwards pour the TEOS acid solution into the bottle containing the P84. Swirls once or twice by hand to mix the contents thoroughly then seal the bottle and age in the dark without further motion. The pH of the gel upon mixing should be 1.76. The final molar ratios for the gel are:

TEOS: 1
HCl: 0.0092

Water: 6.02

Ethanol: 9.96

The amount of P84 depends on the structure you are aiming for and the molar ratios compared to TEOS are accordingly:

Lamellar: 0.0263

Double gyroid: 0.0219

Hexagonal: 0.0202

4. You must now wait for the gel to age. The length of time you wait depends on the structure you are aiming for. The following times are how long you should leave the gel to age in the dark before dip coating:

   Lamellar: 3 Days

   Double gyroid: 3 Days

   Hexagonal: 2 h

5. Use a clip to affix the substrate to the dipping arm. Make sure it sits as vertically as possible in the clip so that the substrate will be perpendicular to the gel bath. This is important so that the contact angle is maintained at the surface during drawing from the solution.

6. Once the gel has reached the correct age for the intended structure pour it out of the bottle and into the beaker being used as the dipping bath. Fill the beaker 2 mm shy of the rim. Seal the bottle and save the remainder in the dark.

7. Reseal the dipping chamber and allow the relative humidity to return to 40%. You may then turn on the dip coating machine and lower the substrate into the gel at a rate of 1 mm/s and then withdraw immediately at the same rate.

8. The substrate will now be coated with silica sol which will already be beginning to condense due to the evaporation of ethanol. Leave the sample hanging in place for a further 15 min for the initial surfactant assembly and film condensation to occur.

9. Taking care not to let the humidity drop much below 40% for too long, transfer the freshly coated slide quickly into the 40% relative humidity ageing box. You may wish to lay it flat or keep it in the upright position and the film should be solid enough that it will not deform during the ageing step. Leave the coated sample to age for 12 h.

10. If you have finished dip coating the gel can be decanted back into the HDPE bottle for further ageing if required.

11. Once aged, a porous silica film with the desired structure should have formed. The surfactant will still be present though and this needs to be removed by calcination. The film can be analysed at this point to check for the structure by powder x-ray diffraction.

12. To remove the surfactant, place the film in a furnace and heat to 400 °C in air for 4 h. The rate at which you heat and cool the sample is very important. Upon heating there will be a significant contraction in the silica structure. For a hexagonally packed structure the ramp rate should be 0.5 °C/mine. For the lamellar and double gyroid structures the ramp rate should be °C per minute. Upon cooling the film is ready for analysis.

The quickest method of assessment is to hold the film up in the light and move it around. You should see a blue iridescent sheen due to interference in the film. The silica film will be between 300−350 nm thick and this can be assessed by scanning electron microscopy of a cross section of the film if you afford to break a piece off. If a surfactant has been used to produce regular porosity then grazing incident or powder x-ray diffraction can be performed which should show reflections correlating to the spacing and lattice parameters of the structure. For porous films with regular orientations the powder X-ray diffraction (PXRD) pattern obtained should show a peak between $2\theta = 1-2°$. This is a fairly low angle for some machines and theoretically some of the peaks should normally only be obtainable by a 2D X-ray technique like grazing incidence instead of only the 1D scan of PXRD. However there are off-specular reflections that can be seen by the PXRD detector if the slit width used is large. For the three different phases the major reflections will shift between uncalcined and calcined films due to a 40% volume contraction upon heating. These are:

Phase structure	2θ	Index plane
Hexagonal	1.17°	02 (two-dimensional)
Calcined hexagonal	1.80°	02 (two-dimensional)
Double gyroid	1.31°	221
Calcined double gyroid	1.92°	211
Lamellar	1.17°	001
Calcined lamellar	No peak	Amorphous

Scanning and transmission electron microscopy is the most definitive method to observe the porous structure in the film. Samples for TEM can be prepared by scraping thin sections of the film from the substrate surface. It is likely that these sections will be too thick to get the electron beam to pass through if placed directly onto a TEM grid. Crushing the film shavings could potentially damage them so if getting a thin enough sample is a problem try sonicating a shaving in ethanol and then place the dispersion onto the grid to dry. In addition to a highly oriented porous structure you should see that the silica walls are about 5 nm in thickness (Fig. 4.16).

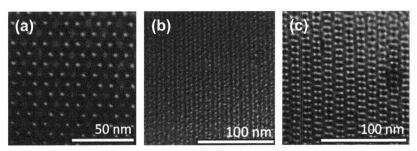

FIGURE 4.16   (a–c) Unprocessed transmission electron micrographs of calcined double-gyroid nanoporous silica films looking down the (a) [111], (b) [211], and (c) [311] lattice planes. *(Adapted with Permission from Tate et al. (2010). Copyright 2010 American Chemical Society)*

MAKING A CUBIC PHASE IN A DIP COATED SILICA FILM USING CETYLTRIMETHYLAMMONIUM BROMIDE

Adapted from Lu et al. (1997)

By some minor modifications to the above recipe you can obtain a cubic phase pattern in a silica film using self assembled CTAB as the template. The TEOS is partially hydrolysed by reflux with a very low level of acid to slow the hydrolysis reaction before the further addition of catalytic amounts of acid and the surfactant precursor.

For a CTAB templated silica film you will need:

1. The dip coating setup as mentioned elsewhere. You will be using everything as with the previous recipe except that instead of P84 you will be using CTAB ($CH_3(CH_2)_{15}N^+(CH_3)_3Br^-$) along with different ratios of the precursors. Before you start make sure the relative humidity has reached above 40% and reprogram the drawing apparatus so that it will pull the substrate out at a rate of 1.26 mm/s.
2. A screw top HDPE bottle (~100 mL) for mixing and storing the dip coating solution in.
3. A 200 mL two necked round bottom flask with a quick fit connector and a reflux condenser with the correct tubing and clips connected to a tap and drainage. The second neck should be fitted with a glass stopper. You will need a clamp stand to hold everything in place and a cork ring for the round bottomed flask.
4. A heating mantle with an oil bath capable of reaching 60 °C fitted with a magnetic stir bar. You might be tempted to use water to fill the heating bath but 60 °C is still a high enough temperature to evaporate away the water while you are not looking. Use mineral oil and have plenty of paper towels in hand to take away the oil on the round bottomed flask when you remove it.
5. A clean substrate for coating such as (100)-silicon. Follow the procedures for cleaning that can be found elsewhere in this book.
6. 14.91 mL (14.02 g, 0.067 mol) of TEOS measured out using a volumetric pipette. As before, it is important that the precursor has not already partially hydrolysed. Make sure there is no sediment in the bottom of the bottle.
7. Two aliquots of pure ethanol. One must be 31.44 mL in volume and the other 55.0 mL. A 5 mL volumetric pipette can be used for accuracy in measuring out even if it takes a little longer!
8. Two solutions of hydrochloric acid which will be at different strengths. The first is a 9.3×10^{-4} mol/L solution of HCl which can be prepared by serial dilution of 37 wt% (11.85 mol/L) stock HCl. Add 2.04 mL of the stock solution to 20 mL of water to form a 0.93 mol/L HCl solution. Now take 1 mL of this solution and make up to 100 mL using a volumetric flask. Similarly, for the second acid solution you require, dilute 1 mL of 37 wt% HCl up to 74.66 mL with water to make a 0.1587 mol/L concentration.
9. 2.453g (0.0067 mol) of cetyltrimethylammonium bromide to use as the cubic phase surfactant template.

To make a CTAB templated silica film:

1. Sit the round bottomed flask in the oil bath and fit the reflux condenser. *Ensure that all clips and clamps are in place and set the water flow on slowly.* Turn on the heating mantle and allow it to equilibrate to 60 °C.
2. Using a volumetric pipette add 3.63 mL of the 0.92×10^{-3} mol/L HCl solution to 31.44 mL of pure ethanol in the round bottomed flask through the side neck. Add 14.91 mL of TEOS to the solution and a magnetic stir bar. Reflux the solution at 60 °C for 90 min under very gentle stirring. The molar ratios for this step are given below:

   TEOS: 1
   Ethanol: 3
   Water: 8
   HCl: 5×10^{-5}

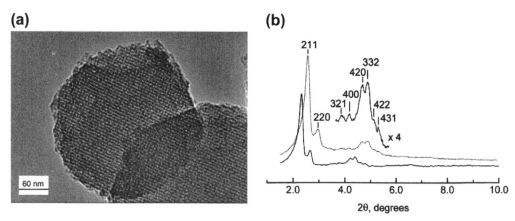

FIGURE 4.17   (a) A transmission electron microscopy image of calcined MCM-48 viewed along the [100] direction. (b) The expected reflections observed for a powder X-ray diffraction pattern of MCM-48. *(Images adapted with permission from Xu et al 1998, copyright 1998 The American Chemical Society)*

3. Set the mantle temperature to 50 °C and lift the round bottomed flask out of the bath and off of the heat. Add 2.45 mL of the 0.13 mol/L HCl solution. Stir for a further 15 min at room temperature.

4. Lower the round bottomed flask back into the oil bath with the reflux condenser still on and age the solution for a further 15 min without stirring.

5. Through the second neck in the round bottomed flask add 55 mL of pure ethanol and stir briefly.

6. Add the 2.453 g of CTAB to the solution and stir until it has dissolved. It is important to get the molar ratio of CTAB to silica, 0.1 to 1 respectively, correct so that the self assembly is well ordered throughout the film. If the molar ratio is above 0.11 to 1 then the cubic lattice structure will collapse during the calcination step. The final molar ratio of the reactants is given below:

   TEOS: 1
   Ethanol: 22
   Water: 5
   HCl: 0.004
   CTAB: 0.1

7. Use the gel immediately and pour it into the dipping bath filling it almost to the rim. Store unused sol in an HDPE container in the dark if it is to be used further but keep in mind that ageing will change the coating properties.

8. With the humidity set to 40% dip the substrate into the bath and draw out at a speed of 1.26 mm/s.

9. Leave the film above the dipping bath for 15 min before transferring to the other humidity box for ageing. This film is now ready for analysis but will still be thick with surfactant.

10. The CTAB can be removed by calcination to 400 °C in air for 4 h using a heating ramp rate of 1 °C/min. For better results, the film can be detached from the substrate before heating which will eliminate some of the surface stress from contraction. It will be easier to remove the film from a silicon wafer than a porous electrode such as a tin oxide layer. Be careful at this stage as the heating can destroy the structure if the CTAB ratio was slightly too high.

You should now have a porous film ready for analysis. Transmission electron microscopy of a calcined film should reveal a square periodic lattice with a spacing of 7.9 nm. This will be the [100] orientation of the cubic phase. Powder X-ray diffraction should also show a strong [100] reflection centred at $2\theta = 2.8°$. If the film is uncalcined then you will see a lamellar structure which will give a reflection for the 001 plane around $2\theta = 2.6°$ then another weaker reflection at $2\theta = 5.2°$. The film is transformed into the cubic phase during the heating step (Fig. 4.17).

DIP COATING TO FORM A DOPED TIN OXIDE FILM

Adapted from Chatelon et al. (1994), Terriera and Rogera (1995)

Tin oxide (SnO_2) is a transparent conductive oxide with interesting thermal and gas sensing properties. The pure mineral form is known as cassiterite and the mineral is a wide band gap semiconductor (~3.6 eV) but exhibits conductivity due to oxygen deficiencies in the crystal lattice (Batzill & Diebold, 2005). In a purified form it is poorly conductive and SnO_2 is often doped with antimony or indium to introduce donor centres near to the conduction band making it an n-type semiconductor. The tin is in a 4+ oxidation state and has a full set of d-orbitals which are largely thermally stable against promotion into the conduction band. A dopant, such as antimony, has loosely bound outer shell electrons which it can readily donate into the conduction band to act as charge carriers. As the conductivity of doped tin oxide films is related to the oxidation states of the metal cation centres, films of this type can be used for gas sensing. Redox reactions on the surface of a doped tin oxide layer will show a change in resistivity with a change in the level of ambient oxygen. Though transparent in the visible frequencies, SnO_2 is reflective in the infra red and it is often used as a coating on glass to produce thermally insulating windows. In this recipe you will make a tin oxide film doped with antimony on glass. One interesting effect of the dip coating approach is that a pure SnO_2 film will exhibit a far poorer conductivity than a film made by spray pyrolysis. This is due to the spray pyrolysis method introducing more oxygen vacancy defects into the formed layer and therefore boosting the observed conductivity. It is possible to make an undoped tin oxide film simply by not using any of the antimony salt in the procedure. The level of dopant used will be 10% Sb by molarity as this will result in the lowest resistivity. If more dopant is added then a large amount of Sb^{3+} which acts as an electron acceptor will compete with the electron donor Sb^{5+}.

The tin chloride forms an alkoxide species when dissolved in the excess ethanol. Though it reforms to the chloride if the solution is allowed to dry out the tin will form an insoluble hydroxyl species in the presence of water. The formed tin hydroxide layer is then dried immediately at 100 °C and then oxidised in dry flowing oxygen at 500 °C to produce the metal oxide film. The water for the reaction is provided by the ambient humidity during dip coating and nothing is added to the dipping solution to promote gelation. The level of humidity will have a drastic affect on the produced film and in contrast to other recipes, the best results will be produced by having the humidity below 40% during the drawing of the substrate. At higher humidity the particulate that makes up the film will be large enough to produce scattering effects which will make the film look cloudy.

To dip coat a tin oxide film onto a substrate you will need:

1. The dip coating apparatus in the humidity box. Ensure the humidity is equilibrated between 30 and 40% and set the drawing speed to 0.13 mm/min. If the humidity is too high then setup an argon flow through the apparatus to bring the humidity down.
2. 8.37 g (0.0370 mol) of $SnCl_2 \cdot 2H_2O$ dissolved in 100 mL of ethanol. This should be prepared in a 200 mL schott bottle with a screw top lid.
3. 0.844 g (0.00370 mol) of $SbCl_3$ dissolved in 20 mL of ethanol. Similarly this should be in a 100 mL screw top schott bottle.
4. A two necked 200 mL round bottomed flask with a quick fit aperture for a reflux condenser and a glass plug for the side neck. The reflux condenser must have all tubes and clips as appropriate.
5. A heating mantle with an oil bath capable of magnetic stirring with a magnetic stir bar.
6. 50 mL of ethanol and a 10 mL aliquot of ethanol for rinsing out the flask.
7. A clean substrate for coating. Ideally use a piece of glass or tempered borosilicate glass which can take heating to 500 °C.
8. A glass funnel for pouring solutions into the round bottomed flask.
9. A drying oven setup to handle the evaporation of large amounts of ethanol with an exhaust or in a fume hood.
10. An oven set to 100 °C with a wire rack. This will be used for drying the films immediately after dipping.
11. A tube furnace capable of reaching 500 °C with a flow of pure dry oxygen. *This could be a potential hazard so ensure the oven is safety tested and nowhere near anything inflammable.*

To dip coat a tin oxide film onto a substrate:

1. Set the heating mantle to 80 °C and place a magnetic stir bar in the round bottomed flask. Turn on the reflux condenser.
2. Pour the $SnCl_2$ in ethanol solution into the round bottomed flask through the side neck using the funnel.
3. Reflux for 2 h at 80 °C. Decant the solution into the Schott bottle and allow the ethanol to evaporate in a drying oven to produce a white powder.
4. Rinse the round bottomed flask with ethanol and then add the $SbCl_3$ in ethanol solution to the round bottomed flask using the funnel.
5. Reflux for 1 h at 80 °C then decant into a Schott bottle and place the drying oven to evaporate.
6. Set the heating mantle to 50 °C.
7. Add 50 mL of ethanol to the Schott bottle containing the dried $SnCl_2$. Once it has dissolved pour this solution into the Schott bottle containing the $SbCl_3$.
8. Once the oil bath has reached 50 °C pour the tin and antimony solution into the round bottomed flask and reflux for 2 h under stirring.
9. Lift the reflux setup out of the oil bath and allow the solution to cool to room temperature. Turn off the heating mantle and *turn off the water going to the reflux condenser.*
10. Decant the solution into the dip coating beaker. It should be used immediately.
11. Affix your substrate in the dip coating arm. Once the humidity has equilibrated to between 30 and 40% lower it into the solution and draw it back out at 0.13 mm/min. Once used, the antimony doped tin solution can be stored in a sealed container. The reproducibility of films using older solutions may vary.
12. Immediately remove the film from the dip coating setup and place horizontally onto a wire rack in the oven at 100 °C for 15 min.

13. Place the films into the tube furnace under flowing oxygen and heat to 500 °C for 15 min. The heating rate can be as quick (>10 °C/min) though you may find less cracking if you heat at a rate of 1 °C/min.

14. Remove the films from the oven and allow them to cool. You should now have a transparent conductive film ready for analysis.

The dip coating process can be repeated up to six times to build up the film thickness. Each layer should be around 100 nm thick. You may see an increase in cracking as the level of antimony doping is increased. This is because the higher valence of the Sb^{5+} ion introduces more stress into the lattice from crosslinking. In films prepared without any doping the cracking observed should be minimal. The resistivity of the film can be quickly tested using a hand held conductivity meter and a figure around 3×10^{-3} Ω (ohms) per centimetre should be observed. If the resistance is very high or the sample is not conductive then the doping ratio is wrong. It is possible that reactions occurring in the dipping solution reduce the amount that gets incorporated into the film. The actual doping in the produced oxide can be a lot lower than the value calculated from the reaction molar ratio. The actual ratio of antimony to tin can be calculated using X-ray photoelectron spectroscopy (XPS). A spectra taken of the film using this technique should show characteristic peaks for the Sb 3d and Sn 3d orbital. Comparison of the peak areas relative to one another gives a ratio of doping in the produced film. Doping levels above 10% result in more Sb^{3+} centres and these reside within the band gap where they compete with the donor sites and essentially 'cancel out' the doping effect. The easiest indication of the doping will be a blue tinge to the film which becomes increasingly strong at higher doping levels. Transmission electron microscopy will reveal a grain size of between 3 and 25 nm and the distribution should be skewed towards the lower sizes. If the grains are larger then this is an indicator that the humidity may be slightly high. Twinning may also be observed which is known to occur for the tetragonal cassiterite structure of SnO_2.

DIP COATING TO FORM AN ANATASE PHASE TITANIA FILM

Adapted from Sonawane et al. (2002)

The uses of anatase phase titania have been outlined extensively previously and so in this recipe we will concentrate on the use of a titania precursor that is not an alkoxide. In this recipe hydrogen peroxide is used to dissolve a hydrolysed, but not condensed, titania. At a very low pH the Ti^{4+} forms into a monovalent triangular $Ti(O_2)OH^+$ cation. At higher pH or with time this will form into a dinuclear unit Ti_2O_5 which will undergo further polymerisation and condensation until a large metal oxide framework has formed (Schwarzenbach et al., 1970) (Fig. 4.18).

The gradually increasing viscosity of the gel due to the slow polymerisation of the titania lends itself well to coating a wide variety of different substrates. This is attractive as many prefabricated or oddly shaped objects can be made self cleaning by applying a coat of this complex followed by heating to produce the photoactive phase of titania. Stainless steel, glass wool, aluminium and ceramics can all be dip coated using this technique. Furthermore, the colour of the complex changes from a deep orange when initially formed (indicating the presence of the monomeric cation) and then changes to yellow as the polymerisation proceeds allowing for an easy visual identification of how rapidly your gel is ageing.

In this recipe we will break away from preparing a flat film and attempt to coat something non-uniform. A stainless steel metal washer will do to begin with but whatever you choose to coat it should be heat stable to at least 500 °C.

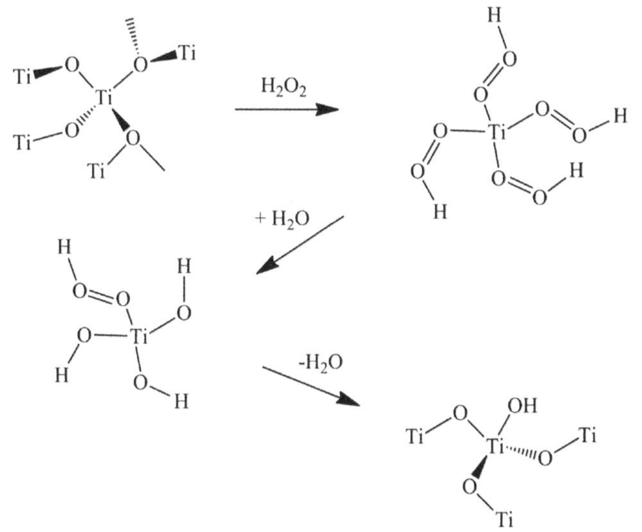

FIGURE 4.18 Schematic of the reaction of the dissolution of amorphous titania with hydrogen peroxide to form a complex and the subsequent polymerisation reaction.

To dip coat an anatase phase film on a non-uniform substrate you will need:

1. A substrate for coating that will not warp or degrade at temperatures of up to 500 °C in air. It must be clean and often soap and a solvent such as ethanol makes a good degreaser for ensuring the surface is free from debris.
2. 4.8 mL (4.8 g or 0.0141 mol) if titanium tetrabutoxide measured out using a volumetric pipette. This is water sensitive but not overly so and can be handled under ambient conditions for this recipe.
3. 100 mL of water in a 150 mL beaker. This will be used to hydrolyse the titanium alkoxide to form a hydroxyl species. You will also need excess water for dilution of the gel.
4. A Buchner funnel and filter paper large enough to handle roughly 5 g of solid. You will need excess water for washing. A tap mounted aspirator will provide enough suction to draw through the moisture. You will need a clamp to hold the receiving flask in place. It can be tempting not to use one but are prone to falling over.
5. 75 mL of a 30% H_2O_2 solution in a 200 mL beaker. This solution should be fresh as hydrogen peroxide solutions will degrade over time. Keep stock solutions in the dark and in a fridge when not in use.
6. The dip coating setup. This procedure can be performed at ambient humidity as the viscosity of the titanium peroxo complex is the major factor in determining the thickness of the film. The drawing speed should be set to 1 mm/s.
7. Somewhere to hang your coated samples from. After dipping in a viscous solution non-uniform substrates should be suspended during the ageing process.
8. A viscometer for determining the viscosity of the gel so you know when it is ready to use. There is a range of ways to determine the viscosity and these will vary in price. More cost effective methods such as timing the passage of a bubble through the solution will give you a reasonable idea of the gel viscosity.
9. A furnace capable of reaching at least 500 °C.

To dip coat an anatase phase film on a non-uniform substrate:

1. Using the volumetric pipette add the titanium butoxide to the 100 mL of water. You should see a white precipitate form. Leave this hydrolyse for 5 min.
2. Place the filter paper in the Buchner funnel and turn on the tap to draw air through. Decant the precipitate into the funnel and let the water get pulled through. Rinse out any remaining precipitate from the beaker with water and rinse through the funnel three times with water to remove the butanol formed during the hydrolysis.
3. Add the precipitate to the 30% hydrogen peroxide solution. As it dissolves to form the titanium peroxo complex you should see the solution turn into a transparent orange. From this point the formed sol will begin to age and gel and will turn more yellow as it does so. You will have a few hours before the solution begins to thicken enough for dipping.
4. If we consider that essentially the entire alkoxide precursor has been successfully converted, the as prepared peroxide gel has a titanium peroxo concentration of 0.188 mol/L. The more concentrated the solution is then the faster the gel will thicken. Dilute the gel using water to give a 0.1 mol/L solution. This can be done by adding 66 mL of water to the 75 mL of titanium complex in hydrogen peroxide solution. The concentration has a large effect on the thickness of the produced film as it relates directly to how quickly the gel thickens from the polymerisation of the titanium peroxo complex.
   The as prepared system will be 4–5 cps and this will increase to 12,000 cps over the course of about 12 h. You will need to perform the dip coating when the viscosity is somewhere between 4000–12,000 cps and this will produce films ranging from 30 to 100 nm respectively with an almost linear relationship. If the solution is too thick when you perform the dip coating then there is a tendency towards the film fracturing and falling off after the calcination step. If it is too thin then the film will be patchy in places and not very thick (Fig. 4.19).

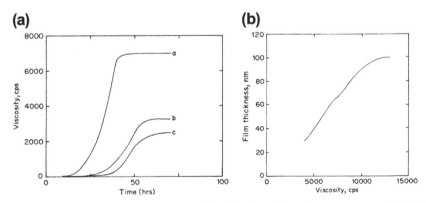

FIGURE 4.19 Plots of (a) Variation in viscosity with time of titanium peroxide sol having titanium ion concentration: (i) 0.01, (ii) 0.005 and (iii) 0.001 M and (b) Variation of film thickness with viscosity of sol. *(Adapted with Permission from Sonawane et al. (2002). Copyright 2002 Elsevier)*

5. Pour the gel into the dip coating bath. Hang your substrate in such a way that it will remain as still as possible during the dipping. Dip the sample into the bath and drawback out immediately at a speed of 1 mm/s.

6. If you are dip coating a flat substrate then allow the film to dry in air and then heat in an oven at 100 °C for 2 h. After this transfer the sample into a furnace and heat it to 400 °C for an hour so that the titania films are transformed into the anatase phase. Non-uniform substrates should be hung to dry in air for 6−8 days for the surface coating to fully solidify. These can then be heat treated in the same way as a flat film would be. As with most sol−gel dip coating techniques, the slower the ramp rate is then the lower the stress on the condensing gel will be and you should see less cracking.

7. You should now have an anatase phase titania coated substrate ready for analysis.

The titania coating will appear optically transparent though substrates can develop a white tint if multiple layers have been built up. Coated materials should show some photocatalytic response and this can be observed by monitoring the photodegradation of an organic dye such as methylene blue. The method for this is outlined elsewhere. A powder X-ray diffraction pattern of the coating should show peak broadening of the major reflections which can be used to estimate the grain size of anatase particles using the Scherrer equation. If you scrape some of the film off the surface you can confirm the size using transmission electron microscopy. If the XRD pattern shows peaks for the rutile phase then the sample has been heated at too high a temperature.

REPLICATION OF ODDLY SHAPED MORPHOLOGIES USING SOL GEL TECHNIQUES

The previous recipe touched on one method to coat 3D objects using dip coating. The following recipes will look at expanding this to replicating larger but more finely detailed structures using examples of complex templates of biological origin. Though there are more templates of biological origin than might be comfortably worked with in a few years, or a lifetime, of laboratory work the following recipes should demonstrate the principle of nanoscale coating to macroscale architectures. This section also serves as a cross over point between making particles and films using sol−gel techniques and into making large monoliths and aerogels with nanoscale patterning and structure.

REPLICATING POLLEN WITH TITANIA USING A SOL GEL APPROACH

Adapted from Simon et al. (2006)

Pollen can be micrometres in size but is possessed of intricate structural features all the way to the nanoscale in the form of ridges and spines which have evolved to aid it in nature to stick to bee legs or be carried in the air. The structure varies by species and so there are a variety of different templates to choose from with most choices exhibiting a high level of symmetry. It is a hardy biological substrate as the outer exine layer is composed of tough biopolymers and other macromolecules known collectively as sporopollenin. This allows the pollen to survive relatively rough chemical procedures and is therefore a good test subject for the refinement of a replication technique. In this recipe the pollen is soaked in water so that the biopolymer exterior becomes hydrated. The hydrated pollen is then isolated and dispersed into a solution of titanium isopropoxide in isopropanol. Titania precursor in contact with the surface of the pollen grain will hydrolyse and condense to build up an amorphous layer of TiO_2 on the surface. The coated pollen is sedimented from the solution by centrifugation and then calcined in an oven to remove the pollen and leave behind a replica of the original structure in titania. One of the limitations with this technique is that there is some shrinkage within the formed oxide layer upon calcination. In native pollen the spines protruding from the surface are 4 μ in length and 700 nm in width. Upon replication and calcination the morphology is maintained but the dimensions are reduced to 0.5 μm long and 100 nm in width. This shrinkage highlights an important difference in working with sol gel materials at length scales on the order of microns and above. Metal oxide films cast onto substrates do experience shrinkage but can remain in place due to adhesion or by the fact that they are very thin. Similarly, spherical nanoparticles experience a shrinkage upon drying that is normally only a few percent of the total particulate volume. As the total volume of the formed metal oxide becomes larger the percentage shrinkage becomes larger and can introduce defects. When working at larger scales this shrinkage must be factored in either by allowing for the volume shrinkage or by maintaining the drying conditions so that the oxide condenses very slowly with minimal capillary forces. Techniques such as critical point drying with supercritical CO_2 can be used for preserving structures sensitive to drying stress.

To make a titania replica of pollen you will need:

1. Roughly 0.01 g of pollen. This may be any type you wish but the source reference uses dandelion (*Taraxacum*) pollen.

2. 1 mL of a 100×10^{-6} mol/L titanium isopropoxide in isopropanol solution. A stock solution can be prepared by dissolving 284 μL of titanium isopropoxide into 10 mL of isopropanol.

3. 1 mL of a 50 wt% ethanol and water solution in a small centrifuge tube. This will begin to soak the pollen grains and the ethanol will solubilise any contaminants on the pollen exterior. You will also need 1.5 mL of water for washing and soaking the grains within the centrifuge tube.

4. A centrifuge suitable for handling 1−2 mL volumes.
5. A sonic bath with a piece of foam to sit the centrifuge tube in. This will be used to disperse the pollen in water and help get it absorbed into the biopolymer surface.
6. An agitator capable of a speed of at least 1000 rpm. This is a device that you can sit a centrifuge tube in and it will get shaken automatically. Depending on what model you have you must ensure that the lid of your sample vial is sealed firmly shut and you may even need to tape the tube in place.
7. A furnace capable of reaching at least 500 °C and a small crucible. Select a crucible with smooth wall as this will make it easier to isolate the rather small amount of product produced.
8. A vacuum oven or vacuum line. This is for drying the samples prior to heating and you do not need the oven component. If you are using a pump to evacuate a bell jar or even the vacuum oven ensure it is fitted with a cold trap to catch the alcohol and alkoxide removed. The amounts will be small but it is good habit to cultivate!

To make a titania replica of pollen:

1. Add the pollen to the water/ethanol mixture and sonicate for 15 min.
2. Centrifuge the sample at 13,200 rpm for 5 min. A pellet of the pollen should form in the bottom. Decant off the top solution leaving the pellet.
3. Redisperse the pellet in the centrifuge tube with 1.5 mL of water then sonicate it for 15 min.
4. Place the sample in the centrifuge tube into the agitator. Set it to run at 1000 rpm for 12 h. This will give the water time to permeate into the surface of the pollen grain.
5. Replace the sample tube into the centrifuge and spin it down at 13,200 rpm for 5 min to once again form a pellet. Pour away all the water on top to leave behind a damp pellet.
6. Add the titanium isopropoxide in isopropanol solution to the centrifuge tube containing the pellet. Leave the sample to stand for 15 min. At this stage hydrolysis and condensation of the titanium precursor will begin to occur on the pollen surface. Do not be tempted to leave the sample for too long in the precursor solution as the reaction could go too far and you will get unwanted precipitation.
7. Centrifuge the sample at 13,200 rpm for 5 min. Pour away the top solution into the correct solvent waste container.
8. Place the sample in a vacuum oven and dry under vacuum for 2 h. Once finished, make sure that you empty the cold trap into the correct solvent waste container. The vacuum step will not only remove the excess solvent and precursor but will also drive the condensation reaction helping to solidify the formed oxide layer.
9. Transfer the dry, titania coated pollen into a crucible and heat in a furnace in air to 500 °C for 3 h at a heating ramp rate of 1 °C/min. If possible affect the transfer by holding the tube upside down over the crucible and tapping it. Trying to dig the dried sample out with a spatula can damage the delicate structures. This step will burn away the pollen leaving behind a hollow titania replica ready for analysis. The slow heating rate will allow the CO_2 formed by the combustion of the template to diffuse slowly out of the oxide coating without damaging it. The step also drives the densification of the titania at the surface resulting in shrinkage and converts the titania into a crystalline form.

Scanning electron microscopy is the most expedient method of checking your results. The replica pollen can be tapped directly from the crucible onto a carbon sticky pad atop a sample stub and then coated with carbon or platinum. A comparison with uncoated control pollen should reveal that the replicas are up to 5 times smaller and this is a result of shrinkage. Despite this, the morphology should be very similar in the replica and the fine structure will remain whilst the larger geometry may appear 'dried out'. The fine structure can be analysed under transmission electron microscopy by sonically dispersing the replicas in ethanol and then drying the resulting dispersion onto a carbon backed copper grid. If your TEM has the facility, then a diffraction pattern will likely reveal that the crystal polymorph of the titania is brookite. This structure is normally expected to form at higher temperatures and is only weakly observed in this synthesis. That is to say, the replicas are not well ordered crystal structures. Further heating to 800 °C does not force a change of phase to anatase and results ultimately in the phase transition to the rutile polymorph.

MAKING A CUTTLEFISH BONE REPLICA

Adapted from Ogasawara et al. (2000)

The cuttlefish belongs to the cephalopod family which includes squid and octopi. In contrast to its marine cousins, the cuttlefish has a large porous internal bone structure called the cuttlebone. Cuttlebone has evolved to act as a ballast tank for the animal and is composed of a mineral $CaCO_3$ phase biomineralised within a well ordered macroporous polysaccharide scaffold. The $CaCO_3$ is nucleated in the aragonite crystal polymorph and their plate like growth is directed by nucleation within the spatially well ordered structure of a β-chitin matrix. This nanocomposite structure makes the cuttlebone very strong. This structure is 95% porous and this allows the animal to adjust the gas to liquid ratio internally to provide lift or drop. From a materials point of view the cuttlebone can

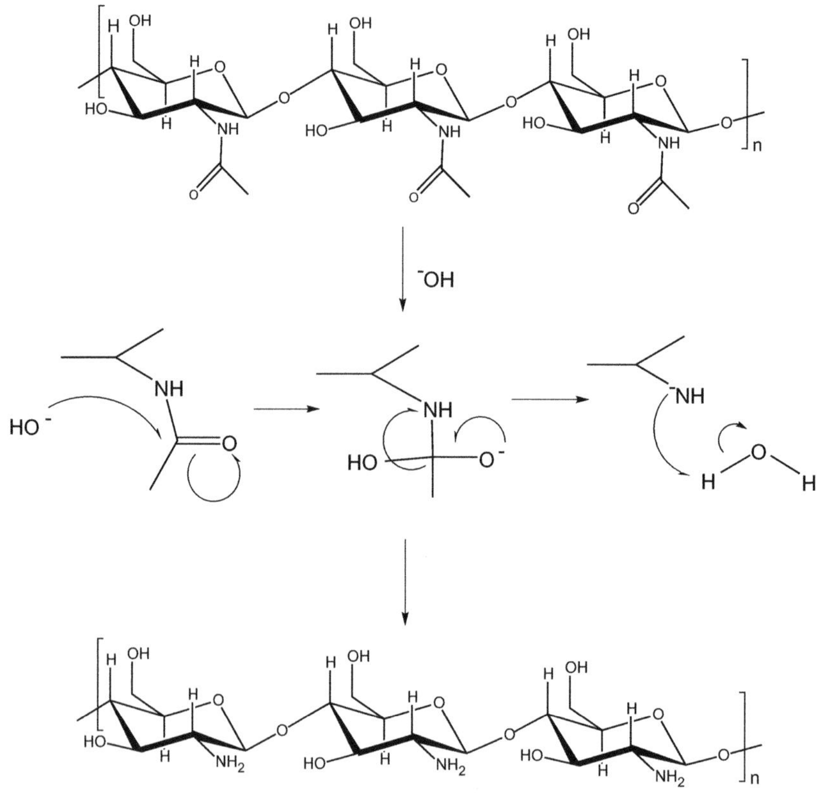

FIGURE 4.20 The deacetylation reaction of chitin to produce chitosan.

be used as a template to produce a well ordered high surface area structure which can be used for filtration, catalyst support and chemical storage and release. By removing the $CaCO_3$ using acid, the structured β-chitin matrix will be left behind and this can be remineralised using the material of our choosing. In this case it will be glass once again though the preparation is not limited to this. β-chitin is a glucose derived N-acetylglucosamine and though it makes for a tough framework this does not always make the best template material due to the low binding ability of the acetyl ($COCH_3$) group. Under ideal conditions the biotemplate will have some form of specific affinity for the mineral precursor to be laid down upon its surface so that the end material is formed at the interface and not free in solution. Normally, most metal salt or sol−gel hydrolysis species will behave as ions free in solution and will interact with surface bound carboxylic acid or amine groups. With this in mind it is possible to boost the level of surface affinity in a chitin scaffold by chemically modifying the polysaccharide and substituting something more 'sticky' as a functional group in the place of the $COCH_3$. The deacetylation of the chitin can be achieved by boiling the demineralised scaffold in a strong sodium hydroxide solution. This procedure cleaves off the $COCH_3$ group and leaves behind the free amine which has a lone pair capable of donation to positively charged metal centres or else, can be protonated to give a positive charge under acidic conditions (Fig. 4.20). If the deacetylation reaction is prolonged then the chitin will be converted to chitosan. Chitosan can be produced in large quantities from waste chitin and is used in waste water reclamation as it forms a gel readily in the presence of waste metal ions. This gel can be removed from the water leaving it cleaner. The deacetylation process preserves the underlying structure of the template so that now a replacement mineral or compound can be nucleated in the place of where the aragonite mineral used to be.

In this recipe we will perform both the removal of the existing mineral in the cuttlebone as well as performing the deacetylation of the chitin. The deacetylation will not be taken to the point that the scaffold is converted to chitosan so as to preserve the native structure as much as possible. The precursor for glass formation will be an aqueous solution of sodium silicate. This technique provides the basic outline for the use of other polysaccharide derived materials as templates including cellulose and chitosan.

To make a glass replica of cuttlebone you will need:

1. A cuttlebone cut into cubes 1 cm^3 in dimension. You can obtain a cuttlebone from a pet shop where it is sold as something for budgies to peck on. Each cuttlebone will be about 15 g and from this you should get at least ten cubes.
2. At least 1.5 L of 2 M HCl in water. This can be prepared by diluting 168.7 mL of concentrated 37 wt% (11.85 mol/L) stock solution up to 1 L in a volumetric flask. *Wear gloves and prepare in a fume hood while working with the concentrated acid.* You will need a 500 mL beaker to use as an acid bath for soaking the cuttlebone samples in.
3. A vacuum oven or vacuum line with a large bell jar. This does not have to be overly powerful as it will be used to degas the cuttlebone as the mineral component dissolves and evolves CO_2.

4. Litre amounts of water for rinsing and soaking steps. A large beaker will also help for rinsing as the cubes can be placed in the beaker and rinsed quickly. You will also need ~100 mL of a 95% ethanol in water solution.

5. 500 mL of a 1 mol/L solution of NaOH. This can be prepared by weighing out 20 g of solid NaOH and dissolving it in 500 mL of water. Add the NaOH slowly to the water under gentle stirring to make sure it does not become too hot. At this strength the basic solution will get rid of associated proteins bound in the chitin matrix.
   Should you wish to deacetylate the chitin then make the NaOH solution five times stronger to give a 5 mol/L solution. This will be used in the place of the 1 mol/L solution during the reflux step.

6. A large 1 L round bottomed flask with a quick fit aperture and a large reflux condenser that fits it. You will need all the correct tubing and clips to hold it all in place as well as a clamp each for the flask and reflux condenser. This should be setup in a fume hood and will be used to reflux the NaOH solution.

7. A large heating mantle and oil bath upon which the reflux setup can sit. You will not need the capability to perform magnetic stirring as this could damage the cuttlebone.

8. A large Buchner funnel and flask with filter paper that will be used for rinsing the samples.

9. 500 mL of bench ethanol which will be used for dehydrating and rinsing the samples.

10. Liquid nitrogen for rapid freezing and a vacuum line or pump for freeze drying the scaffold for analysis. If you have a dedicated freeze dryer then use this. You will need a shallow dewar flask for dipping samples into and a plastic spoon or similar for removing them with. *Liquid nitrogen can cause cryogenic burns, ensure you follow your laboratory guidelines and wear gloves and safety glasses.* Freeze drying will preserve the structure in a way that normal air drying might not. The vacuum lines will need to be fitted with a cold trap to collect water.

11. 5 mL per cuttlebone cube of a 1.72 mol/L solution of sodium silicate. This can be prepared by diluting a 28 mL of a 27 wt% (~6.15 mol/L glass) stock solution to 100 mL with water. The pH should be around 11.5 and this is *not* adjusted using a cationic exchange column.

12. 60 mL of a 25% volume ethanol in water solution. 3 mL aliquots of this will be used to soak the cubes prior to condensation in the presence of an ammonia catalyst.

13. 60 mL of a 0.03 mol/L solution of ammonium hydroxide in water containing 25% volume ethanol. This can be prepared by diluting 2 mL of a 28 wt% (14.5 mol/L) stock ammonium hydroxide solution with 750 mL of water. 75 mL of this solution can then be made up to 100 mL using bench ethanol in a volumetric flask. The pH of the solution should be 10.5.

14. A pH meter for checking the pH of solutions.

15. A furnace capable of reaching 800 °C with a heating ramp rate of 5 °C/min. You will also need a crucible large enough to hole the replicated cuttlebone cubes.

To make a glass replica of cuttlebone:

1. The cuttlebone first needs to be demineralised to remove the calcium carbonate and leave behind the chitin framework. Soak the cuttlebone cubes for 12 h in 500 mL of the 2 mol/L acid solution. Because of the carbon dioxide evolved by a reaction with the acid, gas can accumulate in the porous structure which can block the passage of more acid deeper into the structure. The acid soak therefore needs to be performed at reduced pressure so that the gas is forcibly drawn from the sample as it is formed. The beaker containing the acid can be placed into a vacuum oven or bell jar and exposed to vacuum. The vacuum should not be high enough that the acid solution begins to boil.

2. Decant the acid solution out of the beaker and dispose of in the correct waste solvent container. Add another 500 mL of fresh acid to the beaker and repeat the soaking process under low vacuum.

3. As before remove the acid and rinse the demineralised cuttlebone with copious amounts of water until the pH reaches 7 after the cubes have been sat in water for at least 5 min.

4. Setup the reflux condenser in the oil bath on top of the heating mantle. *Ensure all of the clips and tubes are secured properly!* Add 500 mL of the 1 mol/L NaOH solution to the large round bottomed flask. Place the demineralised cuttlebone cubes into the solution. Secure the reflux condenser in the round bottomed flask and set the temperature of the mantle to 100 °C. Reflux the cubes for 4 h without stirring. This strength of base will denature the remaining protein in the structure.
   If you want to perform a deacetylation of the chitin then use the stronger 5 mol/L NaOH solution and run the reflux for only 3 h. This will destroy remaining protein and deacetylate the chitin to produce amine groups capable of coordinating with the silicate anion.

5. Turn off the heating mantle and the water flow and allow the solution to cool to room temperature. Remove the cubes from the reflux setup and transfer into a Buchner funnel lined with filter paper. Draw off the excess sodium hydroxide solution and rinse the cubes with water. Don't let them dry out.

6. Place the cubes into a beaker with a great excess of water and swirl gently by hand for a few minutes. Check the pH of the water and exchange it for more until the pH reaches 7.

7. Exchange the water for the 95% ethanol solution and allow the cubes to soak for 3 h. Afterwards wash again with water. If the cubes are not going to be used immediately then you can store them in a fridge in water.

8. For every cube of the deacetylated chitin scaffold add 5 mL of the 1.72 mol/L sodium silicate solution into a beaker. Leave the scaffolds to soak at room temperature for 5 h.

9. Remove the cubes from the sodium silicate solution and place into a 25% ethanol solution for 30 min. Use 3 mL of the solution for each cube. In the water and ethanol solution the silicate is supersaturated and begins to precipitate out of solution.

10. After 30 min exchange for fresh 25% weight ethanol solution and leave to stand for another 30 min. The final pH of the solution will be around pH 10.

11. Place the samples in the 25% ethanol and ammonium hydroxide solution. Use 3 mL of solution for every cube. Allow the samples to stand for 1 h. At this stage the ammonia will drive the condensation and solidification of the silica.

12. Wash the cubes by repeated rinsing in excess water until the pH reaches 7.

13. At this point drying the cubes in air could lead to the replica collapsing so they need to be freeze dried to preserve the structure. This is achieved by dipping the still wet cubes rapidly into liquid nitrogen. While the cubes are still frozen expose them to vacuum so that the water will sublime away to collect in the cold trap.

14. The silica coated cubes can be analysed immediately or calcined to remove the chitin scaffold. The mineralised cubes are placed into a crucible and heated to 800 °C for an hour at a rate of 5 °C/min.

Scanning electron microscopy of the glass replica should reveal an organised porous structure with an interval spacing of 100 μm to 300 μm identical to that of the native cuttlebone. If ethanol concentrations higher than 30% are used for the rinsing stages the silicate will precipitate out to give a rough texture to the glass. If the ethanol concentration is greater than 40% then the silica is not soluble enough to remain in solution and will precipitate out of its own accord. The level of deaceylation also helps to bind the formed glass to the chitin surface and the more amine groups there are in the scaffold the greater the concentration of glass that will bind to the surface. Samples of demineralised scaffold before and after the NaOH treatment can be analysed by FT-IR to determine the level of deacetylisation. The characteristic stretches for the demineralised β-chitin and the silica coated β-chitin are given in the table below:

Sample	Bond and stretching mode	Wavenumber/cm^{-1}
β-chitin	OH vibration	3450 and 3095
	NH vibration	3270
	C=O	1620
	NH stretch	1560
Silica coated β-chitin	Si−O−Si stretch	468
	Si−O−Si vibration	795

MAKING A SUPERCONDUCTING CUTTLEBONE

Adapted from Culverwell et al. (2008), Kareiva et al. (1994), Sun (1996)

A superconductor is most simply defined as a material that shows zero resistance to electrical current passing through it. In broad terms the attributes of a superconductor are defined by two parameters known as the Curie temperature (T_c) and the critical current density (J_c). T_c is the temperature below which a material exhibits superconductive behaviour. J_c defines the maximum current that can flow with zero resistance per unit volume per unit time before the superconductivity breaks down and resistance is observed. The materials that demonstrate superconductivity are placed into two categories, type I and type II. Most pure metals are type I superconductors and at temperatures of only a few degrees Kelvin will display zero resistance to charge. Of far more technological interest are the type II superconductors which are mostly ceramic oxides and super conduct at tens of Kelvin with generally higher critical current densities than those observed in type I materials.

The main goal of superconductor research is to produce a type II material with a T_c as close to room temperature as possible or even higher, with a large J_c value. Currently, cuprate perovskite ceramics are the leaders with T_c values above the boiling point of liquid nitrogen (77 K). Liquid nitrogen is seen as a bench mark as this is a commercially available cryogenic and is the minimum temperature range for practical use. Yttrium barium copper oxide ($YBa_2Cu_3O_7$) was the first perovskite formulation to demonstrate superconductivity above liquid nitrogen temperatures (77 °K) and mercury barium calcium copper oxide has been shown super conduct at 135 °K (Chu et al., 1993).

Traditionally, ceramic superconductors have been made by a 'heat and beat' method where mixed metal carbonates and nitrates are ground together before heating to form an oxide. The resulting powder is then pressed together to form a pellet of superconducting material. There are problems with this method in that grinding the oxide powders together does not always result in a homogenous stoichiometry. Upon heating this can lead to heterogeneous phase formation and domains of single metal oxides as opposed to mixed ones. There is also a lack of morphological control and the powders are composed of crystallites of varying size. This causes problems when the powder is pressed into pellets for use as they can be mechanically very brittle. Small crystallites of uniform size minimise this issue and have the added advantage of decreasing grain boundary misalignment and therefore improve the critical current density attainable.

Many of these formulation problems can be overcome by using the sol gel method to form a precursor gel which can then be heated to produce a homogenous oxide with high phase purity. The method is advantageous in that the gel can be formed into films and coatings. In this recipe we will use cuttlebone as a template and coat it with the superconductor $YBa_2Cu_3O_{7-\delta}$. This is also known as Y123, reflecting the molar ratio of one yttrium, two barium and three copper per mole of the oxide. The delta in the formula represents a slight oxygen deficiency in the structure. Cuttlebone makes for an interesting template as the porous structure lends itself well to the diffusion of cryogen throughout the structure. This is important as it allows for faster cryogenic cooling of the entire superconductor, thereby minimising the risk of resistive heating by areas which have not yet achieved the superconducting critical temperature. Additionally, the ordered periodicity of the cuttlebone and the homogeneity of the coating through well aligned crystallites gives an increased J_c not observed in the unstructured bulk material. You do not have to use cuttlebone as the template in this recipe but it is easy to obtain and demonstrates the versatility of the technique well. During the cacination step to form the superconducting phase, the porous template structure allows ambient oxygen to percolate in the sample and this eliminates the requirement for a pure oxygen flow as some preparations require. To form the gel precursor yttrium, barium and copper salts are dissolved in a 1:1 ratio of water:ethylene glycol. The key to the homogeneity of the gel is the ethylene glycol which acts as a chelating agent for the metal cations. The lone pair electrons on the glycol oxygen atoms coordinate to the multivalent metal centres and prevent them from contact with one another and this helps to prevent precipitation from the solution if the salts are poorly soluble. Each molecule of ethylene glycol has two hydroxyl groups at either end meaning it can coordinate to two cations to bridge them together. As the cations are multivalent this means an extensive metal chealate network is formed that dries down to a homogenous blue solid upon drying (Fig. 4.21).

Upon heating to around 300 °C, the organic component of the gel is burned away and with further heating to higher temperatures the metals are oxidised. As the precursor gel is a homogenous mixture with tightly controlled molar ratios the oxidised species produced at above 900 °C will convert into the desired Y123 perovskite structure. Depending on the heating rate, the grain sizes produced will be in the sub-micron domain. In order to produce a coating of Y123 on the cuttlebone template we will simply soak the template in a premade sol of Y123, allow it to dry and then calcine it in a furnace. In this recipe the chitin matrix will not be demineralised as heating the aragonite breaks it down into lime and slaked lime. This impacts on the mechanical strength of the superconducting replica but can be offset in part by the introduction of silver salts into the gel. The silver does not act as a dopant in the sense that it is incorporated in to the superconductor structure as a type of defect. Instead it behaves as a flux, promoting the assembly of the perovskite like structure from the metal oxide moieties and linking them electrically. As mentioned previously, a larger crystal size is detrimental to the critical current density and so methods to strengthen the structure in this way are only required if the product needs to be robust. The method for adding silver to the gel is provided in this recipe though it is possible to substitute this with sodium, cerium lithium and calcium nitrates with similar effect (Walsh et al., 2009).

To make a cuttlebone superconductor you will need:

1. An accurate three figure balance. One of the major difficulties in producing a superconductor is that the molar ratios must be very tightly controlled.
2. A magnetic stirrer and stir bar.
3. 0.958 g (0.0025 mol) of $Y(NO_3)_3 \cdot 6H_2O$.
4. 1.307 g (0.005 mol) of $Ba(NO_3)_2$.
5. 1.744 g (0.0075 mol) of $Cu(NO_3)_2 \cdot 2.5H_2O$.
6. 25 mL water and 25 mL ethylene glycol mixed together in a beaker.
7. Cuttlebone cubes 1 cm^3 in dimension. As mentioned previously this may be obtained from a pet shop.

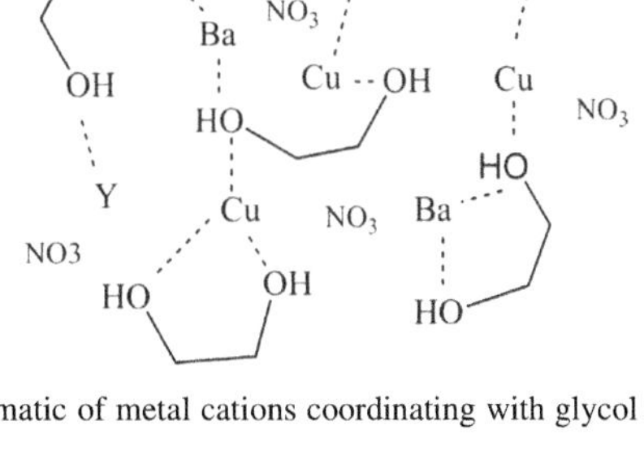

FIGURE 4.21 A schematic of metal cations coordinating with glycol molecules to form a gel.

8. A dish capable of holding 15–20 mL deep enough that solution will cover the cube when filled.
9. A vacuum oven or vacuum line with a bell jar and tap. This does not have to be powerful and will be used for degassing the cuttlebone under solution.
10. A drying oven set to 40 °C.
11. A furnace capable of reaching at least 920 °C with a programmable ramp rate of 1 °C/min.

To make a cuttlebone superconductor:

1. Add the stir bar to the ethylene glycol and water mixture and then add the $Ba(NO_3)_2$. Wait until it has dissolved then add the $Y(NO_3)_3 \cdot 6H_2O$ and after this too has dissolved add the $Cu(NO_3)_2 \cdot 2.5H_2O$. A blue transparent solution should form. If you see any precipitation then increase the proportion of ethylene glycol.
2. Pour 15 mL of the blue solution into the dish and add the cuttlebone cube. It should be entirely immersed.
3. Place the dish containing the cuttlebone into a vacuum oven/bell jar and reduce the pressure to the point that you see the air evacuating from the porous cuttlebone. This will aid the infiltration of the precursor into the structure. Leave under low vacuum for 2 h.
4. Remove the cube from the bath and place into the drying oven at 40 °C on a petri dish. Allow the cube to dry and you should see a blue veneer over the surface of the sample.
5. Place the cube back into the dish full of fresh precursor and repeat the low vacuum soaking and then drying cycle two more times.
6. Once the cube is dried the third time it should look blue all over. Place it into a crucible and heat to 920 °C for 4 h at a heating rate of 1 °C/min. Slow heating rates allow the oxide and phase change to occur slowly so that the fine structure is retained. At higher rates the thermal shock will cause the sample to disintegrate into powder and the different metals will oxidise at different rates causing unwanted mixed phases. After calcination and cooling you should have a black looking replica of the cuttlebone ready for analysis. *It will be very brittle!* The delicacy of the product can be offset by following the variation outlined below.
7. If you wish to have a more robust replica and do not mind the resulting lower critical current density then add 0.4 g (0.0023 mol) of $AgNO_3$ into the gel solution. Keep in mind that silver nitrate is photosensitive and will reduce to native silver if exposed to even weak light. Solutions containing silver nitrate should be kept in the dark at all stages of the preparation. The replica will contain silver after calcination but this does not interfere with the crystal Y123 structure and will actually result in stronger Y123 reflections if powder X-ray diffraction studies are performed.

You should be able to see the replica structure using an optical microscope. Perform powder X-ray diffraction to confirm that the Y123 phase has been formed. The major reflections can be indexed to JCPDS card number 79–1229. There will also be reflections present for CaO which is formed from the decomposition of the aragonite. This can be indexed to JCPDS card number 48–1548. The expected X-ray pattern is given below (Fig. 4.22).

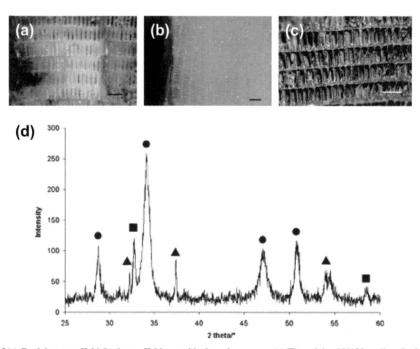

FIGURE 4.22   Photographs of (a) Cuttlebone scaffold (b) the scaffold coated in the gel precursor (c) The calcined Y123 replica. Scale bars in (a) and (c), 500 mm, in (b) 1 mm. (d) The expected powder X-ray diffraction pattern obtained from the replica. Peaks marked with circles are $Ca(OH)_2$, peaks marked with triangles are CaO and peaks marked with squares are from Y123. *(Image adapted with permission from Culverwell 2008)*

To determine the superconductive behaviour you will need to use a SQUID (Superconducting Quantum Interference Device) magnetometer. This type of magnetometer consists of a superconducting coil into which a sample is placed. The sample is cooled to a few degrees Kelvin from room temperature in a magnetic field of a known strength generated by the coil and fluctuations in the magnetic and electronic behaviour of the sample are monitored. Though the operating theory of a SQUID magnetometer is beyond the scope of this book in brief; a plot of temperature against magnetic susceptibility will allow you to determine the T_c value and is the temperature at which the magnetic susceptibility reaches zero. For the Y123 cuttlebone replica the T_c should be 93 °K. It is also possible to determine the critical current density by varying the field strength at different temperatures but this will require some calculation using the critical state model and you will need to know the grain size of the Y123. The grain size of the superconductor can be determined by scanning and transmission electron microscopy.

MESOPOROUS INORGANIC POWDERS

Alumina (Al_2O_3) and silica (SiO_2) are the two most abundant minerals on earth and together form various compositions including clays and zeolites. The first thought at the mention of clays might be of pots and bowls but inorganic minerals are employed in various ways for nanotechnology. In this section we will focus on zeolites and the various routes available for their synthesis. Naturally occurring zeolites are aluminosilicate compounds formed in volcanic rock and have been studied for hundreds of years. The various octahedral and tetrahedral alumina and silica units pack together to form a wide array of cage like structures with long range order. The pores within the framework are of the order of 5–10 Å and this microporosity means that a zeolite can be used as a molecular filter for the separation of small molecules like water or nitrogen gas from bulk phases. Zeolites are the active components in ion exchange columns where one type of cation may be exchanged for another in solutions passing through the column. In a single unit cell of an ordered porous aluminosilicate the silicon is in a +4 oxidation state whilst the aluminium is in a +3 oxidation state. The silicon is charge balanced for every two oxygen present in the lattice but aluminium in the same position is one charge short for keeping the net unit charge zero. The extra positive charge is made up for by the presence of cations residing within the pores of the zeolite framework. These are loosely coordinated and may therefore be exchanged out with other different cations if a strong salt solution is passed through the porous material (Fig. 4.23).

Mesoporous zeolite materials were developed because of the need for larger pore sizes than could be found to occur naturally. Before the 1990s the production of larger pores in synthetic aluminosilicates was technically challenging and an iron aluminium phosphate mineral known as cacoxenite set the bench mark for pore diameter at a maximum of 1.4 nm (Beck et al., 1992). It was not, at the time, possible to coax the assembly of the inorganic constituents into larger ordered pore arrays by simple alteration of stoichiometry or by changing the synthetic conditions. These limitations were sidestepped by taking a sol gel approach to the formation of the zeolite structure and using a self assembled liquid crystal assembly as a template about which an aluminosilicate could condense. In 1991 a team at Mobil laboratories were the first to publish the technique (in the academic literature at least, see the box for more) and named the class of mesoporous materials prepared using it as 'M4S1' (Kresge et al., 1992). The most common preparations are known as Mobil Composition of Matter (MCM)-41 and MCM-48 (Schumacher et al., 2000). MCM-41 forms into a hexagonally packed pore array and MCM-48 forms a cubic/gyroid pore array. The materials are typically very low density compared to the same volume of bulk material and display high surface areas around 700–1000 m²/g. The condensation of the solid phase relies on the capacity for the reaction to take place without disruption of the surfactant template and as long as this condition is

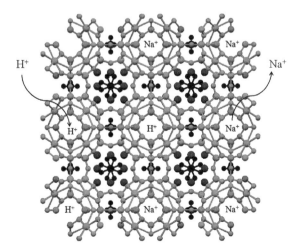

FIGURE 4.23 The porous structure of zeolite A. Cations can be infiltrated into the open spaces and exchanged with other ions. Crystal structure obtained from Depla et al. (2011).

satisfied a wide range of compositions can be used to form the zeolites. It is therefore possible to do away with the alumina component altogether and have a pure silica zeolite structure. The MCM materials are made by taking a silica or alumina precursor, normally a mixture of aluminium nitrate and sodium silicate solution, and mixing it with a base and a large amount of surfactant. The surfactant must be above its critical micelle concentration so that the surfactant molecules form a self assembled phase. The hydrophobic component of the assembly will eventually form the pores as the sol−gel precursor is delivered in the aqueous phase and will condense in association with the hydrophilic head groups. The pore size produced can therefore be controlled by having a larger tail group in the surfactant or by bulking out the hydrophobic phase with an oily molecule such as 1,3,5-trimethylbenzene (Beck et al., 1992). The packing of the porosity is a direct result of the surfactant phase behaviour and in addition to traditional hexagonal, cubic and gyroid structures more esoteric configurations can be achieved by the choice of surfactant (Huo et al., 1994). After condensation and solidification the surfactant phase can be removed by extensive washing or calcination to leave behind a porous ordered powder or monolith.

There is an interesting aside to the development of mesoporous materials which highlights the occasional happenstance in science of rediscovery (Renzo et al., 1997). In 1971 the Sylvania Electric Products company patented a low density silica formulation it had come across in the pursuit of a zinc doped orthosilicate phosphor. The patent outlined the condensation of tetraethylorthosilicate in the presence of ammonia and the surfactant cetyltrimethylammonium bromide. The low density was in fact due to the presence of ordered porosity though it was not until a French team in 1997 looked at the patent and repeated the synthesis that this was discovered to be a mesoporous material preceding the development of the MS41 materials. The scientists at the Sylvania electronics company were not looking for mesoporous materials and so missed out on being the first to publish. Though in hindsight it is possible to see the links between two different research themes this is not so obvious for those focused on achieving specific research goals at the time. It is very difficult, if not possible, to know if an innocuous discovery in one field might have larger implications in another unfamiliar one. The only lesson that can be learned from this is that publication and good communication between all areas of the scientific community can increase the chances of breakthroughs being made earlier rather than being discovered again much later.

MAKING A POROUS ALUMINOSILICATE ZEOLITE MCM-41

Adapted from Beck et al. (1992), Kresge et al. (1992)

This recipe is adapted from the original MCM-41 preparation as outlined by the Mobil research team in 1992. Subsequent development of this recipe has yielded simplified routes towards production and a few of these will be outlined elsewhere. The procedure mixes alumina and silica powders with a strongly basic tetramethylammonium silicate solution. The solid powders partially solubilise during the high temperature hydrothermal conditions of synthesis before re-condensing into a porous lattice. The surfactant template is a quaternary ammonium chloride ($C_{16}H_{33}(CH_3)_3NCl$) which is partially converted to a hydroxide by anionic exchange in a column prior to use. The cationic alkylammonium surfactant self assembles into a hexagonal phase about which the aluminosilicate will condense and the pore diameter can be changed by selecting a longer tail group. The MCM-41 will resemble a powder when removed from the hydrothermal bomb reaction vessel which can then be filtered. Removal of the surfactant from the pore channels is done by calcination to above 500 °C.

To make aluminosilicate MCM-41 you will need:

1. 2 g (0.02 mol) of Al_2O_3 powder. The smaller the particulate grain you can find the better the results will be. It is also possible to substitute an aluminium nitrate salt if required.
2. 100 g of a 10 wt% tetramethylammonium silicate (($CH_3)_4N(OH)\cdot 2SiO_2$) in water solution in a 500 mL beaker. This is available commercially as a 15−20 wt% solution and can be diluted with water depending on the strength of the stock solution obtained.
3. 25 g of fine silica powder. As with the alumina make sure the powder is as fine as possible and nanometre grades are available. *When working with ultra-fine powders be aware of the inhalation risks.* Work with powders in an isolation box away from anything that could disturb it like an opening door! Once the powders are in solution they should be safe to work with on an open bench.
4. 48 g (0.15 mol) of cetyltrimethylammonium chloride ($C_{16}H_{33}(CH_3)_3NCl$). This will be used to make a 29 wt% solution in 200 mL of water which will be run through an anionic exchange column to replace ~30% of the chlorides with hydroxides.
5. An IRA-400 anionic exchange column large enough to accommodate at least 100 mL of fluid running through it. This will be a polymer gel packed into a glass column and will most likely come charged with chloride ions. The column will need to be flushed through with an excess of 1−2 mol/L sodium hydroxide solution to charge it with hydroxyl ions before use in the recipe. If the column is small then you can always run aliquots of the surfactant solution through it and then recharge the gel with sodium hydroxide. You will also need a clamp to hold the column and a large beaker to receive the solution.

6. A static autoclave safe to use at 150 °C.
7. A large Buchner funnel and filter paper.
8. A large ceramic crucible and a furnace capable of reaching at least 540 °C fitted with a nitrogen flow. After an hour of heating this will be switched over to air.
9. A magnetic stirrer and stir bar.

To make aluminosilicate MCM-41:

1. Setup the ion exchange column in the clamp and place a large beaker underneath. Charge the column by running sodium hydroxide solution through it. Dispose of the excess solution that runs out of the column in the correct waste disposal container.
2. Dissolve the cetyltrimethylammonium chloride in 200 mL of water. Pass the solution through the column so that around 30% of the chlorides will be replaced with OH anions. Collect the solution in a 500 mL beaker.
3. Place the magnetic stir bar in the 500 mL beaker containing the surfactant and set the solution to stir. Pour in the 10 wt% tetramethylammonium silicate solution and add the silica and alumina powder. Mix thoroughly.
4. Pour the mixture into the autoclave and seal it. Heat the mixture at 150 °C for 48 h. At this stage the solid zeolite structure will form.
5. Decant the contents of the autoclave into a Buchner funnel and rinse it out with water. Rinse the solid precipitate thoroughly with water. After drying the structure of the zeolite can be checked by powder X-ray diffraction or electron microscopy though the pores will still be blocked with surfactant.
6. The surfactant can be removed by calcination. Place the dried precipitate in the large crucible and calcine in the furnace at 540 °C under flowing nitrogen. After 1 h switch the nitrogen over to an air feed and heat for a further 6 h. After the sample has cooled it will be ready for analysis.

Small angle powder X-ray diffraction of the non-calcined product should give a major reflection at ~4.3 nm corresponding to the [100] reflection of a hexagonal array. After calcination this will shrink by a few angstroms due to contraction in the aluminosilicate. Transmission electron microscopy should reveal a well ordered mesoporous material with wall thicknesses of around 1 nm. The calcined powder will have a very high surface area around 600−1300 m^2/g and this can be determined by performing BET adsorption isotherms with nitrogen though more accurate porosimetry measurements can be obtained using argon physisorption.

The pore size can be expanded by the addition of 1,3,5-trimethylebenzene to the mixture which bulks out the lamellar oil phase. It is a combustible volatile compound so it must not be used near any sources of ignition. The amount to add to the solution mixture prior to heating and the pore sizes expected are given below:

Pore diameter required/nm	Volume of 1,3,5-trimethylebenzene to add/mL
4.0	0
4.5	6.94
5.0	10.42
6.5	25.47

It is possible to alter the recipe further by the incorporation of other materials into the solid framework. This is of particular interest if the zeolite is to be used as a catalyst where a particular metal provides an activation site. By selection of cationic and anionic precursors in the presence of a surfactant you will be able to make an incredible range of mesoporous materials with cubic, hexagonal and lamellar structures. A comprehensive source of compositions including s, p and d-block elements formed as zeolites can be found in the work of the Stucky group (Huo et al., 1994).

AN ALKOXIDE APPROACH TO MCM-41 PREPARATION

Adapted from Hall et al. (1999)

This recipe is more easily implemented and does not require the use of an autoclave or furnace to achieve a templated material. The beauty of using a low temperature synthesis is that organic groups can be incorporated into the silica network structure. The interior pores can be engineered to have a chemical functionality like amides or mercapto groups covalently bonded at the silica surface. Organic groups can also impart structural flexibility required in end applications that need to be a bit tougher than standard brittle glasses. In this recipe a silicone alkoxide is used as the precursor and this can be modified so that an organic unit is incorporated into the condensed phase. Pure tetraethoxysilane can be used to produce a pure silica phase or a chemically modified silane can be employed where one of the alkoxide groups is substituted by a functionalised alkyl chain covalently bound to the silicon atom. Substituted alkoxides will still undergo hydrolysis and condensation to form a gel and then a solid and the functional alkyl group will remain bound to the silicon centre throughout resulting in an organosilane after reaction. Cetyl-trimethylammonium bromide (CTAB) is used as a surfactant template in the presence of sodium hydroxide. There is no need for a column in this preparation and a hexagonal phase will still form about which the organosilane can condense. The surfactant will be removed from the porous solid by refluxing in an acid and ethanol rinse although this procedure will not guarantee complete removal.

To make MCM-41 from a silicon alkoxide you will need:

1. 8.03 mL (7.55 g, 0.0362 mol) of tetraethylorthosilicate. If you are going to functionalise your MCM-41 you will need to substitute the TEOS for another functionalised precursor. The source reference uses a number of organotrialykoxysilanes of the form $(RO)_3Si-R'$. The recipe is known to work for: Phenyltriethoxysilane ($R'=C_6H_5$), 3-aminopropyltriethoxysilane ($R'=(CH_2)_3NH_2$), allyltrimethoxysilane ($R'=CH_2CHCH_2$), 3-mercaptopropyltriethoxysilane (($R'=CH_2)_3SH$). These can be substituted up to 20 molar% with the TEOS and will still form into a mesoporous material. Don't feel limited to these precursors alone though results may vary with anything drastically chemically different.
2. 1.6 g (0.00439 mol) of cetyltrimethylammonium bromide (CTAB).
3. 20 mL of 1 mol/L NaOH. A litre of this can be prepared by dissolving 40 g in 1 L of water in a volumetric flask. Add the solid NaOH slowly as it gets hot.
4. 70.2 g (3.9 mol) of water in a clean 150 mL beaker.
5. Enough parafilm to cover the top of the 150 mL beaker.
6. A Buchner funnel and flask with filter paper and excess water for rinsing. A glass petri dish large enough to sit the paper on is also helpful.
7. A vacuum oven capable of reaching at least 100 °C. This will be removing water from the sample so *make sure you have a cold trap fitted between the pump and the oven.*
8. A 500 mL round bottomed flask with a quick fit aperture and a corresponding reflux condenser. You will need a clamp to hold the glassware in place and all the correct tubes and clips. This will be used for rinsing the surfactant out of the silica pores. You may wish to use up to a 2 L or larger round bottomed flask but these, and the appropriate glassware and mantles, are harder to come by.
9. A large oil bath on the top of a heating mantle capable of reaching at least 75 °C. It must be large enough to receive the 500 mL round bottomed flask.
10. Around 2 L of a 1 mol/L HCl in ethanol solution. This will be used to extract the surfactant from the porous silica. A litre can be prepared by diluting 84.38 mL of a 37 wt% (11.85 mol/L) HCl in water solution up to 1 L with ethanol in a volumetric flask. The ethanol does not have to be reagent grade for this step.
11. A magnetic stir plate and stir bar.

To make MCM-41 from a silicon alkoxide:

1. Dissolve the CTAB in the 70.2 mL of water under stirring. Once this has dissolved add the 20 mL of sodium hydroxide solution.
2. Add the silicon alkoxide reagent to the stirring solution. Place parafilm over the top and poke a few holes in it to let evolved alcohol evaporate away. Leave the solution to stir for 24 h. At this point the molar ratio of the reagents is 1 Silicon:0.12 CTAB:0.5NaOH:130H$_2$O. A white precipitate will begin to form.
3. Decant the solution and precipitate out of the beaker and into the Buchner funnel onto filter paper with the vacuum on. Rinse any remaining precipitate out of the beaker with water and then rinse the collected silica powder in the funnel with copius amounts of water. Allow it to dry in the funnel for ten minutes.
4. Transfer the silica from the funnel onto the large glass petri dish. You may wish to lift the entire filter paper out as this will have much of the product stuck to it. The powder will come off it much easier once it is dried. Place the silica in a vacuum oven and dry under vacuum at 100 °C for 10 h. *Remember to empty the cold trap after use!*
5. Setup the reflux condenser above the oil bath. Set the mantle temperature to 75 °C and allow it to equilibrate. The silaceous product should have the right morphology at the mesoscale but requires an acid and ethanol rinse to remove surfactant from the pores. For optimum results, 1 g silica is refluxed in 1 L of HCl/ethanol solution. This recipe will produce around 2 g of solid so to rinse all the product the reflux is done in batches. This can be scaled up if you have the appropriate equipment. In the 500 mL round bottomed flask add 0.25 g of the silica powder to 250 mL of the HCl/ethanol solution. Affix the reflux condenser to the top and clamp in place with the round bottomed flask sitting in the oil bath. Allow the suspension to reflux at 75 °C for 24 h.
6. Filter the suspension using the Buchner funnel and rinse with large amounts of excess ethanol.
7. As before transfer the product into a large petri dish and dry in the vacuum oven at 100 °C for 10 h. After this you should have a silica, or organosilica, MCM-41 ready for analysis.

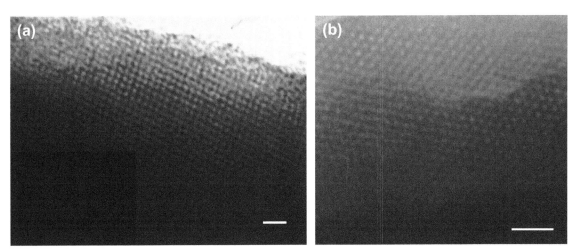

FIGURE 4.24 Transmission electron micrographs of (a) a bi-functionalized phenylamino-MCM-41 mesophase (10 mol% PTES/10 mol% APTES) showing hexagonally ordered channel structure, (b) phenyl-functionalized MCM-48 mesophase (10 mol% PTES) showing cubic structure viewed down the <110> zone. Scale bar (both micrographs) = 10 nm. *(Images courtesy of Simon Hall)*

As with the previous MCM-41 preparation electron microscopy should reveal that the powder is composed of well ordered hexagonally packed arrays. Powder X-ray diffraction can be used to determine the unit cell length of the packing and for CTAB templated silica the 100 reflection should be between d = 4.4 and 4.6 nm depending on the level of organic group substitution in the lattice. Diffractometer scans should run from between 0.5 and 10° to obtain full patterns (Fig. 4.24).

The presence of the organic groups can be confirmed by ^{29}Si and ^{13}C solid state nuclear magnetic resonance spectroscopy. The ^{29}Si NMR spectrum will show different shifts and splitting between Si-OH and Si-OR groups. In organo-substituted samples the signal strength for the Si-OH will be much stronger, denoting that the organosilica is less condensed that a pure silica. This would be expected as for functionalised precursors there are only three hydroxyl groups that may be condensed to form an Si−O−Si bond as opposed to four. As the level of organic substitution in the network increases the capacity of the silica to fully condense by forming these bonds is impaired. This results in an inability for the formed solid phase to adequately replicate the surfactant template. The ^{13}C spectrum should show that the organic moieties are intact within the inorganic framework although infra red spectroscopy might be a quicker way to confirm they are present.

Functional group	13C-NMR chemical shift in ppm	Infra red absorbance wavenumber in cm^{-1}
Phenyl	127, 129, 130, 134	3100−3000, 1668, 1430, 702
Allyl	114, 135,	1400, 800
Thiol	23.0, 41.4	2550−2590 (weak)
Amine	16.8, 44.8	3400, 1620−1650

BET surface area measurement can be used to determine the surface area and pore sizes on acid/ethanol rinsed samples. For the silica only MCM-41 preparation, the area will be near 1200 m^2/g and the pores will be between 3 and 4 nm in diameter though these figures will change depending on what you use as a dopant. Grids for transmission electron microscopy can be prepared by grinding the product and sonicating in ethanol. A droplet of the suspension can then be dried down onto a carbon coated grid. Depending on the orientation of the fragment you are looking at you should see either hexagonally packed pores or well ordered lines denoting the tubular pores adjacent to one another.

AN ALKOXIDE APPROACH TO MCM-48 PREPARATION

Adapted from Hall et al. (1999), Schumacher et al. (2000)

The pores structure formed in an MCM-41 preparation is a hexagonal array of tubes, much like a pack of straws packed together in a box. By modifying the previous method a composition called MCM-48 is produced which results in a bicontinuous framework pore structure arranged in a gyroid structure. The gyroid structure has a space group of Ia3d which is a short hand way of describing the unit cell symmetry and how it behaves in 3 dimensions. The gyroid structure resembles lots of interconnected pipes and is advantageous over traditional MCM-41 for filter applications as this interconnectivity means that the filtrate can move around blockages in the structure.

In this recipe the template surfactant is cetyltrimethylammonium chloride (CTAC) and the condensation of the solid material is driven by a hydrothermal synthesis. It may seem like only a very small tweak to change the halide anion in the surfactant but it has a large effect. The gyroid structure normally self assembles at a much higher concentration but the ethanol generated in the hydrolysis reaction induces a phase transformation in a lower concentration of CTAC from hexagonally packed arrays into the gyroid phase. The other major change is in the choice of reaction vessel as the reaction takes place under hydrothermal conditions. This is like a more advanced form of pressure cooking where the solution is taken to its boiling point or beyond but the reaction volume is maintained at a constant. For this to work properly the condensation step must take place in a poly(tetrafluoroethylene) (PTFE) container and not a glass beaker. The PTFE is chemically inert meaning that it will not provide surface sites for unwanted silica nucleation which could change the dynamics of the system. Because the liquid inside the PTFE container will be near, or over the boiling point the vessel will need to be sealable and strong. This is achieved by using a hydrothermal bomb which consists of a stainless steel sleeve into which the PTFE container can be sat and a lid screwed down on. This will ensure that nothing boils over or pops due to excessive pressure during the heated reaction. As with the MCM-41 preparation an acid/ethanol rinse will be required to remove the surfactant from the product. If you are not going to substitute any organic functionality into the lattice then a calcination step can be used in its place.

To make MCM-48 from a silicon alkoxide you will need:

1. 7.73 mL (7.26 g, 0.0348 mol) of TEOS. As with the previous recipe you may wish to include organic moieties and this is achieved by using organosilane precursors. This can be measured out in two parts using a 5 mL volumetric pipette.
2. 28.7 mL (27.8 g) o f a 25 wt% solution of cetyltrimethylammonium chloride solution in water. This concentration is normally the solution strength provided by suppliers.
3. 0.67 g (0.017 mol) of NaOH dissolved in 17 mL of water. This will produce the required amount of a 0.1 M solution of NaOH.
4. A magnetic stirrer and stir bar.
5. A vacuum oven capable of reaching at least 100 °C. It will need to be fitted with a cold trap.
6. A strong, sealable PTFE container that can hold at least 50 mL, preferably using a hydrothermal bomb. This is going to be heated to 100 °C containing the reactants and so needs to be able to contain the pressure of the liquids in the vapour phase. *Depending on the design the bomb must be assembled correctly to ensure the safety valves will work.*
7. A 500 mL round bottomed flask with a quick fit aperture and a corresponding reflux condenser. You will need a clamp to hold the glassware in place and all the correct tubes and clips.
8. Around 2 L of a 1 mol/L HCl in ethanol solution made up as in step 10 of the previous recipe.
9. A heating mantle capable of reaching at least 75 °C.

To make MCM-48 from a silicon alkoxide:

1. Add the 7.73 mL of TEOS to the 17 mL of NaOH solution. Add the magnetic stir bar and stir vigorously for 5 min.
2. Add the CTAC solution and stir for a further 15 min. The molar ratio of the reagents is: 1 siloxane:0.25 NaOH:0.64 CTAC:62 H_2O.
3. Remove the magnetic stir bar and pour the solution into the PTFE container. Seal it shut by screwing on the PTFE lid (ensuring it is sturdy enough) or sealing it shut in a hydrothermal bomb.
4. Place the container in the oven (At ambient pressure, vacuum is *not* needed at this point) and heat without stirring at 100 °C for 72 h. During this time the silica will condense around the surfactant gyroid structure to form a white precipitate.
5. Remove the container from the oven and allow it to cool. Isolate the formed precipitate using a large Buchner funnel and wash with copious amounts of water. Rinse out the PTFE container into the funnel to make sure the entire product is obtained.
6. From this point onwards the preparation is exactly the same as the previous recipe from step 3 and onwards. In brief, dry the precipitate in the vacuum oven at 100 °C for 10 h. The dried product is then refluxed in the HCl/ethanol solution at 75 °C for 24 h. The product is then decanted into a Buchner funnel and rinsed with excess ethanol. It is then dried once more in the vacuum oven. You will now have a silica or organosilica MCM-48 powder ready for analysis.

The characterisation procedure is identical to those performed for the MCM-41 preparation. The major difference will be in the observed pore structure which should appear very different from the MCM-41. Under an electron microscope you should be able to see a cubic arrangement of pores but the system may also look like it is composed of wavy lines as the gyroid phase intersects with itself. A low angle x-ray diffraction pattern should show a major reflection at d = 3.9 nm for the 100 plane. The difference between MCM-41 and MCM-48 should be visible by inspection of the X-ray diffraction pattern and some examples are provided below. BET surface area analysis should give a figure somewhere in the region of 1000 m^2/g.

The recipe below and the ones that follow are concerned with making large free standing monolithic slabs of silica and similar sol–gel derived materials as opposed to particles and powders. As mentioned previously most sols will begin to gel as the hydrolysis and condensation mechanism occurs. If the rate of gelation is homogenous throughout the sol then it is possible that the forming sol particles will link together to form a network. The inorganic network can extend throughout the solution leaving an interconnected solid phase filled with liquid–normally the alcohol evolved in the reaction and water. This is called a gel and it will conform to the dimensions of the vessel in which it was formed. Because of this property, gels are highly useful as they can be cast as films and blocks in any shape desired for the end application. We have already explored aspects of the hydrolysis and condensation mechanism in the production of particles, powders and thin films. In sol–gel derived monolithic gels, the ageing and drying mechanisms are incredibly important to the end result. If a gel is not aged long enough to allow the network to adequately condense it will be brittle or fall apart. If a gel is dried too quickly then the stress of liquid leaving the pores will crack it apart. Gels are commonly thought of as being wet but even after drying they are still called gels. There is a nomenclature associated with gels to describe their properties and the mechanisms by which they were formed.

Hydrogel: The gel volume is filled with water. This is perhaps the most common gel experienced in everyday life.

Alcogel: The gel pore volume is filled with an alcohol. This is not necessarily the alcohol evolved during hydrolysis. In order to prepare a gel for drying the water phase is removed by soaking the gel in an alcohol with a low vapour pressure. The gel can then be allowed to dry out to air, which will cause shrinkage, or by supercritical removal which has minimal shrinkage. Working with supercritical alcohols is dangerous unless you are in a laboratory dedicated to it as they require relatively high temperatures and pressures.

Xerogel: This type of gel has been allowed to dry out in a controlled way to atmosphere. This is normally from an alcogel state as opposed to a water containing gel and is done under a controlled humidity so that the evaporation rate is slow enough not to crack the gel. Xerogels can demonstrate considerable shrinkage from the original volume due to network densification.

Aerogel: This is a very low density dry gel which is formed by drying out the gel in a way that does not cause shrinkage or densification. An aqueous gel can be flash frozen in liquid nitrogen then placed in a vacuum to freeze dry. A more refined method is to solvent exchange out all the water for a volatile solvent such as ethanol or acetone and use supercritical CO_2 drying.

MAKING A MONOLITHIC SILICA STRUCTURE FROM NON-IONIC SURFACTANTS USING A DIOL BASED SILICA PRECURSOR

Adapted from Brandhuber et al. (2005)

In this recipe we will prepare a large monolithic gel of porous silica using a modified silane precursor and a non-ionic surfactant as the liquid crystal template porogen. In contrast to the MCM-41 and MCM-48 preparations, a solid free standing block can be formed as opposed to a powder exhibiting ordered long range hierarchical porosity. The pore structure is achieved by using a non-ionic surfactant which readily forms into a liquid crystal with long range order at the meso and macroscale. The precursor and template are mixed and allowed to gel. As the product is a block, the drying of the gel is important in preserving the shape and structure. The larger the sample is then the more risk there is that the cumulative drying contraction and pore stress during this process will cause fractures. The drying of the gel is done by sequential exposure to volatile solvents which serves to remove the template and by-products of the condensation reaction and does not generate the same capillary pressure as water upon evaporation from the pore structure. The final drying step is done by very slowly increasing the temperature. The heating is so slow that the condensation and solidification stresses are kept minimal. Other methods for the non-destructive drying of gels are discussed further on in this section.

The major driving force behind a self assembled ionic surfactant phase comes from the relatively attractive and repulsive forces the molecule experiences due to a hydrophobic tail and a hydrophilic charged head group. The pH and salt concentration of the continuous phase in which an ionic surfactant is dispersed will influence the strength of the ionic interactions. Similarly, these factors also influence the strength of repulsion in the tail groups and determine if there is enough free energy to pack the molecules into more energetic lyotropic crystal formations. One problem with using an alkoxide based precursor for templating is that an alcohol is formed during hydrolysis. Alcohol is miscible with water and this decreases the polar strength of the water as a solvent causing the aliphatic surfactant tail groups to be less repelled and therefore more loosely pressed together. The evolution of alcohol limits the amount of alkoxide precursor that can be added to the system and also the selection of surfactants that can be used as a liquid crystal template. It is possible to get around the problems caused by *in-situ* generated alcohol by replacing the alkoxide moiety with a diol or polyol group in the precursor.

The silica precursor is tetrakis(2-hydroxyethyl) orthosilicate (THEOS) and it is made by exchanging the four ethoxide moieties in a molecule of TEOS for ethylene glycol by transesterification (Meyer et al., 2002). THEOS has a number of advantages over TEOS for sol gel synthesis and is especially relevant in procedures where alcohol cannot be used such as those involving biological materials. Although much of the previous section has described mixing TEOS into water with acids or bases in the presence of a surfactant, TEOS is very poorly soluble if mixed directly with water. For many applications it is not suitable for use without additives that allow it to solubilise. THEOS is fully miscible with water and this makes it ideal as a one pot sol–gel precursor. It can be added directly to water where it will hydrolyse to form a short lived silicic acid before condensing to produce a solid silica phase.

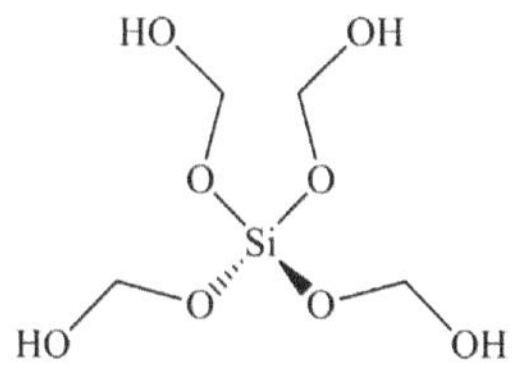

FIGURE 4.25   The chemical formula of tetrakis(2-hydroxyethyl) orthosilicate.

The only by-product generated is ethylene glycol which can rinsed away although we will be using an acid catalyst in this preparation also. Ethylene glycol and a large number of polyols are also benign towards biological entities such as cells, proteins and enzymes and are often used in preservative and storage solutions for bacterial cultures! Interestingly, the glycolated silane is not as moisture sensitive as alkoxide based precursors and a solution of THEOS in ethylene glycol can be left open to air on a bench for many days without visible solid precipitation. The use of a polyol modified precursor to encapsulate cells and enzymes is discussed elsewhere (Fig. 4.25).

Another difference in this recipe is the use of a non-ionic surfactant as the liquid crystal template. Uncharged polymeric surfactants are known to assemble with ordering in the meso and macroscale (Cameron et al., 1999). A limitation with the MCM-41 and similar approaches is that the mesoscale morphology is highly ordered locally but breaks down over long ranges. Both the changing ionic interaction of the charged head groups and the alcohol evolution serve to break the long range structure down over time as the reaction occurs. This often results in the solid product being in the form of a powder as opposed to a large block. Non-ionic surfactants are normally composed of di- or tri-block co-polymers where the constituent polymeric units have a known length and hydrophobic or hydrophilic interaction. For this preparation we will use the triblock polymer P123 which is composed of poly(ethylene oxide)$_{20}$ − poly(propylene oxide)$_{70}$ − poly(ethylene oxide)$_{20}$. The poly(ethylene oxide) tails are slightly hydrophilic whereas the poly(propylene oxide) unit is hydrophobic. This arrangement causes the polymer molecules to behave like a surfactant in water without the need for a charged head group. Non-ionic surfactants of this type are incredibly useful as they can be tailor-made to specific lengths allowing the liquid crystal phases they form to be controlled. Like normal charged surfactants they will form micelles and liquid crystal phases and in this recipe the lack of charge interaction with the condensing silica phase means the template structure will maintain a long range order. Furthermore, the diol generated by the reaction has almost no effect on the template allowing a large monolithic silica structure to form with meso and macroscale porosity.

To make a silica monolith using THEOS and P123 you will need:

1. 10 mL (10.63 g, 0.051 mol) of tetraethylorthosilicate (TEOS). This is the starting point for making a stock of around 10 mL of THEOS solution for use. You will need a large syringe with a wide bore needle for delivering the TEOS through a rubber septum into the reaction vessel.
2. 11.18 mL of ethylene glycol (12.41 g, 0.2 mol). This will be reacted with the TEOS at 4 molar equivalents to replace the ethoxide groups coordinated to the silica. The ethylene glycol needs to be dry before use so add some Na_2SO_4 to the stock bottle and let it sediment to draw the water out before use.
3. A heating mantle and oil bath with magnetic stirring capability and a stir bar. It must be capable of reaching a temperature of at least 140 °C.
4. A two necked 50 mL round bottomed flask with quick fit connections. The side neck should be fitted with a rubber septum for injections. The main neck will be fitted to a Vigreux column which is used to draw off ethanol evolved in the reaction by condensation. The Vigreux column must fit to a receiving round bottomed flask and it will require a water flow to cool the ethanol vapour. The column will also need a bubbler capping the top of the column through which argon will flow. You will also need clamps to keep this all in place. See the schematic provided for the setup (Fig. 4.26):
5. A vacuum oven or vacuum line which will be used to draw out the last of the ethanol from the freshly produced THEOS stock. It will need to have a cold trap fitted. You will also need a cork ring to sit the round bottomed flask in.
6. 0.3 g (5.1×10^{-5} mol) of Pluronic P123 triblock polymer non-ionic surfactant. The molecular weight is around 5800.
7. 0.7 mL of 0.01 mol/L HCl solution. A stock solution can be made by diluting 10 mL of a 37 wt% (11.85 mol/L) stock up to 118.5 mL. Take 1 mL of this solution and make up to 100 mL with water to get the 0.01 M HCl solution.
8. A benchtop vortex.
9. A 2 mL polypropylene centrifuge tube. The vessel for gelation must be plastic as this prevents nucleation on the vessel surface. As the tube will act as a mould for the gel it is important to have one that is wider at the top so the gel can be removed.
10. A drying oven or incubator set to 40 °C. This will be used for ageing the gel.
11. 10 mL of a 10% by weight solution of trimethylchlorosilane in petroleum ether. *This will be highly flammable. Use in a fume hood and keep away from sources of ignition.* This is used to dry the gel after it has formed and to draw out ethylene glycol and polymer. You will need a glass test tube (~10 mL volume) and stand to place the gel in for the rinsing step and a glass plug which should fit snugly over the top to seal it.

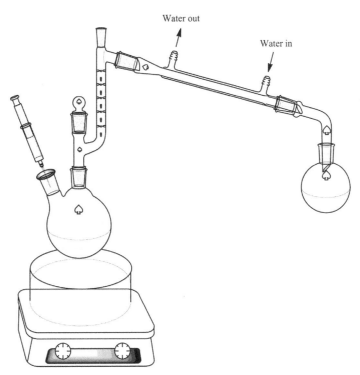

Water out

Water in

FIGURE 4.26 Schematic of the setup for converting TEOS into THEOS.

12. 30 mL of petroleum ether. This will be used to further rinse the monolith. *Highly flammable. Use in fume hood and keep away from sources of ignition.*

13. 30 mL of ethanol for further rinsing of the monolith.

14. An oven capable of reaching at least 500 °C with a programmable ramp rate. The heating rate should be controllable from 1 °C/h to 5 °C/min. This very slow ramp rate can be achieved manually with most furnaces but will, understandably, be time consuming. You will need a crucible for heating the gel in.

To make a silica monolith using THEOS and P123

1. Setup the heating mantle and distillation array for the conversion of TEOS to THEOS. Add the 11.8 mL of ethylene glycol to the round bottomed flask and add the stir bar through the side neck and reseal with the rubber septum.

2. Allow argon to bubble through the system for half an hour before use to ensure there is no moisture in the system.

3. While the system is purged with argon turn on the heating mantle and set it to 140 °C and allow it to equilibrate. Turn on the water supply to the Vigreux column. *It does not have to be fast!*

4. Once the ethylene glycol is up to temperature inject the 10 mL of TEOS through the rubber septum into the ethylene glycol under stirring.

5. A transesterification reaction will now occur and ethanol will be evolved. This will vaporise, move up the column and through the Vigreux column where it will condense to be collected by gravity. Somewhere around 10 mL of ethanol will be produced in the reaction. Ethanol is less dense than TEOS so the volume of alcohol extracted will be slightly more than the volume of precursor put in! Allow the reaction to proceed until no more condensation is observed within the column.

6. Turn off the mantle, argon flow and water feed. Remove the round bottomed flask from the system and stand it upright on a cork ring in a vacuum chamber. Expose the solution to vacuum to remove the last of the ethanol. You should not see excessive bubbling and if you do lower the strength of the vacuum. Leave the solution under vacuum for one hour.

7. You will now have a stock solution of THEOS. The exact concentration of silica can be determined by thermogravimetric analysis in air. This precursor should oxidise to glass comprising 20 wt% of the pre-burn precursor weight. The product can be further checked by ^{29}Si NMR which should show a single peak at around −83 ppm.

8. This recipe is enough to make a gel roughly 1 cm^3 suitable for laboratory scale analysis though if you wish to make larger gels then simple scale up the amounts of THEOS, P123 and acid solution used. Dissolve the P123 surfactant in the 0.7 mL of HCl solution in a 2 mL polypropylene tube. This may take a while to dissolve so seal the lid and vortex the tube.

9. Add 0.71 mL of THEOS precursor to the tube and seal it. Vortex the tube for 1 min. The contents will turn white just before the gel forms. This will take about 4 h.

10. Place the sealed tube in a drying oven and age the gel at 40 °C for 7 days. During this time the gel will solidify into a monolith in the shape of the tube interior.

11. Once the gel has aged a delicate process of solvent exchange begins in order to minimise drying stresses. Remove the solidified gel from the tube by gentle tapping and place into the small test tube. Add enough of the 10% trichlorosilane in petroleum ether to the test tube so that the still wet gel is covered to a depth of three times its height. You should see a surfactant and liquid phase drawn out of the gel immediately. Seal the beaker and leave the gel to stand in the solution for 24 h *in a fume hood away from ignition sources*.

12. Decant off the trichlorosilane solution and dispose of it in an appropriate waste solvent container. Refill the test tube with 10 mL of the pure petroleum ether, seal and leave to stand for 8 h. After this time decant off the solution into the appropriate waste container and refresh with more petroleum ether. Repeat a total of three times.

13. Repeat this rinsing step 3 more times using ethanol.

14. The gel is now ready to be heated to remove the ethanol. There should be only minimal traces of surfactant still remaining within the pores. In contrast to ionic salts, the total organic nature of the triblock polymer template means that calcination will convert it entirely to CO_2 without leaving residual traces of salts. The source reference uses a slow calcination step but you may opt to use drying in supercritical fluid. This technique is discussed further on in this section. Heat the gel to 200 °C at a heating rate of 1 °C/h. Once it has reached 200 °C heat the sample further to 500 °C at a rate of 5 °C/min to remove any residual organics. The porous silica monolith is now ready for analysis.

Small angle X-ray scattering should result in four major reflections corresponding to the [100], [110], [200] and [300] planes of a two-dimensional hexagonally packed structure with p6 mm symmetry. The repeat unit distance of the structure will be around 11 nm. BET nitrogen adsorption isotherms can be used to determine the micro- and meso-porosity volumes and diameters along with the internal surface area. This should be around 1000 m^2/g with total pore volumes of up to 2.5 cm^3/g. Grids for electron microscopy can be prepared by crushing scrapings from the gel and sonicating them in ethanol then drying them down on the grid. You should observe an interconnected pore structure as part of a larger rod particulate. Scanning electron microscopy will show the hierarchical macroporosity generated by the smaller rods aggregated as a network. Mercury porosimetry can better characterise the macroporosity and should reveal a total macropore volume near 0.5 cm^3/g with pore diameters of ~0.4−0.5 µ in diameter.

AEROGELS AND SUPERCRITICAL DRYING

With thanks to Dr. Stephen A. Steiner III and Dr. Micheal Grogan. Adapted from www.aerogel.org

Aerogels are a class of low density, nanoporous solid materials comprised of networks of interconnected nanoparticles or nanostructures. Aerogels typically begin as wet gels physically similar to edible gelatine. Gels are colloidal systems comprised of a sparse, nanoporous solid framework permeated by a liquid throughout its pores. Isolation of this sparse solid framework from the gel's liquid component in a way that substantially preserves the framework's structure and porosity results in an aerogel. In this regard, an aerogel is not a substance per se but rather a geometry that can be taken on by a substance. Virtually any gel that can be prepared can be rendered into an aerogel. Aerogels of silica (Brinker and Scherer, 1990), transition metal oxides (Baumann et al., 2005), organic polymers (Pekala et al., 1992; Mulik et al., 2007), and even quantum dots (Mohanan et al., 2005) can all been prepared, as well as aerogels of substances for which wet gels are not readily attainable, such as carbon (Lu et al., 1993) and metals (Nicholas Leventis et al., 2009), through chemical conversion of other aerogels. Aerogels may also be doped with other nano-materials such as gold nanoparticles to produce composites (Grogan et al., 2011).

Because liquids exhibit the property of surface tension, evaporation of liquid from a gel imparts substantial capillary stresses on the gel's solid framework. Such stresses cause the framework to contract upon drying, usually in an irreversible way, resulting in a low-porosity, densified solid called a xerogel. Thus, conversion of a gel into an aerogel requires a drying process that minimises contraction of the gel's solid framework as the liquid is removed or, alternatively, allows collapse of the framework to be reversed. Following from this, aerogels exhibit extremely low bulk densities; in fact, the world record for the lowest density solid material ever produced is held by a silica aerogel made at Lawrence Livermore National Laboratory in 2003, which exhibited a density of only 1.1 mg/cm^3 (Guinness Book of World Records, 2012).

The most reliable method used to transform gels into aerogels is called supercritical drying. In this approach, the liquid in a gel is heated and pressurised past its critical point, at which the liquid loses its surface tension and behaves like a gas while still possessing the density and thermal conductivity typical of a liquid. If the liquid contained in pores of the gel is, for example, an organic solvent, temperatures in excess of 200 °C and pressures in excess of 45 atm are required for supercritical extraction—conditions at which organic solvents are highly flammable. To avoid this, the gel liquor can instead be exchanged with liquid carbon dioxide, a non-flammable solvent with a critical temperature of only 31.1 °C and critical pressure of 72.8 atm. In typical process for making an aerogel, a gel is formed by performing some sort of polymerisation reaction or particle agglomeration process in a liquid solvent. Upon gelation, the gel may be optionally aged to allow its solid framework to strengthen. The gel is then purified by soaking in

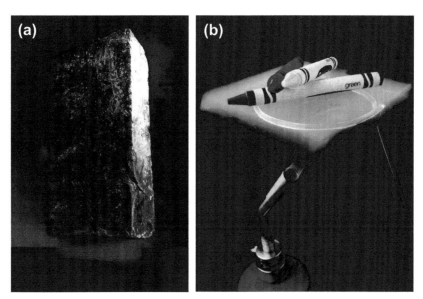

FIGURE 4.27   Photographs of (a) A silica aerogel supporting a brick and (b) An aerogel insulating crayons against the heat of a blowtorch. *(Image Adapted with Permission from NASA Jet Propulsion Laboratory)*

a pure solvent and allowing for diffusive exchange of its liquor for the pure solvent. This solvent must be miscible with both the gel's existing liquor and liquid CO_2 so that the existing gel liquor will be successfully displaced and so that it can be subsequently displaced by liquid CO_2. At no point is the gel allowed to dry out. The purified wet gel is then placed in a gas-tight fixed-volume high-pressure chamber and the solvent exchanged for liquid CO_2 a number of times until as much of the gel liquor as feasible has been displaced. The liquid CO_2 is then heated past its critical point thereby transforming it from a liquid to a supercritical fluid without imparting capillary stress on the gel's solid framework as would be the case if it were dried evaporatively. The gas is then slowly released leaving behind only the low density solid framework of the gel.

APPLICATIONS OF AEROGELS

Their apparent ethereality belies the fact that, by weight, aerogels are very strong—a centimetre-thick layer is capable of supporting a building brick placed (gently) on top. Additionally, the air within the network is extremely slow to diffuse through the tortuous pore network of the monolith which imparts to it a capacity for high thermal insulation, which is the primary use of aerogel-based materials for industrial applications. The best thermal insulating solids ever developed are, in fact, aerogels. Low density silica aerogels can exhibit thermal conductivities as low as 10 mW m^{-1} K^{-1} at room temperature and are stable up to ~700 °C. Silica aerogels have also found use in a wide range of exotic applications. One such application was on NASA's Stardust probe, which used gradient-density silica aerogel tiles to capture high-velocity microparticles ejected from the comet Wild-2 and successfully returned them to Earth for scientific analysis (Brownlee et al., 2006). Monolithic plates of low density silica aerogels with very low refractive indices are also routinely used as Cherenkov radiation detectors in high-energy physics (Poelz and Riethmuller, 1982). The speed of light propagating through a medium, such as an aerogel, is slower than that of light in a vacuum. In a particle accelerator, atoms and electrons are routinely taken to near light speed and collided. The sub-atomic particles generated by these collisions are ejected, often near light speed themselves, and detectors are needed to observe these reactions. A fast moving charged particle passing into the aerogel may find itself exceeding the 'local' speed of light. This causes the local electromagnetic field to distort strongly, like a shockwave, before returning to its ground state by the emission of a photon. This does not normally happen unless the particle is above the speed of light in the given dielectric medium. It is the glow generated by these photons and their wavelength that can be used to track particle collisions in an aerogel-based detector if the detector plate is placed near the collision point. In addition to silica, aerogels of other substances are also useful in many applications (Aegerter et al., 2011) (Fig. 4.27).

MAKING SILICA AEROGEL MONOLITHS

This recipe focuses on the production of silica aerogel for optical applications employing a single-step base-catalysed reaction with the silicon alkoxide tetramethoxysilane (TMOS, also referred to as tetramethylorthosilicate). It is possible to substitute the less-toxic alkoxide tetraethoxysilane (TEOS) for TMOS, however doing so requires use of a two-step catalysis or addition of ammonium fluoride (Cao and Hunt, 1994) to properly catalyse gel formation. This said, for optical applications TMOS typically produces more reliable results. TMOS hydrolyses quickly and the faster the reaction proceeds, the smaller and more uniform the particles comprising the sol will be. This results in a more optically transparent aerogel. Optical transparent is also a useful proxy for general material quality as transparency correlates with a high degree of nanostructural homogeneity throughout the silica framework. Though it is relatively easy to get a gel to form using TMOS using standard grade reagents, high-quality transparent aerogels suitable for optical applications require high-quality reagents. Use the freshest TMOS you can obtain. Old TMOS may become contaminated by moisture over time causing partial hydrolysis before it is even used to form a gel in turn causing a shift in gel time and resulting aerogel properties. Analytical reagent (AR) grade water should also be used as solvated ions in lesser grades of water can cause cloudiness and weakening of the gel framework. Degassing organic solvents and water prior to use, while not necessary, may be helpful in improving transparency in the resulting aerogels.

To make a silica aerogel you will need:

1. 10 mL (10.2 g, 0.067 mol) of fresh tetramethoxysilane. Use a new bottle wherever possible.
2. 20 mL (7.92 g, 0.0247 mol) of pure methanol in two 10-mL aliquots. Keep these in a sealed vial or bottle after the degassing step (see below).
3. A stock solution of 28−30 wt% ammonium hydroxide in water.
4. A large sealable bottle containing 1 L of analytical reagent grade water. *Do not use tap water.*
5. A 40−50 mL PTFE, polypropylene, glass, stainless steel, or other suitable container which will act as a mould for the gel. The choice of container material will affect how easy it is to demould the gel for the soaking stage. The mould should have smooth surfaces. Cling wrap can be used to improve demoulding from a particular mould, however may impart an irregular surface texture to the gel. You will also need a stir bead, stainless steel, or solvent-compatible plastic rod for stirring the mixture together.
6. A 500-mL to 1-L beaker or dish with a liquid-tight cover. This will be used to purify the gel with methanol to prepare it for solvent exchange with liquid CO_2.
7. 2 L of methanol. This will be used for removing water from the gel and subsequent soaking ready for supercritical drying. This does not need to be degassed as this methanol is not involved in formation of the gel.
8. A supercritical dryer and cylinder of liquid CO_2 with a siphon tube. *This equipment requires training to use! Gas under high pressure is involved!* A supercritical dryer (sometimes called a critical point dryer) is a special piece of equipment required to perform solvent exchanges with liquid CO_2 and subsequently perform supercritical drying. It is possible to build a supercritical dryer on your own with a basic level of technical inclination provided careful attention to detail. Schematics, designs and instructions for building and operating such a supercritical dryer are available on the Internet from www.aerogel.org. Samples are placed into the supercritical dryer which is then filled with liquid CO_2 to replace the existing pore liquor in the gel with liquid CO_2. Once this is done the liquid CO_2 is converted to a supercritical fluid by heating the vessel which will simultaneously result in an increase in the pressure of the vessel. Once above the critical point, the vessel is then repeatedly heated and depressurised while maintaining conditions above the critical point for several hours. Finally, the vessel is slowly depressurised back to ambient conditions. As such, the vessel must be capable of operating at ~100 atm at 60 °C. The chamber must also be large enough to accommodate the dimensions of your gel with several gel volumes (at least three) of room to spare. For this recipe the volume of the gel will be 35 mL. You can scale the recipe as needed. If you are using a small dryer then you can try carefully cutting your gel into smaller pieces while the gel is under the alcohol and drying these separately. To prevent the gel from moving around during filling, draining and supercritical processing in the supercritical dryer, the gel should be placed into a support container of some sort. This can be done by taking a fine wire mesh or perforated aluminium foil and fashioning it into a cage into which the gel can be placed. Be careful to avoid stray wires or sharp edges poking into the interior of the cage as these can easily slice into or scrape the gel.

 Some supercritical dryers will feature a bleed valve on the top of the vessel for venting pressure and driving siphon action from the CO_2 tank and a drain valve on the bottom of the vessel for draining off liquid. Both are highly desirable for reliable control over the CO_2 solvent exchange and supercritical drying processes. If your supercritical dryer has a window you will be able to observe liquid mixing and phase changes near the critical point directly. Be advised that although miscible, methanol and liquid CO_2 may not mix well on contact and may require mixing to blend efficiently. If your vessel does not have stirring capability, convective heating or simply allowing gels to soak for 12−24 h periods will facilitate blending of the two solvents.
9. An argon line and piping fitted with a clean syringe for degassing water and methanol. This is a somewhat rough approach to removing dissolved gases from liquids but it is relatively easy. This step is optional and you may find you get good results without performing it. You will need a lid with holes for the needle or some Parafilm® to stop solvent from being ejected.

To make a silica aerogel:

1. **Degassing solvents (optional).** Bubble argon gently through the water and the methanol aliquots to be used in the gel mixture through a clean needle for 30 min. Make sure the needle tip is at the bottom of the bottle so that the gas bubbles through the

length of the container. It is a good idea to check the gas flow in a vial of water beforehand as it is very easy to blow the water out by accident. Place some film over the top of the container to prevent solvent loss, particularly with methanol.

2. **Preparation of catalyst stock solution.** Add 5.40 mL of the 26−30 wt% ammonium hydroxide solution to the 1 L of water to make a stock ammonia solution for use as a catalyst.

3. **Preparation of alkoxide solution ('Solution A').** Mix the 10 mL of TMOS with the 10 mL of methanol in your mould(s). Alternatively, you may mix this solution in a beaker and pour the mixture into your mould(s) after Step 5.

4. **Preparation of catalyst solution ('Solution B').** Add 5 mL of the stock catalyst ammonium hydroxide solution to the other 10 mL of methanol and mix.

5. **Preparation of a gel.** Add the catalyst/methanol solution 'B' to the TMOS/methanol solution 'A' in your mould(s). Stir for about a minute to ensure mixing. Allow the solution to gel. This will take around 8−15 min. Alternatively you may perform this step in a beaker and pour into your mould(s) after mixing if it is too difficult to do this in mould.

6. **Ageing.** After the gel has set pour methanol over the top of the gel to the brim of the mould. If possible you can also place the entire mould into a methanol bath. Allow the gel to age for at least 24 h.

7. **Demoulding and solvent exchange.** The gel should be firm enough to remove from the mould. Handle it using a large spatula or even a spoon. Tweezers or similar will damage the gel. Transfer it into a bath of excess methanol around three to four times the gel volume and seal to prevent evaporation. Exchange the methanol four times or more over the course of a week. This should gradually remove all of the water, unincorporated silica oligomers, and other reaction by-products from the gel so that supercritical drying with CO_2 can be performed.

8. **Loading into supercritical dryer.** When you are satisfied that the gel has been sufficiently purified, transfer the gel directly into the supercritical dryer. Delicate gels should be placed in a mesh holder for support and to dissipate stresses resulting from liquid influx. Shower methanol over the gel to keep it from drying out during transfer if necessary. It should not stand in open air for longer than 20 s. Fill the supercritical dryer with enough methanol so that the gel is submerged. This is to protect the gel from stresses arising from rapid pressurisation and rapid influx of liquid CO_2. Seal the chamber and fill with liquid CO_2. Once liquid is in the chamber, it will produce its own vapour pressure proportional to its temperature. CO_2 will remain in the liquid state below a temperature of 31.1 °C and a pressure of 7.29 MPa (72.8 atm). Draining liquid CO_2 from the vessel will result in a discharge of both CO_2 gas and dry ice. At this point if there is a window you should see the methanol and liquid CO_2 phase separate. CO_2 is less dense than most solvents at 0.59 g cm^{-3} so it will separate as the top phase. Next, drain as much of the methanol out of the chamber as you can while keeping the system under pressure so that more liquid CO_2 is drawn in. You can either watch the methanol level fall in the window or vent the expelled gas into polypropylene or other compatible solvent container to keep track of how much solvent is removed. A methanol/dry ice slush will evolve from the drain valve. Allow the dry ice/carbonated methanol to warm and take note of the volume of the remaining methanol. Once you have retrieved ~70−80% of the methanol initially put into the supercritical dryer you will have removed essentially all of the added methanol from the dryer. Dispose of the methanol in the correct solvent waste container. If you are performing this step by looking at the fluid levels through a window make sure the exhaust is channelled into a fume hood. Ensure the chamber is filled with liquid CO_2 and leave the sample to soak for at least 12 h (Fig. 4.28).

9. **Solvent exchange with liquid CO_2.** Depending on your setup you might have the option to stir liquid inside the chamber. This can reduce the amount of time the gel needs to soak in the supercritical fluid. To exchange a stirred chamber *slowly* allow the

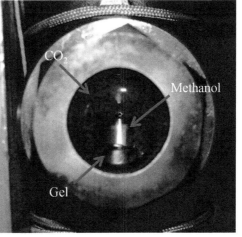

FIGURE 4.28 A critical point dryer setup for the removal of solvent to produce aerogels. The chamber is pressurised using liquid CO_2 through the inlet valve. Methanol is drained from the pressurised vessel through a valve in the bottom of the chamber. *(Photos Adapted with Permission from www.aerogel.org. Copyright Will Walker)*

chamber to vent and let the fill level fall to just above the gel. Let more liquid CO_2 into the chamber and repeat three times. If you are using a non-stirred chamber, drain liquid out of the bottom of the chamber while keeping the intake valve to the chamber open to refill the vessel by siphon action. Take note of the removed methanol volume. In both cases after the CO_2 has been exchanged refill the chamber with liquid CO_2 and leave to soak for 12–24 h. Depending on your setup, refilling may be aided by the help of a cooling loop or freezer or by cracking open the bleed valve to drive siphon action. Repeat this step at least three times.

From a practical point of view valves may get clogged with dry ice formed from expansion cooling of the egressing CO_2. This can be controlled to a limited extend by gently throttling the valves back and forth and releasing pressure slowly enough that dry ice does not build up. *Always be careful opening valves on a pressurised vessel.* Do not feel tempted to try and blow the ice out by opening up the release valve hard. You can lose a lot of pressure very quickly if a plug of ice is being particularly difficult and you over open the valve. *This will damage the gel and is unsafe to the operator as well.*

10. **Supercritical drying.** The next step is to convert the liquid CO_2 to supercritical CO_2. Close all valves (intake, bleed, and drain). Raise the temperature of the chamber to 45 °C–55 °C. As you heat the vessel the pressure will increase. The pressure must increase to greater than about 7.5 MPa but preferably greater than about 10 MPa. During this process, the liquid in the chamber will transform into a supercritical fluid. Do not allow the pressure to exceed 10 MPa. You will need to bleed pressure off from the vessel periodically in order to get to the target temperature range. If your vessel has a window, you may find that starting this process with the vessel only 66–75% full with liquid CO_2, provided that any gels in the vessel are still submerged under liquid, will reduce the frequency with which you will need to vent as the vessel is heated. While the critical point of CO_2 is 31.1 °C and 7.29 MPa, the gel liquor may still contain some methanol which will raise both the critical temperature and pressure of the mixture. By overshooting the critical point of CO_2, we guarantee that the gel liquor, even if it still contains some methanol, will go supercritical. If your vessel has a window, you may observe turbidity (swirling) in the chamber as it heats up and as the liquid in the chamber passes above its critical point you may observe the liquid-vapour meniscus transition from a sharp line to a gooey, thick boundary, and finally disappear. Once at the target conditions, allow the vessel to remain at these conditions for 1–4 h. Next, release the pressure slowly at a rate of 0.7 MPa h^{-1}. This should be slow enough to prevent damage to the newly formed aerogel. Also ensure that the exhaust valve fitted to release the gas does not become clogged with dry ice, although if the fluid in the vessel is >45 °C this generally should not be a problem. Once the vessel has reached ambient pressure you can open it and remove the aerogel(s). They can be delicate and friable so use a spoon or nudge them onto a support of some sort for handling. The aerogels are also hydrophilic (that is, they readily absorb moisture) and keep them in a dry environment and free from contact with water and other polar liquids. Avoid using tweezers or your fingers to pick them up. The aerogels are now ready for analysis. The aerogels will be extremely light and for this particular preparation the bulk density should be somewhere around 0.09 g cm^{-3}. BET surface area from nitrogen adsorption isotherms should reveal an internal surface area of around 700 m^2 g^{-1} with a calculated pore volume of 95%–96% of the total volume. The gels are mesoporous with an average pore diameter around 20 nm.

The as-produced aerogels should appear optically transparent to the point that you should be able to see printed letters through slices a few centimetres thick. The refractive index will be very low, somewhere between 1.002 and 1.05. Aerogels are sometimes called 'frozen smoke' as they can appear to have a bluish cast. This is due to Rayleigh scattering from the silica nanoparticles which comprise the aerogel framework. Similarly, light sources viewed through the aerogels will appear to have a yellow tint due to Mie scattering (Mie scattering is slightly different from Rayleigh scattering in that it describes the behaviour of photons scattered by spherical bodies over a wider size domain; Rayleigh scattering is predominantly scattering from particles smaller than the wavelength of incident light) (Fig. 4.29).

FIGURE 4.29   A photograph of a silica aerogel. The printed matter beneath the gel can be clearly seen. (*Image Used with the Permission of Micheal Grogan*)

TROUBLESHOOTING

Cracking

Cracking of aerogel monoliths is a very common problem, While many researchers attribute cracking to improperly executed supercritical drying (for example, too rapidly depressurising from supercritical conditions), in fact *it is rarely the supercritical drying step that causes cracking* but rather the solvent exchange with liquid CO_2. The volume of a methanol/liquid CO_2 mixture will be greater than or less than the sum of the volumes of the unmixed components, and by how much changes as a function of the composition of the mixture. This phenomenon is due to entropy of mixing effects. As liquid CO_2 diffuses into a methanol-containing gel, differential expansion and contraction throughout the gel monolith can occur. Depending on the geometry and density of the gel, this differential expansion and contraction may result in cracking. Slowing down the exchange of methanol with liquid CO_2 helps to mitigate this by allowing the gel to more gradually accommodate volume changes due to entropy of mixing. This can be done by a gradient solvent exchange. First, partially charge the supercritical dryer with liquid CO_2 and if possible use mechanical or convective stirring to homogenise the mixture. Next allow the gel to soak for several hours. Then, drain a fraction of the mixture and replace with fresh liquid CO_2 and repeat this process until the liquid in the supercritical dryer is all liquid CO_2. Protruding features and sharp corners tend to exacerbate cracking. Lower density aerogels (<0.06 g cm^{-3}) are also more prone to cracking than denser aerogels as their backbones are weaker. Dust, debris, ions from water, and bubbles can also serve as points of crack initiation in an aerogel monolith.

CLOUDINESS

If the gels appear cloudy there could be a number of contributing factors. Water remaining behind prior to the supercritical drying will produce cloudiness so try more methanol exchanges during the soaking step. Similarly, residual methanol following super-critical drying can contribute to cloudiness so try additional liquid CO_2 solvent exchanges and make sure that supercritical drying is performed carefully. If the gels are particularly flimsy this may indicate a lack of condensation so try adding more ammonia catalyst (note that ammonium hydroxide tends to release ammonia gas and may become less concentrated with heat and time). It is also possible to exchange the methanol during the soaking stage for acetone which is slightly more miscible with supercritical carbon dioxide than methanol.

SPHERICAL OR COIN-SHAPED BUBBLES

You may observe spherical or coin-shaped voids in your aerogels. These are a result of dissolved gases that evolve out of solution and become trapped during gelation. The coin-shaped bubbles result when spherical bubbles form just prior to gelation and are compressed into flat coin-shaped inclusions as the gel network strengthens. Since the gelation process is exothermic and the solubility of gases decreases with increasing temperatures, an overcatalysed reaction solution may cause the reaction solution to heat too quickly forcing ammonia and/or dissolved gases to bubble out of solution and become trapped in the silica network upon gelation. Degassing your reagents and not overcatalysing your reaction solution will help to avoid the appearance of bubbles.

STICTION

Stiction is the phenomenon of one material sticking to another by weak interactions. Gels and aerogels, being nanoporous materials, can grip onto surface roughness that may otherwise be imperceptible to the human eye thereby making it difficult to demould them. Using smooth, clean moulds of materials with low surface energies such as PTFE will help prevent stiction of gels and aerogels.

WAVY SURFACES

Gels that start out smooth may exhibit wavy, uneven surfaces many hours following gelation. This can be caused by premature submersion of the gel under methanol. If the gel network has not adequately strengthened by the time it is submerged in methanol, swelling and restructuring of the gel may occur, resulting in buckling of an edge of the gel monolith. Allow gels to age in a sealed container with a minimal skin of liquid methanol over the exposed surfaces of the gel.

ADJUSTABLE PARAMETERS AND ALTERNATIVE METHODS

This particular recipe is a one-step base-catalysed approach. Using a higher amount of ammonia will reduce the gel time but may result in the loss of some transparency in the final product and as mentioned may result in bubbles. You can also try using NaOH but this is a stronger base and so a lower molar strength solution than the ammonia must be used. Additionally, catalytic amounts of ammonium fluoride (NH_4F) can be used to accelerate hydrolysis by a different mechanism than acids or bases. The density of the gel can be adjusted by altering the volume of solvent (methanol) used in the initial gel preparation. Similarly, the recipe can be scaled up or down to produce larger or smaller volumes of aerogel.

For applications requiring even lower refractive indices and densities, a two-step approach can be taken in which a prehydrolysed silica oil is produced and used as the precursor (Tillotson and Hrubesh, 1992). This is achieved by adding aqueous HCl (in the 5 to 10×10^{-5} M range) to the solution and letting it gel prior to adding the base catalyst for around half an hour. To achieve densities below 0.020 g cm^{-3} direct supercritical extraction of the alcohol (as opposed to supercritical extraction with CO_2) is necessary. This is in fact how the first aerogels synthesised by Kistler in 1931 were produced (Kistler, 1931). In high-temperature supercritical drying, the gel is placed into a pressure vessel filled with methanol and is pressurised with nitrogen to 12 MPa. This prevents the methanol from boiling as it is heated past its critical point. The temperature is then raised to 270 °C over the course of 24 h by the end of which the methanol has become a supercritical fluid with no change in volume. The supercritical vapour is then released over the course of six hours while the temperature is maintained. This method requires strict safety practices as hot methanol vapour is a fire and health hazard. Silica aerogels produced through high-temperature supercritical drying have the added property of hydrophobicity (water resistance), as methanol at high temperatures reacts with surface hydroxyl (Si—OH) groups on the backbone of the aerogel to form a hydrophobic methoxylated (Si—OCH_3) surface.

Hydrophobic silica aerogels can also be made without high-temperature supercritical drying by chemically replacing surface-bound silanol (Si—OH) groups with hydrophobic groups while the material is still a wet gel (Smith, 1993). This is done by diffusing in a reactive agent such as trimethylchlorosilane (TMCS, Si(CH_3)$_3$Cl) that, upon contact with a hydorxyl group on the gel framework, will result in a hydrophobic trimethylsilyl group (Si—O—Si(CH_3)$_3$). Since trimethylchlorosilane will react indiscriminately with hydroxyl groups of many sorts including those found on methanol (CH_3OH), a non-reactive intermediate carrier solvent such as dry acetone, acetonitrile, or hexane must be used to perform this step.

RECIPE FOR MAKING HYDROPHOBIC SILICA AEROGEL MONOLITHS

To make a hydrophobic aerogel you will need:

Large quantities of flammable solvents are used. Keep away from sources of ignition and keep in a fume hood in an appropriate container. Trimethylchlorosilane is very reactive and hazardous to your health. Its vapours are also extremely corrosive and will corrode nearby metals over time.

1. Everything listed in the first silica aerogel recipe.
2. 1 L or more of anhydrous acetonitrile.
3. 1 L of 6 wt% trimethylchlorosilane in anhydrous acetonitrile solution. This can be prepared by adding 70 mL (60 g, 0.552 mol) of trimethylchlorosilane to 1190 mL acetonitrile (940 g).
4. A liquid-tight sealable polypropylene container.
5. An oven capable of heating to at least 60 °C.
6. A magnetic stir bead and stir plate.
7. (Optional) 1 L or more of hexane.

To make a hydrophobic aerogel:

1. **Prepare a gel.** Follow the steps in the previous recipe up to step 7. In step 7 of the recipe use acetonitrile instead of methanol for the soaking and solvent exchange. Remember to use 5—10 times the volume of the gel of solvent.
2. **Solvent exchange into acetonitrile.** Exchange the acetonitrile 4 times over the course of a week. For all soaking steps ensure your container is sealed to prevent evaporation.
3. **Solvent exchange into TMCS/acetonitrile solution.** Place your gels into the sealable container and fill the container with the 6 wt% trimethylchlorosilane in acetonitrile solution. Use a volume of solution of at least three to four times the volume of the gels. Place a magnetic stir bead in the bottom the container and seal the container. Proceed to the next step without waiting for this solution to diffuse into the gel(s).
4. **Hydrophobic treatment.** Place your sealed container into the oven with magnetic stirrer and start stirring the solution. Heat the oven to 60 °C. Allow the gels to remain at these conditions for 48 h. In this step lone pair electrons in surface bound OH groups will coordinate to the central silicon in the trimethylchlorosilane releasing an HCl molecule and generating a Si—O—Si(CH_3)$_3$

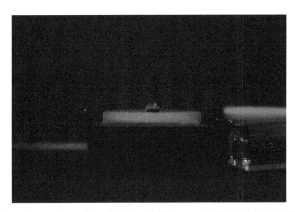

FIGURE 4.30   A photograph of a water droplet resting on the surface of a hydrophobic aerogel. *(Photos Adapted with Permission from www.aerogel.org. Copyright Will Walker)*

surface moiety. Only ~30% of the surface hydroxyl groups need to be converted to achieve hydrophobicity. Be advised that the boiling point of acetonitrile is 82 °C so make sure to use a suitable sealed container and not to overheat. *Take all appropriate safety measures.*

5. **Solvent exchange into acetonitrile.** Take the container out of the oven and allow it cool to ambient temperature. *In a fume hood*, remove the used TMCS/acetonitrile solution and replace with fresh acetonitrile. Leave to stand for 24 h. Repeat this twice more.

6. **Supercritical drying.** Supercritically dry your gels as in the previous recipe. You may but do not need to replace the acetonitrile with methanol prior to exchanging with liquid CO_2.

   Alternative Step 6. **Subcritical drying.** Exchange the gel(s) into hexane 4 times over the course of a week. For all soaking steps ensure your container is sealed to prevent evaporation. Hexane has such a low surface tension and the hydrophobic gel backbone has such a low surface energy that you can actually subcritically dry the gels and obtain aerogels. To do this, place the hexane soaked gel(s) in a jar in which the bottom contains hexane a few millimetres deep. Screw a lid loosely on top and let the sample dry out over the course of a few days. If the gel shrinks substantially place it in a vacuum oven and heat under vacuum to ~50 °C to get it to spring back into shape.

An aerogel prepared by this method should be strongly hydrophobic and should sit on top of water without wetting. The internal surfaces of the silica should be coated in trimethylsilyl groups which repel the water and the large internal surface area makes it energetically unfavourable for water to soak into the gel (Fig. 4.30).

MONOLITHS AND GLASSES CONTAINING FUNCTIONAL BIOLOGICAL MATERIALS

Adapted from Harper et al. (2011), Gill and Ballesteros (1998), Gill (2001)

   Biological materials are true nanoscopic machines, assembled at the molecular level to perform the specific functions required to continue life (Stephen Mann, 2008). In the pursuit of functional nanomaterials the first stop is often to Nature where an abundance of ideas and ready-made solutions can be found. Millions of years of frugal adaption and evolution have yielded proteins, enzymes and much more as solutions to chemical problems encountered in the environment. Very often a function found in Nature can be fitted to a specific problem in the laboratory. For example enzymes, which lower the activation energy of specific biological reactions, are routinely employed as biological washing powders. Once a protein, enzyme or cell is found to be of use then the next step is to design your reaction so that you can get the most activity out of it. This is not easy as biological entities are adapted to very specific environments and deviations from established parameters can cause all kinds of complications. These could include a loss of activity for an enzyme, unfolding for a protein or simple apoptotic or necrotic death for a yeast cell. Complications of this type are mostly overcome by using a buffer or maintaining reaction conditions as close to a natural state for the biological reagent as possible. Typically, a solution phase reaction can be performed in a beaker but there are many applications for which an enzyme or protein needs to be immobilised for use. This could be because the enzyme is expensive and you do not wish to lose it or because the enzyme or protein free in solution will interfere with the UV-vis absorbance of a product. This is where a sol−gel approach can be used to encapsulate and protect a biological entity. Bioinorganic composites are comprised of a functional biological component supported within, or coated with, a ceramic layer. The key feature of these hybrid materials is that they retain biological function whilst gaining mechanical strength or increased resistance to solvent denaturation. Conceptually the idea of mixing a protein into a large lump of glass may make it seem as though the protein will be sealed off, unable to interact with reactants or even move. Although a gel can be formed which does this, it is ultimately of no use and so there are a few design rules for selecting a precursor for use in biological applications. Firstly, the gel must be porous enough to allow diffusion of reactants to the bound protein, enzyme or cell. Secondly, the matrix in which the biological component is bound must allow a degree of flexibility so that the quaternary structure of the protein can deform if this is a part of its function. Thirdly, any by-products of the sol−gel reaction must not destroy the biological component.

To encapsulate a biological material a sol−gel precursor matching the design criteria is added to a buffered solution of the protein, enzyme or cell. If an alkoxide based reagent is used then the alcohol evolved by hydrolysis can denature the biological component. This can be negated partially by prehydrolysis of the precursor to form a silicic acid or similar species or by using diluted amounts of the pure alkoxide. However, making such compromises is not ideal and it is better to use a polyol based silica precursor. As mentioned previously in the recipe for making silica monoliths from a diol based precursor, polyols are not harmful to biological materials and are often used in protective solutions for storage. Polyol based siloxane precursors are fully water soluble and provide a 'one step' option for the formation of silica monoliths without having to go through prehydrolysis. The water solubility of the precursor means it can be mixed readily with other polymers such as PVA and polysaccharides or precursors such as an amino functionalised silane prior to gelation (Shchipunov, 2004; Collins et al., 2009; Lee et al., 2011). These additives impart flexibility in the formed monoliths and can act as engineered binding sites for the protein or enzyme being mixed in. Activities for biological materials encapsulated in gels formed from polyol siloxanes retain far higher activities than alkoxide based procedures.

In the next few recipes we will prepare a poly(glycerol silicate), a 3-aminopropylglyceroxy silane and a 3-glucanamidopropyl siloxane for use in the encapsulation of a range of common biological entities for use as bioinorganic devices. The hardest part is making the polysiloxane precursor and after that it is a case of mixing the precursor with buffer and biological part and a gel will form quickly. The concern with encapsulated biologicals is their long term stability. Though using analytical devices based on 'live' bioglasses is fine in the laboratory they are still not available commercially. The functional activity of encapsulated cells lasts for only a few months and even then only with particularly hardy organisms such as yeasts. The situation is much better for proteins and enzymes as these are less complex than an entire cell and do not rely on nutrients or the need to excrete to remain active. Bioglasses containing these can stay active for years if stored correctly.

MAKING A GLASS CONTAINING YEAST

Adapted from Harper et al. (2011)

Yeast (*Saccaromyces cerevisiae*) is a single celled eukaryote organism and has been used for hundreds of years in the bio-fermentation of alcohol and, marginally more recently, bread. In the last few decades the advances in genetic engineering have seen yeast taking a leading role in biovat fermentation processes where a strain can be tailored to convert sugars into biofuels (Naik et al., 2010). Because yeast is a simple cell it is widely used as a model for eukaryotic cellular processes with genetic modifications allowing more complex studies to be made (Sherman, 1991). In this recipe we will prepare a poly(glycerol silicate) and use it to encapsulate yeast. The source reference used yeast that was genetically modified so that it would express a fluorescent protein when the bioglass containing the yeast was exposed to galactose (Flemming et al., 2006). In this way the long term activity of the yeast could be monitored by how strongly fluorescence was detected in the glass. You may wish to try with your own genetically modified yeast and a general protocol for doing this can be found elsewhere in the biology section of this book. For now we will focus on getting the yeast encapsulated in a monolithic gel using a poly(glycerol silicate). This compound is formed by the substitution of ethanol moieties in tetraethylorthosilicate for glycerol ($C_3H_8O_3$). The preparation method is very similar to the preparation of THEOS detailed earlier in this section. Each glycerol molecule has three hydroxyl groups capable of coordination such that the ligand will form bonds to other silicon centres in other molecules. This is why the compound name is given the poly suffix. This particular preparation will give a 1:10 ratio of silicon to glycerol and is similar to the preparation of the diol based siloxane THEOS in an earlier recipe. The compound itself is doped with a small amount of Ti^{4+} cation which behaves similar to silicon chemically but will catalyse the gelation as this ion hydrolyses more rapidly. Using a titanium dopant is one way to catalyse the gel when you are unable to use acids and bases which could damage the yeast (Fig. 4.31).

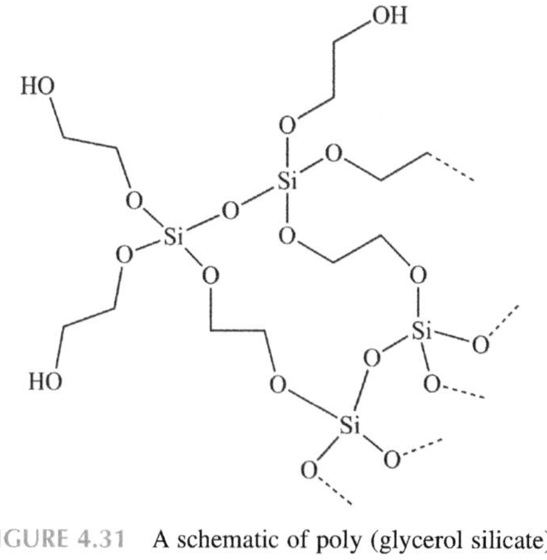

FIGURE 4.31   A schematic of poly (glycerol silicate).

To make a glass containing yeast you will need:
For the poly(glycerol silicate):
These quantities will make far more of the siloxane than you will need but it is very stable and can be used as a stock solution.

1. 50.27 mL (63.4 g, 0.688 mol) of glycerol.
2. 10.86 mL (10.21 g, 0.049 mol) of tetraethylorthosilicate.
3. 1.08 mL (1.02 g, 0.003 mol) of titanium (IV) isopropoxide.
4. A heating mantle with a large mineral oil bath capable of reaching at least 130 °C. It must be fitted with magnetic stirring and a stir bar.
5. A 100−200 mL round bottomed flask with a quick fit connection. You might want to use a two necked flask as this will save you having to remove a reflux condenser to add the silicon and titanium alkoxides.
6. A reflux condenser that sits in the round bottomed flask and all the appropriate tubing and clips. You will need a water supply to run through it. Use clamp stands to hold everything secure.
7. A vacuum line (~1.3 Pa) running through a cold trap. The end of the line should have a quick fit connector the same dimension as the round bottomed flask or the top of the reflux condenser. The cold trap will need an ethanol and dry ice bath in a dewar flask to keep it cool enough to condense out the ethanol. This will be used to remove the ethanol and small amounts of iso-propanol from the produced poly(glycol silicate). An alternative is to use a rotary evaporator but ensure the heating bath is able to run at 130 °C.

For the yeast in glass:

8. A culture of *S. cerevisiae* cells. You can try using simple bakers yeast to start with or you may wish to use genetically modified cells. Details on growing and maintaining a culture are found elsewhere. You will need enough to form a concentration of 1×10^6 to 1×10^7 cells/mL in the buffer solution given below. This can be prepared by dipping a sterile wire in the initial culture media and then dipping it in the buffer. Track the optical density to a given point to get the right amount of cells.
9. A 0.0119 mol/L solution of phosphate buffer for mixing the cells with the silicate precursor. This is prepared by dissolving 1.42 g (0.0119 mol) of NaH_2PO_4 or 1.68 g of H_2NaPO_4 in to 1 L of water. Add 10.21 g (0.1370 mol) of NaCl and 0.15 g (0.0027 mol) KCl. Adjust this solution to pH 6.0 using drops of 1 M HCl or NaOH and a pH meter under stirring after all the salts have dissolved.
10. A 1.5 mL polypropylene centrifuge tube and a benchtop vortex.
11. A volumetric pipette with a range of 100−500 µL.

To make a glass containing yeast:

1. Place the glycerol in the round bottomed flask with a magnetic stir bar and setup the reflux condenser over the oil bath on the heating mantle. Make sure the tubing on the reflux condenser is firmly in place.
2. Turn on the heating mantle and set it to 60 °C. Lower the flask and reflux condenser into the bath and allow the glycerol to equilibrate to 60 °C with stirring. The glycerol level should be level with the heating oil.
3. Add the 10.86 mL of tetraethylorthosilicate and 1.08 mL of titanium isopropoxide to the round bottomed flask.
4. Set the temperature on the heating mantle to 130 °C. Reflux the solution under stirring for 3 h.
5. By this point the product will have formed in an excess of glycerol. Ethanol evolved in the reaction must be removed. Fit the vacuum line to the top of the reflux condenser and place the heated sample under vacuum for 30 min. You should see roughly 12 mL of ethanol condense into the cold trap.
6. After the ethanol has been removed disconnect the vacuum, turn off the heating mantle, magnetic stirrer and the water feed to the reflux condenser. Allow the product to cool and store in a bottle or by plugging the round bottomed flask.
7. This reaction should produce around 60 g of the silicon polyol product with the formula $Si(C_3H_7O_3)_4 \cdot 10C_3H_8O_3$. It should appear as an opalescent white liquid. This can be kept as a stock solution in a dessicator though it should be resistant to hydrolysis by atmospheric moisture if left out on the bench for a while.
8. To prepare the bioglass add 175 µL of the yeast in pH 6.0 phosphate buffer 175 µL of the poly(glycol silicate) solution in the 1.5 mL poly(propylene) centrifuge tube.
9. Vortex immediately for 60 s to mix the solution. Allow the sample to gel for 24 h at room temperature. The gel should be strong enough not to flow when the tube is inverted after around 2 h. After this time a solid silica phase will have formed about the yeast cells. The gel is now ready for analysis.

You should be able to remove a self supporting free standing gel from the centrifuge tube. The viability of the encapsulated yeast cells needs to be determined. The easiest way to do this is to break a portion of the gel up and place it in 200 µL of the buffer solution adjusted to pH 7.4. This is agitated to form a suspension and some of the cells will be released into the solution. Add 100 µL of this suspension to 5 mL of YPD medium in a 15 mL culture tube and incubate at 30 °C for 24 h. YPD is an acronym for yeast extract, peptone and dextrose and these three ingredients act as food for the yeast to grow with. It should be prepared freshly by adding 10 g of yeast extract, 20 g of peptone and 20 g of dextrose to 1 L of water. At regular intervals over the 24 h measure the optical density of the solution using a microplate well reader. You should see the optical density increases as the yeast proliferates.

The health of the yeast cells can be tested using an assay kit which comes with markers that fluoresce in the presence of damaged cells. A basic live/dead staining kit for yeast will contain 5-carboxy-fluorescien diacetate (CFDA) and a complementary acetoxymethyl ester (AM). When mixed together the AM allows the CFDA to pass across the membrane of the yeast cell and get inside. Once inside a live cell, non-specific esterases will cleave the diacetate group on the CFDA resulting in a fluorescent dye that will emit light at 517 nm (green) when irradiated with 492 nm light. As the esterases are only produced by living cells, anything that shows up green is a viable cell. In order to show up dead cells propidium iodide is used. This is a dye that will only permeate into leaky and damaged dead or non-viable cells. Dead cells stained in this way will emit light at 635 nm (red) when irradiated at 490 nm. This means that by imaging the cells under a fluorescent light microscope you should be able to see a mixture of green (alive) and red (dead) cells. You will want to see more green than red. A more in depth look at staining methods is explored in the biology section.

If you have genetically engineered the yeast to express a protein then you can 'feed' the gel and monitor the amount of protein produced. With fluorescent proteins such as YFP this can be used as an indicator of how readily a gene is expressed in the yeast and how active it is compared to non-encapsulated cultures. In the source reference the yeast was engineered to express YFP in the presence of galactose. 250 μL of a yeast extract, peptone and galactose buffer is added to the gel while still in the centrifuge tube and incubated for up to 32 h. The galactose buffer is prepared identically to the YPD buffer solution but replacing the dextrose for galactose and adjusting to pH 7.4. Thin sections of the gel are taken at various time points and examined under a fluorescent microscope. The intensity of the yellow fluorescence is an indicator of how well the cell is functioning in comparison to non-encapsulated controls incubated in the same way. Remember that cells demonstrate auto fluorescence and all measurements must be taken against a negative control of unaltered yeast and silica gel.

As the yeast cells are so large and correspondingly complex regarding the number of biochemical processes occurring, the viability of the encapsulated cells will drop off to around 40−60% after only 9 days. Cells encapsulated while they are in a quiescent state will fare better than ones that are fully active and this is attributed to the stresses imparted by the gel process upon the cell. The next encapsulation recipes work with smaller biological components and so the shelf life of the products is markedly increased.

MAKING A MICROTITER PLATE DIAGNOSTIC FOR GLUCOSE, GALACTOSE, LACTOSE AND LACTATE

Adapted from Gill and Ballesteros (1998) with thanks to Dr. Iqbal Gill

The means to immobilise active biological materials within a substrate means that we can isolate useful components from a particular biochemical pathway and incorporate them into functional materials. In this recipe various enzymes will be immobilised into a gel coating the inside of a 96 well plate. Reactions of interest can be run in the wells and monitored by a well plate reader. After testing, the reaction solution can be thrown away and the plate rinsed out before being used again. This is very useful if you have to monitor a large number of reactions over and over again as you do not have to mix in and then separate out sometimes precious enzymes. In this recipe a number of enzymatic oxidases for oxidising simple sugar molecules into metabolites will be immobilised in an organofunctionalised bioglass. A major variation in this preparation is that the alkoxide precursors are partially hydrolysed by reaction with water before conversion to a polyglycerol. Instead of a single silicon surrounded by an excess of glycerol ligands the silicon will have one or more oxygen coordinated to it before gelation and the precursor has an approximate formula of $SiO_{0.5-1.5}(Glycol)_{0.5-2}$.

Glucose is a small monosaccharide that is used as the main energy source for cellular function. The detection of free glucose in the blood or urine is routinely performed by doctors using strips which can be dipped into fluids and change colour to brown if the sugar is present. The strips will only change colour for glucose and this specificity is due to the presence of the glucose oxidase enzyme which converts the sugar, oxygen and water to hydrogen peroxide and D-glucono-δ-lactone. In the strips that a doctor uses the hydrogen peroxide generated reacts with potassium iodide impregnated in the paper to produce brown iodine. A more accurate determination of the sugar concentration is possible by using a dye which is sensitive to the presence of hydrogen peroxide and monitoring the absorbance with UV-visible spectroscopy. The most commonly used dye of this sort is 2,2'-azino-bis(3-ethylbenzothiazoline-6-sulphonic acid) (ABTS) which has a peak absorbance at 415 nm. A plate well with the enzymes for catalysing the oxidation of small molecules of metabolic importance will allow you to accurately monitor a large number of different samples quickly and all at once. To prepare the plate well for testing it will be primed by rinsing with organosilanes to form a layer on which the gel can grip to the polycarbonate substrate. A gel solution containing an enzyme is then coated over the inside and allowed to solidify. This will form a thin transparent layer so that the optical transparency of the wells remains unaffected by coating. An oxygen saturated solution containing known or unknown amounts of glucose or a similar small molecule are added to the well in addition to the ABTS indicator. The level of oxidation is then monitored spectroscopically. A typical plate well might look like this with strips for the different sugars or different concentrations of the same sugars. After testing known concentrations and monitoring the rate for the reaction you may wish to try seeing how accurately you can measure an unknown concentration of a given sugar (Fig. 4.32).

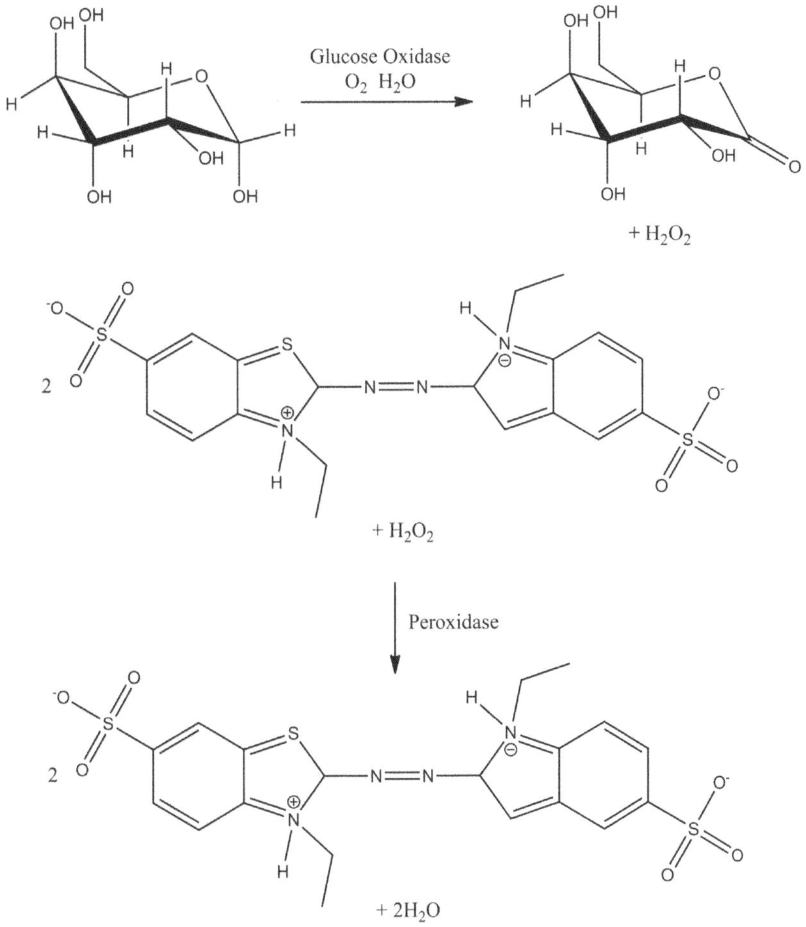

FIGURE 4.32   The conversion of glucose to D-glucono-δ-lactone. Hydrogen peroxide formed in the process reacts with the ABTS dye such that the progress of the conversion can be monitored spectroscopically.

In order to get the highest activity possible the inorganic component is functionalised by the incorporation of an organic group compatible with the enzymes. If an unmodified poly(glycol silane) is used then the activity of the encapsulated enzyme will drop to a fifth of the free enzyme. Both poly(vinyl alcohol) (PVA) and primary amine terminated aliphatic moieties are added to the composition. The PVA introduces flexibility into the structure and allows for conformational movement as well as accommodating the transport of small molecules to and from the trapped enzymes. The amino functionalised groups in the silica act as anchors for the enzymes to stick to within the lattice and secure them in place. To dope the poly(glycerol silane) alkoxide precursors are modified first and then converted to the polyglycerol form. 10% by mole 3-(1'-hydroxy-2'-(2''-hydroxyethylamino)ethoxy)propyl silane (HEAS) which formed in this synthesis by the reaction of 3-glycidoxypropyltrimethoxysilane (3-GPTMS) with ethanol-amine. This product is in turn added to 90% TMOS and the alkoxides are all hydrolysed then treated with glycerol to form the gel precursor. This is then mixed with PVA and the enzymes in a buffer to form the bioglass gel. The molar ratio of PVA:HEAS:PGS in the gel will be 1:2:5 (Fig. 4.33).

To make a microtiter plate diagnostic with an encapsulated enzyme layer you will need:

1. 3.14 mL (3.36 g, 0.1 mol) of 3-glycidoxypropyltrimethoxysilane (3-GPTMS). This chemical is normally provided in a capped bottle to protect it from moisture so you will probably also need a needle and a syringe to get it out. *Make sure you have a sharps bin for disposing of the needle after use.*
2. 6.03 mL (6.10 g, 0.1 mol) of ethanolamine in a 40 mL vial with a screw top lid. *This is corrosive and toxic. Work with it in a fume hood.* You will need a clamp stand to hold the small vial in an oil bath.
3. A heating mantle capable of reaching 50 °C with an oil bath and magnetic stirring. Given the low temperature you might be tempted to use a water bath. The heat will be applied for over 20 h by which time, even at lukewarm temperatures, the water will have evaporated.
4. 25 mL (19.77 g, 0.61 mol) of methanol.
5. 132.7 mL (137 g, 0.9 mol) of tetramethylorthosilicate. This should be placed in a 300 mL round bottomed flask with a quick fit aperture that can connect to a rotary evaporator. You will need a glass plug to seal it and a cork ring to stand it in. Weigh the

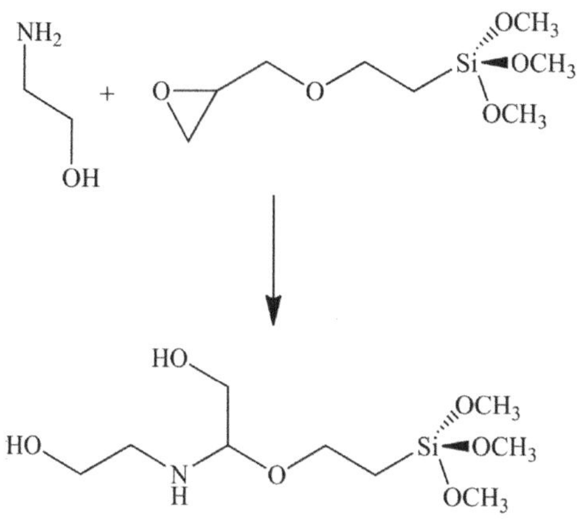

FIGURE 4.33   The reaction of ethanolamine and 3-GPTMS to form HEAS.

    empty round bottomed flask before use and make a note. This will help in determining how much solvent has been removed during the rotary evaporation step.

6. 18 mL (18 g, 1 mol) of water. This will be used to partially hydrolyse the silica precursors before the reaction with the glycerol.

7. 59.15 mL (74.6 g, 0.81 mol) of glycerol.

8. A rotary evaporator with an oil bath that can reach 50 °C and accommodates the 300 mL round bottomed flask. The condenser in the evaporator setup should be cooled by a water/ethylene glycol mixture maintained at −10 °C with a Peltier system and pump. Alternatively you can use a cold finger type condenser filled with dry ice and acetone.

9. 2.5 mL of a 10 wt% poly(vinyl alcohol) (PVA) solution in water. A stock solution can be prepared by dissolving 1 g of PVA (Molecular weight = 85,000−146,000) in 8 mL of water in a small vial. Place the vial on some scales and add water until it weighs 10 g.

10. A stock solution of a 0.05 mol/L Na_2HPO_4 buffer at pH 7.5. This can be prepared by dissolving 7.0 g of Na_2HPO_4 in 900 mL of water and then adjusting to pH 7.5 with 1 mol/L HCl or NaOH. After the correct pH is reached make the solution up to 1 L. This will be used to disperse the enzymes in prior to gelation.

11. A wide pot of crushed ice for sitting samples in. It is important to keep the enzymes ice cold as this prevents them spoiling. Solutions and gels containing enzymes should be kept cold at all times. Have a number of 10 mL glass samples vials sitting on the ice. These will be useful for making up the various gels.

12. A fridge for keeping buffer solutions, water and precursors in. To protect the enzymes they are maintained at 4−5 °C throughout the preparation.

13. 60 units each of glucose, galactose and lactate oxidases and 200 units of horseradish peroxidase enzymes. Double this amount if you need to run unencapsulated gel free controls of some sort. When purchased they will come as a lyophilised powder and should be kept dry and cold in a fridge or freezer until use. The use of 'units' can cause some confusion so if you are in doubt over how much enzyme to put in the gel aim for between 10 and 50 mg of enzyme per gram of final gel solution. This is a rough guess but if you determine you are working lower than 1−0.1 mg you will know that this is too little enzyme and the 'units' definition may be incorrect for the intended use. Each enzyme must be dissolved into 0.5 mL of water shortly before use.

> Because the molecular weight of enzymes can vary massively they are often sold in 'Units'. This is defined as the amount of enzyme that catalyses 1 micromole of substrate (reagent) per minute. This definition can vary from different suppliers and alternatively you may find activity Units commonly quoted in terms of nanomoles of substrate per minute. When you buy enzymes be sure to check the definition of units given by the supplier.

14. A polycarbonate 96 well microtiter plate. This should be supplied as sterile and try to keep it as clean as possible. Although this recipe is not performed under clean room conditions it is good practice to maintain clean conditions with any biological experiments.

15. A priming solution composed of various organosilane alkoxides to make the well plate amenable to gel coating. This is prepared by adding 1 mL each of 3-aminopropyltriethoxysilane, 3-ureidopropyltrimethoxysilane and N-(2′-aminoethyl)-3-aminopropyl-trimethoxysilane to 95 mL of ethanol and then adding 5 mL of water. This should be prepared fresh shortly before use to prime the well plate.

16. Volumetric pipettes and tips capable of measuring in the ranges of 10 μL up to 5 mL.

The following components are required for running the diagnostic.

17. A stock solution of 0.05 mol/L Na_2HPO_4 buffer at pH 6.5. This is prepared identically to the buffer solution detailed in item number 9. To 1 L of phosphate stock add 1.072 g of magnesium acetate tetrahydrate and 0.793 g of calcium acetate hydrate. The calcium acetate hydrate adsorbs water so it may require drying out in an oven before weighing.

18. 0.823 g (0.015 mol) of 2,2′-azino-bis(3-ethylbenzothiazoline-6-sulphonic acid) in 100 mL of the phosphate buffer at pH 6.5 (as given in component item 15). This is normally supplied as a tablet that can be dropped directly into 100 mL of buffer to give a working indicator solution. This should be enough solution to run with the plate. Shortly before use add 100 units of horse-radish peroxidase and a known concentration of glucose or other small polysaccharide. The concentration of sugars should be around 0.005−0.01 mol/L. For glucose a control can be prepared by adding 1.8 g of glucose to 100 mL of the indicator solution.

19. An air line which is fed through a tube and a small nozzle or needle. This will be used to bubble air through the indicator solution so that it is oxygen rich. The oxygen is required for the peroxidase to work. You may also want to have a nitrogen line which is useful for drying the gel coated well plates. If you do dry the well plate using a nitrogen stream, make sure it remains at 5 °C. This can be accomplished by sitting the plate on ice and clamping a nitrogen nozzle in place so that nitrogen flows over the wells.

20. An UV-visible spectroscopic microplate reader monitoring the dye absorbance at 415 nm.

To make a microtiter plate diagnostic with an encapsulated enzyme layer:
To make the 10% organosilane doped poly(glycerol silane) precursor:

1. Place a small magnetic stir bead in the ethanolamine in the 40 mL vial. Clamp it in place on a magnetic stirrer so that it won't fall over. This should be setup over the oil bath but without any heating yet.

2. Add the 3.14 mL of GPTMS to the stirred ethanolamine slowly over the course of 5 h. Replace the cap each time you add more GPTMS.

3. Turn on the heating mantle and set the oil bath to 50 °C. Add 25 mL of methanol to the vial and screw on the lid. Make sure it is secure. Lower the vial into the oil bath leaving about half a centimetre of the methanol solution above the oil line. Leave the solution stirring and heating for 20 h. The ethanolamine will react with the glycidoxypropyl group to produce the HEAS moiety.

4. Remove the vial from the oil bath and set to one side. Keep the oil bath at 50 °C. Clamp the 300 mL round bottomed flask containing the TMOS in its place and sit the flask in the oil bath. As before the top of the TMOS should be about half a centimetre above the oil line. Pour the methanol and HEAS solution into the TMOS under stirring.

5. Over the course of 1 h add the 18 mL of water (~0.3 mL/min). Place a quick fit stopper into the neck of the flask to prevent any solvent getting out.

6. Replace the stopper and let the solution stir at 50 °C for 20 h.

7. Remove the round bottomed flask from the oil bath and remove the magnetic stir bar from the flask. Connect the flask to a rotary evaporator and set the heating bath to 50 °C. Make sure your cooling system is operating at −10 °C or lower.

8. Rotary evaporate the sample until no more liquid is collected. Dispose of waste solvent in the cold trap or bulb in the appropriate waste container. A viscous yellow liquid should remain in the flask. This product is partially hydrolysed but still has methanol groups coordinated to the central silicon atom which we will replace with glycerol in the next step.

9. Place a clean magnetic stir bar into the round bottomed flask and clamp it once more sitting in the oil bath. Set the oil bath to 50 °C and turn on the magnetic stirring. Add the 59.15 mL of glycerol over the course of 1 h (1 mL/min). Replace the stopper and leave to stir for 40 h.

10. Remove the flask from the oil bath and turn off the heat. Remove oil from the bottom of the flask with some tissue. Weigh the round bottomed flask containing the product and make a note of what the product weight is after subtracting the weight of the flask you took earlier.

11. Connect the flask to the rotary evaporator as before with the same temperature setting. This step will remove the methanol displaced by the glycerol. Allow the evaporation to proceed and periodically weigh the flask. Once the solution has concentrated down to 70% of its original weight the product is ready to be used. You should now have a yellow gel with the rough formula $(HEPSiOGlycerol)_{0.1}(SiOGlycerol)_{0.9}$. This can be stored to use as a stock solution.

The following steps are for preparing the microtiter plate. **You will need to make up a different gel for each enzyme**:

12. Prepare the priming solution as outlined in item 13.

13. Fill the wells with the primer solution and leave at room temperature for 1 h. Empty out the wells before use and allow to dry. You do not want any ethanol remaining in the wells as this will damage the enzymes. Sit the empty well plate on the ice.

14. Make up the enzyme solutions by dispersing the powdered enzyme (kept in a vial on ice) into 0.5 mL of pH 7.5 buffer. This recipe will outline encapsulating only the glucose peroxidase for clarity. For the other enzymes repeat the following steps to produce a fresh gel. For the glucose, galactose and lactose oxidases disperse 60 Units of enzyme into the 0.5 mL of water. For the horseradish peroxidase disperse 200 units into 0.5 mL of water. Keep these solutions in vials on ice and prepare immediately before use. Any remaining enzymes should be stored in a fridge or freezer immediately after use.

15. Place 0.4 mL of water in a 10 mL sample vial and place it on ice. In to this ice cold water dissolve 1 g of the $(HEPSiOGlycerol)_{0.1}(SiOGlycerol)_{0.9}$.

16. To the vial containing the silica precursor add 2.5 mL of the 10 wt% in water poly(vinyl alcohol solution). Then add the 0.5 mL glucose oxidase enzyme in water dispersion to the vial. Mix the contents with gentle agitation for a few seconds and move to the next step quickly as the gel will form within 15 min.

17. Depending on the experiment you are running you must think about which wells you would like to coat with a particular enzyme or which ones you will leave uncoated as enzyme free controls. Using a volumetric pipette coat the inside of each 'active' well with 30 μL of the gel solution. Keep the well plate on ice until the solutions have gelled which should take about 15 min.

18. Place the lid back on the well plate and store it in a fridge (~5 °C) for 10 h to give the gel time to age. After this period dry the well plate on ice using nitrogen for 4 h.

19. Give each well 4 rinses with the pH 7.5 phosphate buffer. Replace the lid and store until use. This plate is now ready for use in running microtiter experiments.

The following steps are an example of using the microtiter plate well:

20. Depending on what you are testing you may wish to run different concentrations of sugars and measure the response of the indicator. Just to test the plate and assess the functionality of the enzymes you will need a range of known concentrations. Using a test for glucose as an example; prepare a number of different glucose concentrations in the indicator solution. These should range from 0.005 to 20 mmol/L. How many different concentrations you run will be up to you. Run a duplicate of each concentration three times for accuracy. You can run a range of unknown glucose concentration solutions in parallel to see if you can determine their concentration correctly.

For running titrations on the galactose, lactate and lactose (galactose oxidase, lactate oxidase and the horseradish peroxidase enzyme coated wells respectively) a linear concentration relationship to absorbance should be seen for the ranges 0.005−0.03, 0.005−0.025 and 0.005−0.040 mmol/L respectively.

21. Add 100 μL of the indicator and glucose solution to the plate well. The reaction to oxidise the sugar molecule should be complete after 12 min. You should see the stronger sugar concentrations appear greener than the weaker ones.

22. Place the plate into a plate well reader measuring absorbance at 415 nm to get quantifiable comparisons. You should see a straight line relationship between the absorbance of the solution from dye production and the concentration of sugar put in.

23. Empty the plate and rinse it four times using the pH 6 phosphate buffer. You can now use it for another analysis. You should be able to reuse the plate for analysis up to 60 times with only a small drop off in the enzyme activity. Keep the well plate in the fridge when not in use.

The interesting aspect of this type of sol−gel coating is that an almost identical preparation using a methoxide based precursor instead of a polyol one results in a 50% loss of enzyme activity even though chemically the product is the same−organosilica doped with PVA. The activity retention in this preparation demonstrates the efficacy of using a precursor that evolves its own bioprotective agent upon hydrolysis. The activity compared to native protein should be near 90% when compared to free enzyme in solution.

The poly(glycerol silicate) precursor can be used on most enzymes and proteins for encapsulation with minimal damage to biological function. You may find that for some applications it is better to use an organosilane modified form. The following recipe gives a modification to the procedure to form a free standing bioglass strip which can be used in a spectroscopic application.

MAKING AN OPTICAL THICK FILM SENSOR BIOGLASS FOR THE DETECTION OF D-GLUCOSE-6-PHOSPHATE

Doping poly(glycerol silicate) with an amount of poly(3-aminopropyl glyceroxysilane) will give a bioglass with even higher retention of biological function. This might lead us to think that using only poly(3-aminopropyl glyceroxysilane) will give even higher efficacies but gels made in this way do not solidify correctly as one in four of what should be Si−O−Si bonds are Si−CH(CH$_2$)NH$_2$ bonds which do not contribute to the stability of the silica lattice network. Therefore it is necessary to have a large proportion of precursor that will react completely to form a glass. If you are already setup to produce a poly(glycol siloxane) from the previous recipes outlined in this book then the modifications you will need to make are very minor. You could add a given amount of 3-aminopropyltriethoxysilane directly to the tetramethyltrimethoxysilane during the preparation of the glycerol form but then you will have a stock solution with a fixed ratio of silane to organosilane in the produced glass. For the sake of experimentation it is better to have a stock solution of both silane and organosilane precursors so that you can mix them as required. The methods for producing a stock batch of both are presented below. The preparation method for each is exactly the same so for clarity the process for making the undoped poly(glycerol silane) is described. For the poly(3-aminopropylglycerol silane) replace the tetraethylorthosilicate with 3-aminotriethoxysilane in the amount given.

The biological component in this recipe is D-glucose-6-phosphate dehydrogenase. This enzyme is a component of the pentose phosphate metabolic pathway and aids the conversion of the coenzyme nicotinamide adenine dinucleotide phosphate (NADP) to the hydrogenase form (NADPH). During the conversion the D-glucose-6-phosphate is converted to 6-phosphogluconolactone which is then processed further in the pathway. The conversion of NADP to NADPH can be monitored spectroscopically in a UV-visible spectrophotometer and therefore it is possible to monitor the activity of the encapsulated enzyme.

To make an aminopropyl silane doped bioglass you will need:

1. 106 mL (100 g, 0.48 mol) of tetraethylorthosilicate. This is the precursor for the pure poly(glycerol silicate) solution. Alternatively to produce a stock solution of poly(3-aminopropyl siloxane) use instead 112.3 mL (106.2 g, 0.48 mol) of 3-aminopropyltriethoxysilane.
2. 50 mL (39.4 g, 0.85 mol) of ethanol.
3. 10.4 mL of a 0.25 mol/L solution of HCl in water. A stock solution of the correct strength can be prepared by diluting 2.5 mL of a 37% (11.85 mol/L) HCl solution up to 118.5 mL.
4. 27.75 mL (34 g, 0.38 mol) of glycerol.
5. A heating mantle with an oil bath capable of reaching 70 °C with magnetic stirring. You will also need a clamp for holding a round bottomed flask in the bath.
6. A 300 mL round bottomed flask with a quick fit aperture and a glass plug. This should fit with a rotary evaporator. Pre weigh the flask so that you can easily weigh the amount of product inside it.
7. A rotary evaporator with an oil bath that can accommodate the round bottomed flask. The condenser in the evaporator setup should be cooled by a water/ethylene glycol mixture maintained at −10 °C with a Peltier system and pump. Alternatively you can use a cold finger type condenser filled with dry ice and acetone.

The following items will be needed for making the thick film sensor for D-glucose-6-phosphate:

8. A wide pot of crushed ice for keeping all the biological components cold while out on the bench. Have a number of 10−20 mL vials to hand for working with the various solutions and keep them on ice.
9. A fridge.
10. A magnetic stir plate and a pH meter for adjusting the buffer solutions.
11. 4 mL of a 15 wt% poly(vinyl alcohol) (PVA) solution in water. A stock solution can be prepared by dissolving 1.5 g of PVA (Molecular weight = 85,000−146,000) in 8 mL of water in a small vial. Place the vial on some scales and add water until it weighs 10 g.
12. 220 Units of torula yeast D-glucose-6-phosphate dehydrogenase. This will come as a powder and should be kept frozen and dry until use.
13. 0.5 mL of a 0.1 mol/L phosphate buffer at pH 8 containing 10 mg/mL of NADP for making a dispersion of the enzymes with. This can be prepared by dissolving 14.0 g of Na_2HPO_4 in 900 mL of water and then adjusting to pH 8 with 1 mol/L HCl or NaOH. Take 5 mL of this solution and dissolve 50 mg of β-nicotinamide adenine dinucleotide phosphate hydrate to it. You will only need 0.5 mL of buffer for each gel. Keep the bulk phosphate solution as this will be used later.
14. A silicone mold or small plastic petri dish and lid for the gel to set in. You will produce about 10 mL of the gel solution which will need to be aged so something that can be sealed is ideal. You will need to be able to remove the gel after it has solidified. For cutting the gel you need a scalpel.
15. A nitrogen line and nozzle which can be used to stream nitrogen over the gels to dry them.
16. A stock solution of conditioning and loading 0.075 mol/L phosphate buffer at pH 7.5. This can be prepared by taking 75 mL of the bulk phosphate solution prepared in item 12 and further adjusting it to pH 7.5. Make the volume up to 100 mL. Add 100 mg of NADP to the solution and stir. This solution is used to rinse and condition the gels. It will also be used as the medium in which D-glucose-6-phosphate and NADP are added to the cuvette.
17. A quartz UV-visible cuvette and UV-visible spectrophotometer.
18. For monitoring the reaction you will need a number of different concentrations of D-glucose-6-phosphate. In order to work properly the correct amount of NADP coenzyme must be added for the reaction to be properly catalysed by the dehydrogenase enzyme. For testing you will add 1.5 mL of the phosphate buffer prepared in item 15 to the cuvette and monitor the reaction by measuring the absorbance at 340 nm. The sensor should give a linear response over a range of 0.005−0.05 mol/L of D-glucose-6-phosphate. For every mole of D-glucose-6-phosphate in the buffer solution add 1.5 moles of NADP.

To make and aminopropyl silane bioglass:

1. Turn on the heating mantle and set it to 70 °C.
2. In the large round bottomed flask add the 106 mL of tetraethylorthosilicate and 50 mL of ethanol. Clamp it in place and lower the flask into the oil bath. Add a stir bar and stir the solution vigorously.
3. Over the course of 30 min add the 10 mL of 0.25 mol/L HCl solution. Replace the glass plug when you are not adding the acid.
4. Allow the solution to stir for 15 h.
5. Take the flask out of the oil bath and remove the stir bar. Connect it to the rotary evaporator and reduce the solution at 35 °C. This will produce a clear viscous liquid with the rough composition $SiO_{1.1-1.2}(Ethoxide)_{1.8-1.9}$ which will have some trace amounts of free ethanol still present.

6. To convert this alkoxide precursor to the glycerol form clamp the round bottomed flask back in place in the oil bath and set the temperature to 50 °C.

7. Under vigorous stirring add 27.5 mL of glycerol to the solution over the course of an hour. Replace the glass plug and allow the solution to stir at 50 °C for 40 h.

8. Remove the flask from the oil bath and weigh it so that you know the weight of its contents.

9. Fit the flask to the rotary evaporator. Set the temperature to 50 °C and evaporate the solution until the contents are at 70% of their original weight. You should now have a clear gelatinous poly(glycerol silane) with the rough formula of $SiO_{1.2}$ $glycerol_1$. Keep this in a sealed container in a dessicator until needed.

10. Repeat steps 1−9 using 3-aminopropyltriethoxysilane to form a stock poly(glycerol organosilane).

11. Disperse 220 units of the Torula yeast D-glucose-6-phosphate dehydrogenase in 0.5 mL of the 0.1 mol/L phosphate buffer solution as prepared in item 13. Keep the dispersion cold on ice. Put the stock enzyme back in a freezer once it is used.

12. Add 1 g of poly(3-aminopropylglyceroxysilane) to 2.4 g of poly(glycerolsilane) dissolved in 1.5 mL of ice cold water in 20 mL vial.

13. To the vial containing the silane precursors add 4 mL of the 15 wt% PVA solution and then 0.5 mL of the enzyme dispersion. Mix briefly and pour the contents into the mould. Keep everything on ice. The gel will set within 10 min.

14. Seal the mould and age the gel in a fridge for 10 h.

15. Remove the aged gel from the fridge and dry the gel sat on ice under a nitrogen flow for 5 h.

16. Cut the gel into strips large enough to fit into a UV-visible quartz cuvette using a scalpel. The dimensions of the strips should be $2 \times 5 \times 20$ mm to fit into a $5 \times 10 \times 30$ mm cuvette. Your sizes may vary. The gel strips can now be stored for use in a fridge until needed.

17. In a petri dish soak the gel strips 3 times with 10 mL aliquots of the conditioning 0.074 mol/L phosphate buffer as prepared in item 16. Each soaking step should last for 30 min.

18. Setup the UV-visible spectrometer to monitor absorbance at 340 nm. Place a strip into a UV-visible cell and then add 1.3 mL of a D-glucose-6-phosphate/NADP solution of a known concentration. You should now be able to see the absorbance at 340 nm increase as NADPH is evolved from the enzyme catalysed reaction. Larger concentrations will give larger absorbances and the response time should be about 10 min per 0.01 mol/L D-glucose-6-phosphate in solution.

GROWING ZINC OXIDE NANORODS

With thanks to Christa Bünzli

Zinc oxide nanorods are useful due to their optical transparency, semiconductor nature and piezoelectric properties. Nanorod arrays grown onto ITO substrates are being investigated for use instead of titania as the light harvesting semiconductor layer for photoelectric cells (Law et al., 2005). The band gap is larger for n-type ZnO than anatase phase titania but still small enough for photovoltaic applications. When arranged as wires, the ZnO arrays can be infiltrated with a hole forming conductive polymer to act as a counter electrode. Because the surface area of the vertically arranged nanowires is so high compared to the counter electrode any light generated electrons in the ZnO will be passed rapidly into the polymer. This gets around some of the problems in titania cells where the electron/hole pair generated under illumination can recombine quickly or else experiences electrical states in the titania that behave as traps. As promising as this is, ZnO solar cells are not yet as efficient as titania based solar cells. ZnO nanorods arrays have been shown to operate as very sensitive humidity and gas detectors with rapid response times (Zhang et al., 2005). The high surface area of aligned forests of ZnO nanorods means that in the presence of water moisture or ethanol the conductivity of these arrays can increase greatly. Additionally, the bio-inert nature of ZnO makes it a good candidate material for biomolecular sensor systems as might be integrated with a laboratory on a chip device (Wei et al., 2006). One exciting application is the idea of using oriented ZnO arrays for the conversion of mechanical energy into electrical energy. This principal has been demonstrated by depositing ZnO nanorod arrays onto flexible substrates like plastic films and Kevlar fibres so that bending or friction against the material induces bending in the rods (Gao et al., 2007; Qin et al., 2008). Two fibres coated with ZnO nanorods wrapped around one another were agitated so that the rods on the surface brushed together. This process generates a charge which can be turned into useful electrical energy. The theoretical output of this type of fabric based energy scavenging generator is estimated to be in the range 20−80 mW/m^2. Though this may not sound like much, many devices only need the barest trickle of power to work. If scaled down further this approach can be used power micro and nanoscale devices. The rods are formed by a simple hydrothermal method where a solution of zinc salt precursor is heated under specific conditions so that the rods nucleate on the surface of the substrate.

In this section we will focus on using hydrothermal methods for making films of ZnO nanorod arrays on a wide range of substrates. Nanoscale ZnO is produced by growing a film of vertically aligned rods onto the surface of a conducting substrate such as aluminium foil or indium tin oxide (ITO) electrodes. ZnO arrays can be formed by template procedures but this leads to the potential complication of having to get rid of the template afterwards. Other template free synthesis methods include chemical vapour deposition, laser ablation, electrodeposition and hydrothermal growth (Yi et al., 2005). All of these methods normally require the deposition of a seed layer beforehand. Without a seed layer only a few unaligned nanorods will be observed. The choice of seeding method is determined by whichever gives you the best crystal orientation for perpendicular growth although the time and effort is also a large consideration. Laser ablation and RF magnetron sputtering to produce seed layers are quite technically involved and should only be performed if you have the relevant specialist equipment setup (Fuge et al., 2009). Good results can easily be obtained by dip or spin coating the substrate with a colloidal dispersion of ZnO nanoparticles or alternatively by thermally decomposing zinc acetate on the substrate surface (Govender et al., 2004; Greene et al., 2005). The thin film layer can be further annealed by heating in air to better adhere the oxide particles prior to the rod growth step. During subsequent hydrothermal growth in solution, ZnO crystals are deposited at the seeded nucleation sites forming the tetrahedral wurtzite polymorph and will naturally tend to orient on surfaces of a lower energy. For a ZnO nanocrystal this will be structure faces that have an equal number of zinc and oxygen atoms present such that there is no overall polarisation in the presented facet. These facets are the [211] and [010] faces which result in a hexagonal rod like ZnO crystal growing along the [001] axis of ZnO. During the initial stage of growth, the ZnO can grow in a range of directions but rods that encounter each other will be prevented from growing further. Only those rods that grow perpendicular to the substrate will continue to grow unhindered such that after a prolonged reaction time all of the nanorods appear to be standing up straight and packed together tightly. Nanorods grown in this way form a thick coating on the substrate next to the polymer or fibre layer then grow outwards into oriented arrays of regularly spaced hexagonal wires. Where the aspect ratio of the rods is high, and if they have enough space between them, bending one of the ZnO wires induces a polarisation in the crystal that generates a current. The bases of the nanorods are tightly packed together such that the continuous semiconductor interface layer can serve as a ZnO electrode if doped correctly or under the correct bias. The current generated by deformation can be raised an order of magnitude by first depositing a metallic electrode layer, like gold, onto the substrate or by depositing the rods on to a ZnO seeded layer of an indium tin oxide coated glass. This is a much more efficient charge carrier and also allows nucleation of the ZnO rods by the hydrothermal process (Fig. 4.34).

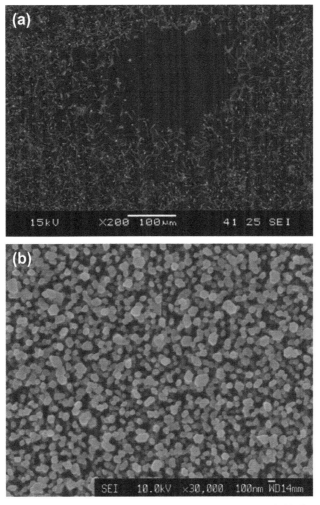

FIGURE 4.34 Scanning electron micrographs showing ZnO nanorods grown (a) hydrothermally on an unseeded indium tin oxide and (b) hydrothermally on indium tin oxide seeded with ZnO by pulsed laser deposition. *(Image used with the permission of Christa Bunzli)*

MAKING SEED LAYERS FOR THE HYDROTHERMAL GROWTH OF ZINC OXIDE NANORODS

Electroplating on a Conductive Substrate

Adapted from Yangping et al. (2010)

Electroplating a metal surface with zinc is widely used to protect metal components, mostly anything with iron in, from corrosion. It is also valued for the metal sheen the technique imparts. The process has been around since the nineteenth century and is easily implementable on copper, nickel and stainless steel metal foils. In this recipe zinc metal is deposited by the reduction of Zn^{2+} on a foil used as a cathode in a solution cell reaction. The corresponding anode in the cell is pure zinc which ensures the cation concentration remains constant during deposition. The foil will be coated with a layer of zinc which will further oxidise in air or else can be placed directly into a solution for ZnO nanorod growth. The rate at which the zinc layer grows will depend on the current and voltage placed across the cell. A slow deposition rate will achieve a smoother more homogenous film. A current of 20 mA/cm^2 applied for 30 min to an hour should give coatings around 10 μ thick.

For the electrodeposition of zinc onto a metal foil you will need:

1. A section of copper, nickel, aluminium or stainless steel foil. This will form the cathode of the cell. The foil dimensions are not that important so long as the entire foil section can be immersed into the deposition bath. Because we will want to analyse the results of the experiment by scanning electron microscopy we will describe using a foil piece of only a few centimetres square. You can use the same process to make much larger areas if you wish.
2. A zinc metal anode. This does not have to be the same dimensions as the foil but you will want to use approximately the same weight. If the entire anode disappears during the process then you used too little.
3. 1 L of a 0.5 mol/L Zn deposition solution to use as a stock. This can be prepared by dissolving 68 g (0.5 mol) of $ZnCl_2$ in 500 mL of water. To this solution add 190 g (2.5 mol) of KCl and 25 g (0.4 mol) of H_3BO_3. Adjust the pH of the solution to 5 using 1 mol/L HCL solution.
4. A beaker large enough to put the cathode and anode in so that they are immersed. A 250 mL beaker will do for this recipe. You may need a clamp or two to hold the cathode and anode in place.
5. An adjustable power supply with wires and crocodile clips. *Risk of electric shock. Make sure this is turned off before you begin.*

For the electrodeposition of zinc onto a metal foil:

1. Attach the cathode wire (negative terminal on the power supply) to the substrate foil using a crocodile clip. Attach the anode wire (positive terminal on the power supply) to the zinc foil.
2. Fill the beaker up with the Zn electrodeposition solution to a few centimetres short of the rim.
3. Dip the cathode and anode foils into the Zn solution making sure they do not touch one another.
4. Turn on the power supply to the desired current and leave for 30 min to 1 h. You should be able to observe the cathode foil becoming lighter in colour (especially if you are using copper) as the zinc is deposited on the surface.
5. After you are happy that a film has formed turn off the power supply and remove the sample from the deposition bath. Rinse the coated foil in water and dry before use.

Though you should be able to visually see the zinc coating you may wish to characterise how thick the zinc layer is. Under scanning electron microscopy the zinc will appear as a smooth film composed of packed hexagonal platelets. X-ray diffraction of the film should show reflections that can be indexed to hexagonally packed zinc metal (JCPDS number 04-0381).

THERMAL DECOMPOSITION OF ZINC ACETATE

Adapted from Greene et al. (2005)

If a substrate is thermally stable to high temperature then a seeding layer can be grown easily upon it by decomposing zinc acetate salt on the surface. This is effective for substrates like conductive indium tin oxide electrodes. A solution of zinc acetate in ethanol is drop cast onto the substrate and left for a few seconds before being rinsed with ethanol and dried. This process is repeated so that a very thin layer of the zinc acetate is adsorbed at the surface and then crystallises upon drying. The sample is heated to 350 °C whereupon the zinc salt oxidises. This substrate can then be used in a range of processes for ZnO nanorod growth. Interestingly the best results are obtained using only zinc acetate and not alternatives like zinc nitrate or zinc acetylacetonate. Although both of these salts decompose to form a ZnO layer, the rods that sprout from them are unaligned and randomly oriented. This hints at a possibly preferential alignment of the zinc acetate crystals on the substrate surface prior to the heating step which results in the oxide produced having the desired crystallographic orientation conducive to perpendicular nucleation. The seeding layer will form with the [001] planes parallel to the substrate surface as long as it is flat independent of the material. The temperature of decomposition is influential on the morphology of the rods and the salt begins to sublime at 175 °C and melts at 250 °C. If the salt is exposed to temperatures of lower than 350 °C, then thinner nanorods will be formed although the seeding mechanism fails completely if the temperature is not maintained above 200 °C.

To seed a substrate layer by thermal decomposition of zinc acetate you will need:

1. A substrate for coating. In this example let us use an indium tin oxide electrode though you could still use foil if you wanted. As before, the dimensions are not important but most ITO electrodes will be a few centimetres square.
2. 50 mL of a stock 0.005 mol/L zinc acetate in ethanol solution. This can be prepared by dissolving 0.054 g of $Zn(CH_3COO)_2 \cdot 2H_2O$ in 50 mL of ethanol.
3. Ethanol for rinsing the substrate with.
4. Clean tweezers for handling the substrate.
5. A pipette for applying solutions.
6. An argon line for drying the substrate surface with. Alternatively a dry air or nitrogen line will work. This should be fed through a tube possibly tipped with a disposable pipette tip.
7. A furnace pre-heated to 350 °C. You will need to be able to open this up and take samples in and out. It is recommended that you have a crucible, tongs and a ceramic plate for getting the hot ITO sample in and out of the furnace. Some laboratory furnaces are monitored by heat sensing fire detectors and opening the door while the oven is hot can sometimes trigger these. If this is the case you may need to temporarily turn them off.

To seed a substrate layer by thermal decomposition of zinc acetate:

1. Turn on the furnace and set it to 350 °C. Allow it to get to temperature before you proceed.
2. Use a pipette to coat the surface of the ITO substrate with drops of the zinc acetate in ethanol solution. Let the solution sit on the surface for 10 s then tip it off into a waste container.
3. Use another pipette to coat the surface with ethanol and then immediately tip it off into a waste solvent container.
4. Pass a steady and gentle stream of argon over the substrate to dry it.
5. Repeat steps 2—4 another 3—5 times. This should build up a very thin layer of zinc acetate crystals on the surface.
6. Place the dried substrate flat into a crucible using tweezers. Transfer the crucible into the hot furnace, close the door and leave for 20 min.
7. Use the tongs to remove the crucible from the furnace and place it on a ceramic plate to cool down.
8. Repeat steps 2—7 once more to ensure a seeding layer is built up. You should now have a ZnO seeded layer on the ITO electrode.

Scanning tunnelling microscopy can be performed to image the seeding layer and you should be able to observe plate like ZnO particles packed together around 5—20 nm in diameter. If you decide to perform STM make sure the substrate is very flat before you treat it. X-ray diffraction analysis of this layer should show a very strong reflection for the 002 plane as the seed crystals tend to orient this way upon decomposition. X-ray scan times will need to be relatively long in comparison to bulk ZnO due to the thinnes of the film. Rod arrays grown hydrothermally from the seed layer should be perpendicular to the surface and well aligned. Should you find that subsequent growth of the rods is patchy or that it does not happen at all then you may need to expose the substrate to the zinc in ethanol solution for longer. Alternatively you may attempt to skip the ethanol rinsing step. This should thicken the oxide layer eventually formed in the oven but you might experience slightly less film adhesion or it may end up too thick.

TO SEED A SUBSTRATE LAYER USING A COLLOID

Adapted from Pacholski et al. (2002)

This method is almost the same as the salt deposition except instead of a salt you will be adding preformed ZnO nanocrystals. The ZnO particles are formed by dissolving a zinc acetate salt in methanol under heating. A solution of methanol-containing potassium hydroxide is added which hydrolyses the zinc which eventually precipitates to form ZnO nanoparticles. Because the reaction is performed in the absence of a stabilising agent, the particles can sediment out of solution. This makes it relatively easy to exchange or filter out the methanol in the reaction solution which must be replaced and also contains potassium acetate as a by-product. When a dispersion of these particles is dried onto a substrate it will form a seed layer that can be used for hydrothermal nanorod growth. Colloid treated seed surfaces can sometimes result in a more disoriented rod growth direction because the nanocrystals may not be aligned in any particular axis with respect to one another after sintering. As with the salt treatment, heating the seeded substrate will form metal oxide bonds between the seeds and the under layer so that when the rods are formed they are adhered strongly to the substrate.

The concentration of zinc acetate used in the colloidal preparation method has a large effect on the morphology and size of the seed particles. The concentration in this recipe is low enough that the ZnO seeds formed will be nearly spherical in shape. When higher (ten times) concentrations of zinc salt are used in the reaction then rod structures will begin to form free in solution. If the as formed colloid is refluxed for many hours in an ageing process then the particles will remain roughly spherical but become slightly larger with an average diameter of 5 nm via Ostwald ripening. However, if the as formed colloid of spherical particles is concentrated by evaporation first and then refluxed for a number of hours then an Ostwald ripening process will occur that generates long, anisotropic wurtzite crystals. The concentration of the particles allows them to aggregate enough to form chains of particles. The Ostwald ripening process then fills in the 'links' to form solid rod. The dimensions of these rods can be as long as 100 nm and 15 nm wide.

To seed a substrate layer using a colloid you will need:

1. 125 mL of methanol in a 500 mL three necked round bottomed flask with a quick fit aperture for a reflux condenser in the main neck and a rubber septum and a glass stopper in the side necks. Place a magnetic stir bar inside.
2. A heating mantle with an oil bath on top and magnetic stirring.
3. 2.195 g (0.01 mol) of $Zn(CH_3COO)_2 \cdot 2H_2O$.
4. 65 mL of a 0.03 mol/L solution of KOH in methanol. This can be prepared by dissolving 1.6833 g of KOH into 65 mL of methanol.
5. At least 200 mL of extra methanol for rinsing and redispersing the colloid in. You will need a Pasteur pipette for removing methanol with.
6. A 100 mL dropping funnel with a tap and a quick fit end that can fit into the side neck of the round bottomed flask. This will be used to deliver the KOH in methanol solution dropwise. You may need clamps to hold this in position. Alternatively you can use a needle and syringe injected through the rubber septum.
7. A reflux condenser that fits in the main neck of the round bottomed flask and all the appropriate tubes and clips.
8. The substrate to be coated.
9. A furnace capable of reaching at least 350 °C. You will also need tongs and a crucible for transferring the sample in and out.
10. A pipette for drop casting the colloidal solution onto the substrate. Optionally, a spin coater can be used for depositing a thin film of colloidal particles on the surface.

To seed a substrate layer using a colloid:

1. Clamp the round bottomed flask containing the methanol in place above the heating mantle and sitting in the oil bath. If you are using a dropping funnel to add the KOH in methanol then fit this to a side neck, with the tap closed, in the place of the rubber septum.
2. Fit the reflux condenser to the round bottomed flask and clamp it in place. Make sure the water intake and out take tubes are clamped in place. The water flow through the condenser does not need to be fast.
3. Turn on the heating mantle and set it to reach 60 °C.
4. Turn on the magnetic stirrer and add the zinc acetate salt through the side neck into the methanol under vigorous stirring. Let it dissolve.
5. Slowly open the tap from the dropping funnel so that drops of the KOH in methanol solution begin to drop into the zinc solution. Alternatively inject the solution drop by drop through the rubber septum. This should be added over the course of 10 min.
6. Once the base has been added allow the solution to stir for 2 h at 60 °C. During this time it should become a milky white colour.
7. Turn off the heating mantle and magnetic stirring and allow the sample to cool. Once it is cool, turn off the reflux condenser. Leave the solution to stand overnight so that the ZnO particles sediment to the bottom of the round bottomed flask.
8. Potassium acetate salts are soluble in methanol and these must be removed before the colloid can be used. Once the white ZnO particles have sedimented, use a Pasteur pipette to gently remove the top solution which will contain the soluble potassium acetate. Dispose off this in the correct solvent waste container.
9. Add 100 mL of clean methanol to the round bottomed flask and redisperse the ZnO particles. Allow them to sediment once more and then remove the top solution. Add another 100 mL of clean methanol.
10. You will now have a stock colloid of ZnO ready to be used as a seeding layer on a substrate. It can be used as formed or you may wish to concentrate it. This can be done by simply letting the methanol evaporate away. It is advisable to check the quality of the as produced colloid by transmission electron microscopy before proceeding further. You should see particles around 3 nm in diameter that may look slightly elongated.
11. Use a pipette to drop cast the colloid on the substrate. Allow the alcohol to evaporate away. Alternatively a spin coater can be used to produce the thin layer. The sample can be stored until needed. You may wish to try performing the hydrothermal growth of the nanorods at this point without the annealing step.
12. Heat treat the coated substrate in a furnace at 350 °C for 20 min to affix the particles. The furnace does not need to be on when you start the process. You may find different heating rates or exposures produce variable results in the further growth of the nanorods. Turn off the furnace and allow the sample to cool to room temperature.

HYDROTHERMAL GROWTH OF ZINC OXIDE NANORODS

With thanks to Christa Bünzli. Adapted from Vayssieres (2003)

Once you have a seeded layer on a substrate, then you can begin to grow the nanorods from them. Hydrothermal synthesis normally involves heating a solution containing both precursor and substrate in a liquid under a fixed volume. It is similar to a normal reflux procedure only instead of using a reflux condenser to allow liquid to return to the vessel we seal everything in tightly. This can produce conditions of a slightly (or greatly depending on the heat) elevated pressure which can shift the solubility product of crystal formation. Under an assortment of pH, temperature, pressure and solvent mixtures unique crystal morphologies can be formed. Often an additive is also added to the reaction solution to influence the direction of crystal growth. In a hydrothermal synthesis of ZnO a zinc salt precursor is encouraged to oxidise in the presence of an amine under basic conditions. In the presence of water the Zn^{2+} ion becomes hydrolysed to form a transient short lived species of zinc hydroxide. At a pH greater than 9 and in favourable conditions this intermediate can undergo a condensation reaction to form a water molecule and solid ZnO (Fig. 4.35).

$$C_6H_{12}N_{4l(l)} + 6H_2O_{(l)} \rightarrow 6CH_2O_{(l)} + 4NH_{3(aq)}$$

$$NH_{3(aq)} + H_2O_{(l)} \leftrightarrow NH_4^+(aq) + OH^-_{(aq)}$$

$$Zn^{2+}_{(aq)} + 2OH^-_{(aq)} \leftrightarrow Zn(OH)_{2(i)} \leftrightarrow ZnO_{(s)} + H_2O_{(l)}$$

A wide range of amine precursors can be used including ethylenediamine, triethanolamine, diethylenetriamine and hexamethylenetetramine (HMT). The latter is the most often used in this type of synthesis and in aqueous solutions the HMT breaks down to form formaldehyde and ammonia which increases the concentration of hydroxyl ions in the solution. The rate at which the HMT decomposes in water is heavily influenced by the pH and temperature and ranges from minutes to hours. Because of this the length of time an HMT and Zinc solution is left to stand before adding the substrate will determine the end structure you produce. Over time the growth of the rods will deplete the reactants in the reagent solution and no further growth will occur. To get around this the solution can be exchanged for a fresh batch which will induce further growth. This comes at a cost of rod density on the surface as well as increasing the diameter of the rods. Eventually this will lead to the rods fusing together so repeated immersion cannot be performed more than a few times.

The aspect ratio of the rod can be tuned by adjusting the pH and temperature of the reaction. Performing the hydrothermal synthesis at 40 °C produces shorter rods while performing the same reaction at 90 °C results in rods of the same diameter but with a greatly enhanced length. Alternatively, further additives or surfactants can be used to adsorb to specific crystal surfaces and direct the crystal growth. Adding 5–7 mmol/L poly(ethyenimine) will result in rods with aspect ratios of 125 (Law et al., 2005). Rods grown without seeding will tend to be a few hundred nanometres in width and can grow between 1 to 25 μ in length under the right conditions. With a seed layer on the substrate the diameter of the nanorods is decreased dramatically into the tens of nanometre region. In this recipe, using a seeded substrate, you should get a layer of zinc oxide nanorods 50–300 nm in diameter and between 1–2 μ in length. The rod density can be as high as 130 million rods per centimetre squared.

To Grow ZnO Nanorods Hydrothermally you will need:

1. The ZnO seeded 1×2 cm^2 ITO coated substrate. You can attempt to treat more than one substrate at a time. It depends how many you can place comfortably in the solution.
2. 40 mL of a 0.1 mol/L solution of hexamethylenetetramine. This can be prepared by dissolving 0.56 g of HMT into water and making up to 40 mL.
3. 40 mL of a 0.1 mol/L Zinc Nitrate solution. This can be prepared by dissolving 1.18 g of $Zn(NO_3)_2 \cdot 6H_2O$ in water and making up to 40 mL.
4. A 100 mL Schott bottle with the lid and a magnetic stir bar.
5. A heating mantle with magnetic stirring. A mineral oil bath should be sat on top.
6. A clean glass microscope slide and a roll of PTFE tape.

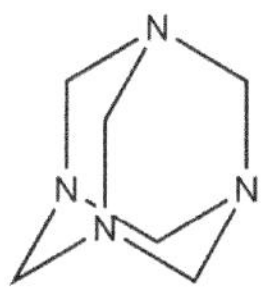

FIGURE 4.35 The chemical structure of hexamethylenetetramine (HMT).

FIGURE 4.36   Schematic of the setup for the hydrothermal growth of ZnO nanorods on a seeded ITO substrate.

7. Water for rinsing and an argon line for drying the slides.

   To Grow ZnO Nanorods Hydrothermally:

1. Turn on the heating mantle and set it to 90 °C.
2. Add the zinc nitrate solution and the HMT solution to the Schott bottle and add the stir bar. Screw on the lid.
3. Turn on the magnetic stirring and allow the solution in the bottle to equilibrate to 90 °C.
4. While you wait place the substrate to be treated on to the glass slide with the seeded side facing up. Position the sample about 2 cm from one end of the slide. Use the PTFE tape to affix or strap the substrate in place making sure the seeded areas are exposed.
5. Once the solution is up to temperature turn off the magnetic stirring and open the Schott bottle. Dip the glass slide into the hot solution. Rest the bottom on one side with the slide tilted over so that the top rests on the opposite side. The substrate should have the seeded side pointed at around a 45° angle towards the bottom as in the diagram. Reseal the bottle (Fig. 4.36).
6. Keep the magnetic stirring off and allow the sample to sit in the heated solution at 90 °C for 4 h.
7. Remove the sample from the reaction solution and remove the substrate from the glass slide. Rinse the sample with distilled water and then dry under an argon flow.

   ZnO nanorods can continue to grow up to 1200 nm in length in one immersion and even further by repeated immersion in fresh reagent solution. After a number of cycles they will begin to coalesce together. Lowering the initial concentration of the zinc reagent will produce rods of a lower diameter but they will be correspondingly shorter. The same analysis methods that you used with the seeding methods can be employed here. An X-ray diffraction pattern of the film should give strong reflections for the ZnO wurtzite structure. You should see that the [002] reflection is much stronger due to the preferential growth along this axis from the surface. For scanning microscopy analysis you may have to fracture or cut a substrate so that it may be viewed side on. Comparing the diameters and lengths of the grown rods will let you determine the aspect ratio. Further thermal annealing of the nanorods arrays to 350 °C should not induce any observable morphological change if this is required. Conductive atomic force microscopy and piezo response force microscopy can be performed which will allow you to view electrical activity by deforming the rods. This can sometimes be tricky if you have to image in contact mode as long rods will deform and can generate charge at the same time (Fig. 4.37).

CADMIUM SULFIDE, SELENIDE AND TELLURIDE QUANTUM DOTS

With thanks to Bo Hou

   The difference in behaviour of materials observed at the nanoscale is very clearly demonstrated in the size dependant optical effects of semiconductor nanocrystals formed from group II and VI elements. As a bulk powder, cadmium selenide appears red but if formed into a colloid it can appear as a range of colours. This effect is even more pronounced if the colloid is viewed under strong light of a particular wavelength as the particles will fluoresce in a colour related to the particle size. CdSe, CdS, CdTe, InP and InAs

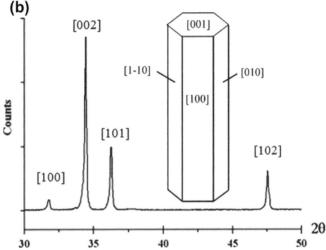

FIGURE 4.37   (a) A scanning electron micrograph of aligned and oriented ZnO nanorods grows on an ITO substrate. (b) The expected powder X-ray diffraction pattern for the rods which can be indexed to the ZnO wurtzite structure (JCPDS card number 36-1451). *(Image adapted from Vaysierres 2003)*

all demonstrate fluorescence under various wavelengths depending on the composition and nanoscale morphology. The relationship between the optical interaction of a semiconductor and its size is related to the band structure and how this alters as the size of a particle is changed. A bulk crystal of cadmium selenide arranged in the hexagonal wurtzite structure is a wide band gap semi-conductor which requires a photon of around 1.7—2.4 eV (red light of around 700 nm) in energy in order to promote an electron from the valence band into the conduction band (Bawendi et al., 1990; Biju et al., 2008).

Valence and conduction bands are a way of describing the bulk molecular orbital structure of a semiconductor. Valence bands are composed of the electronic orbitals in the crystal that are occupied by electrons. The uppermost valence band is called the highest occupied molecular orbital (HOMO) and for cadmium selenide this will be a $4p$ orbital on a selenium atom. The conduction band is a collection of unoccupied orbitals that can receive an excited electron and 'hold it' until it falls back into the valence band. The lowest unoccupied molecular orbital (LUMO) in a semiconductor will be the bottom of the conduction band and in cadmium selenide this is a cadmium $5s$ orbital (Fig. 4.38).

In a semiconductor the band gap is the energy level separation between the LUMO and the HOMO. Band structures are determined by looking at the molecular orbital diagram of an individual molecule and then adding more molecules and seeing what happens. Though the electronic transitions of an individual molecule are discrete, collections of them together and in different orientations blur the allowed transitions so that they are spread over a well defined range of allowable energies when in bulk. This is defined as the band structure and it can be seen by working in the opposite direction from a bulk collection of orbitals to discrete molecules that clusters of CdSe nanoparticles of a few hundred or thousand molecules have quantised electronic transitions. This is a major contributing factor to the quantum confinement effect and the size dependant optical behaviour of the quantum dots.

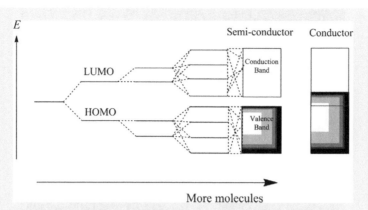

FIGURE 4.38 Molecular orbital energy levels in single molecules form bands as the number of molecules builds up. Eventually this forms bands consisting of a valence band with the HOMO at its top and a conduction band with the LUMO at its bottom.

This interaction with light produces an electron-hole pair in the crystal called an exciton which can be viewed analogous to a hydrogen atom in terms of the energy levels it can posses. In a similar manner to the derivation of a $1s$ orbital, the exciton can be described by a wavefunction much like an electron orbit where the atomic nucleus is replaced by a positive hole. The distance of separation between the electron and the positive hole that remains is called the exciton Bohr radius and may be pictured as a temporary nucleus surrounded by an s-orbital conforming to the Schrödinger equation. The exciton Bohr radius, α_b, can be calculated from the following formula where m_e and m_h are the effective mass of the electron and the hole respectively. These effective masses are quoted as a fraction of an electron mass and for CdSe this is 0.19 for m_e and 0.81 for m_e:

$$\alpha_b = \frac{h^2 \epsilon_{CdSe} \epsilon_0}{\pi e^2} \left(\frac{1}{m_e} - \frac{1}{m_h} \right)$$

For a large crystal there is no real limitation on the behaviour of the exciton or on the wavefunction it assumes as an electron-hole pair. An observer generating excitons in a crystal using a strong light of the correct wavelength may see conduction in the material as the electrons occupy the conduction band or else the exciton recombines into its ground state. As the size of the crystal approaches the exciton Bohr radius an excited electron will find itself confined within the dimensions of the particle and the allowable number of energy states the exciton can have is decreased. The result is a select, or quantised, number of allowable energy states as opposed to a continuous band structure and these will be higher in energy than the LUMO in the conduction band and very discrete. Effectively the particle itself behaves as though it were an atom when it gets close to or below the exciton radius in size. In contrast to an atom, a semiconductor particle can be elongated in dimension and shape so that the electronic interactions with incident light at a given angle to the particle can be controlled. Normally the peak light absorbance of a zero dimensional (spherical) semi-conductor colloid becomes blue shifted and very sharp in comparison to the bulk material. More pronounced and complex effects can be observed in rods, wires and thin films and this makes semiconductor nanomaterials of use in optoelectronic applications such as solar cells, laser diodes and light based logic operations (Fig. 4.39).

A mathematical description of quantum confinement effects was first explored by Brus who related the blue shift seen in the absorbance of semiconductor quantum dots to the size of the particle (Brus, 1986). Using this relationship the peak adsorption wavelength can be used to give an estimate of the particle size according to the following formulas:

The energy of a photon absorbed by the nanoparticle is:

$$E_{Photon} = \frac{hc}{\lambda_{max}} = E_{Bandgap} + \Delta E_{Confinement}$$

where $E_{bandgap}$ is the normal band gap energy of the semiconductor material and $\Delta E_{confinement}$ is the change in energy observed in a nano-particle form. $\Delta E_{confinement}$ is therefore the shift in energy compared to the bulk band gap as the radius of the particle decreases. This term is related to the radius by the formula:

$$\Delta E_{confinement} = \frac{h^2}{8r^2} \left(\frac{1}{m_e} + \frac{1}{m_h} \right) - \frac{1.8e^2}{\epsilon_0 r} + \text{small terms}$$

where h is Planck's constant, r is the particle radius and the term after the subtraction describes a small contribution from charge interactions where ϵ is the permittivity of free space and e is the charge of an electron. The smaller terms describe negligible electronic interactions. By inspection of the formula it can be seen that as the particle radius decreases the band gap gets larger (Winkler et al., 2005).

FIGURE 4.39 Photo demonstrating the size-tunable fluorescence properties and spectral range of quantum dot dispersions having identical concentrations of CdSe versus the particle core diameter. All samples were excited at 365 nm with a UV source.

The size and shape related photoluminescence of nanoparticles makes them extremely useful as fluorescent probes in biological microscopy and spectroscopy (Bruchez, 1998; Chan et al., 2002; Medintz et al., 2005). For fluorescence microscopy a photo-luminescent dye is added to a cell which will bind specifically to a target material. This may be a protein, DNA or enzyme that may be the marker of interest which will bind to the fluorescent probe. This allows the target to be visualised by irradiating the sample with a high energy light, normally in the near UV or blue range, which promotes the electronic structure of the dye into an excited state. When this state decays the energy is released as a lower energy photon of a specific colour. Organic chemical fluorophores experience problems with photobleaching as the powerful UV light required to generate light emission strong enough to be imaged often results in the dye molecule falling apart. In the excited electronic state these dyes are also more reactive with the surrounding environment and are chemically decomposed to non-emissive chemicals the longer a sample is imaged. If the contrast in the image is low between the emissive dye and the background fluorescence of the sample then increasing the intensity of light may degrade the dye with little increase in the light output. This means that cells stained with a chemical dye might only be imaged once and then cannot be used again. Another problem with chemical dyes is that they have broad emission and excitation wavelengths meaning that you cannot image more than one or two target substances at once. A good example of this is in live/dead staining where green and red fluorescent dyes denote the presence of live cells and dead cells respectively. Red and green can easily be distinguished but it might be more useful to have a full spectrum of colours that can be used to probe a multitude of biological species all at the same time on the same side. These complications can be overcome by the use of quantum dots instead of chemical dyes. Semiconductor nanoparticles are highly resistant to chemical degradation under illumination such that the particle will not bleach from strong UV exposure. For the experimentalist this means no limit on how long can be spent getting the imaging setup correctly and that slides can be stored and returned to at a later date. As the emission wavelength is size dependant a full spectrum can be engineered so that multiple entities can be imaged on the same slide as long as the surface of the quantum dot probe can be functionalised to stick to them.

This is not to say that using nanoparticles as fluorescent biological probes is not without some complications. As bright as they are compared to chemical dyes much research is undertaken to maximise the quantum yield of the fluorescence. As a bulk material approaches the nanoscale the surface area to volume ratio becomes huge meaning that interactions with the surface of the nano-crystals significantly influence their behaviour. For emission to occur the surface of the nanoparticle must be free of any chemical 'traps' that might adsorb the energy of the exciton and quench the fluorescence. Surface crystal defects will also result in quenching as the energetic electron can decay down non-emissive pathways back to the ground state. This problem has been compensated for by making a shell around the nanoparticle which acts as a shield to the loss of exciton energy and prevents unwanted chemical surface interactions by passivation of the particle surface. The shell is composed of another semiconductor material, normally zinc sulfide, with a wider band gap than the fluorescent core. The wider band gap means that the mobile electron generated in the exciton cannot tunnel directly into the ZnS conduction band where it could be lost to a non-radiative process. This type of coating is normally performed only in the high temperature synthetic routes for highly monodisperse particles.

For applications where the particles are to be used in-vivo the toxicity of using cadmium, selenium and tellurium components is of concern. Though the particles are photolytically stable they may not be proofed against chemical attack from normal cellular pathways or variations in cellular pH. Once more, bio-inert shells have provided some alleviation of this issue but for anything that is to be used in the body the eventual result of breakdown and the associated risk must be considered before use. Polymer coating with poly(ethylene glycol) has been shown to greatly increase the retention time of quantum dot probes used for in-vivo imaging (Gao et al., 2004). Polymer coatings can be used instead of ZnS coatings without significant loss of particle intensity. Some discussion of the differences in coatings is provided later on.

In this section we will explore a range of methods for the production of semiconductor quantum dots and the various methods of stabilising them including the attachment of targeting groups for biological imaging.

A QUICK WATER BASED METHOD FOR THE PREPARATION OF CADMIUM SULFIDE NANOPARTICLES

Adapted from Winkler et al. (2005)

This method is a simplified process for the formation of cadmium sulfide dots in the presence of a polymer in methanol. It is a good starting point for getting a quick idea of how these materials behave or as a laboratory demonstration of quantum size effects. The method is advantageous in that the reaction is optimal at 10 °C although the particles produced are larger than those produced at high temperature. Particle diameters of around 20 nm are achievable though this can vary depending on the polymer used to stabilise the colloid and they will not be very monodisperse. In this recipe cadmium nitrate is added to sodium sulfide in the presence of a low molecular weight branched polyethenimine as given in the formula below.

$$Cd(NO_3)_{2(aq)} + Na_2S_{(aq)} \rightarrow CdS_{(s)} + 2NaNO_{3(aq)}$$

The amine groups present in the polymer act to cap the cadmium sulfide particles and render them immediately soluble in aqueous solutions. Furthermore the amine coordination at the surface fills in the electronic 'traps' that would otherwise cause an exciton to decay by non-radiative processes. Therefore using a polymer with a large number of amine groups can boost the quantum yield and the overall fluorescence of the particles produced. Though this reaction is performed in methanol it can also be run in water meaning that you can attempt this nucleation process in the presence of biological templates. Should you choose to use water then it will need to be thoroughly degassed to remove oxygen before use.

To prepare a poly(ethylenimine) stabilised cadmium sulfide colloid in methanol you will need:

1. At least 50 mL of methanol that has been thoroughly degassed using an argon or nitrogen line. *This methanol is to be used for making up the following solutions.* If you decide to use water as the reaction medium then this too should be degassed before use. A nitrogen line that can be run through a needle is also required for degassing the reaction flask.

2. 10 mL of a 4×10^{-4} mol/L solution of poly(ethylimine) in methanol. The amount you dissolve will depend on the molecular weight of the polymer used. In this recipe we will use a PEI with a molecular weight of 750,000 which means you will need to dissolve about 3 g of the solid polymer into methanol. This could take some time. Another problem you may encounter is that the polymer will be provided as a 50% stock solution in water. If you are performing this synthesis in water then you only need to dilute the stock further. Otherwise you will need to use a rotary evaporator to dry the polymer before redissolving it in methanol.

3. A 50 mL three necked flask with a magnetic stir bar. Two of the necks should be fitted with rubber septums and the main neck plugged with a PTFE stopper. You will need a clamp for holding the flask in position.

4. A water bath maintained at 10 °C. The reaction will proceed as high as 25 °C but the brightness of the illuminated product is optimal at 10 °C. This should be placed on top of a magnetic stirrer.

5. 5 mL of a 2×10^{-3} solution of cadmium nitrate in methanol. This can be prepared by dissolving 0.030 g of $Cd(NO_3)_2 \cdot 4H_2O$ into 5 mL of methanol in a sealable vial. This means there will be a small amount of water in the cadmium solution but will not be detrimental to the reaction. Prepare the solution fresh prior to reaction.

6. 5 mL of a 2×10^{-3} solution of sodium sulfide in methanol. This can be prepared by dissolving 8 mg of Na_2S (or 24 mg of $Na_2S \cdot 9H_2O$) in to 5 mL of methanol. It is probably easier to make a stock in larger quantities to get more accuracy in the amount of Na_2S weighed out.

7. Two graduated syringes with needles for the controlled injection of the precursors. You may need to use a clamp for each syringe to hold them steady for the injections.

To prepare a poly(ethylenimine) stabilised cadmium sulfide colloid in methanol:

1. Add the 10 mL of poly(ethylemimine) solution to the 50 mL three necked flask. Replace the PTFE stopper in the top and clamp in place in the water bath. Turn on the magnetic stirring and allow it to stand for 5 min so the solution cools to temperature.

2. Fill one syringe with the cadmium in methanol solution. Fill the other with the sulfide solution. Push the respective needles through the septums in the side neck. **Do not inject the solutions yet!**

3. Add 0.5 mL of the cadmium solution to the stirring poly(ethylenimine) in methanol. Count to 5 then add 0.5 mL of the sulfide solution and count to five again. Repeat this process adding each solution respectively until both the syringes are empty. Leave the solution to react for a further five minutes during which time you should observe a colour change. If you add all of the cadmium to the solution at once the particles will be far larger and may result in a less stable colloid. After this point you will have a colloid ready for analysis.

The visible colour of the colloid depends on the presence of water and the speed at which it was added as this will influence the size of the nanoparticle. The colloids will appear visually yellow immediately after synthesis but will redshift over time and may appear different after 30 min as the particles continue to grow. A redshift in the photoluminescent spectra will also be visible for both the broad excitation and adsorption bands after ageing. The broadness in the adsorption spectra indicates that the product polydisperse compared to more high temperature synthesis routes. The general intensity of the quantum dots will be low due to the presence of defects at the particle surface. This can be improved by the use of a dendrimer in the place of a polymer where the high level of branching forms a denser shell about the particle which prevents quenching. You can also try using poly(vinyl alcohol) as a stabilising agent which has the added advantage of allowing you to dry the solution down into a free standing film. The colloid can also be encapsulated in a glass monolith by the direct mixing 1 part of polymer stabilised colloid with 1 part pH 3 HCl in water and 1 part tetramethylorthosilicate. Mix them in a disposable cuvette and allow the gel to age in the dark for 24 h. The glass that forms should emit blue or yellow light quite strongly under UV illumination.

COATING THE TMV VIRUS WITH CADMIUM SULFIDE

Adapted from Shenton et al. (1999)

The formation of quantum dots under aqueous conditions and at room temperatures means that a biological template can be present. In this variation of the water based synthesis the tobacco mosaic virus will be used as a template in the formation of cadmium sulfide. The virus is soaked in a buffer solution of cadmium chloride before being reacted with hydrogen sulfide gas. The cadmium cations coordinate to the surface of the virus through glutamate and aspartate groups on the surface which forms a nucleation site for the formation of a cadmium sulfide particle upon reaction with the sulfide gas. Depending on the length of time the cadmium coordinated virus is exposed to the gas you will observe small particles coating the surface or a granular shell in the shape of a tube under an electron microscope.

To make a CdS nanoparticle coated tobacco mosaic virus you will need:

1. A 30 mg/mL stock solution of tobacco mosaic virus in a 10×10^{-3} mol/L sodium phosphate buffer adjusted to pH 7. This can be stored in a fridge or freezer when not in use.
2. 1 mL of a 10×10^{-3} mol/L cadmium chloride solution in a solution buffered with 10×10^{-3} mol/L solution of (2-([tris(hydroxymethyl)methyl)]amino-1-ethane-sulfonic acid) (TES) and adjusted to pH 7.2. This can be prepared as a stock by dissolving 0.183 g of $CdCl_2$ in 90 mL of water. Then add 0.229 g of TES to the solution and adjust the pH to 7.2 using a pH meter. Make the resulting solution up to 100 mL.
3. A bell jar with a gas tap fitted and an exhaust. *This experiment must be performed in a fume hood! You will be working with H_2S which is an extremely toxic and flammable gas.* The exhaust from the bell jar can be fed by a pipe through an excess of 1 mol/L sodium hydroxide solution which will not only let you monitor the flow rate by bubbles but will also remove H_2S leaving the bell jar by forming Na_2SO_4.
4. A bottle of H_2S gas with a pressure valve and flow meter fitted. *This is highly toxic and flammable. Hydrogen sulfide smells like rotten eggs and will bind to haemoglobin in blood in a similar way to cyanide. This must be worked within a fume hood.* A very small amount of this gas is required and it is also possible to generate it *in-situ* by adding Na_2S to an excess solution of 1 M HCl in a petri dish. Keep in mind that using the sodium salt will afford you less control over the nucleation rate.
5. A glass slide with parafilm stretched over the surface. The parafilm will maximise the surface area of the droplet containing the virus template so that sulfide gas can permeate in.
6. Volumetric pipettes capable of microlitre amounts for making up solutions and for adding droplets of virus suspension to the parafilm. You will need a number of small microcentrifuge tubes for mixing small volumes in.
7. Although optional it is recommended to have an air or inert gas line to flush through the bell jar to get rid of any remaining H_2S.

To make a CdS nanoparticle coated tobacco mosaic virus:

1. Take 10 μL of the stock TMV solution and add this to 290 μL (0.29 mL) of the $CdCl_2$ solution. This solution has a final virus concentration of 0.1 mg/mL. The $CdCl_2$ will not degrade the virus and you may find that the solution can be used after many months of storage.
2. Place the glass slide into the bell jar and add 20 μL aliquots of the virus solution to the surface using a volumetric pipette. Add as many as you feel you might need. They should form droplets standing proud of the surface.
3. If you are using sodium sulfide in acid then stand a beaker of acid in the bell jar next to the slide and add the sodium sulfide. Do not add it rapidly. Seal the bell jar so that the atmosphere inside become saturated with the evolved H_2S. If you have a direct feed of gas then connect it to the bell jar and slowly add the gas until you can see the gas begin to bubble slowly through the sodium hydroxide solution.
4. Depending on the flow rate you should begin to see the droplets begin to turn yellow as the cadmium sulfide forms. There should be no visible precipitate within the droplets. Allow the droplets to age in the H_2S atmosphere for up to one hour.

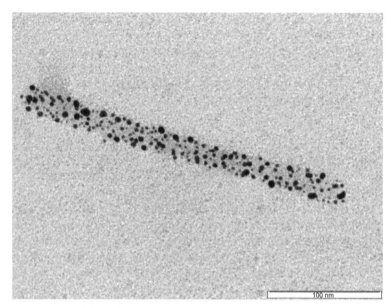

FIGURE 4.40 Transmission electron micrograph of CdS-mineralised TMV particles showing dense mineral coating of the external surface of the virion, scale bar = 50 nm. Inset: Corresponding EDX spectrum (Ni and Cu peaks arise from TEM grid and specimen holder, respectively).

5. Turn off the H_2S gas feed or alternatively remove the dish containing the acid and Na_2S. Before you open the bell jar it is a good idea to flush it through with some air. You should now have droplets containing CdS coated virus ready for analysis. The droplets can be removed and collected in a microcentrifuge tube for further storage.

Photoluminescence and UV-visible spectroscopy on the harvested droplets are quick methods to asses the optical properties of the formed CdS cylinders. Carbon coated nickel grids must be used for transmission electron microscopy as microscopes fitted with elemental X-ray analysis will observe a similar emission peak for the copper and cadmium on standard copper grids. You should be able to see rods of the TMV 18 nm in diameter coated in an electron dense crust of CdS composed of 5–10 nm particles around 15 nm thick. The formed rod structures will be much longer in length than the original virus particles and the act of coating the virus promotes further end to end assembly. A selected area electron diffraction pattern of the crust should give the following d-spacings which can be indexed to the following reflections for a cubic zinc blende crystal structure (Fig. 4.40):

Index plane	111	220	311	311	422
d-spacing/Å	3.36	2.06	1.76	1.33	1.18

PREPARING CADMIUM SELENIDE OR CADMIUM TELLURIDE QUANTUM DOTS WITH WELL DEFINED SIZE CONTROL

With thanks to Bo Hou, Carmen Galen, David Benito Alfonso and David Fermin. Adapted from Peng et al. (2000); Peng and Peng (2001); Dabbousi et al. (1997); Murray et al. (1993)

This method of producing cadmium chalcogenide nanoparticles is more involved than the previous recipe and involves the use of a Schlenk line and high temperature. The increase in experimental difficulty is offset by the excellent size control and monodispersity of the product. This high temperature route provides quantum dot semiconductor crystals that will exhibit photoluminescence in a tight excitation and emission range. A coating of zinc sulfide further enhances the quantum yield by preventing the exciton being quenched. These well defined size dependant optical properties make this type of quantum dot extremely attractive for biological tagging applications. The exterior of the nanoparticles can be functionalised to stick to specific proteins, strands of DNA or cellular materials so that they can be tracked or imaged. Furthermore the emission spectrum of the particles can be engineered such that a range of different coloured particles can be used as tags at the same time. This means that many different biological targets of interest can be investigated at once using a fluorescent microscope. In this recipe cadmium selenide nanoparticles are formed by dissolving selenium in trioctadecylphosphine and then reacting this with a solution of cadmium oxide dissolved in trioctadecylphosphine oxide at 300 °C. As the cadmium selenide nucleates the bulky unoxidised trioctadecylphosphine molecules coordinate to the surface and stabilise the particle. Because the trioctadecylphosphine is easily oxidised the reaction must be performed under argon. Some variations of this recipe use tributylphosphine as the stabilising agent but this has a nasty habit of catching fire spontaneously in air. For the sake of increased safety we will be using the octadecyl form which can be exposed to air without the same fire risk. The size of the nanoparticles is determined by how long they are allowed to age for in the heated solution. The longer you leave the system after the addition of selenium then the larger the particles will be. Rapid quenching of the system stops the ageing process and will halt particle growth. Quenching the reaction after a minute will produce sub 5 nm particles and quenching after 5 min will produce particles of 20 nm. You will be able to monitor the progress of the reaction by the colour that forms. 20 nm CdSe particles will appear black and by this point you should quench the reaction. Determining the exact size during the reaction is difficult though a rapid assessment can be made semi quantitatively by taking an absorbance spectra. Sadly this is not without complications as it can take a minute or so to run the scan during which time the particles are still growing, even worse you might be trying to take an aliquot at a temperature of 300 °C with a syringe and could get burned! The impracticality of this can be set aside and instead you should try the method a few times along with reviewing similar preparations in the literature and see which quenching time suits you. After the quenching procedure the formed quantum dots are precipitated out of the trioctadecylphosphine oxide solution by the addition of methanol and then collected by centrifugation. The methanol induces charge on the surface of the sterically stabilised quantum dots and induces flocculation. Any triocytlphosphine not coating the surface of a nanoparticle will be converted to the oxide form at this stage and is removed with the rest of the solvent by centrifugation. The collected particle sediment can then be washed and redispersed in chloroform.

The second stage of the reaction is the coating of the formed colloidal particles with a zinc sulfide coating. The CdSe particles are once more mixed with trioctadecylphosphine oxide and the chloroform removed by vacuum. Separately zinc stearate dissolved in toluene is reacted with the sulphur containing liquid hexamethyldisilazane. Mixing this solution with a dispersion of CdSe nanoparticles under argon and stirring at 180 °C results in the formation of a zinc sulfide layer over the particle surface. Once more the product is centrifuged and redispersed in chloroform. At this point the particles are still stabilised by trioctadecylphosphine and are stable as a colloid in hydrophobic non polar solvents. In order to make them suitable for use as biomarkers the surface ligands need to be exchanged for a molecule that will make them water soluble and that has a functional group to which target receptors can be attached. This is accomplished using mercaptoacetic acid which binds to the zinc sulfide layer displacing the trioctadecylphosphine. The carboxylic acid moiety can then deprotonate in water to charge stabilise the particle or can be chemically modified further. A general route for the coupling receptors and proteins to mercaptoacetic acid stabilised quantum dots can be found at the end of this section.

This recipe can be adapted to produce cadmium telluride which generally gives redshifted emissions under photoluminescence in the red and near infra red ranges. Simply replace the molar amount of selenium in the recipe with tellurium. The tellurium will react faster in the formation of crystals and so it must be diluted before adding to the cadmium solution to prevent the particles becoming excessively large. A good dilution agent is octadecane which is a non coordinating solvent in the reaction. The exact amount of oleic acid used for dilution will depend upon the system so trial and error will let you work out how much produces desirable results.

The key to getting high-quality quantum dots in this reaction is the heating of the reaction solution. You will not be able to get an oil bath to reach 300 °C. Similarly it is highly impractical to run the reaction in an oven. To get the reaction temperature right you will need to use a woods alloy heating bath into which a round bottomed flask can be placed. Woods metal is a eutectic metal alloy which is liquid above 70 °C and is essentially a liquid metal heating bath. The eutectic must be contained in a (non melting) metal enclosure. When the bath is placed on a heating mantle it provides a very high efficiency of heat transfer in to the solution which drives the ageing and growth of the forming colloid. Metal foil is also placed over the flask being heated to keep the heat in. **A liquid metal at this temperature is a severe burn risk—keep the bath clamped in place and stable at all times.** In a similar manner you can use a sand bath on top of a hot plate which can be packed around the round bottomed flask for efficient heat transfer.

For the sake of simplicity this method is broken up into three sections which may be performed independently as time allows.

Using a Schlenk line to keep the system oxygen free can be tricky and this will be further complicated if your laboratory is not set up for Schlenk line chemistry. To construct your own Schlenk line you will need a good vacuum diffusion pump fitted with a cold trap and an argon gas source. Essentially you will need two taps that can be linked to your reaction vessel as outlined in the schematic and will be used to control the evacuation and repressurisation of the system respectively. Of special consideration is that you must be able to isolate the cold trap for cleaning without breaking the vacuum in your line. Additionally you must ensure that the argon will be fed into the system at ambient pressure and that safe guards like pressure valves are in place to stop glass ware from exploding under pressure. For the manufacture of quantum dots a slightly higher than ambient pressure tends to give better results (Fig. 4.41).

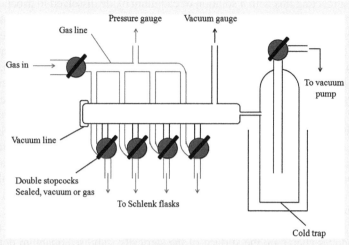

FIGURE 4.41   Simplified schematic of a Schlenk line setup.

To make cadmium selenide nanoparticles dispersed in chloroform you will need:

If you are planning to use tributylphosphine as a stabilising agent be aware it is extremely flammable in air. Techniques for working with it are provided but observe every precaution. Fires of this type must be extinguished with a carbon dioxide/foam extinguisher. Also keep in mind the liquid metal bath presents a severe burn risk.

1. A Schlenk line setup fitted to a vacuum pump and an argon gas line. This reaction is to be performed in a fume hood. A number of separate argon feeds is also useful if you will be working with the tributylphosphine. The vacuum line should run through a cold trap.
2. A heating mantle capable of reaching at least 300 °C with magnetic stirring. On top should be a woods metal eutectic heating bath into which a round bottomed flask can fit. This should be clamped or held in place. The melting point of the metal eutectic is around 70 °C so for some of the lower temperature points early on in the synthesis you will also need an oil bath. Have the heating mantle and bath mounted on a stage that can be raised and lowered. This will allow for an easier transition when changing samples on the line. Alternatively you can use sand packed around the round bottomed flask which may take longer to heat but will be simpler than changing the baths around. This preparation will include instructions for using the woods metal for completeness.
3. 39.6 mg (0.5×10^{-3} mol) of selenium in a three necked 50 mL round bottomed flask. The first side neck will need a rubber septum fitted for injecting liquid and the main neck must be connected to a valve allowing the bulb to be removed from the shlenck line under gas. The third side neck should be fitted with a thermocouple temperature probe so that the reaction temperature can be monitored closely. Place a PTFE coated magnetic stir bar in the flask before connecting it to the Schlenk line.
4. A stock bottle of trioctadecylphosphine. You will need 5 mL (0.011 mol) for making up the selenium solution. The bottle will most likely be fitted with a septum seal and in order to transfer the contents you will need a glass syringe and a needle. In addition you will need a second needle fitted to a balloon for extracting the trioctadecylphosphine under argon as outlined in the box entitled 'The Balloon Technique'.

The Balloon Technique
With thanks to Matthew Hesse

Air and moisture sensitive chemicals normally come in packaging that keeps them under an inert atmosphere. For liquid compounds of this type the chemical is provided in a bottle with a septum cap. The contents are extracted using a needle syringe pushed through the cap but this action presents a problem. If a given volume is drawn out of the bottle then the negative pressure this creates will want to suck either the chemical back into the bottle or draw in external atmosphere to compensate for the pressure difference. Air can also find its way into the reagent bottle because the hollow needles used to penetrate the septum will be full of air before they are pushed in. To get around this problem you can use a balloon to act as a gas reservoir as you draw the liquid out of the bottle.

For extracting a reagent from a septum capped bottle you will need:
1. A large balloon. A fairly thick and robust one will fare better.
2. A feed of the gas you wish to use, setup to flow though a gauge of tubing that you can use to fill the balloon. Most inert chemicals will be packaged under nitrogen. Argon is also commonly used.
3. A needle and syringe for extracting the chemical. Ensure the needle is long enough to get into the liquid in the bottle. It is inadvisable to use a short needle then try to invert the bottle. Caps have been known to fall off bottles treated in this manner and if the chemical is pyrophoric then a spill can quickly become a fire.

4. A small needle and syringe without a plunger fitted. The syringe is only going to be used for attaching the balloon and delivering the gas so a large volume is wasteful. The needle should not be attached when you begin as it is safer to put the needle on after you have fitted the balloon.
5. An elastic band for affixing the balloon to the syringe without the plunger. A snug fitting piece of tubing can also be used.

To extract a reagent from a septum capped bottle:

1. With the needle unattached, place the nozzle to the extraction syringe in the gas feed pipe. Flush the syringe with gas a few times to remove air. The final time you fill the syringe with gas reattach the needle and purge it by depressing the plunger.
2. Fill the balloon with gas and then keep the neck shut with your fingers. There should be enough of the balloon lip free for you to ease it over the plungerless syringe. Still holding the balloon shut with your fingers use the elastic band to hold the neck in place with your free hand. This may require practice!
3. Place a needle on the end of the balloon and syringe. Use your fingers to control the flow rate and let the balloon deflate enough to purge air out of the needle.
4. Penetrate the septum cap of the bottle with the balloon full of gas. Open your fingers slowly and check there are no hissing noises which would indicate a leak. If there is a leak check the condition of the cap or use a smaller gauge of needle. The full balloon should sit happily on top of the bottle.
5. Penetrate the septum using the extraction needle and syringe. Keep a finger on the plunger as the positive pressure from the balloon might push it out. Draw the desired amount of the liquid reagent from the bottle.
6. Remove both the needles from the bottle and use your reagent. Remember to cap any used needles or place them in a sharps bin (Fig. 4.42).

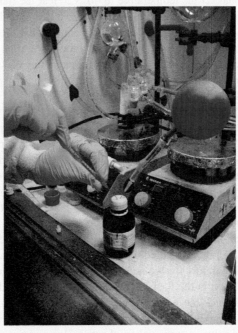

FIGURE 4.42   A photo showing how to use the balloon technique to remove air sensitive chemicals using a needle and syringe.

5. 37 mg (0.3×10^{-3} mol) of cadmium oxide in a 50 mL twin necked round bottomed flask. One neck should be fitted with a rubber septum and the other with a valve so that the flask can be removed from the Schlenk line while maintaining the sample under argon. This flask will also need a magnetic stir bar inside before you connect it to the vacuum line. *The flask must be resistant to thermal shock—it will be cooled very rapidly from 300 °C with water.*
6. 3.6 g (9.3×10^{-3} mol) of trioctadecylphosphine oxide. This is relatively air stable and can be easily weighed out in a normal atmosphere. This will be solid at room temperature but will melt above ~50 °C and can be degassed under vacuum.
7. 0.334 g (1.2×10^{-3} mol) of tetradecylphosphonic acid. This is used to solubilise the cadmium oxide and can be exchanged for oleic or stearic acid if necessary. This acts as a ligand to the cadmium allowing it to remain stable in the solution long enough to nucleate on a crystal surface during growth. In this concentration it promotes the formation of monodisperse particles but at higher concentrations it can promote the formation of rod structures.
8. A large dish with water and a wet towel sat inside. This will be used to rapidly cool the flask and quench the particle growth. You will also need an oven glove or similar so you can handle the hot reaction flask. An oil bath at room temperature can also be used.

9. 10 mL of methanol ready to be dispensed from a glass syringe with a needle.

10. A centrifuge and centrifuge tubes large enough to handle 20 mL volumes.

11. 10 mL of a 3:1 volume methanol:chloroform solution for rinsing the particles after centrifugation.

12. 10 mL of chloroform. The collected CdSe nanoparticles will be redispersed in this.

13. A 220 nm polytetrafluoroethylene filter and syringe for purifying the colloid. This can be optional depending on results. A vial with a lid is also required to collect the colloid in.

14. Silicone grease to ensure the glass joints are gas tight. This needs to be smeared only thinly on ground glass joints.

15. A thermocouple probe mounted in a quit fit plug that can be inserted into the reaction flask. This is vital for ensuring the temperature of the reaction solution is correct before and during the reaction.

To make cadmium selenide nanoparticles dispersed in chloroform:

1. Connect the flask containing the selenium to the shlenk line and evacuate it. Refill the flask with argon and then evacuate again. Repeat this three times.

2. Use a needle and syringe to extract 5 mL of trioctadecylphosphine from the stock bottle. Use the balloon technique to keep the contents dry. Inject this through the septum into the flask containing the selenium. Stir magnetically until the selenium has dissolved to produce a 0.1 mol/L solution. Leave this under argon on the Schlenk line or close the valve on the neck and disconnect it.

3. Add the 0.334 g of tetradecylphosphonic acid and 3.6 g of trioctadecylphosphine oxide to the flask containing the cadmium oxide. Connect the flask to the Schlenk line, clamp it in place and evacuate it.

4. Put the oil bath on the heating mantle and set the temperature to 40 °C. Heat the flask containing the cadmium and other components for 15 min under vacuum. After this time raise the temperature to 60 °C to melt the trioctadecylphosphine oxide and leave the brown solution for another 15 min. This step should dry the reagents so that there is no risk of water contamination in the reaction. If you are using tributylphosphine then this is particularly important. After heating then refill the flask with argon.

5. Lower the oil bath from the flask and use some tissue to completely remove any oil from the bottom of the flask. You may need to wet the tissue with ethanol or methanol to do this. Replace the oil bath with the woods metal bath. Set the temperature of the heating mantle to 300 °C. As the metal melts raise the stage so that the flask is sat in the metal with the contents just lower than the fill line. Turn on the magnetic stirring.

6. Monitor the temperature of the metal bath and ensure it has reached 300 °C before proceeding. The temperature at which the reaction occurs will greatly influence the size of the particles produced. Though a reaction temperature of 230−260 °C using a $CdMe_2$ can produce particle in the range 1−10 nm the size control is better by performing the reaction at 300 °C and leaving the particles to age for longer. Wait for the cadmium solution to turn clear which indicates that the oxide has fully dissolved. If you are taking aliquots during the synthesis and find the size distribution is too wide then lower the temperature of the reaction solution by ~10 °C. The cooling will encourage the focussing process by dropping the solubility of the crystallites in solution. Conversely, if the particles have stopped growing before they are at the correct size then increase the reaction temperature. Smaller particles have a higher surface energy which means they are less stable in a solvent. Heating increases the solubility of the smaller particles and initiates Ostwald ripening. The temperature required to encourage growth increases as the size of the crystallite increases.

7. Once the cadmium solution is up to temperature flush a needle and syringe with argon a few times to make sure there is no air inside. Poke the needle through the septum in the flask containing the selenium and draw up to 2.5 mL of the selenium trioctadecylphosphine solution. Quickly transfer the needle to the heated flask containing the cadmium solution and inject the contents through the rubber septum. You should immediately see a colour change in the solution as the cadmium selenide particles nucleate. Very generally, when the solution turns yellow the particles will be around 1 nm in size and by the time it turns a deep orange they will be around 3 nm in diameter. 1−3 nm particles will form within the first minute of the reaction and thereafter the particle size will grow. If left for up to 5 min the particles will be 20 nm in diameter. Immediately after the addition of the selenium seal the valve in the neck of the flask and prepared to remove it from the clamp.

8. After leaving the reaction to age for the desired amount of time lower the metal bath out of the way and remove the flask from the system. Quench the particle growth by rapidly cooling the reaction solution using the wet cloth in the bowl. Avoid burning yourself by using an oven glove to hold the flask by the neck as you do this.

9. Using a syringe inject 10 mL of methanol into the cooled solution. This will cause the nanoparticles to precipitate out and sediment. The addition of methanol will oxidise any remaining trioctadecylphosphine (or, more importantly from a safety view, tributylphosphine if used) and the flask can now be opened to air. You may observe the presence of some grey sediment which is Cd and Se by-products. These will be removed by filtration or can be removed in a size selective precipitation.

10. Decant the solution and sediment into a large centrifuge tube. Centrifuge the sample at 6000 rpm for 4 min. Discard the top solution and then add another aliquot of the methanol/chloroform solution to rinse the particles. Centrifuge again and discard the top solution. Redisperse the cadmium selenide nanoparticles in 10 mL of chloroform.

11. Use a syringe and pass the colloidal product through the PTFE filter into a clean screw top vial. You now have a cadmium selenide colloid which will contain 0.3×10^{-3} mol of cadmium selenide or roughly 50−60 mg. This can be analysed or used in

the next procedure where the particles will be given a zinc selenide coating. Store these particles in the dark to prevent photolytic degradation. If the size distribution of your particles needs to be very precise then you can perform a size selective precipitation at this point.

Size Selective Precipitation

This technique can be used to isolate a desired size of particle from a polydisperse colloid (Wilson et al., 1993). Particles are dispersed into a solvent where they are stable such as the trioctadecylphosphine oxide in this recipe. To precipitate the CdSe particles from solution a non-solvent miscible with the main solvent is added to the point that the particles can no longer remain stable as with the addition of methanol to the trioctadecylphosphine oxide solution. For size selective precipitation the non-solvent can be added slowly so that a 'tipping point' is reached where the particles are barely stable. Larger particles which have larger van der waals or other attractive forces will precipitate out of solution before smaller, more stable particles. In this way fractions of larger particles can be forced out of polydisperse solutions. Below is an outline for performing a size selective precipitation on the collected CdSe colloid. This method uses chloroform and methanol though you could also use pyridine and hexane or butanol and methanol.

1. Take the CdSe colloid in chloroform and place it in a centrifuge tube.
2. Using a Pasteur pipette add methanol dropwise until the colloid appears opalescent. The opalescence should remain even after stirring.
3. Centrifuge at 6000 rpm for 4 min. Afterwards you should see colour remaining in the solution in addition to a small amount of sediment in the bottom of the tube. The sediment will be larger particles.
4. Decant and keep the top solution for further purification if required.
5. Redisperse the sediment in chloroform and analyse or repeat further precipitation. If all the particle sediment at the same time then the sample is monodisperse.

To coat the cadmium selenide nanoparticles in zinc sulfide you will need:

1. The cadmium selenide in chloroform colloid prepared in the previous section. This contains $0.2-0.3 \times 10^{-3}$ mol of CdSe in 10 mL of chloroform.
2. 3.6 g of trioctadecylphosphine oxide. This should be in a 30 mL three necked round bottomed flask with a magnetic stir bar. One neck should have a rubber septum fitted and the other should have a fitting for a temperature probe. The main neck should accommodate a tap valve leading to the Schlenk line.
3. A Schlenk line with an argon feed.
4. A heating mantle with a mineral oil bath. Depending on the mantle and bath you may have to exchange the oil for a woods metal bath to achieve 180 °C at one point. It should be on a stage that can be raised and lowered.
5. An ultrasonic bath. One that can be maintained at 15 °C is desirable but otherwise this may be cooled with ice.
6. 0.198 g (0.313×10^{-3} mol) of zinc stearate. The stearate ligand will also act as a stabilising agent for the cadmium selenide.
7. 5 mL of toluene in a 10 mL vial fitted with a rubber septum cap. You will also need a small balloon on the end of a needle and a second needle to act as an outlet. This will be used to keep the vial contents under argon.
8. 0.05 mL (0.24×10^{-3} mol) of hexamethyldisilathane. This is the sulphur source for the reaction. This will need to be delivered by a volumetric syringe.
9. A volumetric syringe capable of measuring out 0.7 mL. This will be used to inject the zinc sulfide solution through a septum.
10. 10 mL of methanol ready to be dispensed from a glass syringe with a needle.
11. A centrifuge and centrifuge tubes large enough to handle 20 mL volumes.
12. 8 mL of chloroform for redispersing the formed zinc sulfide coated nanoparticles.

To coat the cadmium selenide nanoparticles in zinc sulfide:

1. Take the rubber septum out of one of the necks and add the 8 mL of cadmium selenide colloid to the round bottomed flask containing the trioctadecylphosphine and the stir bar. Replace the septum.
2. Connect the flask to the Schlenk line and clamp it in place with the valve to the flask closed. Turn on the heating mantle and set it to 60 °C then raise it up until the flask is sat in the oil bath. Once the oil bath is at temperature the trioctadecylphosphine oxide will melt and you can open the valve to the vacuum. Leave it under vacuum for half an hour. This step will get rid of the chloroform.
3. Flood the flask with argon and increase the temperature of the oil bath to 180 °C.
4. Dissolve the 0.198 g of zinc stearate in the vial containing the toluene. The zinc stearate will not readily dissolve so you will need to sonicate the contents of the vial in a sonic bath for a few minutes. Poke two needles through the septum and flush the vial with nitrogen for another 5 min.
5. Add 0.05 mL of hexamethyldisilathane by injection through the septum into the zinc stearate in toluene solution. Leave the solution for 20 min to form a zinc sulfide solution ready to use as for coating.
6. Use a syringe to extract 0.7 mL of the zinc sulfide in toluene solution. Inject this through the septum into the cadmium selenide solution being maintained at 180 °C. Allow the solution to age at this temperature for 10 min under stirring.

7. Lower the temperature of the oil bath to 100 °C and let the solution stir for a further 2 h. During this time the cadmium selenide particles will be coated with zinc sulfide.
8. Turn off the heating mantle and lower it away from the round bottomed flask. Allow the reaction vessel to cool to room temperature.
9. Inject 10 mL of methanol through the rubber septum. This will cause the coated nanoparticles to precipitate out of solution. Transfer the entire sample into a centrifuge tube.
10. Centrifuge the sample at 6000 rpm for 4 min. Discard the top solution. Redisperse the particles in 8 mL of chloroform. At this stage you will have cadmium selenide particles coated in zinc sulfide ready for analysis. They can now be stored until further use. These particles will only form stable colloidal solutions in oil so the next step is a ligand exchange.

MAKING ZINC SULFIDE COATED CADMIUM SELENIDE NANOPARTICLES STABLE IN WATER

With thanks to David Benito Alfonso. Adapted from Gill et al. (2006)

This ligand exchange method will swap out the trioctadecylphosphine coordinated at the particle surface for mercaptoacetic acid molecules. The sulphur head group coordinates to zinc ions at the surface and the exterior facing carboxylic acid groups allow the particles to be dispersed in a polar solvent.

Making zinc sulfide coated cadmium selenide nanoparticles stable in water you will need:

1. 6−8 mL of the zinc sulfide coated cadmium selenide nanoparticles dispersed in chloroform.
2. 10.5 mL (150×10^{-3} mol) of mercaptoacetic acid in a 20 mL sealable vial with a small stir bar. *This must be worked within a fume hood!*
3. A magnetic stirrer.
4. A centrifuge and suitable centrifuge tubes that can handle 20 mL volumes.
5. 6−8 mL (the same volume as colloid used) of a sodium phosphate buffer made up to the desired pH. The concentration might vary depending on the end use but typically a 50 mmol/L concentration adjusted to pH 7.4 is a safe starting point for most biological applications.
6. An ultrasonic bath.

To make the zinc sulfide coated cadmium selenide nanoparticles stable in water:

1. Add the colloid in chloroform to the mercaptoacetic acid and seal the vial.
2. Stir the solution for 12 h.
3. Centrifuge the contents of the vial at 6000 rpm for 4 min. Discard the top solution.
4. Redisperse the pellet into the phosphate buffer using the sonic bath. The sample is ready for further processing or analysis.

You will now have a CdSe colloid solution in water or buffer containing roughly 0.2×10^{-3} mol CdSe. This will be stable for around a week but the stability can be increased to years if the mercaptoacetic acid stabilised particles are modified further with carbodiimide coupling to a protein or bio-receptor. At the visual level you should immediately be able to get a rough idea if the experiment has worked as you should be able to see a stable water soluble coloured colloid. UV-visible spectroscopy should show a significant blue shift when compared to the spectrum of a bulk sample. For a range of semiconductor materials the general shifts in peak absorbance between a bulk sample and nanoparticles in the 2−3 nm range are given below (Yu et al., 2003).

Material	Bulk peak absorbance/nm	Nanoparticle peak absorbance/nm
CdS	512	400
CdSe	716	500
CdTe	827	600

UV-visible spectroscopy can be used to give an estimate of the colloidal concentration but only if the size of the nanoparticles is well known. An estimate of particle diameter is possible from the peak absorbance by using the following formula:

For standard UV-visible concentration calculations:

$$\text{Absorbance} = \epsilon CL$$

where ϵ is the extinction coefficient (L/mol cm), C is the concentration (mol/L) and L is the path length of the light through the cuvette (cm). The figures given here are for a path length of 1 cm.

The extinction coefficient for a semiconductor nanoparticle is size dependant. A polynomial formula for a given material can be used to estimate the particle diameter d (nm) from the peak absorbance wavelength λ (nm):

$$\text{CdTe: } d = (9.8127 \times 10^{-7})\lambda^3 - (1.7147 \times 10^{-3})\lambda^2 + (1.0064)\lambda - 194.84$$

$$\text{CdSe: } d = (1.6122 \times 10^{-9})\lambda^4 - (2.6575 \times 10^{-6})\lambda^3 + (1.6242 \times 10^{-3})\lambda^2 - (0.4277)\lambda + 41.57$$

$$\text{CdS: } d = (-6.6521 \times 10^{-8})\lambda^3 + (1.9557 \times 10^{-4})\lambda^2 - (9.2352 \times 10^{-2})\lambda + 13.29$$

The calculated diameter can then be used to derive the extinction coefficient of the colloidal dispersion:

$$\text{CdTe} : \varepsilon = 10043 \times d$$

$$\text{CdSe} : \varepsilon = 5857 \times d$$

$$\text{CdS} : \varepsilon = 21536 \times d$$

The values calculated in the formula can be off by as much as 20–30% but this can be reduced to around 10–15% if using a figure for d derived accurately from electron microscopy or calculated from peak broadening in X-ray diffraction patterns. The calculations are only valid for particle sizes between 2 and 8 nm.

Photoluminescent spectroscopy should reveal where the peak excitation and emission wavelengths are and this is important if the quantum dots are to be used as fluorescence probes. Ideally the nanoparticles will be monodisperse resulting in very sharp emission and excitation bands so that many different probe colours can be used at the same time. Cadmium selenide particles coated in zinc sulfide should demonstrate visibly stronger emissions under illumination than uncoated controls. The intensity of the emission is directly related to the quantum yield. This can be assessed by a comparison of the quantum dot luminescence compared to a known standard such as rhodamine 640 in methanol. The quantum yield of the fluorescent dye is very near unity, that is to say all of the light going into the dye is being converted into an emission, so a direct comparison can be made to the quantum dot emission spectrum. To do this make up a solution of quantum dots and a solution of the fluorescent dye in matched cuvettes. The concentrations of the two solutions are roughly matched by the optical density of solution as the concentration of a colloid is difficult to determine as accurately by direct spectroscopy using the Beer–Lambert law. An optical density of 0.3 should be sufficient for measurement and this should be measured at the wavelength you intend to excite the samples. For rhodamine 640 excite the solutions at 460 nm and monitor the emission wavelengths from 480 to 800 nm. You will be able to compare the intensity of peak emission of the quantum dots as a percentage of the peak emission of the rhodamine 640. Quantum yields of around 40–50% can be expected for ZnS coated particles.

You may find that an overly thick shell of zinc sulfide actually decreases the fluorescence of the nanoparticles. For the highest quantum yields the surface layer of ZnS should be between 1.3 and 1.6 monolayers thick. This means that the entire surface of the CdSe particle is coated and that surface oxidation to form SeO_2 cannot occur which would act as an exciton trap and quench fluorescence. Between 4 and 6 monolayers provides protection against oxidation in more hash conditions as may be found in a cancer cells if the quantum dots are to be used for imaging (Medintz et al., 2005). As the monolayer increases in thickness beyond this limit there will be an increase in the number of defects produced in the ZnS layer which can also behave as traps for the exciton under illumination. This quenching is reduced somewhat by the further passivation of the ZnS surface by amine terminated organic molecules when making the colloid water stable.

Transmission electron microscopy can be used to determine the exact size of the core and thickness of the shell. If the particles are highly monodisperse then you may see self assembly of the particles into ordered arrays though this also depends on the interactions of the surface bound surfactant. For very small particles you may notice some elongation of the particle in one particular direction. This is indicative of growth along the [00$\bar{1}$] structure of wurtzite. Powder X-ray diffraction should reveal a pattern that can be indexed to the hexagonal wurtzite structure with a strong reflection for the [002] lattice plane. There will be a significant peak broadening in the diffraction pattern which can be used to give an estimate of the particle size using the Scherrer equation. As the particle size moves below ~1.2 nm the peak broadening of the pattern will make identifying any cubic zinc blende structure impurities from wurtzite impossible (Murray et al., 1993) (Fig. 4.43).

SUFACE MODIFICATION OF CADMIUM SELENIDE NANOPARTICLES FOR USE AS FLUORESCENT LABELS AND PROBES

CdSe and other quantum dots are highly advantageous over normal fluorescent chemical dyes in terms of their photo stability and brightness (Chan et al., 2002). In comparison to a single rhodamine 6G molecule a quantum dot is 20 times brighter under the same illumination source. This is due in part to the greater molar extinction coefficient for a semiconductor material over an organic dye

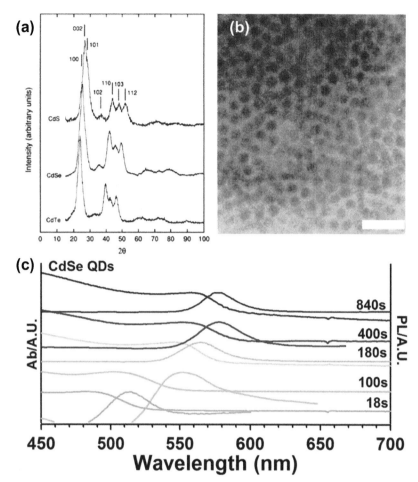

FIGURE 4.43 (a) Powder X-ray diffraction patterns for CdS, CdSe and CdTe with the plane indexes marked on the major reflections. (b) A transmission electron micrograph of CdSe nanoparticles produced by the TOPO method. Scale bar = 20 nm (c) A series of fluorescence spectroscopy plots showing the respective excitation and emission spectra for cadmium selenide quantum dots allowed to grow for various times prior to quenching. *(Images a and b Adapted with Permission from Murray et al. (1993). Copyright 1993 The American Chemical Society)*

molecule; a greater number of photons passing through the nanoparticles will be adsorbed resulting in a higher probability of photonic emission. They also have far higher adsorption cross sectional areas in comparison to the dyes. Chemical dyes are still useful in procedures that require ultrafast responses to illumination but for most biolabelling continuous illumination and emission are used. Of more importance is the stability of the particle in biological conditions and a ZnS or polymer coated fluorescent nanoparticle can last up to 200 times longer than a molecular dye. This means that instead of a single snapshot of a cellular process taking place it can be imaged in real time by marking the various biological components with a nanoparticles probe. One of the first uses of CdSe probes was in tracking receptor mediated endocytosis in HeLa cells by coupling the protein transferrin to the particle exterior (Chan and Nie, 1998). Most commercially available quantum dots sold as probes are produced by the trioctadecylphosphine route because of the tight size distribution. This may be with or without a protective ZnS layer as some of the procedures use polymers to form a protective shell instead. The methods used to convert quantum dots into bioprobes fall into three broad categories:

1. Ligand 'cap' exchange of the trioctadecylphosphine for a thiol with a carboxylic acid terminated chain. The colloid is mixed in a concentrated solution of the thiol which displaces the trioctadecylphosphine by coordinating to surface zinc or cadmium ions on the particle. The carboxylic acid group can be deprotonated under basic conditions to provide an electrostatic charge which stabilises the colloid in aqueous solution. The limitation of this approach is that the colloid is only stable for about a week before it begins to crash out. This can be avoided by coupling a water stable protein or receptor through the carboxylic acid group using a carbodiimide coupling agent (Wu et al., 2003). The lifetime is also extended by using bidentate thiol ligands as desorption of the thiol groups from the surface is minimised which would otherwise result in flocculation. Though gram amounts of particles can be treated in this way the solution must remain under basic conditions if it is stabilised by carboxylic acid groups. This could present a problem if the probes need to be used for imaging a cancer cell where the pH of the cell interior is slightly acidic.

2. Coating the particles with a functionalised inorganic layer (Bruchez, 1998). In a similar way to the thiolated ligand exchange, a silicon alkoxide sol–gel precursor functionalised with a mercapto group is coordinated to the surface of the trioctadecylstabilised nanoparticles in an alcohol. The alkoxide is hydrolysed using an organic base as a catalyst to form a thin organosilane layer over the surface of the particle. This can then be built upon with urea- and aminopropyl-silane precursors to produce an inorganic layer with organic moieties that can be functionalised with suitable receptors. Particles treated like this are well protected from degradation and are highly stable under aqueous conditions though only milligram amounts can be processed at a time. The effective diameter of the particle is about 30–40 nm which is larger than particles treated directly by ligand exchange which tend to be 1–15 nm in diameter. If the glass coating is not fully condensed then surface hydroxyl groups on the surface of adjacent particles can link causing colloidal instability.

3. Polymer coating a trioctadecylphosphine stabilised colloid (Gao et al., 2004). This method involves leaving the hydrophobic phosphine ligands in place on the surface of the particle and builds on it using an amphiphillic di or triblock polymer. Block copolymers can be engineered to have hydrophobic and hydrophilic parts and the hydrophobic components will intermingle with the trioctadecylphosphine ligands on contact. The hydrophillic component is then the outermost layer which renders the colloid stable in aqueous solutions. This self assembly approach is facile and allows for the conjugation of receptors to the surface. Because the original ligands remain in place and the polymer layer is up to 2 nm thick with a 4–5 PEG-receptor shell on top, the CdSe core is well protected against fluorescent quenching. High quantum yields and brightness are seen in preparations of this type with stability at high salt concentrations and a range of pH. One exciting aspect of this approach is that an additional poly(ethylene glycol) coating extends the lifetime of the particles in-vivo so that they can be used as biomarkers in live animals. This means that tumours can potentially be imaged easily in the body though this is currently limited to biological targets near the surface of the skin where the excitation light can penetrate. Particles used in this way must be tuned to respond to IR and Near IR stimulation as these wavelengths penetrate deeper into the skin.

FUNCTIONALISING THE QUANTUM DOTS WITH TRANSFERRIN

Adapted from Chan and Nie (1998)

In the following recipes we will be using the water stable mercaptoacetic acid stabilised nanoparticles prepared in the previous recipe as a starting point. These particles come under the first category of ligand 'cap' exchange for particle modification but this technique can also be applied to the inorganic and polymer coated particles as well. The particles remain stable in aqueous solution due to the electrostatic charge of deprotonated carboxylic acid groups at the surface and it is these groups that make the surface amenable chemical modification. A carbodiimide crosslinking reagent is added to the aqueous solution which forms a reactive O-acylurea intermediate with the carboxylic acid groups. When this comes into contact with an amine group, such as that found on the surface of a protein, it reacts to form an amide bond linking the protein and the particle together. The efficiency of this reaction can be further boosted by the introduction of N-hydroxysulfosuccinamide which forms a reactive succinate ester which also converts to an amide bond by reaction with an amine (Fig. 4.44).

The choice of what to stick on the exterior of the particle depends on what you would like to try and image. Here we will use the protein transferrin to demonstrate the coupling reaction. Transferrin was used in one of the first publications on quantum dot labelling and activates receptor mediated endocytosis. If a coated fluorescent particle comes into contact with a cell with the receptor for transferrin then it will be drawn to the cell interior where it can be imaged under a microscope.

For direct coupling of transferrin to the surface of a mercaptoacetic acid stabilised ZnS coated CdSe nanoparticles using (1-Ethyl-3-(3-Dimethylaminopropyl))Carbodiimide hydrochloride (EDC) you will need:

1. 5 mg of transferrin or other biological receptor with a free amine for conjugation. Biological molecules should remain frozen and dry until use. This should be in great excess for the conjugation.
2. A colloidal dispersion of the mercaptoacetic acid stabilised ZnS coated CdSe nanoparticles. You will need to dilute an aliquot of your stock further using 0.05 mol/L sodium phosphate buffer adjusted to pH 7.4 so that you have around 10 mg of CdSe in 2 mL of the buffer. This should be prepared in a small glass vial with a magnetic stir bar.
3. 1.9 mg (1×10^{-5} mol) of (1-ethyl-3-(3-dimethylaminopropyl)carbodiimide hydrochloride dissolved in 0.5 mL of 0.05 mol/L sodium phosphate buffer adjusted to pH 7.4. This must be prepared fresh shortly before reaction and the EDC should be kept dry and frozen when not in use.
4. A magnetic stirrer.
5. 1 L of stock 0.05 mol/L sodium phosphate buffer solution adjusted to pH 7.4. This will be used for making solutions and for the dialysis step to remove by-products of the reaction.
6. A tube of dialysis membrane and two clips per end to close off the tube. You will need a 200–500 mL beaker for performing the dialysis in.
7. Volumetric pipettes that can measure out 0.1 mL and up to 3 mL in volume.

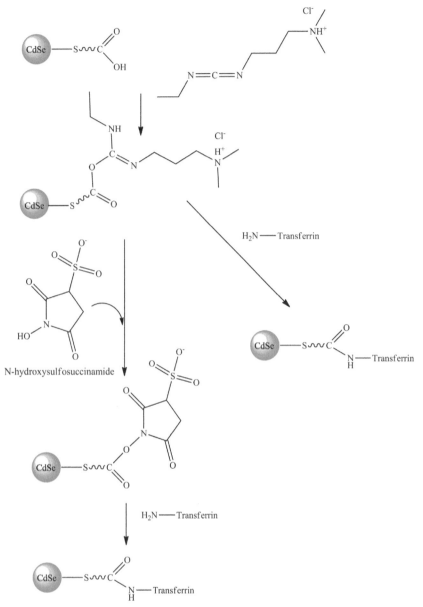

FIGURE 4.44   The chemical reaction for carbodiimide crosslinking. A variation of the reaction using N-hydroxysulfosuccinamide is included.

For direct coupling of transferrin to the surface of a mercaptoacetic acid stabilised ZnS coated CdSe nanoparticles using (1-Ethyl-3-(3-Dimethylaminopropyl))Carbodiimide hydrochloride (EDC):

1. Add the 5 mg of transferrin to the 2 mL of CdSe colloid in the vial and allow it to disperse.
2. Using the volumetric pipette add 0.1 mL of the freshly prepared EDC solution to the protein/colloid solution.
3. Allow the solution to stir gently for 2 h during which time the protein will be covalently bound through the formation of an amide bond to the nanoparticle surface.
4. Add around 200 mL of the phosphate buffer stock solution to the empty beaker and soak a section of dialysis tubing in it. The tubing section should be 2–3 times longer than what you think the sample will require so that the ends can be properly clipped.
5. Double clip one end of the dialysis tube shut and transfer the reaction solution to it using a pipette. Double clip the other end shut and dialyse the solution for 12 h in the buffer changing it twice.
6. Transfer the dialysed colloid back into a glass vial for storage and keep it in the dark in a fridge. The functionalised quantum dot colloid is now ready for use as a dye.

To see if your fluorescent probes work you may try using them to stain a cell culture. The source reference incubated a HeLa cell culture with an excess of the conjugated quantum dots for 12 h at 37 °C. The cells were washed in the petri dish to remove quantum dots that had not been internalised and then transferred onto a glass slide for imaging using a fluorescence light microscope. The procedure for culturing a cell line is given elsewhere. The quantum dots should stand out very clearly against background fluorescence. A good control is to try incubating cells with non biofunctionalised quantum dots as these should not undergo the same endocytosis and should not demonstrate significant fluorescence after rinsing. If you find that only a very small amount of your quantum dots are being coated in the transferrin then you may wish to use the sulfo-N-succinamide variation of the EDC crosslinking reaction as outlined below. This reaction has a higher efficacy as compared to EDC linking alone.

TAGGING THE MERCAPTOACETIC ACID STABILISED ZnS COATED CdSe NANOPARTICLES WITH STREPTAVIDIN SO THEY CAN BE USED FOR TARGET SPECIFIC LABELLING

Adapted from a Invitrogen Qdots and Cytodiagnostics Technical Notes

One of the most useful interactions for biochemists is the very strong binding of the vitamin biotin by the protein streptavidin. In a process known as biotinylation a protein or other target with a primary amine group on it is coupled to a biotin molecule using carbodiimide crosslinking chemistry. The biotin is linked via an amide bond to the surface of the protein where it rarely interferes with the protein function due to its comparatively low molecular mass of 244 as compared to the kilodaltons size of most biological molecules. Once a protein has been functionalised in this way it will have a very high affinity for streptavidin and this property can be used for separating biotin labelled proteins from non labelled ones in assays. A variation of this procedure is to use a quantum dot coated in the streptavidin protein so that it will bind strongly to anything with a biotin molecule on it. This can be exploited to produce a fluorescent probe that will actively bind to only specific proteins or targets that have been biotinylated beforehand. A similar binding interaction is observed if the streptavidin is substituted with avidin. Avidin is a protein that can be readily isolated from egg whites and also binds biotin strongly (Wilcheck and Bayer, 1990).

In this procedure we will use carbodiimide chemistry to tag the water stable quantum dots with streptavidin. There is another step in the complexity of the crosslinking reaction in this method as an activation agent is used to increase the reactivity of the carboxylic acid group. The O-acylisourea intermediate formed from the reaction of the EDC and the carboxylic acid group will hydrolyse rapidly if not reacted with an amine. This is why a molar excess of cross linker and the carboxylic acid must be used for a successful reaction. N-hydroxysulfosuccinamide, also known commonly as sulfo-NHS, reacts with the O-acylisourea intermediate to form a longer lived ester intermediate which increases the chances of a successful coupling to an amine.

The full procedure should be conducted in two parts as you will need to functionalise the quantum dots for use as labels and also functionalise a protein of interest using biotin. A separate general procedure for functionalising a target protein with biotin is provided. The precursor for this is a preformed sulfo-NHS biotin which can be directly added to a protein solution and will react directly. Alternatively you may use the Sulfo-NHS and EDC procedure to convert pure biotin into the sulfo-NHS biotin form during the reaction. The pH for functionalising the particles must be well controlled. The mercaptoacetic acid only stabilises the particles under basic conditions but above pH 7.5 the half life of the reactive EDC and sulfo-NHS intermediates drops from hours to minutes. Once the reaction is complete the colloid should be stable as it will be coated in protein (Fig. 4.45).

For functionalising the mercaptoacetic acid stabilised ZnS coated CdSe nanoparticles with streptavidin you will need:

1. A 0.025 mol/L solution of piperazine-N,N′-bis(2-ethanesulfonic acid) (also known as PIPES) adjusted pH 7.0. A stock of this can be prepared by dissolving 0.75 g of the PIPES in 80 mL of water. Adjust the pH with small volumes of 1 mol/L NaOH or HCl solution then make up to 100 mL.

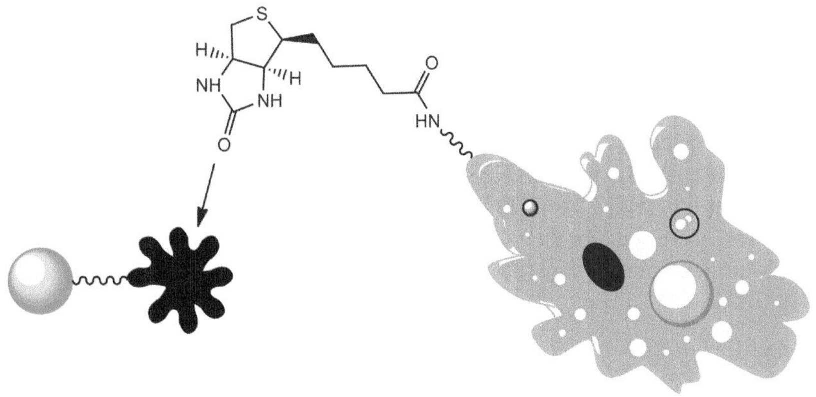

FIGURE 4.45 A picture of streptavidin linked to a nanoparticle. A biotinylated protein will bind strongly to the streptavidin and so the protein becomes fluorescently labelled.

2. The water stable CdSe quantum dot dispersion. You will be using 1 mL of this stock which is roughly 2−3 mg of nanoparticles.

3. A stock solution of tris buffered saline (TBS) solution adjusted to pH 7.4. This reagent can be purchased directly as a powder and made up using 1 L of distilled water. It contains NaCl, KCl and Tris buffer.

4. A stock 10 mL of a 0.02 mol/L solution of (1-ethyl-3-(3-dimethylaminopropyl)carbodiimide hydrochloride in water. This must be prepared fresh before use by dissolving 38 mg of EDC in 10 mL of distilled water.

5. A stock 10 mL of a 0.05 mol/L solution of N-hydroxysulfosuccinamide in water. This can be prepared by dissolving 0.1 g of the powder in 10 mL of water.

6. 0.5 mg of streptavidin. This will be supplied as a powder and can be purchased in amounts as low as 1 mg. Keep refrigerated until use. This is a very small amount to weigh so you may be better off making double the amount of conjugated quantum dots so as not to waste it.

7. A sonic bath, a benchtop vortex and a magnetic stirrer.

8. Volumetric pipettes that can measure 0.1−1 mL and 5−10 mL volumes.

9. A 15 mL glass vial for performing the reaction with a lid and a clean small magnetic stir bar.

10. 0.5 mL of a 0.02 mol/L ethanolamine in water solution. A stock solution can be prepared by adding 0.012 mL of ethanolamine to 10 mL of water. This is used to quench the crosslinking reaction. If ethanolamine is unavailable the reaction can be stopped by dialysing the reaction solution in an excess of the TBS buffer.

11. Ultracentrifugal filter tubes with a 100,000 molecular weight cut off and a centrifuge. These will be used for removing excess streptavidin not bound to the surface.

12. A 0.2 μ filter and syringe and a vial for storing the streptavidin conjugated quantum dots in.

To functionalise the mercaptoacetic acid stabilised ZnS coated CdSe nanoparticles with streptavidin:

1. Add 1 mL of the nanoparticle stock to 9 mL of the 0.025 M PIPES buffer solution in the glass vial with the stir bar. Set the solution to stir gently on top of the magnetic stirrer.

2. Add 0.975 mL of the freshly prepared EDC solution to the vial.

3. Add 0.650 mL of the freshly prepared Sulfo-NHS solution to the vial. At this point the reactive intermediates will be forming with the carboxylic acid groups on the particle surface. The pH should be around 7 but if it is higher then the reaction efficiency will decrease very quickly.

4. In a separate vial dissolve the 0.5 mg of streptavidin in 1 mL of the 0.025 M PIPES buffer.

5. Use a volumetric pipette to transfer 0.5 mL of the streptavidin solution to the gently stirring activated quantum dot solution. The reactive intermediates will react with amines on the surface of the streptavidin to bind them to the particle surface. Leave the solution under gentle stirring for an hour.

6. Stop the reaction by adding 0.5 mL of the 0.02 M ethanolamine solution.

7. Transfer the reaction solution to the ultracentrifuge filtration tubes and spin at 4000 rpm for 10 min. Rinse and repeat with distilled water three times. You may find the filtrate does not redissolve in to water after the first centrifuge filtration. If this happens add a drop of 1 M sodium hydroxide to the water.

8. After the final filtration redisperse the streptavidin conjugated quantum dots in 1 mL of tris buffered saline solution or another buffer of your choice. The functionalised particles are now ready for use in coupling reactions with biotin labelled proteins. For samples that are to be stored before use you may wish to use a buffer containing 1% poly(ethylene glycol) to prevent aggregation.

For attaching biotin to a protein or primary amine containing target you will need:
Adapted from (Bioconjugate techniques by Greg T Hermanson, 2008)

This procedure is very general and a variety of different biotin tags with varying functionalities can be purchased directly from suppliers in kit form. Perhaps one of the most important considerations for which biotin to choose from is to use a molecule with a spacer unit attached. The tail group of a regular D-biotin molecule is around 1.3 nm long so if the biotin is within a deep protein channel it may not have the reach to properly bind to a streptavidin protein nearby. Other common modifications include a section of the spacer unit that can be easily cleaved or a UV-visible chromophore so that the progress of the biotinylation reaction can be monitored. The binding affinity of streptavidin for biotin is so high that is not generally reversible without the use of harsh chemical conditions. Therefore a biotin spacer unit with a 'weak link' can be used so that immobilised proteins captured using this technique can be cleaved from the surface and harvested. For the sake of simplicity this method will use the general term 'sulpho-NHS-biotin'. This preparation is aimed at water soluble proteins but hydrophobic proteins can be treated in the same way by using dimethylformamide or dimethylsulfoxide as a solvent.

1. A stock solution of 0.1 mol/L phosphate buffer containing 0.15 mol/L NaCl adjusted to between pH 7.2 and 7.4. This can be prepared by dissolving 14.49 g of Na_2HPO_4 or 17.8 g of $Na_2HPO_4 \cdot 2H_2O$ in to 800 mL of water and adjusting using 1 mol/L NaOH or HCl solution to adjust the pH. Dissolve 8.7 g of NaCl into the solution and make up the volume to 1 L.

2. A protein for biotinylation. You will be working with this at a concentration of between 1 and 10 mg/mL. For this recipe we will use 10 mg of a hydrophilic protein as this will allow us to use some in an assay for determining the amount of biotin attached. You will need a rough idea about the protein molecular weight so you can estimate how much biotin to use.
3. Between 12 and 20 M excess of biotinylation reagent compared to the molar amount of protein. Use nearer the 20 M excess for a more dilute 1 mg/mL solution. This should be prepared fresh immediately before use in 20−30 μL of water.
4. A small screw top vial or microcentrifuge tube for performing the reaction in.
5. A small magnetic stir bar and magnetic stirrer.
6. A small volume dialysis membrane or cartridge, four clips (if using a membrane) and a 200−500 mL beaker for purifying the product.
7. Volumetric pipettes with a working range 10−50 μL and 1−5 mL.

To attach biotin to a protein or primary amine containing target:

1. Dissolve the protein in 1 mL of the phosphate buffer in a small vial. Set the solution to stir gently on the magnetic stirrer.
2. Add the biotin solution to the protein solution.
3. Leave it to stir for 30−60 min.
4. Soak the dialysis membrane in the beaker with some phosphate buffer so it is ready to receive the reaction solution. Double clip one end of the bag if you are using dialysis tubing.
5. Transfer the reaction solution into the dialysis membrane and dialyse against the phosphate buffer to remove by-products and unreacted biotin.

You should now have a protein that is coated in biotin and will bind strongly to a streptavidin coated nanoparticle. If you have used a type of biotin incorporating a chromophore then you will be able to directly asses the level of biotinylation using a UV-visible spectrophotometer. If you have used a type of biotin without some kind of colorimetric indicator you can instead use the HABA (4′-hydroxyazobenzene-2-carboxylic acid) dye assay to determine the number of biotin molecules per protein. HABA is a dye that absorbs strongly at 500 nm when bound to streptavidin or avidin but is readily displaced by biotin and correspondingly the absorbance at 500 nm decreases as the amount of biotin in solution increases. This procedure can be purchased as a kit and this procedure is adapted from kit instructions (Fig. 4.46).

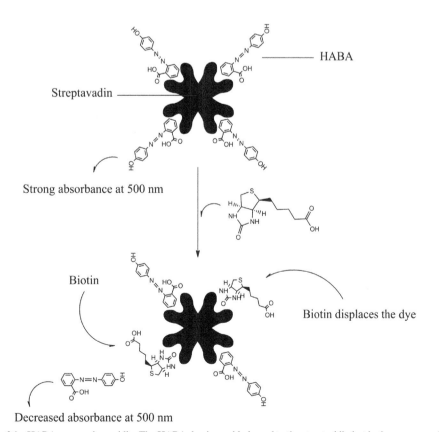

FIGURE 4.46   A schematic of the HABA assay using avidin. The HABA dye is weakly bound to the streptavidin but in the presence of biotin this is displaced and reduces the absorbance of the solution.

To perform the HABA assay you will need:
With thanks to Thermo Fisher Scientific. Adapted from Thermo Fisher Scientific and (Green, 1965)

1. A spectrophotometer with a 10 mL 1 cm path length cuvette. It does not need to be quartz and a disposable one may be used.
2. A stock solution of phosphate buffered saline solution at pH 7.2. This can be prepared identically to the stock solution as outlined in the biotinylation recipe. You will need about 20 mL of the buffer to make up a HABA/avidin solution.
3. A solution of 24.2×10^{-3} g $(0.1 \times 10^{-3}$ mol) of HABA dissolved in 9.9 mL of water with 0.1 mL of 1 mol/L solution of NaOH added. This will help the HABA dissolve and more can be added if required. Only a small amount of this solution is used so this can be kept for performing other assays by storing it in the fridge.
4. 10 mg of avidin. This protein behaves similarly to streptavidin but is harvested from egg white so it is cheaper to use for assays.
5. Volumetric pipettes with a $100-1000$ μL range and disposable tips. You will also need some small vials or microcentrifuge tubes for mixing small volumes in.
6. 100 μL of your biotinylated protein at a known concentration of protein. For the assay the concentration should be around 10 mg/mL and you will need to know the molecular mass of the protein in g/mol.

To perform the HABA assay:

1. Dissolve the avidin in 600 μL of the HABA solution in a small centrifuge tube.
2. Add the 600 μL of avidin and HABA solution to 19.4 mL of the buffer solution. This is now your stock solution for assaying. Take 900 μL of this solution and place into the UV-visible cuvette.
3. Measure the absorbance of the HABA assay solution at 500 nm using the spectrometer. The absorbance should be somewhere near 1. You may need to adjust this using more buffer if it is more than 0.2 absorbance units above 1 but be sure to keep a track of the concentration.
4. Add 100 μL of your biotinylated protein solution to the cuvette and mix thoroughly using the pipette by drawing solution in and out of the tip. Leave to stand for one minute to allow the biotin time to displace the HABA dye from the avidin.
5. Measure the absorbance of the assay solution again at 500 nm.

The Beer–Lambert law can now be used to determine how many moles of HABA have been displaced by biotin molecules. This in turn can be used to give the moles of biotin present and therefore the number of biotin molecules per protein. The calculation is performed in three parts using 34000 M^{-1} cm^{-1} as the extinction coefficient of the avidin–HABA complex.

Firstly the change in absorbance observed at 500 nm after the addition of the biotinylated protein is:

$$\text{Change in absorbance} = (0.9 \times (\text{absorbance value for the initial HABA and avidin solution}))$$

$$-(\text{absorbance value for the HABA and avidin and biotinylated protein solution})$$

The adjustment of 0.9 is to account for the volume change in the cuvette.
Next the concentration of biotin in mol/L is given by:

$$\text{Concentration of biotin (moles per litre)} = \text{Change in absorbance}/(34000 \times \text{path length of cuvette in cm})$$

where the path length for most cuvettes will be 1 cm.
Finally, the molar ratio of biotin to protein can be calculated. The following formula takes into account that the initial protein concentration is diluted by 10 in the assay and that some dilution may have been required for sensible results in the spectrometer. The molar ratio of biotin to protein is given by:

$$\text{Molar ratio of biotin to protein} = (\text{concentration of biotin} \times 10$$

$$\times \text{dilution factor})/\text{known starting concentration of protein in moles per litre.}$$

If you find that no biotinylation has occurred then the biotin molecule may need a longer spacer unit for reaching past the bulk of the protein to the avidin. Alternatively you may need to adopt a different biotinylation procedure or reagent.

CADMIUM SELENIDE NANORODS

Adapted from Peng et al. (2000), Peng and Peng (2001), Manna et al. (2000)
The capacity to control the CdSe electronic structure in two dimensions is possible by the growth of anisotropic nanocrystals. These exhibit unique optical effects not observed in the one-dimensional morphology, having two separate planes of optical quantum confinement (Li et al., 2001). Even more interesting is the ability of semiconductor nanorods to polarise light. When plane polarised light passes through an aligned array of the nanorods the photoluminescent emission of the rods is also plane polarised. This means that anisotropic luminescent tags could be used to look at the orientation of a biolabel linked to a target (Peng et al.,

2000). The quantum yield of the as produced rods is sometimes low but can be boosted by coating with another semiconductor material as in the one-dimensional system. In this recipe we will build upon the trioctadecylphosphine high temperature reaction in trioctadecylphosphine oxide synthesis and adapt it for the production of nanorods.

One of the key stages for the production of highly monodisperse particles is the rapid addition of the precursor. In this recipe the organometallic precursor dimethyl cadmium is mixed with a selenium tributylphosphine solution at low temperature to make a reagent solution. There is no reaction at low temperature as the $Cd(CH_3)_2$ molecule is relatively stable but upon addition of the stock to trioctadecylphosphine oxide at 360 °C the $Cd(CH_3)_2$ rapidly decomposes leaving a high concentration of homogenous Cd^{2+} present throughout the solution which reacts quickly with the selenium to nucleate CdSe crystals of a predominantly wurtzite crystal structure. All crystals have some form of solubility, however low, and this solubility increases as the size of the particle gets smaller and smaller. If the cadmium and selenium precursor concentration is kept high then the driving force for the particles to form rapidly is maintained. In a normal one-dimensional colloid this means that lots of smaller particles will be formed in a tight size regime throughout the solution, an effect called 'focussing'. As the crystal grows the various crystal facets presented to the solution nucleate at different rates depending on the surface energy of that particular plane. In wurtzite the [00−1] lattice plane grows faster than the others so that, in a solution rich with precursor during focusing, an anisotropic rod structure is formed. If the concentration of cadmium and selenium can be maintained above the solubility product of the nanoparticles then this anisotropic growth will continue. Provided the crystal structure is thermodynamically stable after production then the crystals will remain as rods.

If the precursor solution used is more dilute then ions and molecules free in solution will be slower to diffuse to the particle seeds making nucleation and growth more leisurely. Under these conditions the forming crystals will tend towards a spherical shape over a range of sizes. After the initial slow particle formation has ceased the subsequent concentration drop in reagent promotes the more soluble smaller particles to redissolve. The solubilised molecules then nucleate upon the surface a larger particle of a lower solubility. This process continues until a steady state equilibrium is reached between the solid crystal existing as a nanoparticle and a solubilised monomeric unit. This process is known as Ostwald ripening or 'defocusing' and generally results in larger polydisperse colloidal particles. Therefore the production of quantum dots depends on using a high concentration of precursor and getting it to nucleate as quickly as possible (Fig. 4.47).

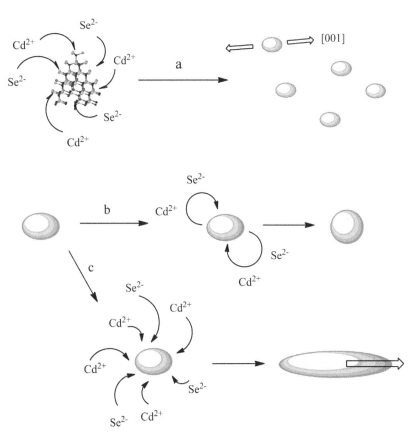

FIGURE 4.47 Schematic showing different conditions that lead to CdSe nanoparticle or nanorod formation. (a) In nanoparticle formation an initial high concentration is rapidly reacted to form discrete slightly eliptical nanoparticles. (b) If the concentration of the precursor ions drops below the solubility product then the Ostwald ripening process occurs and the particles become more spherical. (c) If the concentration of ions is maintained above the solubility product then rods will form as the [00−1] plane (marked with an arrow) grows faster.

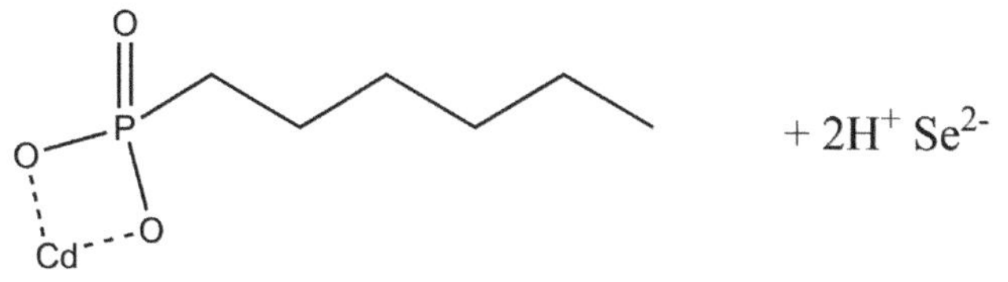

FIGURE 4.48 The hexylphosphonic acid does not stick preferentially to any particular surface on the ZnO wurtzite structure. Instead it forms a complex with cadmium ions which maintains a high concentration of cadmium in the solution and promotes anisotropic growth.

High precursor concentration and fast reaction times are not enough to guarantee the formation of nanorods. Hexylphosphonic acid is added as a surfactant to the trioctadecylphosphine oxide reaction medium. The role of this surfactant in the solution is rather different to the standard mechanism of surfactant directed crystal growth. Normally, in the presence of a crystal nucleating from solution, a surfactant molecule will bind strongly to a particular crystal surface and not so strongly to others. This effectively prevents the deposition of material at sites blocked by surfactant so that material is built up preferentially along another crystal axis. In the growth of cadmium selenide nanorods, the hexylphosphonic acid does not detectably interact with the surface of the forming crystals. Instead it behaves as a ligand and forms a complex with the Cd^{2+} ions immediately after the $Cd(CH_3)_2$ has decomposed (Peng and Peng, 2001). Without the presence of this complex the Cd^{2+} ions will reduce to form a white grey precipitate of cadmium which will react no further. In this respect the hexylphosphonic acid can be viewed as a delivery agent instead of a surfactant, holding the Cd^{2+} in a reactive state long enough for it to diffuse through solution to react with selenium (Fig. 4.48).

The CdSe rods formed after the initial reaction will be grown further by the addition of more of the stock solution at a reduced temperature. Once the reaction solution has been aged for a predetermined amount of time the reaction is quenched by the injection of toluene which drops the temperature rapidly below that required for growth. This is similar in action to quenching the reaction solution using the wet towel method outlined previously. You may find that a range of aspect ratios is obtained so an extra methanol precipitation step is used to isolate out the longest rods (Manna et al., 2000). This is known as size exclusion precipitation where the largest and therefore more unstable particles are forced to crash out of solution so they can be collected. They are rinsed with more methanol to remove the trioctadecylphosphine oxide and can then be redispersed in another solvent for further processing.

To make cadmium selenide nanorods you will need:

This recipe uses tributylphosphine which will catch fire in air. You will need to use gas line techniques as well as a glove box! Take all appropriate safety precautions! The main reaction must be performed in a fume hood.

1. A glove box with an air lock maintained under argon and an accurate set of scales inside. You will also need surplus graduated syringes and needles as the solvents you will be using are kept in septum sealed bottles to protect them.
2. A dry freezer maintained at $-20\ °C$.

For the stock solution:

3. 0.413 mL (0.82 g, 5.73×10^{-3} mol) of $Cd(CH_3)_2$. This should be kept in the glove box and must be weighed out under argon.
4. 2.27 g of a 20% by weight solution of selenium dissolved in tributylphosphine. This should be prepared in the glove box under argon by dissolving 0.453 g (5.73×10^{-3} mol) of selenium in 2.243 mL(1.817 g, 8.98×10^{-3} mol) of tributylphosphine. *Tributylphosphine will catch fire on contact with air!*
5. 16.55 mL (13.41 g, 0.066 mol) of tributylphosphine in a 20 mL sealable gas tight glass vial in the glove box. The vial should be sealed with a rubber septum. This forms the bulk of the stock precursor solution which is stored in a freezer. To ensure that the solution is maintained under argon you will need a larger gas tight container with a valve on the side for connecting to a gas line into which the sealed 20 mL vial can be placed for transfer into the freezer.

For the reaction medium:

6. A Shlenk line with an argon feed and suitable connectors and clamps for fitting to the round bottomed flask used in the reaction. You will need a clamp for holding the reaction flask in place.
7. 0.32 g (1.92×10^{-3} mol) of hexylphosphonic acid. If you have trouble obtaining this surfactant directly then it can be prepared from the more easily obtainable hexylphosphonic chloride by adding it to water. The acid form can then be extracted by washing the water with immiscible diethyl-ether and separating this out. The ether is then evaporated leaving pure hexylphosphonic acid behind. After ensuring this is thoroughly dried out you may keep it in the glove box under argon so that it can be mixed with the trioctadecylphosphine.
8. 3.68 g (9.5×10^{-3} mol) of trioctadecylphosphine oxide weighed into a 50−100 mL three necked round bottomed flask containing a magnetic stir bar. The main neck should be fitted with a valve for connection to a Schlenk line and the other two necks should be sealed with rubber septums. The flask should be stable to rapid thermal changes. This can be done in a glove box but make sure that the hexylphosphonic acid is also in the glove box; it is solid and you won't be able to get it in to the flask very easily without exposing the reaction to air.

9. A heating mantle with magnetic stirring capable of reaching at least 360 °C. On top of the mantle you will need a woods metal alloy heating bath. **This should be setup inside a fume hood near the Schlenk line.**

10. A stopwatch. The progression of the reaction is time sensitive and you will need to time the injections of the stock solution.

11. Glass syringes and long needles for the solution injections. For working with the tributylphosphine remember to flush the needle with argon before drawing in the solution. Use the balloon technique to keep argon pressure maintained in the flask containing the stock solution.

12. Excess toluene for thermally quenching the reaction. This will need to be injected so you need to have it ready in a syringe before use. The toluene must be very dry and there are a few common methods for getting water out of the solvent. The toluene can be refluxed with sodium or a sodium benzophenone mix which will turn the solution blue when the solvent is dry. *Sodium has special safety considerations for handling. Check these before use.* The toluene is then distilled to purify it. It can be easier to distill the toluene directly under reflux with a fractional condenser fitted to the side. The boiling point of toluene is 111 °C so water will be driven off leaving a pure solvent behind. Solid $CaCl_2$ can be added to the solvent which will sink to the bottom and draw water out. Toluene dried like this must be thoroughly filtered to ensure none of the salt remains before use.

For the size selective precipitation:

13. Excess methanol for precipitating out the rods.

14. A centrifuge and centrifuge tubes stable to toluene. They will need to be large enough to hold all of the reaction solution and the methanol used for the precipitation step. Allow for double the volume of reaction solution to be sure.

15. Toluene for rinsing the rods with.

16. A syringe and vials for storing top solution in. Don't throw anything away at this stage as centrifuged solutions will still contain product which can be harvested.

17. Chloroform for redispersing the nanorods in.

18. A syringe and 220 nm PTFE filter with a vial for sample collection.

To make cadmium selenide nanorods:

1. In the glove box under argon prepare the 20 wt% of selenium in tributylphosphine solution. Once the selenium has dissolved add this to the 16.55 mL of tributylphosphine in the 20 mL vial.

2. Add the 0.41 mL of dimethy lcadmium to the selenium solution. Seal the vial with a gas tight rubber septum. Place the vial inside a larger vessel and seal this shut. Make sure it is gas tight.

3. Transfer the cadmium and selenium solution out through the air lock of the glove box. Store the solution immediately in the freezer at −20 °C.

4. In the glove box dissolve the hexylphosphonic acid in the trioctadecylphosphine oxide in the round bottom flask. Plug the septums in place and shut the valve in the main neck. Transfer the flask out of the glove box.

5. Clamp the sealed reaction flask in place over the woods alloy bath on the heating mantle. Connect the flask to the Schlenk line and open the valve to the argon line.

6. Turn on the heating mantle and set it to 360 °C. When the alloy melts then lower the flask in so that the fill line is just below the level of the alloy.

7. Allow the temperature of the reaction solution to equilibrate to 360 °C and get the stopwatch in a position where you can start it easily and see it along with the reaction.

8. Get the stock solution out of the freezer and open up the outer vessel so you can remove the stock solution from within. Flush a needle and syringe with argon and then use the balloon technique to extract 2 mL of the cadmium and selenium stock solution from the vial.

9. Push the needle through a septum in one of the necks. In the exact same moment as you inject the stock solution start the stopwatch. *The stock must be injected as quickly as possible!* You will observe a colour change as CdSe particles are nucleated. The solution temperature will drop to ~300 °C and you want to maintain this. Set the heating mantle to 300 °C.

10. You will now have 1 min and 40 s before you have to add the next aliquot of stock solution to catch the end of the initial focussing reaction. How much you add will depend on the length of rods you are hoping to make. For medium rods add a further 1.5 mL by injection over the course of 380 s (6 min 20 s). This is quite slow and amounts to a few drops every 10 or 20 s. It is worth practicing trying to get the right speed beforehand with some water and a vial.

   For nanorods with very high aspect ratios add 1 mL over 360 s (6 min). Allow the solution to age for an hour at 300 °C then add another 1 mL over 360 s. Repeat this twice more. During the hour wait store the stock solution back in the freezer ensuring that the outer vessel is flushed with argon. In contrast to the spherical preparation the colour will appear to blue shift during reaction. As the rods grow the confinement effect is lessened in the long axis so that the thickness of the rod becomes the dominant factor in determining the colour. As this remains very small the rod solution will turn orange and then shift back to a yellow-green colour as the longitudinal component begins to exhibit bulk behaviour.

11. An hour after the final injection quench the reaction by the rapid injection of toluene. Turn off the heating mantle. Raise the reaction vessel out of the alloy bath and let it cool to room temperature.

12. Using a syringe, inject methanol into the reaction flask until all the cadmium selenide precipitates out of solution and sediments to the bottom. After this point the reaction solution will be stable in air and the tetrabutylphosphine will not ignite.
13. Isolate the flask from the argon feed and disconnect it from the Schlenk line. Slowly open the valve to air then remove one of the rubber septums from a side neck. Decant the entire contents into a centrifuge tube.
14. Centrifuge the tube for 10 min at 13,200 rpm so that all the sediment collects in the bottom of the tube.
15. Pour the top solution into an appropriate solvent waste container. As you near the sediment you may wish to use a teat pipette so as not to disturb the collected CdSe pellet.
16. Add toluene to the tube and redisperse the pellet. Centrifuge the pellet for 30 min at 13,200 rpm. This step will cause the longer and less stable rods to sediment to the bottom of the tube before smaller particles. You might get lucky and all the rods produced are of the same length. If this is the case then the centrifuged solution will appear clear although this could also be a result of centrifuging it for too long. Try slower centrifuge speeds and see if some colour is left behind in the solution. After 30 min you should have some sediment in the bottom of the tube but there should still be colour from rods free in solution. Decant the top solution into another centrifuge tube for further centrifugation and size selection or for storage.
17. The sediment may now be redispersed in chloroform to give a stable colloid of CdSe nanoparticle rods. The longer rods may not be very stable in which case add 1−2 mg of dodecylamine for every 100 mg of CdSe you are trying to disperse.
18. Draw the dispersed colloidal solution into a syringe and purify it through a 220 nm (or similar) PTFE syringe filter into a vial. The colloid is now ready for further processing or analysis. Produced samples should be stored in the dark to protect them from photolytic degradation.

Transmission electron microscopy of the as produced nanorod system may show a number of other shapes in addition to the rods. Commonly you might observe a teardrop morphology indicative of Ostwald ripening having taken place during ageing. This is where the addition of the precursor has been to slow allowing the particle to grow along its thin axis. In samples that have not undergone size selective precipitation you will observe a high number of 'tetrapods' on the grid. These multi armed particles form when the wurtzite rod structure begins to grow on the surface of a small zinc blende impurity which is at the core of these particles. Repeated injections of precursor will allow further branching of the tetrapod arms so that an almost dentritic structure forms. However the branching will only occur two or three times on any one particle. In a similar manner to the quantum dot form a powder x-ray diffraction pattern of the dried powder will show a predominantly wurtzite structure though zinc blende is formed due to the presence of crystal faults (Fig. 4.49).

The CdSe rods will be photoluminescent as in the particles but one of the most interesting things about them is that the emitted light is plane polarised. This is because the fluorescence behaviour only occurs strongly where there is quantum confinement which, for the rods, is in the width dimension. The quantum yield of the unmodified rods will be between 5 and 10% though this can be boosted by

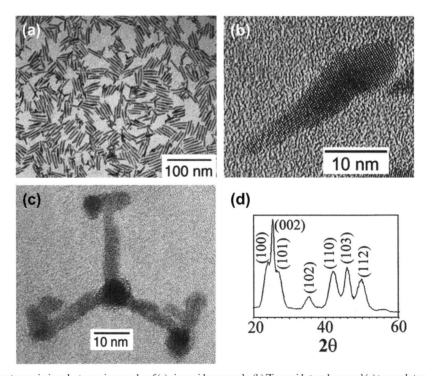

FIGURE 4.49 High resolution transmission electron micrographs of (a) zinc oxide nanorods. (b) Zinc oxide teardrops and (c) terapod structures. Adapted with permission from Manna et al. (2000), Copyright 2000 The American Chemical Society. d) The expected powder x-ray diffraction pattern observed for zinc oxide nanorods which the major reflections indexed to the wurtzite polymorph. *(Adapted with Permission from Peng and Peng (2001). Copyright 2001 The American Chemical Society)*

ZnS coating or by surface functionalisation to block surface exciton traps. The high anisotropy of the rods means that they are relatively easy to orient in the same direction by shear force on a substrate or in a polymer. This is useful if you wish to make a light emitting device for generating polarised light. As a semiconductor material, cadmium based quantum dots have the potential for use in photovoltaic applications for generating electrical energy from light. For one-dimensional dots the recombination of the electron and hole is too fast for the electron to be utilised in practical electrical work. This is not the case in nanorods of the same material where the generated excitons can be transported away in the conduction band down the length of the crystal. This has been demonstrated in the manufacture of very thin solar cell devices which use a mixture of high aspect ratio nanorods with a conductive polymer such as poly(3-hexylthiophene). This polymer can accommodate the electron hole left behind in the CdSe crystal such that under illumination the positive charge can move out of the crystal in the opposite direction to the electron in the crystal conduction band. Because there is a flow of charge a thin layer of the nanorods mixed in with the polymer can be used as a plastic flexible solar cell (Huynh et al., 2002).

MAKING GOLD AND SILVER COLLOIDS

Gold could be considered the earliest used commercial nanomaterial. Gold has been used for hundreds of years in the staining of glass for windows and in the manufacture of rarer items such as the Lycurgus cup. The cup is of roman origin and appears green when light reflects from its surface or red when backlit. When gold was ground extremely finely, or by the addition of salts into the glass melt it was noted that a rich red colour could be produced when light passed through it. Though it was not understood at the time we now know that these colourful effects are a product of plasmon resonance of the electrons in a metal nanoparticle interacting with light (Englebienne et al., 2003). In bulk, gold appears to the eye to have a characteristic yellowish lustre but if a chunk of gold is made small, on the order of tens of nanometres, the 6s electrons in the conduction band can resonate together when interacting with light (Jansen, 2008). Gold nanoparticles will absorb light of a certain wavelength depending on the size of the metal cluster giving the colloid a characteristic colour.

Gold colloids are very useful due to these size and shape dependant optical properties. Perhaps the most common use for gold nanoparticles is in the realm of rapid medical diagnostics and colloids are routinely employed commercially in pregnancy tests (Posthuma-Trumpie et al., 2009). A gold nanoparticle can be coated with antibodies in a way that makes the colloid unstable in the presence of a specific chemical or biological marker. In the case of a pregnancy test this is the protein chorionic gonadotropin which is produced by the placenta of a pregnant woman. This hormone binds to the antibody receptor on the gold particle and causes it to 'crash' out of solution and clump together with other gold particles. This clumping shifts the observed plasmon resonance into the blue and makes the aggregates formed large enough to be caught in a filter paper shaped into a positive or a tick to give a visual indication of pregnancy. There is a wide range of functional receptors and molecules that can be stuck onto the surface of a gold nanoparticle and these recipes will provide a starting point into making the basic spherical and rod shaped particles for use in further experimentation.

In an aqueous colloid a non-soluble solid particle is dispersed within a liquid medium. In this case it will be a solid gold nanoparticle dispersed within water. For a stable colloid to form the solid phase must not precipitate out of solution. This can be difficult to achieve as nanoparticles have a very high surface area when compared to their bulk counterparts. A larger surface area takes energy to maintain and because of this the natural tendency of very small particles is to flock together to minimise the surface area. It is analogous to oil separating out of water. If a mixture of oil and water is shaken up you will see droplets form but then they will separate out. Even though you put energy in to make the oil into smaller droplets the increased surface area is not stable and the oil coalesces and then separates. A solid dispersed in a liquid will experience a similar effect and can sometimes clump together in what is known as a flocculate. By coating the surface of the solid particle it is possible to give it a charge so that when it comes into contact with another particle there is electrostatic repulsion which prevents them aggregating. The same may be achieved by using very bulky molecules adhered to the surface to prevent the particles touching and this is known as steric stabilisation. Whichever method of stabilisation is used it is important that the repulsion between particles under normal Brownian motion is greater than the force of gravity trying to separate out the solid.

The following recipe gives a basic introduction to the formation of metal nanoparticles and with variations of stabilising and reducing agents it is possible to make a wide variety of silver, gold, platinum and palladium nanoparticles. By alteration of the type and concentration of the reducing agent it is possible to control the morphology of the particle formed. In the case of gold, rods and triangles can be formed and these parameters will be explored.

MAKING AN AQUEOUS GOLD COLLOID USING THE FRENS METHOD

With thanks to Dr Mei Li.

There are a number of ways gold colloid can be made but a reliable method which produces quite stable colloids in a short amount of time is the Frens method (Frens, 1973). In this recipe you will take the salt gold tetrachloroaurate and dissolve it in water before heating it under reflux and then adding a reducing agent to form a gold colloid. The reducing agent is sodium citrate which will not only reduce the metal from a soluble ion to form a particle but will also coat the surface of the particle to stabilise it against precipitation. The reduction step in this process takes place over the course of minutes and so the gold particles formed will be slightly larger than those formed over the course of a few seconds. In this preparation the gold colloid produced will be just under 20 nm in average diameter. The speed of metal reduction is key to obtaining smaller nanoparticles. If reduced slowly the nucleating particles have time to collect any unreduced ions in the solution and incorporate them. This leads to larger particles. If a solution of metal ions is reduced quickly and all at once then there is nothing left to add and the resulting colloid consists of small particles. High concentration of reagent, vigorous mixing when adding the reducing agent and a strong reducing agent all contribute to making the formed particles smaller.

After the Frens method an alternative recipe will be given that uses sodium borohydride as a reducing agent. This is much quicker and correspondingly the nanoparticles formed will be smaller.

As the gold nanoparticles formed in this recipe are far smaller than the human eye can resolve they will appear to be transparent. A fun demo to demonstrate that small solid particles are dispersed within the water is to take a laser pointer and aim it through a vial of water and then a vial of colloid, any type of colloid should do. The water has no solids in it and so the laser passes directly through. When it reaches the colloid however the laser beam is scattered by the suspended nanoparticles and can be seen. This is called the Tyndall effect (Figs. 4.50 and 4.51).

FIGURE 4.50   A solution of colloidal gold next to a solution of pure water. The particles in the colloid cause the laser light to scatter.

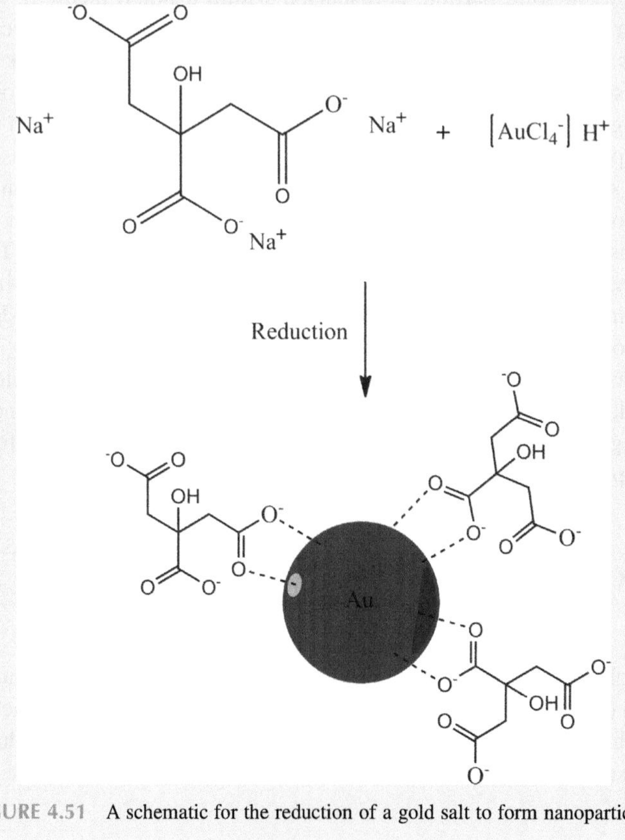

FIGURE 4.51   A schematic for the reduction of a gold salt to form nanoparticles.

Cleaning Glasssware Using Piranha Solution (Aqua Regia)

Having clean glassware before you start any experiment is an obvious standard practice but for the successful synthesis of gold and silver colloids it is vitally important. The glassware used in the following procedures, particularly any vessel where the metal reduction is to take place, must be first cleaned using aqua regia. Even trace amounts of dirt or gold from a previous synthesis will provide a nucleation site at the surface for gold to gather on instead of as free particles in solution.

Aqua regia is a 1:3 by volume mixture of concentrated HNO_3 and HCl respectively. Concentrated generally means the supplied content of the stock solution though nitric acid can vary widely and above 86% by volume in water is known as fuming nitric acid. For this cleaning solution we will be using a 1:3 mixture of 70% HNO_3 and 36–38% HCl. The volume you require depends on the size of the glassware to be cleaned. Ideally you will need enough to totally immerse the glass ware. If you have too little you will need to manoeuvre glassware around in it and, for the sake of safety, you should not need to move things around in the solution more than is necessary. Of course, it is also possible to mix the aqua regia solution directly inside the glassware to be cleaned. The aqua regia is capable of dissolving gold and other noble metals according to the following formula:

$$Au_{(s)} + 3NO_{3(aq)}^{-} + 6H_{(aq)}^{+} \rightarrow Au_{(aq)}^{3+} + 3NO_{2(g)} + 3H_2O_{(l)}$$

$$Au_{(aq)}^{3+} + 4Cl_{(aq)}^{-} \rightarrow AuCl_{(aq)}^{4-}$$

This property was used to hide the Nobel prizes of Max von Laue and James Frank during the Second World War. George de Hevesy (a Nobel prize winning chemist himself) dissolved the medals in aqua regia and placed the resulting solution in a jar. The solution sat on a shelf with hundreds of other chemicals at the Bohr Institute of Theoretical Physics during the German occupation of Copenhagen. It was never discovered and after the war the gold was reclaimed and cast back into medals to be represented to the winners by the Nobel foundation (Hevesy, 1962).

The acid also digests residual organics from the surface of the glass such that only silica should remain after immersion. Do not try to clean large amounts of organic material from the glassware as there is a risk of explosion. Clean glassware as best you can before using the aqua regia as a final step. The solution has to be prepared fresh before each use as it will slowly decompose and lose its cleaning potency. It will appear orange when fresh and will slowly turn more yellow as it degrades. The decomposition reaction will produce chlorine and nitrosyl chloride which will decompose further to nitric oxide. *Because of the evolution of gasses a solution of aqua regia must never be kept in a sealed container.* After cleaning the acid solution must be neutralised by slowly adding sodium bicarbonate before dilution and disposal. Do not add the bicarbonate rapidly as this will heat the solution extremely quickly and may cause it to splash.

To clean glassware using aqua regia you will need:

Every step of mixing the aqua regia and cleaning the glassware is to be performed in a fume cupboard. The acids being used are extremely corrosive. You must wear face protection and suitable heavy duty gloves when handling the acids at any time.

1. A clamp for holding the round bottomed flask securely in place or alternatively, a high walled sturdy glass bath if items are to be immersed in the solution. Any vessel containing the aqua regia should not be more than two thirds full. If you are using a bath of aqua regia then have something with a pouring lip. This will make it easier to transfer the solution out of the bath.
2. A sign on the fume hood stating that you are working with aqua regia. Most laboratories that use this sort of procedure routinely will have a dedicated fume hood.
3. Concentrated HCl and HNO_3 in the amounts required for making up the solution. You should have three times the volume of HCl to HNO_3.
4. A large measuring cylinder and funnel for measuring out the acid volumes.
5. A large High Density PolyEthylene (HDPE) container holding at least 7 times the volume of water as the total volume of aqua regia you intend to make along with enough room for the aqua regia itself. You can use a number of small bottles if this is easier.
6. Large amounts of sodium bicarbonate for neutralising the aqua regia and a spatula.
7. A Pasteur pipette and pH paper for checking the pH of the neutralised solution.
8. Optionally, a long glass rod for stirring the solution.
9. Plastic tongs for handling immersed glassware.
10. A large basin of clean water and a wash bottle for rinsing.

To clean glassware using aqua regia:

1. Measure out the amounts of acid required using the funnel. Add the HNO_3 to the HCl solution slowly. You may see fumes. Stir very gently with the glass rod to mix.
2. If you have prepared this in a glass bath then immerse your glassware into the bath slowly. Ensure there is no trapped air inside.
3. Leave the glass exposed to the aqua regia for 30 min to an hour.
4. Use the plastic tongs to slowly lift the glassware out of the bath letting it drain. Transfer the glassware into the basin of water to rinse and then rinse thoroughly using the wash bottle. If you have filled the round bottomed flask with aqua regia then gently decant it out into a suitable container for neutralisation.
5. The glassware can be dried and used. If you have finished using the aqua regia then you must dispose of it using the following steps.
6. If possible dilute the aqua regia with water. If the volume is not large then you can decant the solution into the HDPE bottle. However, if there is even a remote possibility you might drop the bath then do not entertain the idea of trying. Instead you will need to neutralise the aqua regia in the bath.

7. Add a spatuala full of sodium bicarbonate to the diluted aqua regia. *This will effervesce very vigorously so be careful not to get splashed!* Repeat this process many times. The solution will get hot as it is being neutralised. After a few additions dip the tip of a Pasteur pipette into the solution and dab a small droplet onto pH paper to see if it has been neutralised. Allow the solution to cool each time you add the sodium bicarbonate.

8. Once the solution has been neutralised it may be disposed safely in an appropriate solvent waste container.

To make 100 mL of sodium citrate stabilised gold colloid by the Frens method you will need:
This experiment should be performed in a fume hood although no particular hazards are to be expected.

1. A measuring cylinder that can measure 10−100 mL.
2. A 250 mL two necked round bottom flask and a cork stand. Have glass or rubber stoppers to seal the round bottomed flask with. A larger round bottomed flask will also do. Ideally a round bottomed flask will be roughly a quarter to a half way full when in use. Overfilling a round bottomed flask during a reflux can lead to a variety of complications which could be unsafe or ruin the recipe.
3. A reflux coil condenser that connects to the round bottomed flask. If you get one that is mismatched you may use an adaptor with clips as depicted in this recipe. *It is very important that you have screw clips for the water pipe connections both in and out.*
4. A heating mantle or oil bath fitted with a magnetic stirrer. It should be able to maintain a temperature of 100 °C. One with a fitted clamp stand for holding the glassware in place with is perfect.
5. Depending on the glassware setup you might need an extra stand and clamps to keep everything stable and safe.
6. Tetrachloroaurate salt ($HAuCl_4 \cdot 3H_2O$), Trisodium citrate dihydrate and at least 100 mL of distilled deionised water. The gold salt must be kept cold, dry and in the dark. Although slow compared to silver salts, Au^{3+} can be reduced by light.
7. A magnetic stir bar large enough to ensure the entire solution is being stirred vigorously.
8. A 5 mL volumetric pipette and tips.
9. A four figure balance, weighing boats and a spatula.

To make 100 mL of sodium citrate stabilised gold colloid by the Frens method

1. *Ensure all the glassware is clean!* Dirt on the glassware can act as a nucleation site for the formation of the gold. This can lead to the glass being coated in pink gold nanoparticles but none remaining in solution or even the gold crashing out to form a clumpy black material.
2. Weigh out 0.035 g (8.9×10^{-5} mol) of $HauCl_4 \cdot 3H_2O$. The salt should be a yellow colour so if it is black then this indicates it has spoiled. You may also find that it can appear wet as the salt is very good at absorbing moisture. If this happens you can try placing the wet gold salt in a vacuum to dry it.
3. Measure out 90 mL of water and add it to the round bottomed flask. To this add the gold salt and dissolve it. It should be a transparent yellow colour (Fig. 4.52).
4. Clamp the round bottomed flask by the neck above the oil bath (or heating mantle) which should, at the moment, be off for safety. Setup the reflux condenser as depicted ensuring that you have clipped the water hoses to the reflux condenser. If they pop off it can be a terrible mess and laboratories have been flooded before by only small flows being left on for an entire weekend. Ensure the water inlet hose connects to the bottom of the condenser. For the out hose place it down the sink. It can be a good idea to hold the out hose in a clamp to ensure it will not flail out by accident. The condenser should be able to sit in the round bottomed flask but once you begin the reflux it is always good practice to clamp it in place for safety.
5. Lower the round bottomed flask with the reflux condenser fitted into the oil bath (or heating mantle). In the case of the oil bath you should be able to see about half a centimetre of the reflux solution standing proud of the fill line of the oil.
6. Turn on the water slowly to test for leaks in your system. You do not need a raging torrent passing through the condenser and a flow rate just above a trickle will suffice. In some buildings the pressure can vary so make sure it is flowing enough that it doesn't stop during a drop.
7. Turn on the heating mantle and set it to 100 °C. Turn on the magnetic stirrer and set it to go as fast as possible without the magnetic bead losing control or the gold solution splashing about.
8. While waiting for the gold solution to come up to temperature, weigh out 0.1178 g of trisodium citrate dehydrate (0.4×10^{-3} mol) and dissolve it in 10 mL of water.
9. Once the gold solution has come up to temperature you should see condensation forming in the reflux column and falling back into the solution. It is important that the solution is up to temperature or the reaction may not work very well.
10. Add the sodium citrate solution through the unused neck in the round bottomed flask as fast as possible using the pipette. Replace the stopper into the neck after you have done so.

FIGURE 4.52   A photograph of a standard reflux setup. Inset: a tetrachloroaurate salt dissolved in water.

11. Keep the solution refluxing for 15—30 min. You will see the solution change colour from yellow to a red wine colour. Sometimes you may also see an intermediate stage where the solution appears to be turning black. It could be mistaken for the gold precipitation out of solution but eventually you should see the red colour emerge. The darkening is a result of the forming particle being temporarily elongated before a size focussing process makes them more spherical (Pong et al., 2007) (Fig. 4.53).

12. After half an hour take the round bottom flask off the heat and allow the solution to cool in a cork ring. Turn off the water, heating mantle and magnetic stirrer. You should now have a gold colloid.

The colour will depend on the size of the gold nanoparticle formed and this is the most immediate way to gauge the size of the particles. Dynamic light scattering is also a rapid method for assessing the size and polydispersity of the colloid but this will give the hydrodynamic radius of the particles and not the precise size of the metal core. Due to plasmon resonance UV-vis spectroscopy can be used to gauge the size of the gold nanoparticles based on the position of the absorbance peak. Red colloidal

FIGURE 4.53   The solution appears to darken half way through the reduction process.

dispersions are around 5–20 nm and this will appear increasingly purple then blue as the particle diameter becomes closer to 100 nm. Transmission electron microscopy will allow you to image the colloid and if you are lucky enough to have a high resolution system it is possible to image the crystal facets of the gold. Instead of being perfectly spherical a gold particle will appear to be composed of tetrahedral units. A single particle will often demonstrate a five fold axis of symmetry around the centre and this is a result of multiple twinning where many crystal units of face centred cubic arranged gold clusters have nucleated from a single point. Controlling the relative rates at which these crystal facets form are the key to producing anisotropic gold nanoparticles and this will be explored further on.

If the produced colloid has a black precipitate in it which gets larger over the course of hours or days then the colloid is not stable. If you only wish to use the gold colloid for demonstration purposes then you can run it through a small syringe driven filter. This will eliminate larger debris and you may find the extruded colloid lasts longer than it might otherwise. Try adding more of the sodium citrate to ensure that the formed gold is fully coated. The gold can also reduce on the surface of the glass which will lead to the solution going clear and the glass turning black or red/pink. This can only be solved by cleaning the glassware properly (Fig. 4.54).

Making smaller gold nanoparticles using sodium borohydride

This preparation is similar to the first recipe though it is quicker to run and does not require heating the gold solution. It is based on Jana method and provides seeds for the nucleation or gold nanorods (Jana et al., 2001). Sodium borohydride is a strong reducing agent and by treating a gold salt solution with it in the presence of a stabiliser the nanoparticles are nucleated on the order of seconds as opposed to minutes. This recipe will produce a sodium citrate stabilised colloid of gold nanoparticles around 4 nm in diameter. The temperature of the reaction solution will influence the size of the particle produced. The instructions that follow are performed at room temperature but you may wish to experiment with cooling the gold salt solutions using an ice bath prior to reduction.

To make gold nanoparticles using sodium borohydride you will need:

1. A clean conical flask of 30–50 mL in size and a clamp to hold it with. Generally it has been found that a smaller vial produces better results as the reactants are mixed more rapidly.
2. Tetrachloroaurate salt ($HAuCl_4 \cdot 3H_2O$).
3. Trisodium citrate dehydrate. Other monobasic and dibasic citrates can be substituted. In this recipe the citrate molecules will coat the surface of the nanoparticles and act as a stabilising agent.
4. Sodium borohydride ($NaBH_4$). Ensure this is fresh and has been kept absolutely dry. This chemical must be stored in a desiccator when not in use as it can 'go off' if exposed to water and lose its potency.
5. A magnetic stirrer and large stir bar.

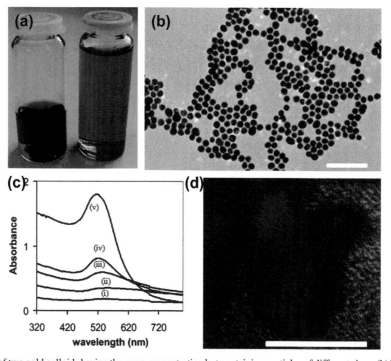

FIGURE 4.54   (a) A photograph of two gold colloids having the same concentration but containing particles of different sizes. (b) Transmission electron micrograph of a sodium citrate stabilised gold colloid. Scale bar = 200 nm. (c) UV-visible spectrum of a gold colloid as it develops (i) colourless (ii) dark blue (iii) dark purple (iv) purple (v) ruby-red. *Reprinted with permission from Pong et al (2007). Copyright 2007 American Chemical Society.* (d) High resolution transmission electron micrograph showing the lattice fringes of atomic gold layers making up the particle.

6. Distilled deionised water.
7. A four figure weighing balance, weighing boats and clean spatulas. Small vials for making up solutions in will also be useful.
8. A volumetric pipette and disposable tips. This will be useful for adding the borohydride solution quickly.

To make the gold colloid using sodium borohydride:

1. Weigh out 0.01 g (2.5×10^{-4} mol) of the $HAuCl_4 \cdot 3H_2O$ and dissolve in 20 mL of water in the conical flask. This will appear yellow.
2. Weight out 0.0735 g (2.5×10^{-4} mol) of sodium citrate and add it to the gold solution. As the solution is at room temperature the reduction of the gold salt by the citrate will not happen straight away. You will have a few minutes to work before you notice anything happening but if left for a long time the gold will reduce and form solids in the flask.
3. Place the magnetic stir bar into the conical flask and clamp the flask in place so it will not fall over. Set it to stir as vigorously as you dare.
4. Make up a 0.1 M solution of sodium borohydride. You will need very little and the concentration does not have to be exact so long as there is an excess of the borohydride to drive the reduction. If you have scales accurate enough weigh out 4 mg (0.004 g) of borohydride and dissolve it in 1 mL of water. Otherwise you can weigh out 40 mg into 10 mL if you can afford to waste some. This solution will begin to effervesce which indicates the borohydride is reacting with water to produce hydrogen according to the following formula:

$$NaBH_4 + 2H_2O \rightarrow NaBO_2 + 4H_2$$

There should not be enough to produce any sort of fire hazard but the bubbling solution does let you know that the borohydride solution is potent. Use immediately before the borohydride is converted to borates and take care to not get any on your skin. To dispose of waste solution leave the vial open to air for at least $3-4$ h by which time the borohydride will have fully reacted and may be poured down the sink.

5. Add 0.6 mL of the sodium borohydride solution rapidly to the vigorously stirring gold/citrate solution using a volumetric pipette. You will see an immediate colour change to a transparent orange red solution which is a result of the plasmon resonance of the formed gold nanoparticles.
6. Leave the formed colloid for about 3 h at room temperature in an open flask or vial. This will allow any excess borohydride time to react with the water.

You may characterise the colloid in an identical manner to the other gold colloid recipe.

MAKING AN OIL STABLE ALKANETHIOL COATED GOLD COLLOID USING THE BRUST METHOD

In this classic preparation a gold colloid is prepared by borohydride reduction in the presence of an alkanethiol (Brust et al., 1994). There are two major differences with the gold particles prepared using this method as compared to the aqueous routes listed previously. Firstly, the particles are stabilised with a long chain aliphatic group strongly bound to the surface of the gold through a terminal thiol and this enables them to be dispersed into a non polar solvent. Secondly, the colloid can be dried down into a waxy residue which can then be redispersed into another solvent phase. Drying down an aqueous colloid will generally result in an irreversible flocculation of the solid material. In this preparation the gold colloid can be redispersed into toluene, pentane and chloroform as well as other solvents. The gold particles are smaller than those produced by the other reduction methods and can be produced over a diameter range $2-5$ nm (roughly $110-4200$ atoms per particle respectively). Below 2 nm in diameter the particles are technically classed as metal clusters composed of only hundreds of gold atoms but will still have an optically observable plasmon absorbance. Clusters consisting of only 200 atoms or less will lose their plasmon band and begin to exhibit quantised size effects. Gold clusters in these size domains exhibit high catalytic activity for the reduction of CO to CO_2, a phenomenon not observed in the bulk noble metal famed for being chemically inert. Additionally, metal clusters show quantised optical effects distinct from plasmon resonance of interest in optical physics experiments. Some discussion of this is given in the section on fluorescent silver nanoclusters.

The synthesis is slightly more involved than just adding a reducing agent and begins with the tetrachloroaurate salt dissolved in water with an alkyammonium salt. The gold exists in the water as the anionic species $AuCl_4^-$ and this coordinates to the alkylammonium cation. Because of the tail groups of the surfactant are so hydrophobic the complex can successfully transfer into a non polar solvent when one is mixed with the aqueous solution. Once the gold species is within the oil phase dodecanethiol is added which will do nothing until a reducing agent is added and the gold begins to nucleate. At this point the thiol group coordinates to the gold surface which caps and stabilised the nanoparticle.

$$AuCl_{4(aq)}^- + N(C_8H_{17})_4 + (C_6H_5Me) \rightarrow N(C_8H_{17})_4 + AuCl_4^- (C_6H_5Me)$$

$$mAuCl_4^-(C_6H_5Me) + nC_{12}H_{25}SH(C_6H_5Me) + 3me^- \rightarrow 4mCl_{(aq)}^- + (Au_m)(C_{12}H_{25}SH)_n(C_6H_5Me)$$

The core diameter of the gold nanoparticles can be controlled by adjusting the molar ratio of gold to decanethiol, the temperature of the solution during the addition of borohydride and the rate at which it is added (Hostetler et al., 2008). The largest effect is from the adjustment of the dodecanethiol to gold molar ratio prior to reduction and variables are given for controlling the size using this parameter. Once the gold particles have been formed by reduction a rotary evaporator is used to remove the toluene and any remaining water. The product will still have unwanted reagents present at this stage so the waxy residue of the colloid remaining after evaporation is redispersed in a small amount of toluene. This dispersion is then treated with a large amount of ethanol which causes the alkanethiol coated gold to precipitate out of the solution. The sample is centrifuged in an excess of ethanol and any impurities will remain in the ethanol toluene mixture. The sample is washed using the ethanol and then centrifuged again so that only organic groups sticking to the surface of gold nanoparticles will remain. Once isolated and dried the particles can be redispersed into toluene, chloroform, pentane and other non polar solvents.

To make an oil stable gold cluster colloid you will need:

This is to be performed in a fume hood. Thiol containing molecules stink. Even in a fume hood it is likely you will smell them so keep any bottles closed when not directly transferring chemicals between vessels.

1. 0.354 g (0.001 mol) of $HAuCl_4 \cdot 3H_2O$ dissolved into 30 mL of water. This will give a 0.03 mol/L solution.
2. 2.187 g (0.0043 mol) of tetraoctylammonium bromide in 80 mL of toluene in a 300–400 mL beaker. This will give a 0.05 mol/L solution. *Toluene is flammable, keep away from flame and work within a fume hood.*
3. 0.479 mL (0.002 mol, 0.4048 g) of dodecanethiol. The molar ratio of dodecanethiol will to gold will heavily influence the size of the nanoparticles. This particular ratio will produce particles around 2 nm in diameter. Higher molar ratios will produce larger particles as the addition of gold to the particle surface is blocked. A 4:1 thiol to gold molar ratio will produce particles around 4 nm and a 12:1 thiol to gold molar ratio will produce particles around 6 nm in diameter.
4. 0.38 g (0.01 mol) of sodium borohydride. You will also need a glass vial containing 25 mL of water. Do not dissolve the $NaBH_4$ in the water until it is required.
5. A 250 mL round bottomed flask with a quick fit neck and a glass plug. This will be used for reducing the gold in the toluene phase and eventually for rotary evaporation to isolate the product. You will need a clamp or cork ring for holding the flask in.
6. A magnetic stirrer and a large stir bar. Alternatively, you may wish to use an overhead stirrer as this is far more vigorous. This may be required for the production of sub 2 nm clusters. An overhead stirrer will probably cause your solution to splash out of a beaker so you may wish to use a large wide necked conical flask instead.
7. A Pasteur pipette and a 5 ml (or larger) volumetric pipette.
8. A rotary evaporator with a fitting for the 250 mL round bottomed flask and an oil bath for heating.
9. 800 mL of ethanol used for inducing precipitation of the colloid from toluene and rinsing.
10. A 600 mL beaker with a clean stir bar. This will be used for ethanol precipitation of the product. You will need a cover for the beaker as it will be sat in a freezer at one point. Clingfilm will suffice.
11. A freezer at ~ -18 to $-22\,°C$ which the 600 mL beaker can sit in.
12. A centrifuge and suitable centrifuge tubes. Falcon tubes are recommended but you will be able to transfer precipitates into smaller tubes if required.
13. A vacuum oven or bell jar fitted to a vacuum pump. This will be used for drying the product.
14. 10 mL of the non polar solvent you wish to have your produced colloid in. We will use toluene.

To make an oil stable gold cluster colloid:

1. **In the fume hood** place the beaker containing the toluene and tetraoctylammonium chloride onto the magnetic stirrer. Set the solution to stir vigorously.
2. Add the yellow $HAuCl_4$ in water solution to the stirring toluene solution. You should see the toluene become an orange brown in colour as the gold ion is drawn into the toluene to form a complex with the alkylammonium surfactant.
3. After a minute or so turn off the stirring. The water should sink to the bottom and form a transparent layer as there should be no gold salt remaining in it. If this is so then use a Pasteur pipette to remove the water from the beaker so that only the coloured toluene layer remains. It does not matter if some of the water remains as you will be adding more to deliver the borohydride. If there is still perceptible colour in the water then continue stirring for another couple of minutes.
4. To the remaining toluene add the 0.749 mL of dodecanethiol with stirring. Allow this to stir for 10 min. For reactions using higher molar ratios of dodecanethiol to gold you will see the solution become pale yellow or even transparent within 5 min.
5. Set the gold containing toluene solution on to stir at as high a speed as possible. Use an overhead stirrer if you have one.
6. Dissolve the 0.38 g of $NaBH_4$ into the 25 mL of water.

7. Add the fresh borohydride solution to the vigorously stirring toluene over the course of ten seconds. The toluene will darken. You may find you wish to add it more slowly depending on results. The rate at which the reducing agent is added has a major influence of the resulting particle diameter. After the addition of the reducing agent allow the solution to stir for 3 h.
8. Transfer the toluene and gold particle solution into the 250 mL round bottomed flask. Set the heating bath to just below 50 °C and reduce the volume of the solution to about 10 mL. If the solvent has been completely removed it will appear as a waxy brown black substance. This can be redispersed by adding 10 mL of toluene.
9. Transfer the product in toluene from the round bottomed flask into 400 mL of ethanol under stirring. If necessary rinse the round bottomed flask out with ethanol. You should observe the product precipitating out of solution. Any dodecanethiol not bound to the surface of a gold nanoparticle will remain in the ethanol phase.
10. Set the beaker containing the ethanol and product into the freezer. Cover the beaker with clingfilm to prevent evaporation. All of the gold particle product should settle to the bottom of the beaker after 4 h.
11. Carefully remove the mixed toluene and ethanol top solution either by gentle pouring or pipetting away until only the sediment at the bottom of the beaker remains in less than 50 mL of ethanol. Pour this into a centrifuge tube making sure all the sediment is transferred.
12. Centrifuge the precipitated colloid at 8000—9000 rpm for ten minutes or until all the sediment has collected in the bottom.
13. Add a drop or two of toluene to allow the precipitate to be redispersed then add enough ethanol to fill the tube. For a Falcon tube this will be about 40 mL. Shake the solution and then centrifuge the sample again.
14. Decant the top solution off and give the sample one more rinse and centrifuge with only ethanol.
15. Decant away the top solution once more and then allow the remaining ethanol to evaporate.
16. Place the pellet sample still in the Falcon tube into a vacuum chamber to remove the last of the ethanol or any remaining solvent. 30 min should be enough.
17. Redisperse the precipitated gold clusters into the 10 mL of toluene. You now have a sample ready for analysis.

Toluene is volatile enough that a drop of the purified colloid can be applied directly to a copper electron microscopy grid and allowed to dry. Transmission electron microscopy should show electron dense particles packed closely together but not touching. The spacing is due to the alkanethiol coating over the surface of the particles but will be slightly larger than the actual length of one monolayer due to disordering. This type of oil soluble colloid can be easily employed in the production of a sensitive chemoresistive sensor (Wohltjen and Snow, 1998). The colloid is directly deposited onto an interdigitated electrode where a dry film of the alkanethiol coated gold will pack together to form an electrically resistive connection. In the presence of specific organic molecules, such as toluene or tetrachloroethylene, the resistivity changes drastically as these molecules are adsorbed into the alkanethiol/gold layer. The detection level of this type of sensor is in parts per million though the thiol groups are susceptible to oxidation in ambient atmospheres. This makes the long term stability of these sensors an issue and despite some progress in using multiple thiol groups on the terminating stabilisation groups attention has turned towards similar sensors using carbon nanotubes as the sensor component instead. More successfully, the capacity of the organo stabilised gold to form a film by casting has allowed films to be formed on carbon electrons and used to catalytically oxidise CO (Daniel and Astruc, 2004). The organic ligand layer can be cross linked to produce a solid thin film. Gold is only catalytically active at the nanoscale and so a thin film assembled on an electrode or metal oxide surface can be used in catalytic redox chemistry reactions.

Along more simple lines, a basic colorimetric test in organic solvent for inorganic anions can be made using the produced colloid (Watanabe et al., 2002). The particles can undergo ligand exchange so that the stabilising molecules are terminated with amide groups. When dispersed in dichloromethane the presence of sulphate, phosphate and nitrate anions reduces the intensity of the observed plasmon peak in concentrations on the order of 10^{-8} mol/L.

HOW TO MAKE GOLD NANORODS

With thanks to Michael Thomas

Gold nanorods add, literally, an extra dimension to spherical gold nanoparticles. By extending a spherical gold nanoparticle into a rod a dual plasmon band is formed. By tuning the reaction conditions during formation, gold nanorods can be produced of various aspect ratios with tailor-made dual plasmon bands. Their anisotropic structure also lends itself to alignment and various projects can be conceived where gold nanorods may be aligned into large wires or assemblies. They may even be functionalised to give visible responses to optical or electronic stimulus.

Gold nanorods are produced by adding gold 'seeds' to a growth solution containing more gold salts, a weak reducing agent and an excess of surfactant which will act as a structure director during the formation of the gold rods (Gole and Murphy, 2004). Tetrachloroaurate is again used as the gold source in the solution but instead of being directly reduced from Au^{3+} to Au^0 the gold is weakly reduced first to Au^+ by ascorbic acid in solution and then only further reduced to Au^0 at the surface of the gold particle. In this recipe the structure directing surfactant is cetyltrimethylammonium bromide also known as CTAB (Johnson et al., 2002). This cationic (positively charged) surfactant will preferentially adsorb onto the 100 plane of the face centred cubic gold nanocrystal in solution. This effectively blocks gold from being deposited on this crystal face from solution and so the particle begins to grow along the 111 crystal lattice face. Eventually this will form a rod (Fig. 4.55a,b).

To make gold nanorods you will need:

1. Tetrachloroaurate salt ($HAuCl_4 \cdot 3H_2O$).
2. L-Ascorbic acid, also known as vitamin C. It will most likely be called L-asorbic acid when you purchase it as a chemical. The L stands for 'laevo' and denotes the specific stereochemical isomer of this molecule. That is to say that it has a right and left hand form.
3. Cetyltrimethylammonium bromide (CTAB) surfactant.
4. One millilitre of a 'seed' solution of citrate stabilised gold nanoparticles. For the best results use particles as small as possible. The 3–4 nm particles prepared in the borohydride reduction prep will work well. The concentration of gold in the solution should be around 2.5×10^{-4} M.
5. A water bath to keep the growth solutions between 27 and 35 °C. It can work just as well to keep the sealed solutions in a drying oven maintained at this temperature.
6. A four figure weighing balance, weighing boats and clean spatulas.
7. Two 20 mL vials and one 100 mL conical flask for the growth solutions. A 20 mL vial will also be required for the ascorbic acid solution.
8. A volumetric pipette capable of measuring out between 50 and 250 μL and disposable pipette tips.
9. Distilled and deionised water and a measuring cylinder that can measure 20–100 mL.
10. A centrifuge for isolation of the produced nanorods and suitable centrifuge tubes.

To make gold nanorods:

1. Make up a 0.1 M solution of CTAB in 100 mL of water by weighing out 3.644 g (0.01 mol) of CTAB and adding it to 100 mL of water. CTAB can be difficult to dissolve so heat the water and stir it until the solution is transparent then allow it to cool.
2. Weigh out 0.001 g of the gold salt and add it to the CTAB solution. It will turn to an orange colour as the $[AuCl_4]^-$ anion forms a complex with the cationic CTAB. This is not a plasmon resonance effect and does not mean the gold has reduced!
3. Make up a 0.1 M solution of ascorbic acid by weighing out 0.176 g (0.001 mol) of ascorbic acid and dissolving it in 10 mL of water.
4. Add 9 mL of the CTAB/gold salt solution to one of the 20 mL vials then mark this 'A'. Add another 9 mL of the CTAB/gold salt solution to a second 20 mL vial and mark this 'B'. Finally add 45 mL of the CTAB/gold salt solution to the 100 mL vial and mark this 'C'. Warm them to 35 °C using a water bath or oven.

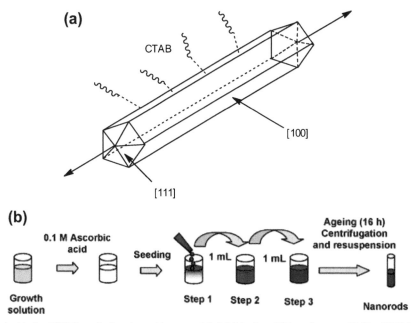

FIGURE 4.55 (a) CTAB adsorbed to the [100] face of a forming gold nanoparticle inhibits the addition of more gold. The [111] facets grow preferentially to form a rod structure. (b) A schematic representation of the seed mediated growth of nanorods. *(Reprinted with Permission from Gole and Murphy (2004). Copyright 2004 American Chemical Society)*

5. Using a volumetric pipette add 50 μL of the 0.1 M ascorbic acid solution to vial A. Add the same amount to vial B. Add 250 μL of the 0.1 M ascorbic acid solution to vial C. The colour from all three solutions will disappear as the gold is reduced from Au^{3+} to Au^+. The ascorbic acid is only weakly reducing so solid gold will not form until the seed particles are added.

6. Add 1 mL of the gold nanoparticle 'seed' solution to vial A. Give it one quick swirl then leave it for 15 s. It will begin to turn a red colour which indicates that the gold seeds are growing.

7. After 15 s remove 1 mL of solution A and add it to solution B. Give it one swirl and then leave to stand. Once more the solution should start to change colour but it should appear slightly slower this time turning from a red orange to a deeper red. The sample from solution A contains larger gold particles than before which will shift any plasmon absorbance towards the blue end of the spectrum. Leave for 30 s.

8. After 30 s take 1 mL of solution B and add to solution C. Give it one swirl to mix it and then leave it to stand at 35 °C for 16 h. This solution will begin to darken to a purple colour as the particles within it get larger. More solution is needed as the seed particles from solution B are bigger and will require an excess of gold to grow. This is reflected by the relatively slow nucleation compared to solutions A and B.

   Larger seeds will lead to low aspect ratio rods with a fairly uniform length which is what might be expected from solution C. Smaller seeds, such as those from the borohydride reduced solution or solution A and B, will lead to high aspect ratio rods which may differ in length.

   Lowering the temperature can lead to high aspect ratio rods but can also lead to higher differences in the relative length when the rods are compared to one another.

9. After 16 h the solution will appear to have separated into a lower purple blue layer and a red/brown top solution. The top solution contains large spherical. Hexagonal or triangular nanoparticles which must be removed before further purification of the gold nanorods. This can be achieved by using a volumetric pipette to gently suck away the top most layer but be warned that it is very difficult to get a gold nanorod sample without the other shapes being present.

10. Once only the purple coloured bottom solution remains decant it into centrifuge tubes. Make sure the centrifuge is balanced and you may add water to do this. Centrifuge the solution at 7000–10,000 rpm for ten minutes. After this time the nanorods will settle to the bottom and you can pour away the clear solution which will contain the excess CTAB surfactant. Resuspend the gold rods into water within the centrifuge tube and then centrifuge again at the same settings. After the second centrifuge step you may collect the gold nanorods for further analysis as a pellet or by suspending in water.

In contrast to a spherical colloid, particle sizing by dynamic light scattering is not easily interpreted as the calculations accounting for size in these machines assume a spherical, not anisotropic, shape. Transmission electron microscopy is a useful assessment method as it allows for visual size distribution analysis as well as crystallographic information if the microscope has a diffraction mode. UV-visible spectroscopy will also allow you to characterise the longitudinal and transverse plasmon absorbance peaks. That is to say the nanorods will have two optical absorbencies in the width and length dimensions of the rod. If the rods are very long or thick it is possible that only one or neither of the bands will be seen as the plasmon absorbance is not in the visible range (Fig. 4.56).

HOW TO MAKE TRIANGULAR GOLD NANOPARTICLES

Adapted from Miranda et al. (2010)

The description of metal nanoparticles as spherical is misleading as a cluster of metal atoms together will pack in a face centred cubic structure. The shape is determined by whatever configuration will present the lowest surface energy during the growth and ageing of the particle. Different facets of the crystal have a different surface energy meaning some will grow faster than others. In the previous recipe the growth of gold along a preferred crystal direction (the [111] plane) was demonstrated. In general the crystallographic [111], [110] and [001] planes (Hornyak et al., 2009) have the lowest surface energy. This means that these are the facets that will grow most easily in a forming crystal like a gold nanoparticle. The equilibrium of facet formation can be altered by the addition of additives during the growth stage, as has been seen with the nanorods, so that a highly anisotropic crystal form can be produced. To produce an anisotropic gold nanostructure, surfactants in the solution are added which will inhibit the growth along a given axis. However, this can be limited in that the thermodynamic driving force behind nucleation will still promote growth in the other axis depending on the strength of the reducing agent. In this recipe we will use a technique for controlling the reduction rate in addition to the use of capping agents to fine tune the reaction equilibrium to produce gold nanotriangles.

The basic reaction template for this procedure is the 'polyol' method which involves adding a metal salt to a refluxing solution of ethylene glycol or other polyol (Li et al., 2008). In a similar manner to the reduction of gold salts by the sodium citrate, the polyols reduce and cap the nucleated gold particles. In order to generate a triangular particle in this synthesis the solution is not heated and instead of a hydrothermal reduction mechanism the reduction is initiated photolytically using a porphyrin and amine redox reaction. In the presence of light the tin centred porphyrin generates an excited state which reacts with ethanolamine in the solution to form a reactive radical porphyrin species. This reacts with the $Au^{(III)}$ cation and reduces it to the unstable $Au^{(II)}$ which is then rapidly

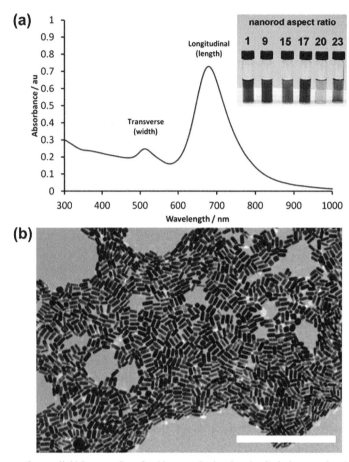

FIGURE 4.56 (a) A UV-visible spectrum for a colloidal suspension of gold nanorods showing the dual plasmon absorbencies along the length and width of the nanorod. Inset: A photograph of various gold nanorod dispersions having different aspect ratios. *Reprinted with permission from Gole and Murphy (2004). Copyright 2004 The American Chemical Society.* (b) A transmission electron micrograph of gold nanorods. Scale bar = 500 nm.

reduced further to solid $Au^{(0)}$ by disproportionation or reaction with water. The irradiation forms seed particles in the solution which are then aged in the dark. The geometry of the seed particle is conducive to slow growth along the edges of the particle which the [111] plane growing comparatively very slowly.

To make gold nanotriangles you will need:

As with the other preparations involving gold you must ensure all the glassware used is clean before use. Solutions used should be syringed through a 0.2 μ filter to ensure purity before use.

1. 20 μL of a 0.02 mol/L tetrachloroaurate solution. This can be prepared as a stock solution by dissolving 15 mg of $HAuCl_4 \cdot 3H_2O$ in 100 mL of water. This solution should be kept in the dark when not being used as it will reduce if left in the light long term.
2. 0.199 mL (1.5×10^{-3} mol, 0.2237 g of triethanolamine. This will be solid below 21 °C so you will need to warm it slightly if you find that the contents of the bottle have solidified.
3. A stock solution of 0.2 mol/L ammonia in water. This can be prepared by dilution of a 28% ammonia stock which is roughly 15 mol/L. Add 2 mL of concentrated ammonia solution to 148 mL of water in a screw top Schott bottle in a fume hood. This is twice the concentration it will be used at but we must make it slightly stronger so that the pH can be adjusted upon the addition of the ethanolamine then made up to the correct concentration by dilution.
4. 100 μL of a 3 μmol/L solution of Sn(IV) meso-tetra(N-methyl-4-pyridyl)porphine tetratosylate chloride. A stock solution can be prepared by dissolving 44 mg of the porphyrin into 1 L of water. Dilute 1 mL of this solution with 9 mL of water to give the correct working concentration.
5. A stock solution of 0.01 mol/L cetyltrimethylammonium bromide in water solution. This can be prepared by dissolving 36 mg of the surfactant into 10 mL of water.
6. A 250 W discharge lamp in a screened area of a fume hood or laboratory. This will be used for driving the photoreduction.
7. A 2−5 mL screw top glass vial for performing the reaction in.
8. Volumetric pipettes capable of delivering 10−100 μL and 1−5 mL volumes.
9. A pH meter that can monitor small (~1 mL) volumes or some pH paper.

To make gold nanotriangles:

1. Add the 0.199 mL of triethanlolamine to 5 mL of the 0.2 mol/L ammonia solution in a vial under stirring. When the volume is doubled after the pH adjustment step you should have a solution of 0.15 mol/L triethanolamine and 0.1 mol/L ammonia.
2. Add 250 μL of the double strength triethanolamine and ammonia solution to a small reaction vial.
3. To the vial add 20 μL of the 0.02 mol/L tetrachloroaurate solution. You should see the yellow colour of the gold solution disappear and the solution will be transparent and colourless. Adjust the pH of the solution to 7 using microlitre aliquots of the 1 mol/L HCl solution dispensed by pipette. Keep a track of how much you add. Check the pH using a meter with a small probe or by dabbing small amounts onto pH paper. When the correct pH has been reached make the volume up to a total of 520 μL by adding 250 μL (minus whatever the volume of acid was added) of water. You should now have a solution of 0.15 mol/L triethanolamine and 0.1 mol/L ammonia. Keep the solution under stirring. If the solution is below pH 6 during the reduction of the metal then the reaction will be too fast to result in the triangle structure.
4. Add 500 μL of the cetyltrimethylammonium bromide solution.
5. Add 2.1 mL of water.
6. Add 100 μL of the tin porphyrin solution. This may seem like a very dilute amount but you should observe a light pinkish colour to the solution upon addition from the presence of the porphyrin. If there is too much the reaction will proceed too quickly and will produce random morphologies.
7. Seal the vial so that the contents will not evaporate. Place under the lamp so that the photoreaction can occur. Leave under irradiation for 2 h. You should see the colour of the solution change to violet as seeds having the correct crystal orientation for further anisotropic growth are nucleated.
8. After 2 h turn off the lamp and place the reaction vial in a dark place for 24 h. During this time Ostwald ripening will take place and the adsorbed smaller particles will lengthen along specific axis of the seed particles resulting in a triangular shape. After 24 h the colloid is ready for analysis.

UV-visible spectroscopy should show a plasmon band at around 540 and, if the spectrometer has the range, another at 820 nm. The visible band is a plasmon oscillation in the thickness of the triangle while the band at 820 nm is for plasmon resonances along the length of the triangles. Transmission electron microscopy should reveal that over 80% of the formed material is in the triangular structure with the rest of the material being roughly spherical. The triangles should be around 15 to 19 nm thick with edge lengths ranging from 133 to 430 nm. As you look at the triangular particle lying flat the [111] lattice plane is exposed which may show some fluctuation in density under the beam. You may be lucky enough to find a triangle lying side on under the beam and this may appear to have a darkened line running through it. This is a twinning plane in the centre where two crystals have grown from the same nucleation point. As demonstrated in the production of a 'zero' dimensional gold colloid a particle of approximately spherical dimensions often has a five-fold twinning axis. Some control over the average edge length of the triangles is possible by using less or more of the porphyrin. More porphyrin will result in faster nucleation times and smaller seed particles. Edge lengths down to 45 nm can be achieved using a concentration of 1 μM phorphyrin. However, the triangular morphology is les prominent with an increase in the reduction speed and many truncated corners might be observed. A powder X-ray diffraction pattern of the plates will show a very strong peak for the [111] plane with other reflections comparatively very weak (JCPDS No. 4-0783) (Huang et al., 2006). Because the face of the triangle is the stabilised [111] plane this is the most strongly reflected plane in the X-ray diffractometer.

USING PLASMON RESONANCE IN GOLD COLLOIDS AS A COLORIMETRIC TEST FOR VARIOUS MOLECULES

The use of gold colloids as diagnostic tests has already been briefly mentioned in determining the presence of pregnancy hormones. A colloidal particle is coated with receptors that cause it to flocculate in the presence of a hormone. For quick home tests these flocculates get caught in a filter shaped like a positive or tick to give a visual affirmation. A more advanced development is to observe shifts in the colour of a colloid when changes in the aqueous environment promote changes in the plasmon resonance. Because plasmon resonance is a surface effect, localised influences on the electron density at the particle surface shift the resonant frequency of the collected electrons when stimulated by light. This can be utilised as a sensitive colorimetric test for a target molecule such as a protein or DNA molecule in solution (Mucic et al., 1998; Fritzsche, 2001). There are a number of analyte detection methods that use gold nanostructures but in general there are two very different approaches in use which must first be clarified to avoid confusion. In this section we will discuss colour changes arising from the flocculation or dispersion of gold nanoparticles in a bulk solution in the presence of a given analyte molecule or protein. That is to say, we will produce a blue or red coloured colloidal solution that shifts its colour in the presence of the chemical trigger of our choice. This section does not cover 'surface plasmon resonance' as described in the box below. Though highly sensitive and widely used the technique requires some specialised equipment and so is only outlined here.

Surface plasmon resonance (SPR) in the analytical detection sense is a quantifiable detection setup routinely used to monitor the presence of biomolecules in a solution. It is a very sensitive technique but limited only to the detection of biomolecules over 20 kDa in size and 1–10 nmol in concentration (Englebienne and Weiland, 1996; Lyon et al., 1998; Englebienne et al., 2003). The basic principle is to monitor the intensity of a beam of plane polarised light reflected off the back of a thin, less than 200 nm, gold film at a given angle through a prism. The metal film is thin enough that the plasmon effect can occur in the dimension of the metal perpendicular to the surface plane. When the polarised light of the right wavelength reflects off the backside of the gold film at the right angle it stimulates the surface plasmon. Because this takes energy the intensity of the reflected light is decreased at this angle. A biomolecule adsorbing to the film surface changes the resonant frequency of the plasmon such that an increase in the intensity of the light will be observed. In a normal setup a gold film is evaporated onto a glass prism into which a visible or infra red light beam passes. The other side of the film is functionalised with affinity receptors for the analyte of choice and exposed to solution. Once the minimum intensity and angle has been determined any biomolecules linking to the tags can be detected by the shift in the reflected light intensity. Various improvements and variations of the technique exist and the sensitivity can be boosted 25 times further by the use of gold nanoparticles in addition to, or instead of, the flat gold film. This is termed as 'enhanced' surface plasmon resonance and this works by adding a gold colloid functionalised with a secondary antibody for the molecule of interest. The gold colloid in solution will bind to the target protein and then this in turn will bind to the primary antibody attached to the gold film. This results in a very large change in the reflected light intensity and can detect picomolar amounts of proteins present in solution. Detection applications involving nanoparticles are termed 'localised' surface plasmon resonance to better delineate the differences between the thin film and particle structure (Fig. 4.57).

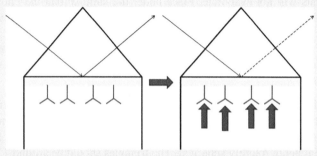

FIGURE 4.57   A schematic representation of surface plasmon resonance. Antibody coated gold reflects light reflected from its backside at a given angle. When a complimentary molecule binds to the receptor on the gold surface the reflectivity at a given angle decreases.

A gold particle can be surface modified to carry an antibody or string of DNA. When the target molecule is present the modified gold particles will bind to it through the complementary species and this leads to bridging and flocculation of the gold particles. This type of diagnostic is very sensitive and accurate as the response is only seen when the target molecule is complimentary to the antibody or DNA sequences used. This is advantageous over more traditional flourophore coupling to biomolecules because the response is more selective and the visible effect is greater at lower concentrations. For the detection of specific DNA sequences two separate aliquots of gold colloid are modified with single strands of oligonucleotides that are complimentary to opposite halves of the target oligomeric sequence. The two differently tagged colloids are then mixed together. In the presence of the target sequence, and only that specific sequence will work, the oligomeric strands bind to the complimentary base pairs on the target sequence causing a large assembly to form. If a single strand DNA of a different sequence is used then the agglomeration will not happen. The close proximity of other nanoparticles when the agglomerate is formed means that the individual plasmons can interact so that the resonant frequency is redshifted to a longer wavelength. For a gold colloid this means a red wine coloured solution will turn to a darkened purple colour. This is just the same as if the gold nanoparticle had a larger diameter except that the cluster is prevented from fusing into one large gold particulate by the surface coatings and target molecules sandwiched between them. Of course, getting this type of test to work requires that the various surface interactions keeping the gold particles stabilised are balanced. For use as a commercial diagnostic test the colloid must remain stable for long periods and possibly over a wide range of temperatures. Conversely the colloid must be sensitive enough to the analyte that it will readily flocculate or disperse. As electrostatic interactions are responsible for the aqueous stability the sensitivity of the analytical gold colloid can be fine tuned by altering the pH and salt concentrations such that the gold colloid/analyte interactions are idealised for a colorimetric response.

A QUICK DEMONSTRATION OF CHANGING COLOUR AND COLLOID STABILITY WITH VARYING SALT CONCENTRATION AND pH

This quick demonstration illustrates how salt concentration can influence the electrostatic interactions in a colloid (Filipponi, 2010). By adding salt to a citrate stabilised gold colloid we will be able to observe the colour shift from red to blue. An aliquot of the colloid can also be mixed with a polymer or protein such that the particles become coated and will not crash out in the presence of a high salt concentration. In this experiment we will use egg white as a source of ovalbumin for coating the nanoparticles with. This protein in particular is useful as it has a number of sulfhydryl groups which act to store metals within the protein structure during normal function. Here these sulphur groups will allow the 45 KDa protein to anchor strongly to the gold surface providing steric stability to the nanoparticles. In comparison to the citrate stabilised control colloid the polymer or protein coated test colloids will not flocculate until a significantly higher amount of salt has been added to the solution. However, all of these variations will demonstrate sensitivity to the solution pH with the colour turning blue as the pH becomes acidic. For citrate stabilisation this is a simple matter of the carboxylic acid groups becoming protonated and losing their charge and the red colloid will turn to blue around pH 2−4. Similar charge effects are also seen for the protein coated gold through the colour change will be observed between pH 4.5 and 6 and this may be interpreted as more sensitive to a pH change than the control colloid. The protein coated colloid is more sensitive because, in this case, the lowered pH is neutralising the charge on the protein as opposed to a molecular species like the citrate molecules. Effectively the flocculation and corresponding shift in absorbance maxima is observed at the isoelectric point of the protein where the net surface charge of the collected aminoacids is zero. When working with gold colloids coated in biomolecules it is important to determine where the stable regions of pH and salt concentration lie. The following method is a simple route to determining this.

To determine pH and salt stability of an ovalbumin coated gold colloid you will need:

1. Prepare a sodium citrate gold colloid as outlined elsewhere. You will need at least six 5 mL aliquots of citrate stabilised gold colloid. Place them each into separate 10 mL glass vials.
2. 10 mL of a 1 mol/L solution of NaCl. This can be prepared by dissolving 0.584 g of NaCl in 10 mL of water.
3. A volumetric pipette in the 0.1−1 mL range.
4. 10 mL of a stock solution of 1% wt poly(vinyl alcohol). This may require heating and stirring to dissolve the polymer. A few seconds in a microwave can help dissolve it rapidly.
5. An egg white in water solution which will act as the source of the ovalbumin. For the purposes of this preparation you do not need to use anything of outstanding purity so simple egg white taken from a chicken egg will do! For optimal protein coverage of the gold colloid you will need a 1000 fold molar excess of protein to gold.
   To make the stock ovalbumin solution add 1 mL of egg white to 1 mL of water. Mix this together trying to avoid foaming and let it settle.
6. A stock solution of 1 mol/L HCl in water.
7. A stock solution of 1 mol/L NaOH in water.
8. A pH meter and magnetic stirrer and stir bar.
9. Optionally you can keep track of the colour changes using a UV-visible spectrometer. You will need cuvettes from placing samples in.

To determine pH and salt stability of an ovalbumin coated gold colloid you will need:

1. Place the control citrate stabilised gold colloid on the magnetic stirrer and add the salt solution dropwise using the volumetric pipette. Keep track of how much you are adding.
   After adding 0.5 mL or less you should observe the colloid change in colour from red to blue as the salt shields the charge of the citrate anions and flocculates begin to form. If you continue to add the salt solution then you will see the solution become transparent and a black precipitate of the gold will collect in the bottom of the vial.
2. Use a pipette to take 1 mL of settled egg white solution from the bottom of the vial. Add this to another 5 mL aliquot of gold colloid under gentle stirring. You may see the colloid turn pink slightly but no large colour change should be observed.
3. Add 1 mL of the 1% poly(vinyl alcohol) solution to another aliquot of the gold colloid under gentle stirring.
4. To the protein and polymer stabilised gold colloids add the same amount of salt solution. You should observe no colour shift due to the extra stabilisation of the additives. Keep adding salt until you observe a colour shift and keep a track of the volume added and final concentration. Repeat this approach with buffers you might wish to use in order to examine the stability conditions for the colloid.
5. Aliquots can be removed from the test vial and placed into a UV-vis cuvette so that the absorbance spectrum can be monitored. You may need to dilute solutions accordingly if the peak absorbance is significantly larger than 1.

Next we will look at the pH stability of the colloids.

6. Take the next three 5 mL aliquots of citrate stabilised gold. Leave one as a control and add egg white and polymer to the other two samples as outlined in steps 2 and 3. Take a UV-Vis spectrum of the samples if required.
7. Place a colloid to be tested on the magnetic stirrer and insert the pH meter. Add the HCl solution in 0.1 mL (or less) amounts and monitor the pH. After each amount of acid is added you may also wish to take the UV-visible absorbance spectrum. Repeat for all three colloids.

You should observe that more acid must be added to the control before it becomes visible blue whereas the protein stabilised colloid will require less. This is a rather simplistic approach towards exploring the stability ranges of a colloid but does serve as a quick and easy laboratory demonstration.

DNA BASED GOLD ASSEMBLIES AND APTAMER COLORIMETRIC TESTS

With thanks to Dr Christopher J. Johnson

When dealing with nanomaterials it is important to have both an ability to produce the materials in question but also to manipulate them. Just as a brick is a single component in a wall, nanomaterials ordered together with hierarchical structure can become more than the sum of their parts. For a gold colloid it is desirable to transform a collection of colloidal particles free in solution into an ordered two- or three-dimensional array. The use of DNA as a template for bottom up self assembly has already been mentioned and will be expanded upon further here. A gold colloid is produced with small amounts of a reactive modifiable functional group present on the surface which may then be further modified. In the first demonstration of this technique by Alivasatos and Shultz groups the functional moiety was an N-propylmaleimide (Alivisatos et al., 1996). To produce basic structures the molar ratio of gold calculated so that only one molecule of the N-propylmaleimide is present on the particle surface. The molecules are then cross linked in solution to thiolated oligonucleotides having a well defined sequence to produce a stable gold colloid where each particle has just one single stranded DNA functionality at its surface. A separate single strand oligonucleotide sequence is added to the solution which contains complimentary pairings that bind the gold conjugated sequences at predetermined positions. Dimeric and trimeric gold nanoaparticle structures can be produced in this way in relatively large quantities. This technique has been extended such that very intricate patterns of DNA templated nanostructures can be produced with not only gold but other metal and semiconductor materials in various mixtures (Mucic et al., 1998; Mitchell et al., 1999). At the same time as the DNA template approach was presented, the Mirkin group published a method using binary mixtures of oligonucleotide functionalised gold colloid. Two colloids are produced separately that are functionalised with non complimentary oligonucleotides sequences on the particle surface. Typically the DNA is linked directly through thiol termination on the oligomer so that a direct DNA to gold conjugate is produced though it may be necessary to introduce a spacer unit where the end application requires. When mixed together with a DNA linker strand complementary to both oligonucleotides (on opposite halves of the single strand) the Watson−Crick base pairing drives the colloids to assemble into arrays. Macroscopically the solution changes from red to purple as the gold/DNA assemblies form. If using different sized particles then a larger particle can be coated with smaller ones to produce intricate patterns. If left to stand for a few hours these solutions form a pinkish grey precipitate but interestingly this process is fully reversible. By heating the solution the DNA base pair bonding is disrupted so that the original colloidal mixture is regenerated and turns from purple to red once more (Reynolds et al., 2000). This is similar to the 'melting' step in PCR where the complimentary DNA strands are thermally disrupted to produce two single strands (Fig. 4.58).

This type of disassembly does not have to be heat induced and gold/DNA aggregates can be produced that will disassemble in the presence of a target analyte in solution as a colorimetric test (Liu and Lu, 2006). In these colloidal dispersions the gold is stabilised by strings of oligomeric DNA called aptamers which can be specifically designed to strongly interact with any target of your choosing. Aptamers are obtained through a combinatorial chemistry approach called SELEX (Systematic Evolution of Ligands by Exponential enrichment) which is relatively simple and avoids having to design your own sequences for the desired interactions. A target molecule is adhered in a column and a library of 10^{14-15} short and random DNA sequences are allowed to flow through the column. Any sequences that interact strongly with the target molecule will be left behind in the column. These bound sequences are

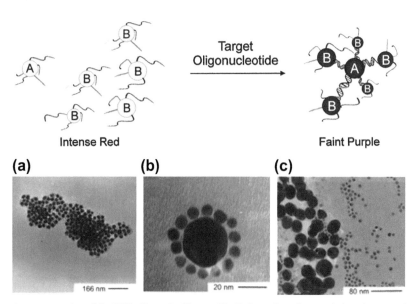

FIGURE 4.58 Above a schematic representation of the DNA directed self assembly. Below selected transmission electron micrographs of self assembled gold structures. *(Adapted with Permission Reynolds et al. (2000). Copyright 2000 American Chemical Society)*

then flushed out and amplified by PCR so that you are left with a large amount of short sequences that will selectively bind to your target. The manufacture and selection of these sequences are beyond the scope of this book but custom services to isolate, sequence and produce them to order are commercially available. In an aptamer based colloidal gold test a binary combination of oligomer stabilised gold colloid is mixed together. The oligomers that stabilise these colloids are designed such that they are complimentary to a specific single strand sequence of linker DNA. This forms a gold/DNA conjugate in a similar manner to the original Mirkin group method of aggregate self assembly but with an important difference. The linker DNA has a long overhanging aptamer strand left free and unbound to anything. This will sit there and do nothing if the purple solution is left to stand but in the presence of a given analyte the aptamer will bind to the target molecule. A small section of the aptamer sequence, a few base pairs long, is directly involved in linking to the thiol terminated oligomer sequence attached to the gold. The conformational change in the linker DNA from the binding event disrupts this short chain of bonds and so the gold located at the aptamer end of the linker DNA cannot stay bound. The aggregate separates to give one gold particle free in solution and the other bound by the linker DNA and aptamer to the target molecule (Fig. 4.59).

This recipe outlines a general procedure for the synthesis of a colorimetric gold colloid test for the small metabolic biomolecule adenosine. The source reference notes that many other biomolecules can be detected by simply changing the aptamer sequence used. The two separate thiol modified oligomers and the aptamer containing linker sequence are purchased beforehand. In some cases the DNA sequences used to stabilise the colloids may be self complimentary meaning the aggregates may not separate as desired. This can be avoided by checking the interactions using a software package. The supplier will also be able to aid you in the suitability of the DNA design. Thiolated DNA can often be provided with some form of protecting group on the end which may need to be removed prior to use according to supplier instructions. For the surface modification to occur efficiently the thiolated DNA is primed with a reducing agent, tris-(2-carboxyethyl) phosphine hydrochloride, which prevents disulfide bonds from forming. If the thiolated DNA were to dimerise then it would not stick to the surface of the gold efficiently. Once a gold aggregate solution has been formed the ideal salt concentration for detection must be determined. The temperature at which the DNA will readily undergo melting is strongly dependant on the salt concentration. As has been demonstrated in previous examples, high concentrations of salt will stabilise aggregate formation because of charge shielding. This means that the sensitivity of the test can be improved by lowering the salt concentration to the point that it will readily disassemble in the presence of the target analyte. To determine the ideal salt concentration the gold/DNA aggregates can be centrifuged out of solution and redispersed in buffers of varying NaCl concentrations. Aggregates in various salt strengths are then heated from room temperature to 60 °C in a thermally controlled UV-visible spectrometer. The absorbance of the solution at 260 nm will increase sharply when the DNA assembly begins to fall apart and the DNA linker is free in solution. This is the melting temperature and can be lowered by lowering the salt concentration of the

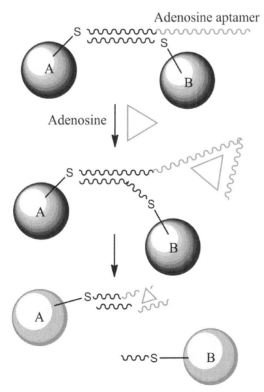

FIGURE 4.59 Schematic diagram of aptamer based gold self assembly and disassembly in the presence of a specific trigger molecule.

buffer used. Because most sensor applications will occur at room temperature you are aiming to tune the salt concentration until the melting transition is around 2–3 °C higher than ambient conditions. This way the aggregates will be stable in solution but bound weakly enough that aptamer binding to a target will cause the assembly to come apart.

To make an aptamer based gold colloid colorimetric test you will need:

1. A stock of the citrate stabilised gold colloid produced as outlined earlier in this section. This recipe will require around 6 mL of colloid.
2. Alkane thiol capped oligonucleotide sequences for stabilising the gold, labelled 'DNA 1' and 'DNA 2', and the corresponding linker sequence labelled 'Linker'. You will need enough of each to make 9 μL of a 1×10^{-3} mol/L solution of each of the stabilising sequences. You will also need enough of the linker DNA to make up 10 μL of a 0.01 mol/L solution. The sequences for the detection of adenosine as outlined in the source reference are given below. The aptamer binding sequence is highlighted in green. The DNA must be desalted and the linker sequence should be purified by gel electrophoresis:

Sequences listed running from the 5′ to 3′
DNA 1: TCACAGATGAGTAAAAAAAAAAAA–(CH$_2$)$_3$–SH
DNA 2: SH–(CH$_2$)$_6$–CCCAGGTTCTCT
Linker: ACTCATCTGTGAAGAGAACCTGGGGGAGTATTGCGGAGGAAGT

---

3. 1.5 μL of a 0.01 mol/L solution of tris-(2-carboxyethyl)phosphine hydrochloride (TCEP). Do not make up the solution until it is required as it needs to be fresh. To prepare add 28.6 mg (1×10^{-4} mol) to 10 mL of water shortly before use.
4. Clean 20 mL glass vials with lids. The vials must be cleaned thoroughly using a strong base prior to use or else the gold may stick to the glass surface. Fill the vials with a 12 M solution of NaOH and leave to stand for an hour. Afterwards rinse the vials out with large amounts of water and leave to dry. *The strong NaOH solution is extremely corrosive and must be worked within a fume hood using appropriate gloves and goggles.*
5. A stock solution of a 0.5 mol/L tris(hydroxymethyl)aminomethane (Tris) buffer adjusted to pH 8.2. This can be prepared by dissolving 60.57 g of the basic salt in 900 mL of water. Adjust the pH to 8.2 using a 1 mol/L solution of HCl then make up the volume to a total of 1 L.
6. A stock solution of 0.5 mol/L acetate buffer adjusted to pH 5.2. A stock solution can be prepared by adding 79 mL of a 0.5 mol/L sodium acetate in water solution to 21 mL of a 0.5 M acetic acid in water solution. To make the sodium acetate solution dissolve 4.101 g of sodium acetate in 90 mL of water then make up to 100 mL. To make the acetic acid solution add 2.86 mL of glacial acetic acid to 90 mL of water and make up to 100 mL. Check the pH is 5.2. If it is not then you can adjust the ratio of the acidic and basic acetic acid solutions respectively.
7. A stock solution of 0.025 mol/L Tris buffer at pH 8.2 containing 0.3 mol/L NaCl. This can be prepared by dissolving 1.753 g (0.03 mol) of NaCl in 5 mL of the 0.5 mol/L Tris buffer (as outlined in item number 5) and diluting to a total volume of 100 mL.
8. A stock solution of 0.025 mol/L Tris buffer at pH 8.2 containing 0.1 mol/L NaCl. This can be prepared by dissolving 0.584 g (0.01 mol) of NaCl in 5 mL of the 0.5 mol/L Tris buffer (as outlined in item number 5) and diluting to a total volume of 100 mL.
9. A stock 1 mol/L solution of NaCl in water. This can be prepared by dissolving 0.584 g (0.01 mol) in 10 mL of water.
10. Test solutions of the target analyte and controls. Here we will use 0.05 mol/L solutions of adenosine, cytidine, uridine and guanosine each in water. Only the solution containing the adenosine should trigger a colorimetric response. You may need to heat these to get them to dissolve in water.
11. A magnetic stirrer and a selection of clean stir bars.
12. Polypropylene microcentrifuge tubes and a microcentrifuge.
13. 0.2 μ syringe filters and syringes.
14. A UV-visible spectrophotometer with temperature control between 15 and 60 °C and low volume quartz cells. The cells should have PTFE plugs or else be sealed with clingfilm to prevent sample evaporation during heating.
15. Volumetric pipettes and tips in the 1–10 and 50–500 μL and 1–5 mL volume ranges.
16. A marker pen for keeping track of solutions.
17. A fridge maintained at 4 °C.

To make an aptamer based gold colloid colorimetric test:

All solutions should be filtered using the syringe filter before use. This experiment is very sensitive to contamination so ensure everything you use is clean.

1. Pipette 9 μL each of the 1×10^{-3} DNA 1 and DNA 2 solutions into separate microcentrifuge tubes. Label the tubes.
2. To each tube add 1 μL of the 0.5 mol/L acetate buffer (pH 5.2) and 1.5 μL of the 0.01 mol/L solution of TCEP. Leave these tubes to stand for 1 h at room temperature. This step will ready the thiol groups so that they will adhere to the gold surface.

3. In two clean and dry glass vials add 3 mL of the citrate stabilised gold colloid to each. Label them 1 and 2 respectively.

4. Transfer the activated DNA 1 solution from the microcentrifuge tube into the gold colloid marked '1'. Repeat this with DNA 2 and the vial labelled '2'. Place the lids on the vials and give them both a gentle swirl to mix the contents. Place these vials in the dark at room temperature and allow them to stand for at least 16 h.

5. After the colloids have been allowed to incubate add 30 µL of the 0.5 M Tris buffer (pH 8.2) drop by drop. Swirl the vial by hand to mix.

6. To each vial add 300 µL of 1 mol/L NaCl solution dropwise with gentle swirling. Leave the vials to stand in the dark for a further 24 h though they will keep for weeks at this point. Because the DNA can possibly degrade over time it is best to use the DNA stabilised colloids after they have been freshly prepared.

7. Pipette 500 µL aliquots of each gold/DNA colloid 1 and 2 into separate microcentrifuge tubes. Label the tubes.

8. Centrifuge the tubes at 13,200 rpm for 15 min. Afterwards you should see the red gold particles collected in the bottom of the tube with a clear top solution. Centrifuge for longer if required.

9. Decant or pipette away the top solution leaving the sediment in the bottom. This will remove any unbound DNA that might interfere with the aggregate formation.

10. Redisperse the pelleted gold/DNA particles in 200 µL of the 0.025 mol/L Tris buffer (pH 8.2) containing 0.1 mol/L NaCl.

11. Centrifuge the colloids at 13,200 rpm for 10 min or until the particles have once more sedimented to the bottom. As before, remove the top solution.

12. Redisperse the pelleted gold/DNA particles in 500 µL of the other 0.025 Tris buffer (pH 8.2) containing the higher salt concentration of 0.3 mol/L NaCl.

13. Add gold colloid 1 to gold colloid 2. To this add 10 µL of the 10 µmol/L solution of linker DNA in water and swirl the contents by hand. The concentration of linker DNA should be 100 nmol/L in the binary gold colloid mixture.

14. Place the gold colloid into the fridge and leave to stand for 1 h. You should observe that the red colour turns blue/purple as the linker DNA binds the two aptamer coated gold colloids together to form aggregates. All of the nanoparticles should have aggregated within 24 h.
    Now that you have the gold/DNA aggregates you will need to optimise the buffer conditions for sensitivity at room temperature. At this point the salt strength is high enough to keep the colloid stable but means that the melting point of the DNA assembly is above room temperature.

15. Take 50 µL of the freshly prepared aggregate and add 150 µL of the 0.025 mol/L Tris buffer containing 0.1 mol/L NaCl. This will give a final NaCl concentration of 0.150 mol/L.

16. Transfer this into a quartz UV-visible cell and place in the spectrometer. Ensure the quartz cell is sealed.

17. Set the temperature to 15 °C and give the sample a minute to equilibrate properly. Record the absorbance spectra and note the peak absorbance at 260 nm.

18. Raise the temperature in steps of 2 °C and record the absorbance at 260 nm. Ensure the sample is equilibrated and shake the sample after each scan to ensure that the particle aggregates do not sediment out of solution. At the melting temperature you should see a sharp increase in absorbance.

19. Depending on where you find the melting temperature is adjust the concentration of NaCl in the buffer. The stock aggregate can be centrifuged and redispersed in a 0.025 mol/L Tris buffer (pH 8.2) with any chosen salt concentrations. Repeat the procedure to check the melting point is correct for the chosen application.

20. After you have determined the optimal salt concentration you can now use the gold/DNA aggregates as a colorimetric test. Take a 49 µL aliquot of the stock gold/DNA aggregate solution and centrifuge it at 13,200 rpm for 15 min.

21. Redisperse the gold/DNA pellet in 0.025 mol/L Tris buffer (pH 8.2) adjusted to the optimal salt concentration. The solution should appear purple.

22. Add 1 µL of the 0.05 mol/L test solution to give a final concentration of analyte of 1×10^{-3} mol/L in solution. When adenosine is added you should observe the solution change colour from purple to red almost immediately. If a more dilute amount of analyte is used then the colour change may occur more slowly and this can be used in conjunction with UV-visible spectroscopy to give a quantitative analysis of analyte concentration.

Perhaps the most important factor in getting this aptamer based gold test to work is clean glassware. The DNA will adhere to any dirt on the surface of the glass vials causing the aggregates to crash out of solution. You may find that the aggregates can sometimes not be redispersed after centrifugation. This may indicate that the gold has not been well functionalised by the DNA and so there might be a problem with the thiol groups on the end. To eliminate this problem use a ten times stronger concentration of TCEP during step 2 then remove the excess TCEP with a desalting column. There is also the possibility that the two component colloids are binding to one another without the linker strand if you are using aptemers of your own design. To test for this try redispersing a centrifuged pellet in pure water and see if the red colour returns. The complete lack of salt should cause the aggregates to disassemble. If the colour remains purple then you may need to check that the DNA sequences are correct (Fig. 4.60).

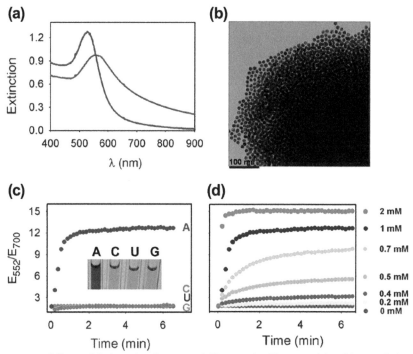

FIGURE 4.60   (a) UV-visible spectrum of dispersed (red trace) and aggregated (blue trace) gold nanoparticles. (b) transmission electron micrograph of aptamer linked gold aggregates. (c) Kinetic plots comparing the ratio of intensity of absorbance at 552 nm against absorbance at 700 nm when various nucleosides are added to an aptamer stabilised colloidal aggregate. The trace marked A is for a colloid treated with adenosine which turns red. (d) A kinetic plot for colour change in the colloid after *treatment with varying concentrations of adenosine. (Reprinted with Permission from Liu and Lu (2006). Copyright 2006 Nature Publishing)*

SILVER NANOPARTICLES AND STRUCTURES

Silver has long been values for its antimicrobial properties and is held up as an early antibiotic treatment. The silver is only biologically active in its cationic form and complexes with the respiratory molecules in bacteria to eventually kill them. Medically, a complex of silver will be used as an external cream or salve to prevent infection on burn injuries. Many commercial products now employ silver in bulk or as a nanomaterial ingredient because of the perceived and real advantages. Clothing is now available with silver embedded in the fibres so that the microbes causing a sweaty wearer to smell are kept in check. The sweat itself in addition to uptake by the microbes generates the small amounts of Ag^+ required for the antimicrobial effect. Conversely, though there is no evidence for any benefit, large amounts of silver colloid are sold for ingestion as an alternative antibiotic medicine. To date there is no evidence of its efficacy as an antibicrobial agent *in-vivo*. Ingestion of too much silver causes agyria, a condition where silver particles build up in the skin causing a blue or grey discolouration. For this reason it serves as a good example of a nanomaterial already interacting with the environment and society and a more in depth look at agyria and silver nanomaterials is outlined in the safety section of this book.

Silver is in the same period as gold and demonstrates similar properties when formed as a nanomaterial. That is to say it packs in a face centred cubic structure and demonstrates similar behaviour in terms of the ways it can be processed to produce nanostructures. It demonstrates a plasmon band in the visible range though gold is more commonly used for surface plasmon resonance experiments as it is easier to modify the surface. Thiols do not stick to the surface of silver in the same way as they do gold. There is also the issue of oxide formation at the surface of the silver particle which can also hinder modification. Despite this silver in the nano particulate form shows excellent potential as a catalytic material and films can be formed easily onto solid supports for this purpose. Furthermore, silver has one of the highest known metallic conductivities which makes it a good choice as an electrical contact material at small length scales. The major difference between gold and silver from a materials chemistry point of view is that silver can be reduced by light in addition to normal reducing agents. This opens up some unique possibilities in terms of synthetic procedures where the use of high temperatures, as with sodium citrate reduction, or harsh reducing agents, as with borohydride reduction, can be eliminated. Instead a template or solution suffused with Ag^+ ions is irradiated with ultraviolet light. Electrons from anionic counter ions, or other sources, can be promoted into the conduction band of the silver where the charge is adsorbed to reduce the silver into the metallic state. Additives in the solution can act as capping agents for the formed silver particles or else provide some form of diffusion limitation preventing the silver from growing too much. In this section various methods to produce silver nanoparticles will be explored with reference to controlling the morphology and the use of light as a reducing agent.

All silver reagents will reduce to metal on exposure to light. Silver solutions will appear colourless but will leave indelible stains after exposure on surfaces and skin. Though this is normally not toxic some people can become sensitised and gloves should be worn when working with silver solutions at all times to prevent contact. Stains on surfaces can be washed with nitric acid which redissolves the silver to form silver nitrate. This must then be rinsed away with copious amounts of water.

MAKING A BASIC SILVER COLLOID BY REDUCTION WITH CITRIC ACID

Essentially the same reaction that works for gold will work for silver. The reduction of the salt by refluxing with citric acid as outlined for gold will also work if silver nitrate is used instead of gold tetrachloroaurate. Instead of a red purple colour developing you should see a yellow or orange colour where the plasmon band peak absorbance is between 380 and 420 nm depending on particle size.

1. All of the equipment as outlined in the citric acid reduction of gold experiment.
2. 0.0151 g (8.9×10^{-5} mol) of $AgNO_3$. This is to be used in the place of the $HAuCl_4 \cdot 3H_2O$.

Then simply run the reaction as outlined. Analysis can be performed in the same way. Should you see the silver precipitate out of the solution and sediment that try adding 1% poly(vinyl alcohol) or 1% poly(ethylenimine) to the solution prior to reduction. If the particulate solution contains large silver clusters then the solution will appear turbid and brown in colour. As with the gold reaction you can try using borohydride or hydrazine as a reducing agent instead. For a quick demonstration it is also possible to produce a silver colloid in small (5 mL) amounts by microwaving the silver and citric acid solution for 10 s. This approach can be useful if you happen to be trying to describe plasmon resonance to students in a lecture!

PREPARATION OF FLUORESCENT SILVER NANOCLUSTERS

With thanks to Jamie Shenston

The colour of a colloidal solution of a gold or silver is dependant on the plasmon resonance. As has been explored previously, this collective oscillation of the conductive electrons in the metal is size dependant on the diameter of the nanoparticle. What happens then, if the diameter is decreased still further? In a similar manner to the quantisation effects seen in a semiconductor material, the behaviour of the bulk electrons in the metal begin to be determined more and more by the electronic structure of its constituent single metal atoms. As the size of the metal particle moves below the wavelength of a single electron at the Fermi level, then the electronic behaviour of the total particle becomes quantised into discrete allowable transitions. Effectively the metal particle now behaves electronically as a molecule and can undergo similar electronic interactions such as excitation. Below 2 nm the term metal 'nanoparticle' is swapped for 'cluster' and in silver clusters it is possible to observe fluorescence when these clusters are excited by UV light. It is important to recap that the fluorescence is not an effect of metallic plasmon resonance but is from the excitation of an electron and the subsequent radiative emission as might be observed for a traditional organic dye molecule in the same conditions. This electronic transition is still influenced by the size of the cluster because the number of metal atoms present will determine the level of quantisation in the electronic band structure. In this respect the electronic band structure of a metal nanocluster has more in common with semiconductor nanomaterials rather than size dependant plasmon interactions. Fluorescent clusters composed of nobel metals are also resistant to oxidation and corrosion making them good for use as markers in conditions that might destroy an organic analogue. Additionally, they do not experience blinking as with semiconductor particles.

In this recipe we will use a facile method to produce fluorescence nanoclusters of silver stabilised by poly(methacrylic acid) (Diez et al., 2009) (Diez et al., 2010). The poly(methacrylic acid) serves a number of purposes in the formation and stabilisation process. This polymer is rich in carboxylic acid groups and it would be normally assumed that the polymer would coordinate through deprotonated hydroxyls to the surface of the silver. However, in this procedure the reaction occurs at pH 3 which means the carboxylic acids retain their protons. In the acidified form the carboxylic acids along the polymer backbone will form hydrogen bonds with one another such that the polymer chains become viscous in solution. Silver ions mixed in with this viscous solution experience very limited diffusion and when reduced photolytically under UV light they stay in place within the polymer matrix. The formed metal clusters are immobilised and prevented from agglomerating to one another. The viscosity also minimises the amount of silver ions that can diffuse to the surface of another particle before they are reduced themselves. These conditions give rise to the formation of sub 2 nm clusters which will behave as fluorophores in solution stabilised by the poly(methacrylic acid). The absorbance and emission spectra can be fine tuned by altering the molar ratio of silver to methacrylic acid groups. This will alter the cluster size and ultimately the quantisation levels observed. Another advantage of using the poly(methacrylic acid) is that it can be readily transferred into a hydrophobic solution of butanol by mixing with the aqueous solution and allowing it to separate. Silver nanoclusters transferred into butanol will demonstrate higher quantum yields under fluorescence. If isolated from the aqueous solution and transferred into other solvents then a solvatochromic shift in the absorbance spectra is seen. Though the effect is common in larger particles having a plasmon resonance the effect here is more closely linked with the effect of a polar or non polar solvent on the stability and energy level of the excited species.

To make silver nanoclusters stabilised by poly(methylmethacrylic acid) you will need:

1. 10 mL of a stock solution of 5×10^{-5} mol/L poly(methacrylic acid) (molecular weight 100,000) in water. This can be prepared by dissolving 50 mg into 10 mL of water. Shake the solution vial with the lid on to dissolve the polymer quicker.
2. 10 mL of a stock solution of 0.0117 mol/L silver nitrate solution. This can be prepared by dissolving 100 mg of $AgNO_3$ into 10 mL of water. Once the salt has been added to water keep the solution wrapped in aluminium foil and in the dark or it will reduce rapidly in light.
3. A number of 5 to 10 mL glass or transparent vials that can be sealed.
4. An automated rotator that can hold the vials.
5. A 1–5 mL volumetric pipette.
6. An 11 W UV (365 nm) or sun lamp that can be positioned in front of the rotator. *If you are using a strong UV source ensure that it is partitioned away so it will only illuminate the rotator.* Alternatively this reaction can be performed using sunlight. The formation of the clusters is not particularly rate dependant so even patchy sunlight will get the job done.

To make silver nanoclusters stabilised by poly(methylmethacrylic acid):

1. Use the volumetric pipette to add 2 mL of the polymer solution to a vial.
2. Using a fresh tip add 2 mL of the silver nitrate solution to the same vial.
3. Seal the lid and place on the rotator.
4. Turn the rotator on so that the solution is constantly mixing.
5. Turn on the UV lamp and expose the solution for 40 min to 1 h. Alternatively leave the solution in daylight for 24 h. In strong direct sunlight the reaction should be complete in around 7 h. The solution should turn a pinkish colour and you should see no sediment. Reaction times will be longer if you choose to use more silver.

The absorbance and emission wavelengths can be tuned by adjusting the molar ratio of silver ions to the number of monomeric methacrylic acid units present. For the poly(methyl acrylic acid) there are approximately 1161 monomer units per mole at a molecular weight of 100,000 g/mol. This means that, for the reaction given above, there is a 1:1 molar ratio of silver ions to carboxylic acid units. Both the absorbance and fluorescent emission maxima can be adjusted by altering the amount of stock solution added to the polymer. If the molar ratio of silver is less than half the molar amount of monomer units then the clusters formed are either too small or the silver remains as Ag^+. In this event you will not see any colour or fluorescence. If the molar ratio of silver is over 8 times higher then the silver particles formed are unstable and crash out of solution. In the table below is a brief outline of the molar ratios required to fine tune the absorbance and emission spectra. The stock solution can be diluted for molar ratios less than 1:1. Use a more concentrated solution for the higher silver cation ratios so that the total volume of the final solution remains 4 mL with a constant concentration of polymer. This ensures the viscosity of the solution remains high enough to form the clusters:

Molar ratio	Absorbance maxima/nm	Emission maxima/nm
0.5:1	504	614
4:1	528	626

The as formed aqueous solution will still have unreacted silver ions or larger silver particles present. Extraction with butanol will specifically draw the desired nanoclusters out of the aqueous solution and into the organic phase. The carboxylic acid groups within the polymer are weakly attracted to the cluster surfaces which leave the hydrophobic methyl groups exposed in solution. The butanol phase is therefore a preferable solvent for the coated clusters and they transfer across from the liquid phase. To make this transfer, simply add an equal volume of 1-butanol to the aqueous dispersion and shake vigorously. For best results place the sample back on the rotator and leave overnight. Afterwards allow the solutions to phase separate from one another. You will see the purple/pink colour denoting the clusters in the butanol which will be floating on the top. At the interface you may observe a darker band which is larger nanoparticles settling out. The top solution can be pipetted off to give a purified solution of the silver clusters. This technique can also be used to concentrate the clusters if you add a smaller volume of butanol (Fig. 4.61).

MAKING SILVER NANOCUBES AND RODS USING THE POLYOL PROCESS

Silver, like gold, atoms pack together most often in the face centred cubic structure. As has been demonstrated with the synthesis of gold rods and triangles it is possible to control the relative rates at which the facets grow in a particular plane in order to produce unique morphologies. In this section the 'polyol' method is used in the formation of cubic particles and rods by the reduction of the

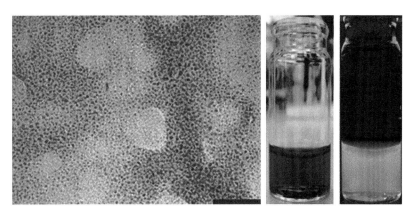

FIGURE 4.61 TEM image of silver clusters protected by a long chain polymer (left — scale bar is 50 nm) and images of clusters in daylight (right) and under 365 nm UV light (far right). *(Images used courtesy of Jamie Shenston)*

metal salt in hot ethylene glycol (Sun and Xia, 2002). This acts as both solvent for the reaction and the reducing agent which reacts with the silver according to the following formula at 160 °C:

$$HOCH_2-CH_2OH \rightarrow CH_3CHO + H_2O$$

$$2CH_3CHO + 2AgNO_3 \rightarrow 2CH_3CO-COCH_3 + 2Ag + HNO_3$$

The structure directing agent is poly(vinyl pyrolidone) which interacts through the oxygen and nitrogen moieties within the heterocyclic ring to cap [100] planes on the metal crystal surface. This can then allow for growth on the [111] planes as the polymer is far less weakly interacting with that particular surface. This can result in rod structures which have a pentagonal symmetry around the centre, like a 'zero' dimensional nanoparticle, but elongated along the axis of the crystal multiple twinning centre. The cube morphology is formed by balancing the growth rate of the [100] and [111] lattice planes by adjusting the silver concentration, capping agent concentration and the growth time. If the growth rate of the [100] face compared to the [111] face is 0.58 times fast then a cube will form. If it is faster, up to 1.73 times, then tetrahedral or octahedral particles will form. These rates are determined by the amount of poly(vinyl pyrrolidone) in the solution compared to the concentration of silver. If the concentration of silver ions is lowered below a certain threshold with the concentration of polymer remaining the same then nanorods will be formed. The silver salt and the polymer must be introduced to the hot solvent at a controlled rate in order to maintain the correct ratio and reaction time.

To make silver nanocubes using the polyol method you will need:

1. A heating mantle with an oil bath capable of reaching at least 160 °C with magnetic stirring and a stir bar. You will need a clamp attached for holding the reaction vessel and syringes in place.
2. 5 mL of anhydrous ethylene glycol in a glass vial. This should have a lid through which needles can be poked. A simple press on cap will suffice. If you are using a septum cap ensure a needle is poked through to allow gas to get out during the delivery of the reactants.
3. 3 mL of a 0.25 mol/L silver nitrate in ethylene glycol solution. This can be prepared by dissolving 0.127 g (7.5×10^{-4} mol) of $AgNO_3$ into 3 mL of ethylene glycol. Keep this solution wrapped in foil or in the dark to prevent reduction.
4. 3 mL of a 0.375 mol (of the monomeric unit) per litre solution of poly(vinylpyrrolidone) (molecular weight = 55,000, monomer weight ~ 110). This can be prepared by dissolving 0.123 g of polymer in 3 mL of ethylene glycol. You may need to vortex the solution to dissolve the polymer and make sure this is fully dissolved before use.
5. A double barrelled syringe pump with a programmable delivery rate with 3 mL syringes. You will need tubing running from the syringes so that the solutions can be delivered to needles. You will also need two needles that the tubing connects to. One of the needles needs to be wide enough that the viscous polymer solution will not encounter problems during delivery.
6. A centrifuge that can accommodate 50 mL falcon tubes. You will also need water for rinsing the tubes. You will also need scales for balancing the samples for centrifuging.

To make silver nanocubes using the polyol method:

1. Add the magnetic stir bar to the 5 mL of ethylene glycol and clamp it in place in the oil bath.
2. Fill one syringe with the polymer solution and fill the other with the silver solution. Fit them into the syringe pump. Cover the silver containing solution with foil.
3. Engage the pump and let both solutions pass through the tubing until they are just about to come out of the needle. Turn off the pump.

4. Poke the needles through the lid of the reaction vial and hold them in position with a clamp.

5. Turn on the heating mantle and set it to 160 °C. Allow the ethylene glycol to equilibrate for 1 h.

6. Set the syringe pump to deliver the silver and polymer solutions at a rate of 0.375 mL/min. Turn on the pump and let the solutions drip into the ethylene glycol under stirring. All of the reactants will be delivered within 8 min and you will see the solution turn from colourless to yellow then red and then a thick dark red green as the silver nanocubes form. Leave the reaction for a further 45 min.

7. Turn off the heating mantle and remove the vial from the oil bath. Allow the solution to cool.

8. Transfer the contents of the reaction vial to a falcon tube and dilute the sample with water. Make up a counterweight tube and centrifuge the sample at full speed for 10 min. Discard the top solution and redisperse in water. Repeat the centrifugation step twice more to get rid of the ethylene glycol. You will now have a dispersion of silver nanocubes ready for analysis.

Scanning electron microscopy should reveal an abundance of cubic particles with an average edge length of 175 nm (around 15 nm deviation). Samples that are highly monodisperse will demonstrate packing to form large arrays. You may observe that about 5% of the sample will be composed of nanorods and if you need to get rid of these then you can dialyse the formed nanocubes shortly after they have formed in the ethylene glycol solution. The temperature plays an important role in the speed of the crystal growth and if the temperature is more than 10 degrees either side of 160 °C then the morphology of the formed silver will not be cubic. Powder X-ray diffraction of the cubes should reveal strong reflections for the [111], [200] and [220] planes for face centred cubic silver (JCPDS card number 04−0783). The intensity ratio between of the [111] plane compared to the [200] will be slightly higher than would be observed in the bulk metal. This is due to the cubes resting flat on the surface of the sample holder such that the orientation of the flat [111] plane is more ordered in comparison to the other facets (Fig. 4.62).

With a small modification to this procedure, long nanowires of silver can be produced (Sun et al., 2003). Adjust the concentration of silver nitrate in ethylene glycol in the reactant solution to 0.1 mol/L. This will produce highly anisotropic rods around 40 nm in diameter and up to 50 μ in length.

Interestingly, both the silver rods and cubes can be used as sacrificial templates for making hollow gold replicas (Sun and Xia, 2002). If tetrachloroaurate salt is added to an aqueous dispersion of the cubes then a redox reaction occurs generating silver chloride and depositing solid gold.

$$3Ag_{(s)} + HAuCl_{4(aq)} \rightarrow 3Au_{(s)} + 3AgCl_{(aq)} + HCl_{(aq)}$$

Take a 250 μL aliquot of the as produced silver nanocubes dispersed in water and add to 5 mL of water. Set this solution on to reflux at 100 °C for 10 min. After this time and add a 1×10^{-3} mol/L solution of $HAuCl_4$ dropwise in 50 μL aliquots. You will begin

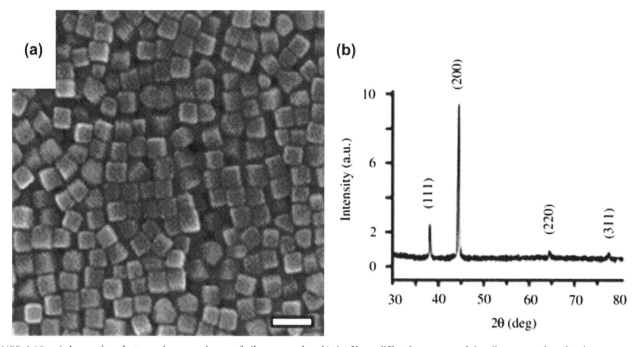

FIGURE 4.62 a) A scanning electron microscope image of silver nanocubes. b) An X-ray diffraction pattern of the silver nanocubes showing a pronounced reflection of the (200) peak due to the way in which the cubes pack together. *(These images are used with permission From Andrew R. Siekkinen, Joseph M. McLellan, Jingyi Chen, Younan Xia, Chemical Physics Letters 432 (2006) 491–496)*

to see a colour change in the solution as the plasmon band of the solidified gold develops. Up to 800 µL of the gold solution can be added at which point there should be no silver solids left in the solution. The gold solids can then be isolated by centrifuging and washing with water. This procedure will also work using platinum chloride to form platinum replicas.

A HYDROTHERMAL WATER BASED METHOD FOR PREPARING SILVER NANOCUBES

In this recipe a variation of Tollen's reagent will be used in the formation of nanocubes of silver (Yu and Yam, 2004). Tollen's reagent is a classical teaching experiment where a complex of $[Ag(NH_3)_2]^+$ in solution is used as a test for the presence of aldehydes. The silver complex reacts with an aldehyde to form solid silver which, if performed in a test tube, gives the inside of the glass a mirrored appearance. Here we will react glucose with the silver diamine complex to form silver in the presence of n-hexadecyl-trimethylammonium bromide. For the successful development of cubic structures the surfactant must be present in larger than 2.5 M equivalents of the silver ions. Any less than this and the particle morphology produced will become more spherical as the concentration of surfactant decreases. The surfactant not only acts as a structure directing agent but also as a source of bromide anions. The anion reversibly forms AgBr in equilibrium with the $[Ag(NH_3)_2]^+$ meaning that most of the silver in the reaction solution is in the form of silver bromide. As the diammonium silver complex reacts with the glucose present to form silver the equilibrium shifts in favour of producing more of the $[Ag(NH_3)_2]^+$. Tollen's reagent normally reduces to silver at room temperature but due to the presence of the bromide can be performed at the 120 °C required for cube formation. The elevated temperature is required because of the relative energetic considerations driving the rate of facet growth. The silver complex is made by the reaction of silver nitrate with sodium hydroxide to form an oxide then further reaction with strong ammonia solution following this mechanism:

$$2AgNO_{3(aq)} + 2NaOH_{(aq)} \rightarrow Ag_2O_{(s)} + 2NaNO_{3(aq)} + H_2O_{(l)}$$

$$Ag_2O_{(s)} + 4NH_{3(aq)} + 2NaNO_{3(aq)} + H_2O_{(l)} \rightarrow 2Ag(NH_3)_2NO_{3(aq)} + 2NaOH$$

The formula for glucose does not, at first glance, seem as though it will react with the silver reagent because there is not an aldehyde group present in the structure. Though this is true the ring opens to form a short lived linear glucose molecule comprising less than 1% of the possible sugar isomers at room temperature. At elevated temperature the linear chain glucose forms long enough for reaction with the silver complex which drives the equilibrium to produce more of the linear form until the reaction is complete. Because 120 °C is well above the boiling point of water the reaction must be performed in a sealed container where the liquid will find it impossible to convert to steam. This approach is called hydrothermal synthesis. The reagents are loaded into a poly(tetrafluoroethylene) container housed inside a stainless steel 'bomb' which seals it shut and will hold it closed even when high pressures are generated inside.

To make silver nanocubes using Tollen's reagent you will need:

1. 6 mL of a 0.1 mol/L solution of silver nitrate in water. This can be prepared by dissolving 0.1 g (6×10^{-4} mol) of $AgNO_3$ in 6 mL of water. This must be prepared fresh before use in a 20 mL glass vial with a lid.
2. 3.75 mL of a 1 mol/L solution of sodium hydroxide. A litre of stock solution can be prepared by dissolving 40 g in 1 L of water.
3. A stock solution of 28% (~15 mol/L) ammonium hydroxide solution. You will need a Pasteur pipette for adding small amounts to solution. *This should be used in a fume hood.*
4. A magnetic stirrer and stir bar.
5. A 100 mL graduated glass beaker and water. You are going to be making the silver complex solution up to 60 mL.
6. An oven capable of reaching 120 °C. One with an automatic timer is preferable.
7. A stainless steel autoclave hydrothermal 'bomb' with a PTFE container and lid. The container should hold 20−25 mL of liquid though we will not be using this much.
8. 10 mL of a 0.0075 mol/L solution of glucose. This can be prepared by dissolving 0.0135 g of glucose into 10 mL of water.
9. 3 mL of a 0.05 mol/L solution of hexadecyltrimethylammonium bromide. This can be prepared by dissolving 0.054 g in 3 mL of water.
10. A centrifuge and suitable centrifuge tubes large enough to handle a 20 mL volume of liquid. You will need water for rinsing the product.

To make silver nanocubes using Tollen's reagent:

1. The first step is to prepare the Tollen's reagent fresh for use and this should be done in a fume hood to avoid fumes from the concentrated ammonia. *The Tollen's reagent must be disposed of after use as it will form a precipitate of silver nitride if left to stand for a long period. Ag_3N is a contact explosive and the solution can be treated with nitric acid to prevent formation after use.* Add the 3.75 mL of NaOH solution to the 6 mL of $AgNO_3$ solution under stirring. You should see a brown precipitate form which is Ag_2O.

2. Add the concentrated ammonia solution dropwise using a pipette to the silver solution until the brown precipitate dissolves. This step will form the $[Ag(NH_3)_2]^+$ complex.

3. In the 100 mL beaker make the final volume of the silver complex solution up to 60 mL. If any brown precipitate forms then add more ammonia. You should now have a 0.01 mol/L solution of $[Ag(NH_3)_2]^+$ that can be used for the hydrothermal synthesis.

4. Place the PTFE container on the magnetic stirrer. Add 5 mL of the silver complex solution, 10 mL of the glucose solution and 3 mL of the HTAB solution under magnetic stirring.

5. Remove the stir bar and place the lid on the PTFE container. Seal the container inside the hydrothermal 'bomb'. *Designs can vary so ensure you have assembled the components correctly before proceeding!*

6. Place the 'bomb' in the oven and heat to 120 °C for 8 h. Afterwards allow the container to cool to room temperature before handling.

7. Remove the PTFE container and decant the contents into a centrifuge tube. Centrifuge at 6000 rpm for 20 min to collect the silver nanocubes in the bottom of the tube.

8. Redisperse in 2 mL of water to concentrate the colloidal dispersion. You should now have a sample of silver nanocubes ready for analysis.

9. Add nitric acid to any remaining Tollen's reagent until the pH is neutral. It can then be disposed of in an appropriate solvent waste container.

Visually the produced colloid should be a transparent yellow from the plasmon absorbance of the silver and should remain stable due to surface bound surfactant. Non cubic particle dispersions will appear nearer orange in colour. The peak plasmon absorbance for the colloidal dispersion will be centred at ~420 nm. If dried onto a flat surface such as a transmission electron microscopy grid then you should see the cubes packing together to form an organised array in the microscope. The cubes should have a side length of around 55 nm with a variance of less than 5 nm. Excess surfactant may make imaging cloudy but this can be removed in the colloid by repeated centrifugation and washing with water. Powder X-ray diffraction of the colloid dried onto a silicon wafer will demonstrate the expected pattern for silver but with a very prominent [200] reflection. This is similar to the enhanced reflections observed in other nanocube samples where the cubes sit flat upon the surface to generate an enhanced [111] lattice reflection. In this case the tight two-dimensional packing in the plane of the wafer is such that the [200] lattice reflection is greatly enhanced by the ordering. The tendency of the cubes to order so well is aided by the surfactant adsorbed to the cube surface. The hydrophobic tail groups are attracted to one another by Van der Waals interactions so that the cubes will sit adjacent to one another upon the evaporation of water.

FERROFLUIDS

With thanks to Dominic Walsh

Magnetic materials are most often mentally pictured as metallic lumps of ferrous ore. Magnetism is a force arising from the alignment of spin within a solid lattice and the size of the lattice can vary widely. In something as large as a bar magnet there are a great many individual magnetic domains pressed together within the larger bulk. In a ferrofluid, the magnetic domains are within a nanoparticle which may be dispersed as a colloid within a fluid medium. In a high enough concentration, magnetic nanoparticles accelerated within a magnetic field can drag the surrounding fluid with them as they move. These colloids can exhibit some unique effects in a magnetic field and have found a number of niche applications such as magnetic oils for engines. Ferrofluids were first developed at NASA in the 1960s by Stephen Papell for use as a rotating shaft lubricant in zero gravity (Papell, 1965). These early ferrofluids were prepared by laborious grinding of the mineral magnetite ($Fe_2^{(III)}Fe^{(II)}O_4$) in a ball mill over the course of several weeks. During the grinding surfactant stabilisers and liquid were added as required until a ferrofluid was ready to be removed from the mill. For the nanotechnologist this method is time intensive and produces particles with a high polydispersity.

Fortunately Nature has a solution which may be copied. Magnetite is produced biologically in a number of organisms and provides these creatures with the capacity to sense and respond to the earths magnetic fields using an ability known as magnetotaxis (Mann et al., 1990; Meldrum et al., 1993). The formation of magnetite in a bacterial cell occurs within chains of small phospholipid vesicles. Iron is transported in the form of a chelate into the cell where it precipitates as amorphous $Fe_2^{(III)}O_3 \cdot nH_2O$ also known as ferrihydrite. The iron oxide hydrate is then sequestered within a lipid vesicle where cellular regulation and confinement make the phase transformation into the nanosized crystalline magnetite favourable. Biologically derived particles have size limitations placed upon them by the behaviour of the magnetic domains. The magnetite crystals must remain above 5 nm in size so that the alignment of electronic spin giving rise to a magnetic domain is not overcome by thermal fluctuations, a condition known as superparamagnetism. A superparamagnetic nanoparticle will align extremely rapidly in the presence of an external magnetic field having a far higher magnetic susceptibility than a normal paramagnetic material. This property can be exploited in a number of medical applications to be discussed further on. While the crystal size is maintained between 5 and 10 nm the particle has one magnetic domain aligned throughout. For magnetite crystals larger than 10 nm multiple magnetic domains can exist within the particle. In this condition some of the spins may not be aligned with one another and a decrease in the net magnetism of the particle is observed.

Synthetic magnetite is prepared from the hydrolysis of Fe$^{(II)}$ (sometimes called a ferrous ion) and Fe$^{(III)}$ (sometimes called a ferric ion) chlorides in the presence of ammonia to produce a metastable iron oxide hydroxide having mixed valence states for the iron. Under the correct conditions the metastable ferrihydrite is converted to nanoparticles of magnetite. This gives the overall reaction:

$$2Fe^{(III)}Cl_3 + Fe^{(II)}Cl_2 + 8NH_3 + 4H_2O \rightarrow Fe^{(III)}_2Fe^{(II)}O_4 + 8NH_4Cl$$

The magnetism in the oxide arises from the difference in the number of electrons in the two different valence states of the iron and the manner in which they are arranged. Magnetite is an inverse spinel structure with the oxygen atoms in a close centre packed arrangement. Fe$^{(II)}$ ions occupy a quarter of the octahedral gaps between the oxygen whilst Fe$^{(III)}$ ions occupy another quarter of the octahedral gaps and an eighth of the tetrahedral gaps. The Fe$^{(II)}$ has 6 d-orbital electrons and the Fe$^{(III)}$ has 5 d-orbital electrons. Both of these are high spin arrangements meaning that the d-orbitals have unpaired electrons around the metal centre. Fe$^{(III)}$ ions in the octahedral gaps align their spin with the Fe$^{(II)}$ ions positioned in adjacent octahedral gaps and so an overall magnetisation is seen. This behaviour is not limited to iron alone and analogues of magnetite such as Mn$^{(II)}$Fe$^{(III)}_2$O$_4$ and Co$^{(II)}$Fe$^{(III)}_2$O$_4$ will also exhibit magnetism.

Magnetism

The majority of materials exhibit either para- or diamagnetic behaviour. When materials are place in a magnetic field they will respond with a magnetisation that represents the bulk magnetic moment of the material (Skomski, 2003; Dujardin and Mann, 2004). This is defined in the equation below as M (cm^3 mol^{-1} Gauss). The bulk magnetic moment is composed of the individual moments of the material, which are related to the spins of electrons within the atomic orbitals. Whether a compound has a high spin or a low spin electronic configuration will determine greatly its magnetic properties. By dividing M by the strength of the applied field H_0 (Gauss) the magnetic susceptibility χ (emu mol^{-1}) can be determined.

$$\frac{M}{H_0} = \chi$$

χ is independent of the strength of the applied field and represents a parameter unique to the sample being observed. A sample that is weakly repelled by a magnetic field is diamagnetic and a sample that is attracted is paramagnetic. Paramagnetism arises from unpaired electrons present in the electronic structure of a compound.

These classifications may be extended to encompass the different types of magnetic behaviour seen for various long range (larger than one molecular unit) structures. For example if looking at a paramagnetic material each molecule will have its own magnetic moment. Looking at a bulk amount of this material it may be seen that all the moments are aligned or opposed. In these materials the magnetic behaviour may be described as ferromagnetic or antiferromagnetic. Ferro- and antiferromagnets exhibit temperature dependant behaviour with respect to magnetic ordering. In paramagnets there is no organisation between separate dipole components at all temperatures. There also exists a behaviour called ferri-magnetism where the dipole moments line up opposing each other but there is a larger moment in one direction as in magnetite. Certain magnetic states can switch around depending on their thermal energy, known as a Curie temperature. Below the Curie temperature of a magnetic material the dipolar regions will align or oppose one another as demonstrated in Fig. 4.63. Above the Curie temperature the random thermal action of electrons will overcome any magnetic alignment. This is why magnets lose their potency if heated.

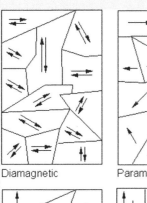
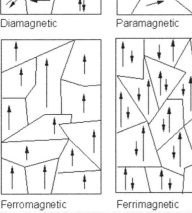

Diamagnetic   Paramagnetic

Ferromagnetic   Ferrimagnetic

FIGURE 4.63   A diagram demonstrating different types of magnetism. Each rectangle represents a cross section of material exposed to a magnetic field. Arrows represent the magnetic moments of the respective domains in a material.

For a colloid to remain stable, the thermal fluctuations within the solution must be of a sufficient energy to push the solid particles around and prevent them sedimenting due to gravity. In a magnetic colloid there is also magnetic attraction between the particles to contend with. A standard colloid, for example 30 nm gold particles, can be forcefully drawn out of solution by artificially increasing the gravity in a centrifuge. In a similar fashion a magnetic colloid can be drawn out of solution by a strong magnet. This property has enabled magnetic colloids to be used in special reverse osmosis bags for water purification processes. These work by having a bag partitioned in two by a dialysis membrane. Into one half of the bag the dirty water is added. In the other half of the bag is a powder of magnetic nanoparticles. The particles draw water across the dialysis membrane by osmotic dilution. Once the maximum amount of water has been pulled across, the magnetic nanoparticles can be removed with a strong magnet and the water is ready to drink.

The core unit of this recipe is the preparation of the magnetite nanoparticles. A few different preparation methods are provided that will offer some control over the particle diameter. The particular selection you make will depend on the end use. For use as an MRI contrast agent you may wish to have a very tightly defined diameter. For use in a school demonstration the size control will be less important. These variations also include a number of different stabilising agents suitable for dispersing the magnetite in both water and oil environments and even how to coat the magnetite with a glass coating.

MAGNETITE COLLOID STABILISED BY TETRAMETHYL- OR TERTABUTYLAMMONIUM HYDROXIDES

Adapted from Berger et al. (1999)

This is perhaps the most widely used preparation for making water stabilised magnetite colloids and forms the most basic recipe. It is a good starting point before attempting more complex preparations. The iron salts are prone to oxidation if left in air or exposed to moisture as they will rust. Store iron salts in a dessicator when not in use and always use a fresh solution each time you make a new batch.

To make a magnetite colloid you will need:

1. 10 mL of 2 mol/L $FeCl_2$ in 2 mol/L HCl in water. To make this, dissolve 39.76 g of $FeCl_2 \cdot 4H_2O$ in 100 mL of 2 M HCl in water. Use immediately.
2. 10 mL of a 1 mol/L $FeCl_3$ in 2 mol/L HCl solution. To make this dissolve 67.58 g of $FeCl_3 \cdot 6H_2O$ or 40.55 g of $FeCl_3$ in 250 mL of 2 M HCl in water. Use immediately.
3. 0.7 M solution of NH_3 in water. This can be made by taking 48 mL of concentrated ammonia and diluting to 1 L with water. *This will need to be handled in a fume hood.*
4. A 25% by volume solution of tetramethylammonium hydroxide (TMAOH) or tetrabutylammonium hydroxide (TBAOH) in water. This is the stabilising agent which will coat the magnetite nanoparticles and prevent aggregation in solution. This sterically bulky molecule coordinates to free iron atoms at the particle surface through the lone pair electron donation on the nitrogen. Using TBAOH will yield a slightly more stable colloid due to the increased steric bulk of the butyl moiety over the methyl. This is only a minor consideration as the colloid is mostly stabilised by electrostatic repulsion. *It is a corrosive and toxic substance. Contact on skin and inhalation is to be avoided. Use with gloves and in a fume hood.*
5. A 100 mL conical flask for mixing the reagents in.
6. A 50 mL burette with a tap and clamp to stand it with.
7. A magnetic stirrer and a large magnetic stir bar.
8. A vacuum aspirator or Buchner flask for removing excess ammonia from the produced colloid.
9. A centrifuge. One that can handle liquid volumes in tens of millilitres is preferable. If this is not available then you can separate samples into smaller centrifuge tubes though it will take slightly longer to isolate the produced magnetite and redisperse it in the TBAOH.

To make a magnetite colloid:

1. Add 1 mL of the 2 M $FeCl_2$ solution to 4 mL of the 1 M $FeCl_3$ solution in the 100 mL conical flask and stir magnetically. The small volume may be an issue here so if you find this amount is too small mix it together in a vial under stirring first. Follow step 2 until you have ~20 mL then transfer to the conical flask. Keep the stirring rate as high as possible.
2. Fill the burette with 50 mL of the 0.7 M ammonia solution and clamp it above the conical flask. Open the tap and add the ammonia solution dropwise to the iron chloride solution. The slower this step is done the better. A black/brown precipitate will form and this is the Fe_3O_4.
3. When all the ammonia solution has been added turn off the magnetic stirrer and allow the formed magnetite precipitate to sediment. This will take about 10 min.
4. Carefully pour off or pipette away the excess top liquid so that a sludge of magnetite remains in the bottom of the conical flask.

FIGURE 4.64   A photograph of a ferrofluid responding to a strong magnet. *(Reproduced with Permission from Berger et al. (1999). Copyright 1999 The American Chemical Society)*

5. Centrifuge the magnetite sludge for 1 min at 1000 rpm. This will form a pellet in the bottom of the centrifuge tube. Once it is finished pour the remaining liquid off the top.
6. Transfer the pellet into a conical flask fitted with a vacuum aspirator. For every 1 mL of magnetite sludge that was originally centrifuged add 0.4 mL of TBAOH or TMAOH solution. Mix it with a glass rod to redisperse the magnetite into the solution and add a magnetic stir bar.
7. Stir the solution for 30 min under vacuum to remove the any remaining ammonia. If the solution begins to bubble vent the flask slightly or turn down the vacuum.
8. After 30 min are up you should have a ferrofluid. There is still a minor problem in that the magnetic stir bar you used will be covered in the colloid. To get this off you can use a weighing boat on top of a stronger rare earth magnet. *With gloves on* remove the stir bar from the flask and hold it at one end above a weighing boat on top of the rare earth magnet. The remaining colloid should be drawn off the stir bar and pool in the weighing boat. Pick up the weighing boat and add the remaining ferrofluid to the rest of the colloid. *It is a good idea to keep the magnetic colloid in a sealed glass vial. It will stain most surfaces and is extremely troublesome to clean off a rare earth magnet!*

Now you can analyse the product. The first check is to make sure it reacts around magnetic fields. If you wave the rare earth magnet near the flask containing the fluid you should be able to see the fluid move. If this is not the case then leave the magnet by the glass for a few minutes after which you should see the colloidal particles pooling near the magnet. If this does not happen then you may have formed a non magnetic oxide. This can happen if the iron salts are not fresh and have partially oxidised. If the fluid is not very reactive then allow water in the system to evaporate until the liquid thickens. Thicker ferrofluids react more impressively in the presence of a strong magnetic field and you should observe spikes breaking out over the surface. This is caused by magnetic alignment of the nanoparticles overcoming both gravity and surface tension. The colloid can be imaged by TEM to determine the size of the magnetite particles. The hydrodynamic radius can be determined by dilution with water and examination by DLS. If this is fitted with a zeta potential analyser you may also wish to determine the surface charge of the particles. If allowed to dry out completely the presence of magnetite can be confirmed by powder x-ray diffraction. The JCPDS powder diffraction card number is 19−0629. Prominent reflections should be seen as outlined in the table below. As nanoparticles the X-ray reflections should exhibit peak broadening and this can be used to further calculate particle size using the Scherrer equation. The magnetic properties of the particles can be quantifiably determined by a magnetometer. For simple assessments of field strengths a Hall effect magnetometer can be used. For accurate measurements a SQUID magnetometer can be used to determine the Curie temperature and field strengths (Fig. 4.64).

d-spacing in Å	*hkl* reflection
2.96	[220]
2.53	[311]
2.09	[400]
1.71	[422]
1.61	[511]
1.48	[440]

MAGNETITE COLLOIDS STABILISED BY A BIOPOLYMER

Adapted from Kim et al. (2009)

In this variation the magnetite particles are stabilised by the biopolymer dextran. Alkyl chain amines are good at keeping magnetic colloids stable but they are of no use if the magnetic colloid is to be used in the biological field. One of the attractions of a magnetite colloid is that it can be used as a contrast agent in magnetic resonance imaging (MRI) (Hu et al., 2009). MRI scans apply a strong magnetic field which flips the spin of hydrogen in water molecules to align with the field. Once the field is turned off the spin decays back to its original unexcited state and gives off a photon in the process. The energy of that photon (we cannot use the term colour as it is not in the visible range) is characteristic of the state the hydrogen atom found itself in within the body. In the presence of soft or hard tissues water molecules which have dipole moments will experience different relaxation times when the magnetic field is turned off. This in turn affects the energy of the photon detected in the machine. Water molecule protons next to magnetic nanoparticles experience faster relaxation times than water molecule protons elsewhere in the body. This is called relaxivity and contrast agents, such as magnetite, have high relaxivities resulting in dark patches in an MRI scan. That is to say water in the presence of a magnetite nanoparticle will emit less photons to be detected than surrounding areas. If a nanoparticle can be tailored to seek a specific site in the body, perhaps by functionalisation with a biological receptor, or even direct injection then the location will look darker in an MRI scan. Perhaps even more importantly, a magnetite nanoparticle can be heated by exposure to an oscillating electromagnetic field (somewhere around the same frequencies used in FM radio transmissions). This is known as hyperthermic treatment and a given amount of colloidal magnetic particles are injected into a tumour where they can be heated up to 44 °C and trigger apoptosis in a cancerous cell. As little as 5% by weight of magnetite in a tumour can generate enough heat to kill the cancer cell but leave surrounding tissue undamaged (Ito et al., 2005). In terms of reducing the impact on a patient, research continues in alternative methods of making these particles self targeting with a view to not only imaging the source of the trouble but also treating it (Berry and Curtis, 2003). To take advantage of these properties the magnetite must be stabilised with a biologically safe agent. Ionic surfactants are unsuitable for this as they tend to insert themselves into the lipid bilayers of cells and disrupt them. Similarly, the alkylammonium hydroxides used in the previous recipe are also toxic to cells. One simple method to overcome these biocompatibility issues is by the use of polymers as delivery agents (Letchford and Burt, 2007). Poly(ethylene glycol) is routinely used as a 'stealth' coating for nanoparticles as it does not trigger a macrophage response (Barrera et al., 2009). In a similar manner the polysaccharide dextran can be used as a cell safe alternative to ionic surfactants. Dextran is a long chain glucose biopolymer linked mostly by $\alpha 1-6$ glycosdic bonds. In this recipe the iron salts are heated in a microwave under basic conditions which not only drives the formation of Fe_3O_4 but also oxidises the aldehyde groups in the dextran to form carboxylic acids. These carboxylic acid groups are capable of losing a proton and coordinating to the positive metal ions on the surface of the magnetite particle. The formed magnetite particles then end up stabilised by long chain dextran molecules coordinated to the surface. This has an added benefit in that surface bound dextran will prevent further magnetite nucleation and particle growth. Due to the diffusion limitation process of growth the molecular weight of the dextran has an influence over the size of the particle formed. By selecting heavier molecular weights the diameter of the particle formed is increased. By adding an enzyme capable of digesting the dextran into small maltoses the long chains can be clipped away from the particles (like a hair cut) leaving only one or two glucose units behind on the magnetite surface to provide an electrostatically repulsive layer which can be chemically modified if desired.

$$2FeCl_{3(aq)} + FeCl_{2(aq)} + 8NaOH_{(aq)} \rightarrow Fe_3O_{4(s)} + 8NaCl_{(aq)} + 4H_2O_{(l)}$$

This equation will also work if another divalent cationic salt such as Co^{2+} or Mn^{2+} are used. It is also possible to replace sodium hydroxide with ammonium hydroxide. An alternative reaction for this preparation may be substituted as (Fig. 4.65):

$$3M(NO_3)_{2(aq)} + 6NH_4OH_{(aq)} + 0.5O_{2(g)} \rightarrow M_2O_{4(s)} + 6NH_4NO_{3(aq)} + 3H_2O_{(l)}$$

To make a colloid of dextran stabilised magnetite you will need:

1. 1.4 g (7×10^{-3} mol) of $FeCl_2 \cdot 4H_2O$.
2. 3.8 g (1.4×10^{-2} mol) of $FeCl_3 \cdot 6H_2O$.
3. 8 g of dextran. The molecular weigh you select will depend on what size of particle you would like to produce. Molecular weights of 6000, 70,000 and 500,000 will produce magnetite particles of ~2.5, 4.5 and 5.5 nm respectively.
4. 80 mL of a 1 M NaOH solution. This can be made by stirring 4 g of NaOH into 100 mL of water under stirring. *When prepared freshly this will get hot and is corrosive so take care.*
5. A stock solution of 6 mol/L HCl solution in water.
6. A 300 mL beaker that can be used in a microwave containing 75 mL of water.
7. An ice bath large enough to cool the beaker.
8. A magnetic stirrer with a magnetic stir bar.
9. A pH meter or pH paper.

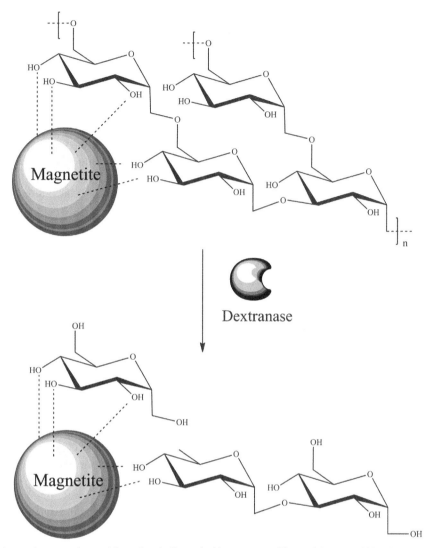

FIGURE 4.65 Dextran coating on the magnetite particle surface is digested with an enzyme. The particles are stabilised in solution by remaining surface bound glucose units.

10. A centrifuge and centrifuge tubes.
11. An 800 W microwave, a kitchen, and a thermometer.
12. A sepharose G-50 size exclusion column or similar and an excess of water for rinsing the column through. You will also need a beaker for collecting the product at the bottom. It should be packed in a glass column large enough to accommodate about 100−150 mL of solution passing through it.

For the enzyme digestion you will need:

1. A water bath or oven maintained at 34 °C.
2. A 1% volume acetic acid in water solution.
3. 250 Units of f.d.dextranase penicillin (1,6-α-D-glucan 6-glucanohydrolase). The measurement 'Units' is a biological term for the activity of the enzyme and reflects how much of the dextran can be digested by the enzyme in a given period of time.
4. A freeze dryer. There are a few ways to approach a freeze drying setup. Unless you have access to a dedicated freeze drying machine the simplest method is to freeze your sample in liquid nitrogen and then expose it to a vacuum. If you do this you will need a cold trap between the vacuum chamber and the pump.

To make the dextran stabilised colloid:

1. Dissolve the iron salts together in 75 mL of water in the beaker under stirring.
2. Once the salts have fully dissolved then add 80 mL of the 1 M NaOH solution rapidly with vigorous stirring. Leave this stirring for 20 min. It will form an iron hydroxide very briefly before converting to magnetite. This will look like a black dispersion.

3. *Remove the magnetic stir bar from the beaker before you perform this step!* Place the solution in the microwave and heat on full power until the reaction solution has reached 80 °C. This may take a minute or two and you should keep stopping to check the temperature with the thermometer.

4. Once the mixture has reached 80 °C, remove from the microwave and cool on ice. Using a pH meter to monitor the pH, add the 6 M HCl until the pH of the solution reaches 7. Allow the solution to cool for 1 h.

5. A black brown precipitate will have formed in the solution but some of this will be unwanted large lumps. Transfer the beaker contents into a centrifuge tube and centrifuge at 5300 rpm for 30 min. *Remember you will want to collect the top solution—not the precipitate in the bottom of the tube!*

6. The colloidal solution (the supernatant from the centrifuge) must be further purified using size exclusion chromatography. This is a column that acts like a sieve and allows the smaller colloidal particles to pass through whilst larger particles are captured. Add portions of the unrefined colloid to the column and allow it to wash through. If it is not already coming out the bottom then add water as an eluent to push the colloid through. Collect the fist 50 mL that comes from the bottom of the column—this is your dextran stabilised magnetite colloid.

7. You may wish to keep it suspended in solution or proceed to the enzymatic digestion step. Alternatively one unique feature of this preparation is that the colloid can be dried down into a magnetic film. This film will redissolve when exposed to water and is fairly brittle. However, should you repeat this preparation with analogous cobalt and manganese salts you will find the composite films formed by drying to be flexible.

To digest the dextran corona on the magnetite nanoparticles:

It may seem that because a stable magnetic colloid has been formed there is no need for further modification. However, the hydrodynamic radius of the magnetite particles can greatly affect the behaviour of the colloid in a biological system and particles having a radius greater than 50 nm will be detected and cleared by the immune system. For use as an MRI agent the corona size of the stabilising agent has direct implications on the induced relaxivity observed in adjacent water protons. The digestion and shortening of the dextran molecules are extremely useful in that it imparts a capacity for fine tuning the colloid for use. Hydroxyl and carboxylic acid groups are relatively easy to modify chemically so that further functionality can be engineered in the shortened chains without having to increase reagent amounts to account for groups in longer and 'unused' dextran polysaccharide chains.

1. Dissolve 0.3 g of a dried dextran magnetite film in 20 mL of water maintained at 34 °C in a water bath.

2. Using a pH meter to monitor the pH add dilute acetic acid under gentle stirring until the pH reaches 5.8. This is the ideal pH at which the dextranase has the highest activity.

3. Add the 250 units of 1,6-α-D-glucan 6-glucanohydrolase and stir the solution very gently for 24 h.

4. Transfer the digested colloid into a vessel suitable for freeze drying. After flash freezing in liquid nitrogen expose to vacuum so that all the water is sublimated. This will produce a fine magnetite powder which will also contain some left over maltoses from the digestion. These may be washed away by repeated centrifugation and rinsing with water. The magnetite powder can be redispersed in water when required to form a stable colloid.

Dried dextran film can be analysed by FT-IR and should show a shift in the dextran carbonyl bands from 1650 cm^{-1} in native dextran to 1635 cm^{-1} in the magnetite dextran films denoting coordination through this group to the particle surface. Minor shifts in the hydroxyl stretching modes can also be observed at ~3300 cm^{-1} commensurate with hydrogen bonding to the metal oxide. The presence of the iron oxide will also show a metal oxide spinel band at ~550 cm^{-1}. As with other magnetite preparations the magnetic properties can be accurately characterised by SQUID magnetometry.

MAKING A POLYOL STABILISED MAGNETITE COLLOID

Adapted from Hu et al. (2009)

In keeping with the theme of making magnetic colloids that can be used under biological conditions the next recipe concerns the formation of magnetite which uses a polyol as the stabilising agent. As discussed previously polyols are useful in that they do not trigger the bodies' immune response as readily as most foreign agents.

One of the problems with making a polyol stabilised magnetite is that it is not always very stable. One way around this is to form the magnetite and then exchange one stabiliser for another. This reaction uses iron (III) acetylacetonate , Fe(acac)$_3$, as the iron source and diethylene glycol, DEG, as both a solvent and a reducing agent. Once the iron oxide particles have formed the DEG is rinsed away and the solid magnetite redispersed in poly(ethylene glycol) bis(carboxymethyl)ether, HOOC—PEG—COOH. The replacement stabiliser is capped with carboxylic acid groups which, as seen in the previous preparation, can coordinate with iron cations at the surface. If dried the HOOC—PEG—COOH coated magnetite particles can be redispersed in phosphate buffer, foetal bovine serum, calf bovine serum and is stable in NaCl concentrations up to 1 M. The colloid will also remain stable over a pH range

5—11. This recipe can be adapted to give particles 3, 4—5 and 6—8 nm in diameter which allows for delicate control over the size dependant magnetic properties.

To make a PEG-stabilised magnetite colloid you will need:

1. 0.4238 g (1.2×10^{-3} mol) of Fe(acac)$_3$.
2. 20 mL of diethylene glycol. This is in a vast molar excess to the iron salt. This solution will be heated to 200 °C but the boiling point of DEG is ~240 °C so setting up a reflux apparatus should be unnecessary.
3. A heating mantle with a mineral oil bath capable of getting up to 200 °C. If you have access to one then you may use a Teflon lined hydrothermal bomb as the reaction vessel.
4. A nitrogen line for purging the DEG.
5. A 50 mL round bottomed flask with a clamp stand that can be sat in the oil bath. This will need a quick fit aperture that can be fitted with a nitrogen line in (through a glass tube is good) and a bubbler out. You want the setup to keep out air!
6. ~200 mL of 1:8 ratio ethanol:ether by volume. *This is highly flammable! Keep away from flame and heat and use in a fume cupboard! Dispose of waste in the correct waste solvent bottle.*
7. 0.084 g of poly(ethylene glycol) bis(cerboyxmethyl) ether (HOOC—PEG—COOH). This should have a molecular weight of ~600 g/mol.
8. 57 mL of a 0.01 M phosphate buffer solution adjusted to pH 6.4.
9. A centrifuge and suitable tubes. You will also need centrifugal filters with a 30,000 Mr size cut off. These are centrifuge tubes with a dialysis membrane within them which allow you to place a sample within a reservoir in the tube then spin out smaller molecules (below 30,000 Mr) to leave behind larger molecules like proteins or nanoparticles. It is possible to use a normal dialysis membrane but this process is faster. If you do use a normal dialysis bag make sure the pore size in the membrane is smaller than the nanoparticles you are hoping to obtain or there is a chance the product will diffuse out of the membrane.
10. An ultrasonic bath.
11. A magnetic stirrer and a stir bar.
12. Weighing scales accurate to 4 decimal places. A 200 mL beaker and two ~40 mL sealable glass vials.

To make a PEG-stabilised magnetic colloid:

1. Turn on the oil bath and set it to 140 °C.
2. Dissolve the Fe(acac)$_3$ in the 20 mL diethylene glycol solution in the round bottomed flask.
3. Set the r.b-flask in a clamp and fit the nitrogen line onto it. Allow N$_2$ to bubble through the mixture through a glass tube which will remove extra oxygen from the DEG. This is important to reduce the chance of other iron oxides forming.
4. Lower the r.b-flask into the oil bath and allow the mixture to heat for 1 h. After this period increase the temperature to 200 °C and allow the mixture to heat for a further 2 h. The solution will turn black as the magnetite nanoparticles are formed. The durations above will give rise to particles 3 nm in diameter and larger 4 and 5 nm diameter particles can be formed by leaving the mixture at 200 °C for 3 h and 6 h respectively. 6 nm particles can also be formed by doubling the amount of iron salt added (0.8476 g or 2.4×10^{-3} mol) and prolonging heating at 200 °C for 8 h.
5. After the heating step, a black colloidal mixture will have formed. Allow this cool to room temperature and transfer to a 200 mL beaker. Add 100 mL of the ethanol:ether mixture and the magnetite will precipitate out.
6. Transfer the solution and precipitate to a centrifuge tube (or more than one centrifuge tube). Centrifuge the solution until a solid pellet forms in the bottom of the tube and the precipitate cannot be seen in solution. Decant off the top liqueur and redisperse the pellet in the centrifuge tube with more of the ethanol ether mix and repeat the centrifuge step.
7. Remove the pellet from the tube and allow to dry in air. This should only take a few minutes given the volatility of the solvent mixture. Weigh the product. At this point the nanoparticles of magnetite are stabilised by DEG. The next step is to perform a ligand exchange reaction to replace this with the modified PEG.
8. Add the solid DEG coated magnetite to water and stir. You will want to add 0.43 mg/mL of water.
9. Dissolve the 84 mg of HOOC—PEG—COOH in the 27 mL of the 0.01 M phosphate buffer and check the pH is 6.4. This should be done in one of the sealable ~40 mL glass vials.
10. Add 3 mL of the DEG coated magnetite solution to the phosphate buffer solution containing the HOOC—PEG—COOH.
11. Seal the vial and place in an active sonic bath for 30 min. This step will shake loose some of the DEG surface coating from the magnetite and allow the HOOC—PEG—COOH to replace it. You may find you have to sit the vial in a piece of foam to stop it tipping over in the sonic bath. Depending on the strength of the sonic bath it can heat up very quickly. Heating at this stage is not a problem but make sure the bath does not get hot enough that the vial pops open.
12. Add a clean magnetic stir bar and allow the colloid to stir for 12 h at room temperature.
13. By this point the HOOC—PEG—COOH should have replaced all of the DEG at the particle surface. You will still have an excess of both the polymers in solution which you need to get rid of. This is achieved by dialysis with water or more quickly by using

a centrifugal filter. The colloidal magnetite will not pass through the interior membrane and can be further rinsed with water or removed from the tube. Depending on what brand of centrifuge filter you purchase the instructions for use may vary. When performing a more traditional dialysis, with the colloid sealed in a dialysis tube placed into a large volume of water, it can be hard to determine when you should stop. A dialysis on a colloid has gone too far when the colloid crashes out within the dialysis bag. This means that the stabilising agent has become too dilute to provide adequate electrostatic or steric repulsion between particles. This is obviously not an ideal marker after you have worked so hard to make the colloid stable. In very general terms; 30 mL of colloid in a dialysis tube can sit in 0.5−1 L of water for six hours while dialysis occurs. Changing the water twice more should ensure most of the excess is gone. This may vary on your setup however so you should determine what works best for you.

14. Remove the HOOC−PEG−COOH coated magnetite from the centrifuge filter and redisperse in the remaining 30 mL of the 0.01 M phosphate buffer.

This colloid is now ready for analysis.

Infra red spectroscopy is a very quick method to determine how well the DEG has been replaced by the HOOC−PEG−COOH. During steps 7 and 14 you will have the opportunity to collect some of the dry polymer coated magnetite. By comparing the spectra against control scans of both DEG and HOOC−PEG−COOH alone you will be able to easily spot if the ligand replacement has been successful. In DEG there are symmetric and asymmetric C−O−C stretches at 1075 cm^{-1} and 1132 cm^{-1} respectively. After substitution with the PEG the C−O−C stretching band manifests as a single peak at 1106 cm^{-1}.

TEM analysis of the magnetite particles coupled with size distribution analysis will give you an excellent indication of the average diameter and the level of polydispersity. Hopefully the size range will be within ±10% of the average. If the colloid is to be used biologically then it is important to determine the hydrodynamic radius as this will be the most influential factor in how it interacts in the body. This is accomplished most easily using dynamic light scattering. A table adapted from the source reference of the average core size against the hydrodynamic radius for both the DEG coated and PEG coated particles is provided below.

Diameter of particle core by transmission electron microscopy/nm	Diameter measured by dynamic light scattering/nm	
	DEG coated particles	PEG coated particles
3	8	16
4	9	17
5	9	19

SQUID magnetometry of the colloid should show superparamagnetic behaviour. Research has shown that the surface modification of the magnetite has a strong influence over the relaxation time of water surrounding the particle. For a PEG corona surrounding the nanoparticle will have water diffusing throughout and this is slow in comparison to the bulk water. This in turn means that the proton relaxation time is extended when sited interstitially between PEG molecules.

MAKING STABLE FERROFLUIDS IN NON POLAR SOLVENTS

Adapted from Ewijk et al. (1999)

So far we have looked at ways of making water stabilised magnetic colloids. What of a magnetite colloid that is stable in a hydrophobic solvent non polar liquid or oil? A magnetic oil is useful as an engine oil component or, as mentioned previously, a vacuum stable lubricant for low gravity conditions (or less excitingly−normal gravity). The most common preparation of this sort is to use oleic acid as a stabilising agent. Oleic acid is a naturally occurring omega-9 fatty acid found in olive and peanut oil and human fat tissue. The long chain aliphatic tail is terminated by a carboxylic acid group. This carboxylic acid can be deprotonated in order to facilitate binding to the particle surface through iron in a similar manner to that seen in the dextran based recipe. The long hydrocarbon tail then allows the particle to be dispersed within an oil continuous phase. In this recipe we will use a slightly different crystal arrangement of magnetite called maghemite. Maghemite is, like magnetite, magnetic and has an inverse spinel crystal structure. The difference is that there is a deficit of $Fe^{(II)}$. This is called the gamma form of magnetite and is written chemically as γ-Fe_2O_3 to reflect the vacancy left behind by the absence of the $Fe^{(II)}$ in the structure. It may be more correct to write $(Fe)(Fe_{5/3} \square_{1/3})O_4$ where the square represents an absence of the $Fe^{(II)}$ cations and the individual irons denote the positions of $Fe^{(III)}$ in tetrahedral and octahedral holes respectively (Haneda and Morrish, 1977).

The maghemite is synthesised without a stabilising agent under aqueous conditions almost the same way as magnetite but with a small extra step to drive the oxidation a bit further to produce an excess of $Fe^{(III)}$ in the structure. The dispersion is then flocculated to form sediment using a magnet and a small amount of oleic acid is added to the solution. This draws out the maghemite from the water phase as a molecule of the fatty acid is more tightly bound by the carboxylic acid to the particle than water molecules. This results in the particle becoming hydrophobic and remaining in the oil phase. The oil phase you choose to disperse the colloid in will depend upon the application and how fluid you wish the product to be. Heavier fatty oils will move slowly and will shown more dramatic effects in magnetic fields. Lighter more volatile organic solvents like cyclohexane will allow the particles to be drawn out of the solution in a strong magnetic field.

To make an oil stable maghemite colloid you will need:

1. 2.08 g (0.0164 mol) of $FeCl_2$.
2. 5.22 g (0.0321 mol) of $FeCl_3$.
3. 5.07 g (0.0210 mol) of $Fe(NO_3)_3$.
4. 20 mL of a 25% volume NH_3 solution. *This will smell strongly so work with it in a fume hood.*
5. At least 100 mL of a 2 mol/L HNO_3 solution in water. *This is a strong acid so take care and wear glasses and gloves!*
6. A measuring cylinder or cylinders that can range from 10 to 500 mL. You will also need ~1 L of water and a few 50–100 mL beakers.
7. A 500–600 mL round bottomed flask that may be fitted with a reflux condenser. You will also need a heating mantle and oil bath with a magnetic stirrer. This will be used to boil an aqueous solution. Depending on the setup you may want to transfer the reaction solution to a smaller round bottomed flask for the reflux step. *Ensure that when you fit the reflux condenser together all the fittings are clipped, that the glassware is clamped in place and that all the connections have safety clips on them to prevent flooding!*
8. A strong permanent magnet.
9. 2 mL of oleic acid or, to use the correct name, *cis*-9-octadecenoic acid.
10. A small separating funnel. It needs to be about 30–50 mL in volume.
11. 30–40 mL of ethanol.

To make the oil stable maghemite colloid:

1. Dissolve the $FeCl_2$ and $FeCl_3$ in 380 mL of water in the round bottomed flask using a magnetic stirrer. It can be best to do this above the heating mantle with the magnetic stir bar running. That way the setup can be lowered easily into the oil bath or mantle when the reflux step is performed. *Don't turn the heat on yet.*
2. Add 20 mL of the 25% ammonia solution quickly and stirring as fast as possible. A black precipitate will form which is, at this stage, magnetite. Leave this to stir for 5 min.
3. Raise the round bottomed flask off the heating mantle and turn off the stirrer. Use a strong magnet to induce sedimentation in the flask. Pour away the top solution. As you do so it can help to use the magnet to keep the sediment mineral in place.
4. Add 40 mL of the 2 M HNO_3 solution and return to stir for 5 min.
5. Dissolve the $Fe(NO_3)_3$ salt in 60 mL of water to make a 0.35 mol/L solution. The next few steps will convert the magnetite to maghemite.
6. Turn on the heating mantle and set it to 100 °C. Place the round bottomed flask containing the magnetite on to heat. You may find that ~100 mL looks rather small in a large round bottom flask and this may impact on the reflux. Transfer to a 200 mL round bottom flask if required. Remember the flask is ideally a quarter to a third full during a reflux. If there is too little liquid in the round bottomed flask then too much of this can vaporise affecting the reaction concentrations.
7. Add the $Fe(NO_3)_3$ solution to the magnetite dispersion and fit the reflux condenser. *Ensure all the clips and fittings are in place!* Leave this under reflux for 1 h with stirring and the black dispersion should take on a red/yellow colour.
8. Turn off the heat and stirrer and allow the formed maghemite to sediment out. Decant off the top solution and add another 40 mL of the 2 M HNO_3 solution to wash the sediment. Allow the precipitate to sediment again using a magnet and pour away the acid. Once the sediment has dried you should have a powder of maghemite nanoparticles.
9. Weigh out 56 mg of the maghemite and disperse in 10 mL of water in a beaker. Then take 2 mL of this solution and dilute with 50 mL of water. You will be working with a very small amount of maghemite—just 11.2 mg in 50 mL.
10. Add a few drops of the 25% ammonia solution which will cause the maghemite to flocculate together and crash out of the solution. Use a permanent magnet to hold the precipitate in place at the bottom of the beaker and add a further 50 mL of water. Remove the magnet and swirl the beaker around to wash the mineral. Use the magnet to sediment out the maghemite and then pour off the water.
11. Redisperse the maghemite in 100 mL of water and add 2 mL of oleic acid. Stir this gently with a spatula and you should see the maghemite collecting in the oil droplets which will turn black. All the maghemite will be taken up by the oil droplets over the

course of 5 min. Decant the whole lot into a separating funnel. If the oil droplets have coalesced (as they should) then it is better to use a pipette to draw the oleic acid phase off the top of the water.

12. Use a separating funnel to get rid of the water phase and keep the black oleic acid phase which will be floating on the surface.
13. Transfer the oleic acid to a small beaker and add 10 mL of ethanol. This will wash out any oleic acid not stuck to the surface of the maghemite particles and remove any remaining water. Use a magnet to hold the sediment in place while the ethanol is decanted away. Repeat this twice more. Three washes should be more than sufficient and there is no need to use an excess of the ethanol.
14. Dry the precipitate in a dry atmosphere. The formed powder can be taken for analysis at this stage.
15. Redisperse the maghemite colloid in a non polar solvent such as cyclohexane. Hopefully it will remain stable.

You should now have a stable magnetic colloid dispersed in a non polar solvent. Should you decide to scale up this preparation you will need ~13 mL of oleic acid for every gram of maghemite you draw out of the water in step 11. Other fatty acids can be employed as stabilising agents and the source reference mentioned dodecanoic acid may also be used. In general the larger the fatty acid is then the higher the melting point will be. This means the temperature of the oil phase must be maintained above the melting point of whichever fatty acid you use. FT-IR analysis of the dried powder free from excess olecic acid should show symmetric and asymmetric carboxylate stretching at 1428 and 1531 cm^{-1} respectively. A peak might also be expected at 1750 cm^{-1} for C=O stretching in the carboxylic acid head group but this may be absent. This suggests that a condensation reaction has occurred at the surface of the maghemite through iron hydroxyl groups to form a Fe—O—C bonds.

MAKING A GLASS COATED MAGHEMITE COLLOID

Adapted from Ewijk et al. (1999)

In this recipe the sol—gel process is adopted as a means of coating the maghemite in a silica shell. This will form a purely inorganic phase for suspension in a liquid medium. The surface charge of an iron oxide particle at a given pH has a massive influence on the stability of the dispersion and on how organic molecules will interact with the surface. Some of the recipes in this section include protocols for the formation of a flocculate of the mineral and this is achieved by reducing the electrostatic surface charge to zero so that the particles can clump together. For magnetite and maghemite the pH at which the surface charge crosses over from positive to negative is around pH 8. The surface of a metal oxide, especially under aqueous conditions, will have surface hydroxyls attached to it. In the case of magnetite this manifests as a surface coating of Fe—OH. At a pH value lower than the isoelectric point or point of zero charge (PZC), in this case between 8 and 9, the hydroxyl groups can be protonated to form Fe—$(OH_2)^+$ which gives the surface a positive charge. At pH values greater than the PZC the opposite will happen and the hydrogen atom on the surface bound hydroxyl group will be removed leaving behind a surface negatively charged with Fe—O^-. The sol—gel reaction to form the glass coating will be performed under alkaline conditions near the PZC of the maghemite. Without modification the maghemite would crash out of the solution before an even coating of silica has condensed around it. Some more difficult procedures negotiate this problem by using a sodium silicate solution in water to first form a thin layer over the surface before using a silicon alkoxide to form a thicker coat. This recipe replaces that more difficult step with a coating of citric acid in the aqueous phase. This will lower the PZC to below pH 7 so that the citric acid coated maghemite will be stable when dispersed in the basic alkoxide solution. The glass shell will have a PZC somewhere about pH 3 meaning the glass coated spheres can be dispersed at neutral pH in a polar solvent. If the glass is modified with organic groups then it is also possible to disperse them in oil phases.

To make a glass coated maghemite you will need:

1. 252 mg of dry uncoated maghemite nanoparticles. These can be prepared by following the recipe 'making stable ferrofluids in non polar solvents' up to step 8.
2. A 0.01 M solution of citric acid in 100 mL water. This can be made by dissolving 0.1921 g in 100 mL of water.
3. Tetramethylammonium hydroxide (TMAOH). This will be used to raise the pH of the acidified solution so you will probably only need a very small amount. It is a good idea to dilute a 25% by volume solution in water so that the addition of reasonable quantities will alter the pH if a citric acid solution. You do not want to completely alter the concentration of the solution you are adjusting. Before you run the experiment have a go at adjusting the pH of a 0.01 citric acid solution using the TMAOH using a Pasteur pipette under stirring. It can be very easy to overshoot the target pH and very often this will ruin the experiment so always have a practice beforehand.
4. A pH meter and a magnetic stir bar with stirrer. You will also need a Pasteur pipette and a 1 mL volumetric pipette.
5. Two 200 mL beakers. Measuring cylinders and water.
6. 75 mL of ethanol.
7. 5 mL of a 25% volume ammonia in water solution.

8. Tetraethylorthosilicate (TEOS) and a 100 μL volumetric pipette to dispense it with. It is possible to modify the alkoxide precursor as discussed elsewhere so that the condensed glass formed has functional groups bound to the silicon centre. For example you might try 3-aminopropyltriethoxysilane so that the formed glass has an abundance of easily chemically modified 3-aminopropyl groups at the surface. If you try something different it is important to remember that the rates of hydrolysis vary widely and there is a chance that the metal oxide may form to rapidly to coat the maghemite. *Even more importantly—some precursors will react with water vigorously so always make sure you have performed all the required safety checks!*

To make a glass coated maghemite colloid:

1. Disperse the maghemite in 45 mL of water in a beaker and monitor the pH using the pH meter.
2. Using a Pasteur pipette add the citric acid solution dropwise until the maghemite precipitates out of the solution and sediments. Give the beaker a gentle swirl each time you add a drop and then leave it to rest.
3. Use a fresh pipette to add TMAOH (or TMAOH in water solution) and adjust the pH back to pH 7 under gentle magnetic stirring. It should not harm the maghemite if you overshoot your target point but it is good practice to adjust the pH slowly and surely. This step should redisperse the maghemite to form a colloid and it should not sediment after you finish stirring.
4. Take 1 mL of the colloidal citric acid coated maghemite and add to a solution of 75% ethanol, 20 mL water and 5 mL of the 25% ammonia solution. Stir magnetically and add 0.1 mL (100 μL) of TEOS. Under these basic conditions the TEOS will begin to condense at the surface of the particles. It is important to mention that this is not through coordination with the surface bound citric acid which is known to repel glass formation. The citric acid is merely present to keep the maghemite particles from flocculating in the high pH environment. Seal the beaker so that nothing evaporates away and leave for 24 h.
5. Add two more 100 μL aliquots of TEOS to the solution. This will thicken the coating on the maghemite. More is needed as the diameter of the particles should have increased and there is now more surface area for the extra glass to cover. Seal once more and leave for another 24 h.
6. Isolate the particles by centrifugation and wash with ethanol if required.

The produced silica coated maghemite colloid should now be ready for analysis. TEM analysis should show the particle diameter to be somewhere around 40 nm though this can vary if the amounts of TEOS used were altered. It may also be observed that the silica shell accommodates two or three particles at once. This is caused by particles coming into contact with one another during the condensation step and sticking together. As there is an excess of forming silica oligomers these clusters are then coated to form a sphere.

ALLOTROPES OF CARBON

Carbon is a uniquely versatile element forming a major component of organic life with the capability to form a diverse range of chemical bonds. Carbon has been used since antiquity as a fuel in the form of charcoal and in more recent history as an additive to strengthen steel or in its graphite form as pencils. Recently the formation of various nanostructures from elemental carbon has created a surge in research to synthesis, characterise and utilise the various allotropes which demonstrate useful properties at the extremes of physical behaviour. The versatility of carbon arises from its electronic structure of $1s^2 2s^2 2p^2$. The s and p orbitals are both in the second quantum energy state (quantum number $n = 2$) and are close enough in energy that they will readily hybridise to produce four sp^3 orbitals (Hornyak et al., 2009). The four sp^3 orbitals are arranged tetrahedrally around the carbon atom at 109.5° to one another. Carbon centred molecules like methane therefore exhibit a tetrahedral configuration when linear sigma bonds are formed with a hydrogen atom capping the ends of an sp^3 orbital. If those hydrogen atoms are removed and replaced with another tetrahedral carbon and we repeat this to build up a network then the hexagonal face centred cubic structure of diamond is observed (May, 2000). Diamond is an insulator because there are no p orbitals to form overlapping π bonds to convey charge through. The degree of hybridisation in the central atom can vary and so carbon will also readily form three sp^2 orbitals (with one p-orbital remaining unchanged) and two sp^2 orbitals (with two p orbitals remaining unchanged). In these lower order hybridisations the bond geometry can be configured so that an overlap of the spare p orbitals on adjacent carbons allow extra electrons occupying the orbital to conduct across them. This is the basis of conductivity in aromatic, or π-bonded, organic molecules and for an allotrope of carbon the conductivity can be extended over the entire carbon framework. This is the case for graphite where the carbon is sp^2 hybridised and the hybrid orbitals are arranged in the same plane at 120° from one another to give the minimum electronic repulsion. Long range carbon to carbon bonding forms a flat hexagonal repeating structure where each of the carbons has a spare p-orbital perpendicular to the plane of the C−C bonding. The p-orbitals on adjacent carbons overlap so that the aromatic system extends across the top and bottom of the carbon sheet. In chemical notation this is represented by drawing double bonds between the carbons. In bulk graphite the sheets stack on top of one another in alternating ABABAB layers known as Bernal stacking. Because there is no direct bonding between the layers individual sheets can be delaminated easily. The sheets can also slide over one another leading to graphite being used as a high performance lubricant. The orbitals in sp hybridisation are at 180° to one another and so form into a linear molecule when bonded to another atom. The two remaining p orbitals are at right angles to one another and can form two orthogonal pi bonds in the axis of the sp bond. This is written in chemical notation as a triple bond and occurs in ethyne

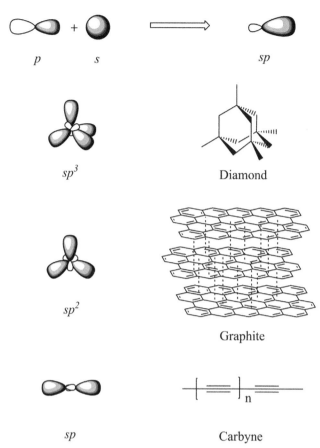

FIGURE 4.66 Hybrid sp^3, sp^2 and sp orbitals result in diamond, graphite and linear chain allotropes. Other carbon allotropes demonstrate a mixture of hybrid character giving rise to unique allotropes and behaviour.

and similar triple bonded functional groups in organic chemistry. The triple bond of sp hybridised carbon is quite reactive and the allotrope related to sp bonding, called linear acetylinic carbon, is not observed naturally (Lagow et al., 1995). It is possible to synthesise it but methods are limited to chains 300 carbons in length and having a fully substituted methyl group terminating the ends.

Diamond and graphite represent bulk networks of carbon where the repeating structure unit can be extended theoretically infinitely. The discovery of the C_{60} buckminsterfullerene demonstrated that it was possible to make other allotropes of carbon where the repeating structure unit is limited in size (Kroto et al., 1985). Similarly, it was initially thought that the graphene sheets of which graphite is composed could not be isolated and manipulated individually. This is not the case and single graphene sheets can exfoliated to form monolayers on surfaces or else induced to coil up to form tubes (Park and Ruoff, 2009; Thostenson et al., 2001). Curved carbon nanostructures like spheres and tubes represent a blend of the three-dimensional sp^3 character of diamond and the two-dimensional sp^2 character of graphite. In a sense making these materials is about controlling the level of hybridisation within the carbon to give rise to a desired morphology. In this section we will look at the various methods developed to produce carbon nanomaterials and why carbon in particular holds so much potential for technological revolution (Fig. 4.66).

THE BUCKMINSTERFULLERENE

With thanks to Dr Jonathan Hare, The Creative Science Centre, University of Sussex. Adapted from www.creative-science.org

Under the right conditions and with more than 40 atoms present, carbon can form into a closed cage arrangement named a buckminsterfullerene. These materials were named after the twentieth century American polymath Richard Buckminster Fuller who became famous for designing geodesic domes for use as strong lightweight buildings. The icosahedral geometry of the domes matches closely the structural arrangement of carbon in the C_{60} structure and so these allotropes inherited the name. Odd numbers of carbon will form into linear structures but even clusters of carbon having greater than 40 atoms will form into a cage fullerene. It is possible to form carbon cages having lower numbers of constituent atoms but these will be increasingly small and so will experience

FIGURE 4.67   A representation of the C_{60} allotrope of carbon also known as the Buckminsterfullerene.

instability due to the increasing strain that curvature puts on the bond angles. A high level of bond strain is observed in the synthetically formed cubane molecule (C_8H_8) as the sp^3 hybridised orbitals are forced to an angle of 90° with respect to one another rather than the standard 109.5° for a four coordinate tetrahedral molecule (Eaton and Cole, 1964) Above a number of 40 carbons the enclosed fullerene structure is preferentially formed as the bond strain is low enough to be offset by the stability of the structure. For a cage structure composed of N carbons there will be $(N/2) + 2$ faces (Bhushan et al., 1993). C_{60} in particular is formed in larger quantities that higher or lower numbered carbon clusters. The C_{60} molecule is a truncated icosahedron formed from a mixture of hexagons and pentagons (Fig. 4.67).

In order to achieve this highly strained curvature the orbital hybridisation is a predominantly sp^2 as would be found in graphite but with some sp^3 character to allow for the curvature. This results in slightly less aromaticity within the structure than you would observe for a purely hexagonal graphene arrangement. The hexagonal units comprising a fullerene have only two double bonds each and the pentagonal units have no double bonds which provide the asymmetry to form the cage structure. Larger cages are formed by the inclusion of more hexagonal units around an assumed centre. This results in elongation until a carbon nanotube is formed capped by two hemifullerenes (Dai et al., 1996). The axis about which the extra carbon is placed into the fullerene will influence the chirality of the formed tube and therefore the conductivity (Thostenson et al., 2001). This will be explored further on in the nanotubes section. Each C_{60} has 30 π bonds which can be employed in chemically functionalising the structure. Addition to the surface of a fullene by chemical modification relieves some of the local strain by imparting extra tetrahedral character to the bonded carbon. Using chemical modification it is possible to link fullerenes together in chains or as superstructures (Briggs and Miller, 2006). C_{60} will naturally pack together in a face centred cubic arrangement held together by weak van der waals interactions between the aromatic bonds. This allows the fullerenes to rotate freely at room temperature despite being within a crystal lattice. It is also possible to insert small molecules and metal atoms into the cage structure. Excitingly, fullerenes containing alkaline metals show superconductive behaviour (Hebbard et al., 1991; Tanigaki et al., 1991). The nomenclature for molecules enclosed within the carbon is $X@C_{60}$ and these are described as endohedral fullerenes. Getting a heterogenous species inside a fullerene can be achieved by forming the fullerene in the presence of a vaporised dopant as with laser ablation techniques. A pulsed laser is aimed at a carbon target and heats it rapidly to over 10,000 °C whereupon carbon is ejected from the surface and reconfigures into various forms. This is the original way in which C_{60} was detected in the laboratory. There are much simpler ways to form fullerenes and encapsulate materials within them. To encapsulate a dopant using preformed fullerenes you can simply seal and heat the dopant and the C_{60} together. The highest superconductivity yet observed is for the $Cs_2Rb_1@C_{60}$ endogenous fullerene which is made by heating the metals and fullerenes together in a glass capillary under helium at 390 °C for 74 h. This compound demonstrates a curie temperature of 33 K and it is hoped that high temperature organic based superconductors similar to the perovskite metal oxide superconductors might be found.

Fullerenes are produced naturally in small amounts in high energy interactions like lightning strikes and combustion. Large amounts of fullerenes have also been detected in space 6500 light years away (Cami et al., 2010). They can be made synthetically in large quantities by arc discharge or laser ablation at reduced pressures under inert gases like argon or helium. More recently a chemical origami approach has been used to fold up a planar aromatic molecule into a closed fullerene cage on a platinum catalyst. This is performed under an ultra-high vacuum within a scanning tunnelling microscope so the yields are low but the efficiency of the reaction is near 100%. A polycyclic aromatic is thermally evaporated into a vacuum chamber and adsorbs to the [111] plane of a platinum surface. The platinum is heated to around 500 °C which induces dehydrogenation in the organic molecule and causes it to fold up on itself to form a cage (Caillard et al., 2008).

In this recipe the arc discharge method will be used. Two carbon rods are gently contacted end to end within a bell jar maintained at a seventh of atmospheric pressure. A voltage of 20−30−50 volts with a large current of around 100 A is placed across the rods which induces arcing across the carbon. Where the arc passes through the rods the carbon is superheated and this ejects carbon

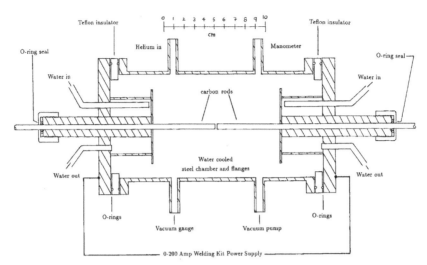

FIGURE 4.68   A schematic design for a fullerene generator. *(Adapted with Permission from www.creative-science.org)*

material from the surface of the rod. Around 5–10% of this material will be fullerenes. The carbon debris is scraped from the inside of the bell jar and dispersed into toluene. All of the non fullerene soot sediments to the bottom while the product remains in the toluene. Fullerenes are soluble in toluene to 2.8 mg/mL and even more soluble in dichlorobenzene to 8.5 mg/mL. Further purification can be performed by column chromatography to remove impurities with a toluene rinse followed by a dichlorobenzene wash to remove any larger fullerene allotropes. Small amounts of soot (less than a gram) as generated in the bell jar experiment can be roughly purified quickly by toluene extraction and will contain both C_{60} and C_{70}. Larger amounts of soot product (more than a gram) are required in order to use column chromatography to separate these allotropes further.

To make fullurenes you will need:

1. A schematic for a fullerene generator capable of yielding large batches is provided below and you will need the expertise of a mechanical workshop to help you assemble this. The design consists of a water cooled steel vacuum chamber 20 cm in diameter and water cooled carbon rod holders. The holders consist of tubes into which carbon rods 0.5 cm in diameter can slide and are isolated from the chamber wall by insulating layers and bolts made from PTFE or nylon. *Given that large currents will be flowing through this setup it is vital that it is checked for electrical safety and that there is no electrical contact with the chamber housing or the water flow!* O-ring seals on the end of the carbon rod holders maintain vacuum while allowing for carbon rods longer than the length of the chamber to be used. This way you will be able to push the rods in as they are consumed and generate more soot from which the fullerenes can be extracted. *The rods should not be pushed or touched while the generator has current passing through it!* It is important that the rods meet directly face to face with one another in the chamber in the same axis and that they touch. Because a spring mechanism would likely melt from the high temperatures that will be reached you will have to keep turning the current off to feed in the rods (Fig. 4.68).

   Alternatively a method that works but produces an order of magnitude less product is to use a large and strong bell jar that can accommodate the arc setup. This is a more simple approach and can be done without having to machine components. It will produce around half a gram of soot from which around 25 mg of fullerene can be extracted. Any type of strong chamber that can be evacuated and re-pressurised will do but you need to be able to harvest the soot from the inside after use. It should hold a vacuum as any air making it into the chamber will stop the nanotube formation. Evacuate your bell jar down to a pressure of around 1 Pa and check that the vacuum holds when the chamber is isolated from the pump. If using this sort of setup you will be limited to having to open up the bell jar to replace the carbon rods after each run. Bell jar systems work better using a DC power supply and having only one rod resting on a large carbon block. The arc in this case will use up only the positive (cathode) terminal which should be the rod. This way the weight of the carbon rod will make good contact with the carbon block even as it is used up (Fig. 4.69).

   Wiring and connections should be designed to allow for large current flows. Thick wire and bolted electrical connections should be used. Small crocodile clips will not work and are likely to melt. Both designs require that the apparatus to be setup on top of an insulating surface. In both cases larger chambers tend to give better results.

2. A vacuum system for removing air and maintaining the helium pressure. A valve should be fitted so that you can isolate the vacuum or allow a small flow should you wish to try flowing helium instead of static helium for the reaction. You will also require a manometer or other pressure gauge connected to the chamber to allow for accurate pressure control. The pressure will be maintained at 13.3 kPa of helium for the synthesis.

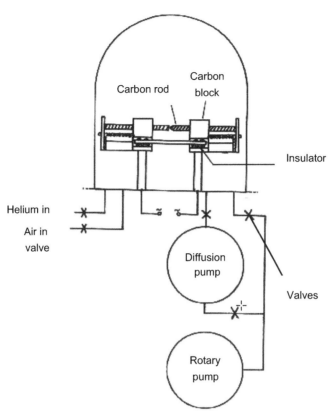

FIGURE 4.69   A schematic of the setup required for generating fullerenes using a bell jar. *(Adapted with Permission from www.creative-science.org)*

A water circulation system will also be needed if you are following the generator design. A simple setup as you would use for a reflux condenser can be used with water being passed in from the tap and out through a drain.

3. Carbon rods 0.5 cm in diameter of a high purity. The length of the rods depends on what will fit in your apparatus but two 5–10 cm rods pressed together will yield around 0.5 g of soot of which 25 mg will be fullerenes. For a bell jar based experiment you will also need a flat carbon block to use as the counter electrode and this will yield half as much as only the one rod will be used up.

   If using the alternating current fullerene generator you will need a method of pushing the rods into the chamber with externally. It is possible to accomplish this by hand wearing thick rubber gloves. However, in the interests of safety you should setup some kind of insulated plunger system. This can be as simple as a piece of wood with a strip of rubber on the end that you mount on a slide so it moves straight back and forth.

4. A helium line that can be fed into the chamber. This should be controlled by a tap valve and you will need to use the manometer to gauge how much to let in after the initial chamber evacuation.

5. For the soot generator two rod design you will need an alternating current power supply with a voltage of 40–50 V and a 100 A current. A welding kit power supply can be used for this.

   For the bell jar design you will need a direct current power supply also operating at 40–50 V with a 100 A current. A car or lorry battery can be used but will need to be recharged after use.

6. A stopwatch or countdown timer. Applying the current must be done in bursts so that the apparatus does not get too hot.

7. A spatula long enough to reach the bottom of the upturned bell jar or the length of the chamber. This will be used for scraping the produced soot from the walls. In both cases you will need a large section of stiff paper or card for collecting the scraped product onto.

8. A fume hood is required when working with the soot to prevent carbon dust getting everywhere and to prevent the risk of inhalation.

9. A large screw top bottle for keeping the collected soot in before the isolation step.

To make fullerenes using the two rod soot generator:

1. Assemble the rod holding components and insert the carbon rods. Push them together so that they meet in the middle between the holders. Separate them slightly so there is only a distance of 1 mm between them. Assemble the chamber so that it is air tight. Check that the O-ring seals on the outside of the chamber sit snugly where the rod goes into the chamber.

2. Turn on the vacuum pump and evacuate the chamber by opening the valve. It will take around ten minutes to reach a pressure of 1.3 Pa. Once this is reached seal the valve to the pump. You might want to try up to 6 Pa if your chamber is particularly large.

3. Open the inlet valve for helium and fill the chamber to just under atmospheric pressure.

4. Repeat steps 2 and 3 twice more to eliminate air and water from the system.

5. Turn on the water cooling to the rod holders and the chamber jacket.

6. Evacuate the chamber once more and refill with helium to a pressure of 133 Pa.

7. Turn on the AC power supply. Because the rods are not yet touching there should be no current flow but you should hear a buzzing coming from the power supply.

8. Depress one of the insulating plungers so that one of the rods moves closer to the opposite rod. When they make contact in the middle of the chamber you will hear the buzzing of the power supply get louder and make a rasping noise. This indicates that the rods have begun to arc.

9. Let the rods arc for 20–25 s then turn off the alternating current power supply. The chamber will be very hot and even with water cooling will need a few minutes to cool back down. If you run the current for too long then you risk damaging any plastic or rubber seals.

10. Turn on the power supply and press the other rod into the chamber until it arcs once more. Feeding in one rod and then the other will keep the arcing roughly centre in the chamber and will help prevent one rod being used up more than the other. Again let it arc for 20–25 s then let it cool. You will be able to repeat this process between 10–20 times.

    If the sound the power supply makes while in use turns from the rasping noise to a smooth hum then carbon deposits have built up on the end of the rod. This will decrease the amount of fullerenes being produced. To stop this turn off the power, open the chamber and remove the excess carbon debris by gently pushing the rods together and twisting them in opposite directions.

11. Once the carbon rods have been used up allow the system to cool then turn off the water cooling and close the helium inlet and allow the chamber to evacuate for 10 min. This will remove any toxic gas produced by the arc and the soot will compact against the wall of the chamber.

12. Close the outlet valve leading to the vacuum and very slowly allow helium to feed back into the chamber to bring it to ambient pressure. Doing this slowly will prevent the soot from being blown loose from the chamber walls and will make collecting the product easier.

13. If it is not already, then place the chamber in a fume hood. Depending on the strength of the air flow you may find keeping the window sash high up will prevent a strong gust from blowing away your product. Keeping all of the valves sealed, remove the inlets and outlets and stand the chamber on its side. Remove the side facing up then place a stiff piece of paper or card on top.

14. Holding the card in place turn this assembly upside down so that the paper is now on the bottom and the other end is pointing up. Remove now the uppermost side of the chamber and use a long spatula to scrape any soot from off the surface letting it collect on the paper at the bottom of the tubular chamber. Continue to scrape soot from the inside of the chamber with a spatula. When you remove the chamber you should have a pile of soot in gram amounts. Take care to ensure the soot product is not blown about in the fume hood as you lift the chamber. The collection step has the potential to be messy!

15. You should get about 5 g of soot but this will depend on how much of the rods you have used up. Transfer this into a screw top bottle ready for the step to isolate the fullerenes.

To make fullerenes using the bell jar:

This procedure is very similar to that using the soot generator. In this preparation there is no water cooling so you can only run the voltage for a short period of time or else components will start to melt. The arcing rod can reach a temperature of 4000 °C. Alternating current can be used in the same way as outlined for the chamber design but this preparation will describe the use of a direct current supply.

1. Setup the arcing apparatus as outlined in the schematic. Make sure all electrical components are isolated from the base of the jar. The carbon rod should be resting upright and perpendicular to the carbon block. The rod must be electrically connected to the positive terminal of the power supply (the cathode). The block is connected to the negative terminal (the anode).

2. Evacuate the bell jar down to 1.3 Pa and then purge it with helium to just under atmospheric pressure. Repeat this 3 more times.

3. Evacuate the bell jar once more and allow the chamber to fill with helium to a pressure of 1.3 Pa.

4. Set the direct current power supply to 40–50 V with a current of 100 A then turn it on for 10–20 s and turn it off again. If the bell jar is transparent you will see the rod glow white hot and begin to eject material which will look like sparks. The interior of the jar should begin to appear black as the soot builds up.

5. Wait 10 min for the system to cool down. Slowly evacuate the bell jar if possible.

6. Open the tap to the helium supply very slowly and return the chamber to atmospheric pressure. You want to disturb the soot as little as possible so nothing escapes when you open the jar.

7. Gently lift off the bell jar and turn it upside down. Transfer this to a fume hood taking care not to lose the product in the air flow. Use a spatula to scrape the inside of the bell jar to collect about half a gram of soot. Transfer this into a screw top bottle ready for further purification.

To purify fullerenes from the soot you will need:

1. 30−40 mg of the carbon soot collected in the screw top bottle.
2. Toluene for solubilising the fullerenes with. Around 500 mL should be more than enough for filling and rinsing the column with. *Flammable liquid, work within a fume hood.*
3. A glass funnel and collecting conical flask. You will need a folded filter paper to fit in the funnel.
4. Two 100 mL conical flasks for collecting separated C_{60} and C_{70} fractions respectively.
5. A 50 mL dropping funnel.
6. A glass separating column with a tap at the base. It should be large enough to hold at least 50 mL of fluid.
7. Clamps to hold the glassware in place with.
8. 30−50 g of carbon granules to be crushed and packed into the column.
9. 200 mL of dichlorobenzene. *Flammable liquid, work within a fume hood.*
10. A solvent waste bottle for toluene and dichlorobenzene.

To purify fullerenes from the soot:

1. Clamp the column in place and pack it with the crushed carbon granules.
2. Place a collection flask underneath the outlet and fill the column with toluene. Open the tap so that the toluene drips out of the column and let it drain until the toluene level is just below the granules. Close the tap.
3. Add 50 mL of toluene to the screw top bottle containing the soot. Replace the lid and shake the contents. You should see the toluene turn a deep red/purple colour. Allow the sample to stand while you perform the next step.
4. Close the tap on the dropping funnel. Once the visible soot has settled in the screw top bottle open it up and carefully decant the purple toluene into the dropping funnel. You can add more toluene to the soot and repeat if required.
5. Clamp the dropping funnel in place above the column. Place a clean receiving conical flask under the column and open the tap. At the same time open the tap on the dropping funnel.
6. Initially the toluene dripping from the bottom of the column will appear colourless. Collect these portions and dispose of them in a waste solvent container. Remember to close the tap when you remove the conical flask. This will take around half an hour.
7. When the colour of the drops turns purple/magenta the C_{60} fraction has begun to filter through. This portion will continue for another 20 min then the toluene will run clear again. Let the toluene level sink to the level of the carbon granules. Remove the flask containing the purified C_{60} and seal it or transfer the contents into another container for further analysis.
8. Add 50 mL of dichlorobenzene to the dropper funnel and allow this to run through the column.
9. As before allow the clear portion to run through and discard it. When a red colour appears then you can begin collecting the C_{70} fraction.
10. When the solvent turns colourless again remove the flask and either seal the flask or transfer into another container for further analysis. Allow the remaining solvent to flow through the column into another container and dispose of it in the solvent waste container.

You should now have samples of purified C_{60} and C_{70} dispersed in solvents. It is also possibly to purify the fullerenes using a Soxhlet extraction procedure. Hot solvent is allowed condense down and drip onto the sample loaded into a porous holder. As the solvent leaches out is takes the fullerene with it and this fills into a collection well. A siphon then removes the product and the process repeats. For fullerenes chloroform or toluene is used as these are less hazardous to work with than benzene as the extraction solvent. Because the solvent is reused in the Soxhlet method and the sample is rinsed many times the process yields the maximum amount of fullerenes from the soot.

The successful synthesis of the fullerene product can be checked quickly using infra red spectroscopy (Hare et al., 1991). This can be performed directly on the soot harvested from the generator or bell jar before the purification step if necessary. The high symmetry of the C_{60} fullerene results in a distinct four peak infra red signature with major absorbances at around 527 cm^{-1}, 576 cm^{-1}, 1182 cm^{-1} and 1429 cm^{-1}. The spectrum for C_{70} is slightly more complex and exhibits more peaks although a very strong absorbance should be observed at around 1430 cm^{-1}. Both the C_{60} and C_{70} allotropes retain solvent through aromatic interactions on the surface even when they appear dry. To dry the fullerenes remove the bulk of solvent by evaporation and then heat the samples at 170 °C under vacuum for 2 h. When working with vacuums and nanoscale powders take precautions to ensure that the sample does not get blown away.

The fullerenes can be imaged by scanning tunnelling microscopy under ambient conditions although atomic resolution may be difficult as the fullerenes rotate at high speeds. To make a sample for STM the fullerenes must be dissolved in benzene to a concentration of 0.01 mol/L (Bhushan et al., 1993). 50 μL of this can then be spin cast onto a flat gold substrate. The gold substrate

FIGURE 4.70   (a) Optical picture of C_{60} dissolved in toluene.

is prepared by evaporating gold onto a cleaved mica surface stuck to a metal base. It is important to ensure the gold on top of the mica is electrically conductive to the metal puck or support mounted in the microscope. The sample is then dried at 50 °C to drive off the benzene from the fullerenes. Exposure to a vacuum can also help this. Although you may not be able to resolve the individual hexagons and pentagons making up the cage you should be able to see dome structures around 7 Å in diameter. Depending on the sample preparation method you may also see some hexagonal packing of the fullerenes (Fig. 4.70).

CARBON NANOTUBES

Carbon nanotubes (CNT's) are perhaps the most high profile nanomaterial in the public mind and are well publicised for having physical properties that outstrip current engineering and electrical component tolerances by orders of magnitude (Baughman et al., 2002). There are two main areas, electronics and materials, where research is hoping to move nanotubes from the laboratory and into a commercial reality. Carbon nanotubes have the strongest tensile strength, weight for weight, of any fibre in existence making them attractive as a material for fabrication of light, strong and flexible structures (Zhang et al., 2002; Gao et al., 1998). Carbon nanotubes can also demonstrate conductivity over 1000 times that of copper coupled with ballistic charge transport making them a potential component in ultrafast electronics (Frank et al., 1998; Vijayaraghavan et al., 2007). Ballistic conduction describes a state where electrons are unimpeded and not subject to the scattering that leads to resistance in other materials. It is not superconductivity as some of the associated field effects are absent but it does make for a highly efficient wire. To date the full potential of carbon nanotubes is unrealised because the methods for its production are not far enough along to afford complete control over the product. The growth mechanism is well understood and a number of reliable methods exist for the mass production of both single walled and multiwalled rods having well defined thicknesses and aspect ratios (Li et al., 2004; Singh et al., 2003a,b). These can have either closed or open ends from growth or by modification post production.

Property	Single walled	Multi walled
Youngs modulus	1,100 GPa	1,280 GPa
Tensile strength	150 GPa	60–150 GPa
Thermal conductivity (along tube axis)	3500 $Wm^{-1}K^{-1}$	200 $Wm^{-1}K^{-1}$
Charge mobility	200,000 $cm^2 V^{-1} s^{-1}$	>100,000 $cm^2 V^{-1} s^{-1}$

Structurally, a carbon nanotube is a fullerene that has been elongated about one of its axis. The cylindrical component can be viewed as a graphene sheet rolled over and joined to itself along one edge. In these graphene sheets there is slightly more sp^3 character in the carbon atoms to allow for the curvature. The angle at which the graphene sheet is rolled up to produce the tube influences the electronic behaviour. This is called the nanotube chirality or wrapping vector and it determines whether the carbon structure has semiconductor or metallic properties (Saito et al., 1998).

For a single walled carbon nanotube there are three main conformations—armchair, chiral and zigzag. Armchair tubes are metallic, zigzag tubes are semiconducting and chiral tubes range from conducting to semiconducting and a formula can be used to calculate which property the nanotube will exhibit. A flat sheet of graphene is laid flat as in Fig. 4.71. Counting horizontally along the top of the sheet gives you the number n, counting vertically down the sheet gives you the number m. When $n-m$ or $m-n$ is a multiple of 3 the nanotube will be a metallic conductor. All armchair tubes are metallic conductors because $n-m = 0$ and all zigzag tubes of the form $n-m = 3q$ (where q is an integer) are also conducting. The values of n and m can also be used to determine the diameter of a single walled nanotubes.

To do this the chiral wrapping vector is described by the numbers n and m where

$$C_h = na_1 + ma_2$$

where a_1 and a_2 are the primitive unit cell vectors of graphite and both are equal to 0.246 nm (a_0). The diameter, d, of the tube as a function of n and m is given by the formula:

$$d = \frac{a_0\sqrt{((n^2 + nm + m^2))}}{\pi}$$

The chiral angle of the carbon nanotube is given by the formula:

$$\theta = \sin^{-1}\frac{\sqrt{3}m}{2\sqrt{(n^2 + nm + m^2)}}$$

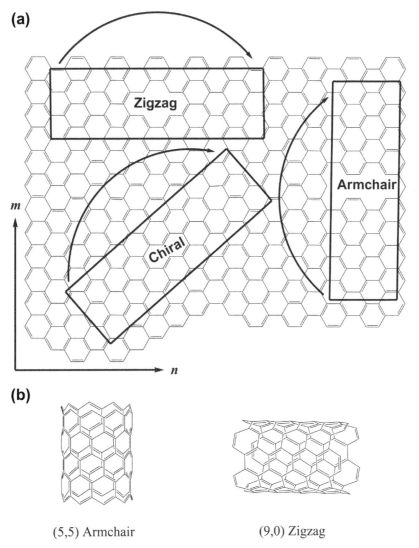

FIGURE 4.71 (a) A graphene sheet laid flat with sections cut into it. These sections will form nanotubes of the marked conformations and chirality when rolled up in the direction of the arrow. The n and m vectors are marked accordingly. (b) Examples of armchair and zigzag conformations.

There are a number of methods for producing carbon nanotubes and a broad overview of some of the more technical approaches is provided. In a similar manner to fullerene synthesis a high current arc discharge can generate single walled nanotubes with a low yield and a high percentage of amorphous carbon by-product (Journet et al., 1997). An arc is generated between two carbon rods in a stainless steel vacuum chamber with a gas inlet. Helium is then maintained under a low pressure flow in the chamber with the rods slightly separated while current is passed between them. Results are better with a continuous rather than static gas flow which would otherwise result in fullerene production. The SWCNTs are formed in a collar on the base of the cathode and they will not form without the presence of a metal catalyst. The advantage of the arc discharge method is that the growth rate is quick and the nanotubes can grow at a rate of several microns per second.

Laser ablation gives far higher yields than arc discharge (Morales and Lieber, 1998; Scott et al., 2001). An Nd:YAG laser is aimed at a metal doped carbon target in a furnace under argon flow. The pressure and flow rate of the argon has a great influence over the morphology of the tubes as this affects the laser heat transfer mechanisms. As with the arc discharge method a metal must be present to catalyse the rod formation. When the target is doped with nickel and cobalt nitrates a higher yield of single walled tubes is produced in contrast to using other transition metal salts. Though high in purity the method is limited by the very low quantities of product.

Catalytic chemical vapour deposition is far easier to do and although nanotube formation is slower than in laser ablation methods the quantities of nanotube produced are orders of magnitude larger (Andrews et al., 1999; Singh, et al., 2003a,b; Fahlman, 2002). A carbon containing gas feed is heated in the presence of a metal catalyst flowing through a furnace. The metal can be mixed with the gas feed in the form of a metallocene which decomposes in a reducing atmosphere to form nanoscopic metal clusters which act as points of nucleation for carbon nanotube growth freely dispersed in the gas flow. This technique is known as 'floating catalyst' synthesis and ferrocene, coboltacene and nickelacene can all be volatilised and passed into a gas phase reaction (Sen et al., 1997). The carbon released by these metallocenes can be used as feedstock for nanotube growth along with the carbon containing feed gas. Carbon dissolves into a metal particles surface and then recrystalises on the surface to form a cap in what is called the Yarmulke mechanism after the Yiddish word for skull cap (Dai et al., 1996). The chemical potential at the carbon-particle interface is lower than the chemical potential over the rest of the particle. This results in the addition of more carbon on the cap edges and so the nanotube grows out from the particle surface. In order to form single walled nanotubes the carbon cap must be in place on the surface before the metal particle 'sinters'. This means that using a metal ion which reduces in-situ is more selective to the production of single walled carbon nanotubes. The deactivation mechanism is not well defined but it is thought that growth may slow down as the tube gets longer which results in cap termination. The choice of carbon feedstock can influence if single walled or multiwalled nanotubes are produced in the flow. Benzene, toluene and xylene are regularly used as feed sources for large scale synthesis although it has been noted that CO tends to form more single walled carbon nanotubes more readily (Lamouroux et al., 2007). Using metal clusters adhered to a substrate surface in the decomposing gas flow results in vertically aligned arrays of carbon nanotubes (Singh et al., 2003a,b). When nanotubes form on a surface they will grow into one another and terminate and so grow together in the vertical direction perpendicular to the substrate. Nanotube forests grown from surfaces tend to be longer than those grown heterogeneously by purely floating catalysis. Nanotube densities of up to 500×10^6 tubes per square millimetre can be achieved with lengths of the order of hundreds of microns. In this recipe the CVD method will be used to produce multiwalled cabon nanotubes by feeding a mixture of ferrocene in toluene under an argon flow through a quartz tube in a furnace. The solution is pre-heated to 200 °C so that the ferrocene and toluene are vaporised before passing into a quartz sleeve in a tube furnace at 760 °C where the ferrocene decomposes to form iron particles on which the nanotubes can grow. Previously the argon flow was mixed with 10% hydrogen to ensure a reducing atmosphere but good results can be achieved without it and the risk of explosion from oxygen getting into the furnace is dramatically reduced. An absence of hydrogen in the argon flow will result in nanotubes up to 90 μ long. Hydrogen reacts with amorphous carbon in the furnace interior to form methane and its presence ensures a high purity of the nanotube allotrope. Enough hydrogen is generated by the decomposition of the organic reagents that you should not need to add further hydrogen to the gas flow unless you find your procedure is generating a large amount of amorphous soot. Hydrogen can also be added to inhibit the growth so that short tubes can be produced. Aligned nanotube arrays will form over the interior of the quartz furnace tube and over any quartz substrate placed inside it and, if trying to coat a different surface material, it is important to use a substrate that is stable at 760 °C. A silicon wafer makes a good substrate as these are extremely flat and tubes growing from the surface will not grow into one another as growth takes place. Nanotubes formed by CVD are predominantly in the armchair configuration making them metallic in character. Shifting this equilibrium to a predominantly zigzag semiconductor conformation can be done by doping the toluene feed with 40% diazane (Ducati et al., 2006). This heterocyclic nitrogen containing ring dopes the forming nanotubes with nitrogen defects which ultimately result in the formation of the zigzag structure. After the formation of the nanotubes they can be cleaved from the substrate surface for further processing (Fig. 4.72).

To grow an array of vertically aligned carbon nanotubes you will need:

With thanks to Krzysztof Koziol. Adapted from Singh et al. (2003a,b)

If you choose to use hydrogen in this experiment there is a risk of ignition or explosion! Ensure your reactor is gas tight before you begin and observe all applicable safety protocols local to your laboratory.

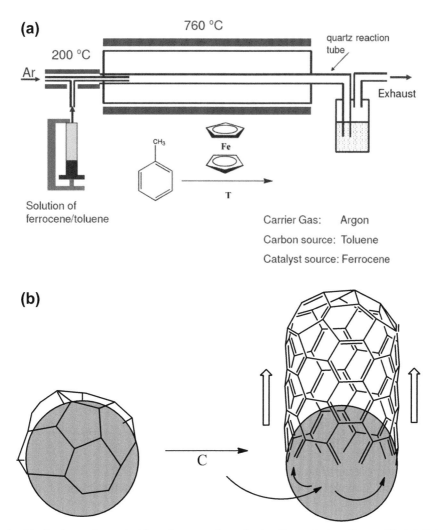

FIGURE 4.72 (a) A schematic of the floating catalyst method for carbon nanotube production. *(Image thanks to Krzysztof Koziol)* (b) The Yarmulke mechanism by which the nanotubes grow from the surface of a catalyst particle.

1. A stock solution of 9.6 wt% ferrocene dissolved in toluene. This can be prepared by dissolving 1 g (0.0053 mol) of ferrocene into 12 mL of toluene. *Toluene is flammable, work with it in a fume hood.* This solution should loaded into a glass syringe tipped with a long steel needle.
2. An automated syringe pump with a programmable injection speed.
3. A quartz tube around half a metre in length and having an interior diameter of between 17−34 mm. The quartz tube will normally have ground fittings on each end that cap the tube. Make sure these ands have fittings to which you can attach the gas inlet and outlet tubing. You will need to ensure these fit tightly as you do not want air getting into the reactor. This is a particular concern if you are to be using hydrogen in the gas feed. Ideally the quartz tube will extend a little way past the furnace.
4. A two stage tube furnace into which the quartz tube can rest. The first half is to be maintained at 200 °C and the second half at 760 °C. The furnace should be a few centimetres shorter than the length of the quartz tube. If you cannot get hold of a two stage furnace then you can try the experiment using a normal tube furnace at 760 °C. The preheating stage can be replicated using a heating element situated around a length of quartz tubing feeding into the main reactor tube.
5. A large oil bubbler some way after the exit of the furnace. This will be receiving hot gas for over an hour so use a volume of oil or similar that can absorb the heat. It is advisable to water cool the gas flow as it leaves the reactor by inserting a water cooled copper finger at the exit of the reactor (Andrews et al., 1999). You may also wish to include a filter to prevent carbonaceous nanomaterials making it out of the reactor although the bubbler should prevent this.
   The exhaust for the gas bubbler must lead out to a fume hood.
6. An argon feed and a gas flow rate meter fitted to it. Rubber tubing can be used to feed the argon into the prestage heater. Make sure any connections are gas tight to prevent air getting into the system. You will use around 60 L of argon per run.

7. A quartz or silicon substrate for coating. Quartz will yield better results and microscope slides can be used. Weigh them beforehand so you will be able to work out how much carbon is deposited. A long piece of wire or a long spatula is required for inserting the sample into the reactor and eventually retrieving it with. This can also be used to harvest nanotubes that have formed on the inside of the reactor.

8. Extra tubing fittings and connectors. Depending on the glassware you have assembled you may find different ways of connecting the various components together. Tubing should slip over a glass or quartz connector and be clamped in place to make it gas tight. Use glass tubing and connections beyond the exit to the reactor. Normal tubing will melt beyond this point unless you have a way of cooling the gas, like a water cooled metal cold finger. The syringe used to inject the catalyst and feedstock can be injected into the prestage via a stainless steel needle inserted through a rubber septum. Ideally the end of the needle will be a few centimetres past the gas inlet and almost half way into the prestage heater.

9. Assemble the components for your carbon nanotube generator as outlined in the schematic below. It is important to ensure the various connections are all gas tight and that ambient air will not get into the reactor (Fig. 4.73).

To grow an array of vertically aligned carbon nanotubes:

1. Load your substrate into the quartz reactor tube. Push it in to the centre of where the tube will be sat in the 760 °C portion of the furnace.

2. Place the tube into the tube furnace and connect the fittings to each end.

3. Turn on the argon flow to 750 mL/min. Allow the system to purge for 10 min though times may vary depending on the volume of your setup. It is important that there is a positive pressure within the reactor to prevent air getting in.

4. When you start the gas flow also turn on the furnaces. The prestage should be set at 200 °C and the main furnace at 760 °C. Turn on any cooling equipment for the reactor exhaust.

5. Fit the syringe containing the ferrocene and toluene solution to the syringe pump. Connect the syringe to the stainless steel needle acting as your feedstock inlet. Set the syringe pump to inject its contents at a rate of 1.2 mL/h.

6. Once the reactor is up to temperature and you are satisfied all the air has been purged from the system turn on the syringe pump.

7. Carbon nanotubes should form on the inside of the quartz tubing and this should appear to turn black. Allow the generator to run for 60 min then turn off the syringe pump.

8. Turn off the furnaces and allow them to cool for a few minutes then turn off the argon flow. A fast cooling rate will not damage the nanotubes.

9. Turn off your water cooling apparatus if fitted.

10. Disassemble the generator and remove the quartz tubing. Slide or push your substrate out or harvest the nanotubes from the inside of the quartz. You should now have nanotubes ready for analysis.

When you harvest the nanotubes take care not to let them blow around and keep them in a sealed container. Carbon nanotubes are known to gather in the lungs and have a similar action to asbestos in that they cannot be readily cleared by the bodies' macrophages. The aligned nanotube arrays can be examined by scanning electron microscopy. Because the tubes will be conductive there is no need to sputter coat them before viewing though you may need to apply some silver paint to the quartz so that the tubes have a path of conduction with the sample stub. You should be able to see a dense forest of nanotubes all aligned perpendicular to the substrate. For an argon only gas feed synthesis you will find them to be many tens of microns in length and around 30–40 nm in width which is consistent with the formation of mutliwalled nanotubes. To get single walled nanotubes you can try dropping the level of ferrocene in the toluene feedstock to around 0.2 wt%. The iron clusters formed in the prestage heating will be significantly small and therefore will only produce a single walled tube during the cap growth stage. One drawback of this is that the weight of nanotubes per square centimetre will drop from around 1 to 2 mg to less than 0.002 mg/cm^2. Another alternative is to iron dope the substrate with surface metal clusters where growth will be initiated. This has been demonstrated on iron doped aluminium oxide (Fahlman, 2002) and a similar approach has been used to grow carbon nanotubes directly onto atomic force microscopy tips (Cheung et al., 2000).

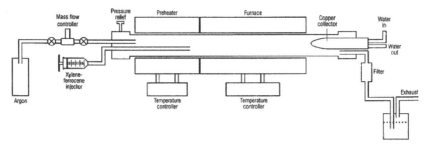

FIGURE 4.73   A schematic for a water cooled CVD carbon nanotube generator. *(Reproduced with Permission from Andrews et al. (1999). Copyright 1999 Elsevier)*

Increasing the reaction temperature will produce wider nanoparticles but this may be due to the deposition of amorphous carbon on the tube surface. Temperatures of above 850 °C will result in a reduction of the nanotube yield.

Carbon nanotubes can be difficult to get to disperse homogenously in water and often require sonication to get them to disperse. Aqueous nanotube suspensions of around 0.5—50 mg/mL can be prepared by adding 2 wt% sodium dodecyl sulfonate to the solution and sonicating it or applying a high shear rate. The anionic surfactant can also be exchanged for 1% solutions poly(vinyl pyrrolidone) or poly(styrene sulfonate) which will coat the nanotubes and separated by centrifugation (O'Connell et al., 2001). The chirality of the tubes can be determined by raman spectroscopy and a good review of this approach is provided (Dresselhaus et al., 2002). Metallic and semiconductor carbon nanotubes can be separated in solution by dielectrophoretic separation in a strong radiofrequency field. By altering the frequency of the applied electric field the metallic nanotubes can be manipulated to deposit out of solution to form a contact between two electrodes (Vijayaraghavan et al., 2007). It is ultimately possible to spin fibres from a dispersion of carbon nanotubes stabilised by a polymer. A composite carbon nanotube based biofibre is wet spun by injecting a carbon nanotube dispersion into a non-solvent. Once in a flowing non-solvent the polymer and nanotubes coalesce to form a strong and stable fibre (Lynam et al., 2007). From an engineering perspective pure nanotube fibres are desirable and this is now possible by spinning fibres directly from a CVD reactor as outlined in this recipe (Li et al., 2004).

GRAPHENE AND GRAPHENE OXIDE

In a similar fashion to mining diamonds from amorphous charcoal carbon, sheets of graphene exist prefabricated in lumps of graphite ready to be extracted. Monoatomic layers of graphene can be generated fairly easily using a micromechanical method (Novoselov et al., 2005). Graphite is rubbed over a flat surface and layers are exfoliated onto the substrate which are extremely thin but can be up to microns in length and width. The position of the exfoliated sheets can be determined by looking at optical interference under a phase contrast microscope then investigating the spot further with an atomic force microscope. It was previously thought that getting a monoatomic layer of graphene was unlikely as, energetically, a thin sheet would be more likely to roll up on itself. There were also issues with finding a single graphene sheet using optical microscopy methods; a monolayer of carbon is optically transparent and does not cause the same identifiable interference effects as a multi-layer stack. This problem was solved by rubbing the graphite on a silicon wafer having a precise 300 nm layer of SiO_2 deposited on the surface. This changes the optical path for even a monolayer and allows it to be visualised using phase contrast techniques. Single sheets found upon the surface can then be isolated and experimented upon further. Other techniques to produce a single sheet thickness crystal include an advanced form of cleaving individual sheets from the surface of highly oriented pyrolytic graphite (HOPG). Squares of the desired area are cut into an HOPG surface using oxygen plasma then a photoresist is poured on top and allowed to set. This is peeled from the HOPG surface and cleaves multi-layers of Burnal stacked graphene sheets with it so that they sit atop the raised section of solid photoresist. Tape is then used to repeatedly peel more layers from the surface until only single sheets remain (Novoselov et al., 2004). Both of these techniques are good enough for performing laboratory based experiments but due to the relative slowness of isolating the sheets optically followed by analysis with AFM these procedures are not useful for large scale production of graphene sheets. Producing large quantities in a controllable manner is a research target as the electrical and physical properties of graphene are of increasing technological importance. The 2010 Nobel prize for physics was awarded to Andre Geim and Konstantin Novoselov for isolating single two-dimensional carbon crystals and for further demonstrating a range of unique electrical and physical properties arising from the thinnest material in existence (Geim and Novoselov, 2007).

Property	Value
Youngs modulus	1,100 GPa
Fracture strength	125 Gpa
Thermal conductivity	5000 $Wm^{-1} K^{-1}$
Charge mobility	200,000 $cm^2 V^{-1} s^{-1}$
Surface area	2,360 $m^2 g^{-1}$

Graphene has a semimetal character and this means that there is a slight overlap of the valence and conduction bands. In the plane of the graphene sheet extremely high densities of current can flow. Furthermore, electrons moving over the surface experience minimal scattering. The periodicity of the carbon lattice gives rise to an unusual effect in the electrons such that they do not behave according to the Schrödinger equation, a purely quantum mechanical description, and can be described more accurately using the Dirac equation. This describes the behaviour of electrons conducting through a graphene layer as though they have lost their mass and are moving at relativistic velocities, that is to say near light speed. The unusual quantum states that arise in graphene can be exploited to monitor quantum electrodynamic effects at room temperatures whereas previously condensed matter physics research of this type has only been possible at ultra-low temperatures. In particular the morphology of graphene makes it attractive for use in scanning probe microscopy experiments where the electronic behaviour can be visualised with the graphene sheets in proximity to various materials and field to see how it behaves. As explored in the fullerene and nanotube sections of this chapter, the conductive properties of graphene can be altered by the introduction of curvature which disrupts the periodicity of the carbon lattice and, by extension, p-orbital overlap in the aromatic system. Therefore the band structure of the graphene can be modified locally by introducing mechanical stress or chemically by absorption of charged moieties at the surface.

Bulk methods of graphene preparation have started to be developed as the chemical research catches up with the physics research (Park and Ruoff, 2009). A facile method of making large dispersions of graphene in solution involves exfoliating sheets by sonicating graphite in a solvent (Hernandez et al., 2008). An organic solvent is used which matches the surface energy of graphene ($70-80$ mJ m^{-2}) such as N-methyl-pyrrolidone or benzyl benzoate. During sonication the solvent molecule infiltrates between sheets as the sonic energy teases the carbon layers apart and remains in contact with the carbon surface by weak van der Waals interactions. Unexfoliated components are removed by centrifugation to leave behind a dispersion of solvated unoxidised graphene sheets which will remain stable for weeks before sedimentation. The concentration of the dispersions is around 0.01 mg/mL. Higher concentrations can be achieved in aqueous solutions by oxidising graphite to form exfoliated graphene oxide then chemically reducing the sheets back to graphene in the presence of a stabilising agent. In this recipe we will use this synthetic route towards producing dispersions of DNA stabilised graphene sheets in concentrations up to 2.5 mg/mL. The DNA stabilised graphene can be used to make nanocomposite structures by drying down the solution in the presence of a guest molecule such as a protein. This forms a free standing film with a lamellar structure composed of the graphene sheets and the guest molecule.

The Hummers–Offerman method is used to form graphene oxide sheets (Hummers and Offeman, 1958). This is a harsh chemical oxidation step whereby graphite is treated with potassium permanganate in the presence of concentrated sulphuric acid. The preparation involves a significant level of risk and the reaction is highly exothermic. Before starting this procedure you must take every precaution and set this up in a fume hood. After the graphene oxide has been formed and isolated, the graphene oxide is further sonicated to ensure exfoliation and mixed with single stranded salmon sperm DNA. The graphene oxide is reduced back to graphene by adding hydrazine to the solution. The purine and pyrimidine bases within the DNA will coordinate to the surface of the graphene sheets by $\pi-\pi$ interactions. This orients the DNA so that the phosphate backbone protrudes from the surface and gives the graphene a negative electrostatic charge. This charge keeps the sheets homogenously dispersed within the aqueous solution (Fig. 4.74).

To make a colloid of exfoliated DNA stabilised graphene you will need:

With thanks to Avinash Patil. Adapted from (Patil et al., 2009)

This procedure uses strong acids and oxidising agents and is, at points, extremely energetic. You will need to perform each step of the exfoliation and following reduction in a fume hood with full protective gear. This includes heavy duty gloves, apron and

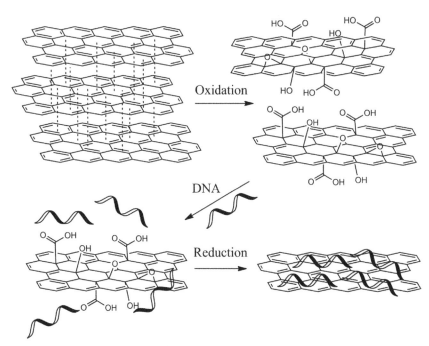

FIGURE 4.74 Schematic showing synthesis of DNA-stabilized graphene sheets. Graphite is oxidised and exfoliated to form brown graphene oxide. This dispersed into a solution containing single stranded DNA and reduced. The DNA coordinates to the graphene and forms a colloidal dispersion.

clothing and a face guard. The reflux with sulphuric acid must be performed behind a polycarbonate blast shield. Check with your safety office before performing this synthesis.

1. 1 g (0.083 mol) of powdered flake graphite.
2. A 300 mL three necked round bottomed flask. One side neck should have a glass stopper and the main neck should accommodate a reflux condenser. The remaining neck should be fitted with a thermometer.
3. A tall Coil reflux condenser which fits into the round bottomed flask. You will need the associated inlet (bottom) and outlet (top) tubing along with clamps to hold it in place. You will need a clamp stand for holding the condenser.
4. A dry ice bath large enough to sit the round bottomed flask in. This can be prepared by filling a pyrex dish one third full with bench ethanol and adding dry ice. When you make the ice bath add the ice slowly to begin with or you will find ethanol gets splashed about.
5. 23 mL of ice cold 18 mol/L (concentrated) H_2SO_4.
6. 3 g of potassium permanganate ($KMnO_4$).
7. A heating mantle capable of magnetic stirring and a large stir bar. You will also need an oil bath which can sit on top of it. The magnetic stirring must be strong as the solution can thicken at certain stages and you do not want it to remain still.
8. Water for adding to the reaction mixture. You will need a measuring cylinder as various 47 mL and 100 mL aliquots must be added at different stages of the preparation.
9. 10 mL of 30% H_2O_2 in water solution.
10. A centrifuge and falcon tubes.
11. A stock solution of 2 mol/L HCl in water. This can be prepared by dilution of 200 mL of a 36% (12 mol/L) concentrated HCl up to 1.2 L with water.
12. 1 L of a stock solution of 0.05 mol/L barium chloride solution. This is used to test for the presence of sulphate anions. You will need a few small vials or test tubes for testing in also.
13. Acetone for rinsing the product with.
14. A drying oven at 65 °C for drying the centrifuged intermediate graphene oxide product with.
    The next items are required for the stabilisation of graphene oxide using single stranded DNA and reducing it to graphene.
15. A 30 mL glass vial with a lid.
16. A sonic bath and a small piece of foam into which the vial can sit.
17. Dialysis tubing having a 12−14 kD cut off and four clips. You will also need a 1 L beaker and excess water for performing a dialysis with. A Pasteur pipette will help with loading the dialysis tubing.
18. 10 mL of a 2 mg/mL solution of single stranded DNA in water. This can be freshly prepared by dissolving 20 mg of salmon sperm DNA into 100 mL of water and heating it to 95 °C for 2 h. Use immediately or store aliquots in a freezer for later use. They will need to be reheated to boiling to get the single stranded DNA but this will not be an issue in this preparation as the DNA solution will be heated near boiling anyway.
19. 20 μL of a 35wt% hydrazine solution in water. *This is carcinogenic. Follow safe handling procedures.* Commercially available hydrazine is provided at a 50−60 wt% concentration so you will need to dilute this to the working concentration with water shortly before use in a microcentrifuge tube in microlitre amounts. Keep this in the dark and in the fridge when not in use.
20. A 100 mL two necked round bottomed flask. The side neck should have a glass stopper and this will be used for adding the hydrazine solution through. The main neck should fit a reflux condenser. Because you will be refluxing only 20 mL use a second smaller reflux condenser.
21. Volumetric pipettes and tips in the 1−5 mL and 10−100 μL ranges.

To make a colloid of exfoliated DNA stabilised graphene:

1. Double check that all of your safety procedures are in place and you are wearing the correct equipment.
2. Make up the dry ice bath and place it on top of the heating mantle. Sit the round bottomed flask into the dry ice bath and clamp it in place. Fit the reflux condenser and check that the clips holding the water tubes are firmly in place.
3. Use a pipette to add the 23 mL of concentrated sulphuric acid to the flask through the side neck. Add a magnetic stir bar and set the solution to stir slowly, just enough to keep everything moving. Adjust the thermometer so that it is in contact with the solution.
4. Add the 1 g of graphite to the stirring acid and let it disperse. Use a spatula to add the potassium permanganate in very small amounts *slowly.* As you add the permanganate the temperature will begin to rise and the solution may effervesce. Let the solution cool back down before adding more and do not let the temperature get higher than 20 °C.
5. Once all the permanganate has been added and the temperature has stabilised lift the reflux apparatus so that you can exchange the dry ice bath for the oil bath. Sit the reflux apparatus back into the oil bath.
6. Set the heating mantle to 35 °C and once the solution has reached this temperature maintain it for 30 min with slow stirring. Turn on the reflux condenser. During this step the solution will begin to thicken into a brown-grey paste.

7. Turn off the heating mantle and lift the reflux apparatus out of the oil bath. Replace the oil bath with the dry ice bath but do not sit the reaction back into the ice. Keep the stirring on.

8. Using a pipette slowly add 46 mL of water dropwise through the side neck in the flask. The reaction solution will begin to vigorously effervesce. The temperature of the solution will climb to 98 °C. If it looks like the temperature will go higher than this then lower the solution into the ice bath until it has cooled. Once all the water has been added allow the solution to keep stirring at 98 °C for a further 15 min during which time it will begin to appear brown from the formation of graphene oxide. Allow the solution to stir for a further 40 min.

9. Dilute the solution further by adding 140 mL of water through the side neck with stirring.

10. Add the 10 mL of hydrogen peroxide solution. This will reduce the remaining permanganate and manganese dioxide to a soluble manganese sulphate salt. The solution will appear yellow.

11. Turn off the heating mantle and raise the reflux apparatus out of the oil bath. Allow the solution to cool and give the bottom of the flask a wipe with tissue to remove excess oil. Decant the solution into large centrifuge tubes.

12. Centrifuge the solution at 5000 rpm for 30 min so that the graphene oxide product sediments in the bottom. Pour away the top solution into a solvent waste container *suitable for strongly acidic solutions.*

13. Redisperse the sediment in the 2 mol/L HCl solution and centrifuge again. Take a few mL of the top solution and add it to an equal amount of the barium chloride solution. In the first wash you will see barium sulphate precipitate form with the residual sulphate anions. Keep washing and centrifuging the sample until you no longer see any precipitate form when you test the top solution.

14. Wash the sediment using the centrifuge three times with acetone. Pour away the waste acetone into a waste solvent container.

15. Air dry the graphene oxide in an oven at 65 °C overnight. The formed graphene oxide can be stored indefinitely at this point until required.
The following describe the formation of a single stranded DNA stabilised graphene oxide colloid and its subsequent reduction to form colloidal graphene.

16. Weigh out 100 mg of the graphene oxide into a 30 mL vial and add 10 mL of water. Seal the vial.

17. Slip the vial into a foam float and place it in a sonic bath. Sonicate the solution for 2 h to completely exfoliate the graphene oxide.

18. While the graphene oxide is sonicating soak a section of dialysis tubing in the 1 L beaker filled with water. It should be a large enough section to hold 10 mL allowing for the end sections to be folded over. Fold one end of the tubing over and double clip it.

19. Use a Pasteur pipette to transfer the exfoliated graphene oxide into the dialysis tubing. Folt the top over and double clip it shut. Dialyse the solution for 2 h with 2 changes of water during this time to ensure there are no remaining salts.

20. Transfer the contents of the dialysis tubing into a centrifuge tube and centrifuge at 3000 rpm for 20 min. Any unexfoliated materials will sediment to the bottom. Transfer the top solution containing the exfoliated sheets into a fresh vial. This is now a stock solution and should contain around 8.4 mg/mL of exfoliated graphene oxide.

21. Take 1.15 mL of the graphene oxide dispersion and make up the volume to 10 mL in a clean vial with water. This will give a 1 mg/mL solution of graphene oxide.

22. Place the oil bath on the heating mantle and setup the 100 mL round bottomed flask for reflux. Set the heating mantle to 95 °C

23. Add 10 mL of the 2 mg/mL of freshly prepared salmon sperm single stranded DNA to the round bottomed flask with a magnetic stir bar. This should already be hot but it is alright to reheat the DNA solution under reflux for an hour before adding the graphene oxide in the next step.

24. Set the heating mantle to 100 °C. Add all 10 mL of the 1 mg/mL graphene oxide dispersion to the round bottomed flask with stirring.

25. Add 20 µL of the freshly prepared hydrazine solution to the round bottomed flask. Allow the reaction to reflux for 1 h at 100 °C.

26. Turn off the heating mantle and raise the solution out of the oil bath. Allow this to cool to room temperature and turn off the reflux condenser.

27. Soak another section of dialysis tubing in the 1 L beaker filled with water. Double clip one end of the dialysis tubing and transfer the contents of the reflux condenser into the tubing using a Pasteur pipette. Fold the end over and double clip the tubing shut. Dialyse the sample for 2 h with three changes of water to get rid of any remaining hydrazine and excess DNA.

28. Transfer the contents of the dialysis bag into a clean vial with a sealable lid. You should now have a black colloidal dispersion of single sheets of graphene stabilised by DNA.

Single stranded DNA stabilised colloids containing graphene in amounts lower than 5 mg/mL should remain stable indefinitely. Zeta potential analysis will reveal the sheets to be negatively charged and the strength of the electrostatic charge will increase with an increase in the ratio of stabilising DNA. The successful reduction of the graphene oxide to graphene can be monitored by UV-vis spectroscopy. Graphene oxide has a peak absorbance in the UV region at 231 nm. Upon reduction to graphene peak should shift to 266 nm though this may be masked by the absorbance of the single stranded DNA. The level of oxides still present in the graphene can be determined by X-ray photoelectron spectroscopy (XPS). The carbon 1s spectrum for graphene should display a peak with a binding

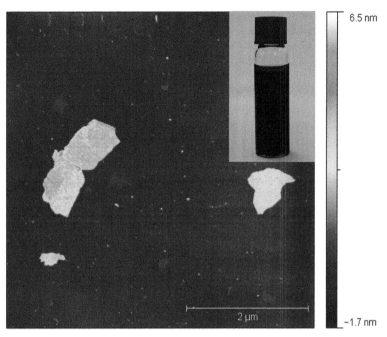

FIGURE 4.75   An AFM image of DNA stabilised graphene sheets resting on a mica surface. Inset is a photograph of a DNA stabilised graphene dispersion in water. *(Image courtesy of Avinash Patil)*

energy of 284.5 eV indicative of sp^2 hybridisation. Deviations from this in the product will be due to sp^3 character from various carbon–oxygen binding modes. Free standing graphene films will form upon drying the colloid down on to a flat silicon substrate. The presence of the stabilising DNA coordinates the lamellar stacking of the individual sheets. X-ray diffraction analysis of the films should reveal a broad reflection centred at $2\theta = 5.8°$. This equates to an interlayer spacing of 1.51 nm for a single graphene sheet coated in a monolayer of single stranded DNA. Individual sheets can be examined more closely using atomic force microscopy under ambient conditions. A sample is prepared by drying a very dilute solution down onto a freshly cleaved mica surface. You may see the flakes folded over on themselves and line profiles taken of cross sections of the image should show the thickness to be around 1– 1.5 nm in height which will hopefully match the interlayer spacing calculated from the X-ray diffraction pattern (Fig. 4.75).

METAL ORGANIC FRAMEWORKS

With thanks to Matteo Lussi, Mairi Haddow and Guy Orpen.

Many of the recipes in this book are concerned with arranging a material or chemical into a specific shape on the nanoscale. The various bottom up approaches to generating nanostructures mostly rely on self-, or driven-assembly of molecular components to form the desired structure. Many of these routes depend on the nanoscopic morphology to alter some key property of the material in contrast to the bulk counterpart. An example of this is the formation of titania nanoparticles with the anatase crystal phase. This polymorph exists in the bulk material but when generated as a small particle the band gap is altered and the surface area is greatly enhanced. In both the large and small forms you can have the TiO_2 molecules arranged in the anatase structure. Similarly, gold particles in solution and bulk gold can both have face centred cubic packing. Both of these nano particulate examples are forms of chemically driven assembly to form nano analogues having the same crystal structure as the bulk matter. That is to say, the way the constituent atomic building blocks pack together to form the long range order is the same for the nanoscale and bulk scale in some materials. Polycrystalline domains can be formed in a material produced by 'bottom up' methods and you can preferentially select which crystal facets grow faster by using surfactants or additives to introduce anisotropy. Even though the shape of the crystal is altered, the underlying crystal arrangement as seen in the bulk phase remains identical.

To add an extra dimension to the bottom up approach, it is desirable to have control over how the constituent atoms making up the material, bulk or nanoscale, are arranged. Crystal engineering is a branch of supramolecular chemistry that has the potential to provide such a solution by assembling crystals from metal cation centres and organic molecules. It is not limited to the inclusion of metal centres and much of the field is involved with the production of purely organic crystal structures. The principle concept is to make crystals through the exploitation of non covalent molecular interactions such as hydrogen bonding, π-stacking interactions, electrostatics and van der Waal forces using multivalent metal centres and bridging bi or tridentate organic ligands to form the lattice with. In this section we will be making crystals using organic ligands and metal centres. These structures are called metal organic frameworks (MOFs) and the aim of research within crystal engineering is to identify the molecular building blocks that will predictably self-assemble and result in a crystal with a specific desired structure. A simplified terminology for MOFs is to call the components that build up the lattice framework tectons. The term tecton is usually linked with geology but in this case derived from a latin word meaning 'build'. The term synthon is used to describe the binding motif within the structure made up from tectons. For example a synthon could describe the coordination of two carboxylate oxygen to a metal centre.

Quite often these supramolecular assemblies will have a functionality imparted to them as a result of their physical structure. This could be Ångstrom or nanoscale porosity which is useful for hydrogen storage and drug delivery, or some electronic property such as thermocromic or thermomagnetic behaviour resulting from divalent metal centre spin transitions across a lattice framework. In the following recipes we will explore some of the routes available for making common examples of MOFs in bulk and as nanoparticles.

COBALT BIPYRIDYL POLYMER

With Thanks to Matteo Lusi, Adapted from Adams et al. (2010)

In this recipe we will make an archetype MOF using cobalt chloride and a bipyridyl ligand. The interesting aspect of this compound is that it can be switched between a hydrogen bonded salt and a true metal organic framework structure reversibly. Most MOFs are prepared by mixing a metal halide salt and a bi or tridentate aromatic heterocyclic ligand, usually an onium cation. Onium cation ligands are organic molecules in which the guest element in the carbon heterocycle is from group 15, 16 or 17 and has hydrogen linked to it. The aromaticity of the ligand tecton contributes to the rigidity of the formed lattice through a combination of π stacking with adjacent molecules and aromatic delocalisation which negates rotational freedom in the organic moiety. Bonding to a metal halide tecton occurs by hydrogen bonding of the onium cation to one of the halogens coordinated on the metal centre. This arrangement can be reversibly modified by removing the hydrogen and chlorine so that the remaining lone pair on the ligand nitrogen coordinates directly to the metal cation. This general mechanism of MOF formation is known as the Anderson rearrangement where inner sphere ligands on the metal centre are exchanged for outer ones (Anderson, 1855). During a dehydrohalogenation reaction, the metal centre exchanges some of the coordinated halides for the more basic nitro group within the ligand. In an established MOF structure this reaction is reversible and for cobalt, zinc, platinum and iron bipyridyl MOFs of the type described in this section, the addition or removal of HCl will induce a structural change in the lattice. For the cobalt centred MOF in this recipe there will be a visible change in the colour as the structure is converted between the acidified and basic form through exposure to HCl gas (Fig. 4.76).

In addition to the direct linkage of metal centres through a pyridine, the longer range packing can be tailored by altering the functional groups in the ligand structure. This is achieved using hydrogen bond donors and receptors to modify the packing parameters as well as the behaviour of the final lattice in accommodating guest species, such as hydrogen gas. Strong hydrogen bonding interactions between molecular units in MOF structures are observed for OH—O, N—H—O, O—H—N and N—H—N. Functional groups giving rise to these synthon motifs are most commonly NH_2, OH and CO_2H groups. Much research in MOFs is concerned with deriving the bond angles and strengths of the hydrogen bonding so that tectons can be synthesised to form predictable crystal structures. X-ray crystallography is used to work out the bond lengths and angles which can be described by a number of parameters as illustrated below. In addition to simple linear hydrogen bonding, it is also possible to have bifurcated and trifurcated hydrogen bonds acting as bridges between more than one ligand (Fig. 4.77).

Classical approaches to growing a crystal are concerned with achieving structural and chemical purity. There are three very general methods for producing crystals but the exact choice of mixing method will influence the final structure. Crystals formed with identical chemical formulas often have more than one crystal structure and this is known as polymorphism. For a given compound there will be a crystal structure that has the lowest energy of formation but when the environment is right the higher energy configurations can be produced. Factors that cause polymorphism involve the pressure and temperature at which it was produced as well as the presence of solvent ions trapped in the lattice. Getting the desired polymorph therefore involves growing a crystal in the right conditions using a variation on one of the three general synthetic methods:

Growth from a melted phase which is cooled slowly to initiate crystal nucleation.
Crystal nucleation and growth from a supersaturated or metastable salt solution.
A new crystal structure can be formed by grinding two reactive crystals together.

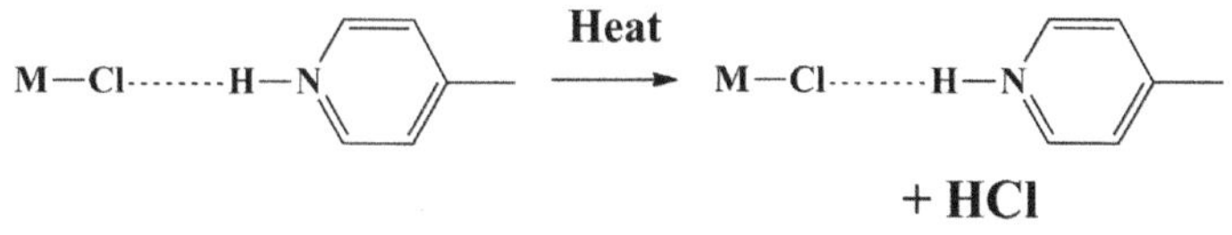

FIGURE 4.76 A schematic of the Anderson rearrangement.

FIGURE 4.77 A schematic representation of a synthon. It is described by the bond angle θ, the distance of the coordination bond d and the length of the solid bond r. Coordination through hydrogen can also give rise to single, bifurcated and trifurcated hydrogen bonds.

The melting method is how the fan blades for jet engines are made. Only a large single defect free metal alloy crystal can stand the extremes of stress and pressure at which a jet engine operate. Although the melt method is good for producing large crystals of a high purity, it is not used for making MOFs as high temperatures would induce thermal decomposition of the organic ligands. Therefore only the solid state mixing and solvent saturation routes remain for the manufacture of MOFs.

For solid state mixing the dry powders may be ground together in what is called a mechanochemical synthesis. This is also known as the heat and beat method used often in the inorganic synthesis of metal oxides. Heat does not have to be involved although friction can generate local heating and often only grinding is required. An extreme example of the difference heating can cause is demonstrated by the reaction between HgCl and AgCl by a heating or grinding method. If they are heated together the solids melt and sublimate. However, if they are ground together they will form a halide gas and metal alloy (Fernandez-Bertran, 1999). By grinding the components together with a mortar and pestle, the surface area of the solid reagents is increased to the point that direct reaction can occur. It may seem surprising that the products of a mechanochemical reaction are homogenous. It might be expect that having no solvation medium in which a molecular exchange can take place would result in unreacted lumps of reagent. When grinding alone is not enough to produce a homogenous product, solvent assisted grinding can be used. This variation entails using a few drops of water or solvent added to the grinding mixture. The solvent is not in an amount to dissolve the bulk reagent crystals but instead possibly provides localised dissolution to aid the reaction. One potential drawback of solvent assisted grinding is that it can result in a solvate instead of a true polymorph. Another solvent-free variation on the solid state reaction is to allow a gas to diffuse into a precursor crystal to form the product. Because the organic ligands involved in MOF assembly can often be volatile it is possible to introduce them as a gas to the crystal whereupon they will react with the metal centres or ligands directly. Gas phase—solid state interactions of this type can also be reversible and this is of interest in the production of gas storage materials. This last example is not strictly mechanochemical, but the reaction is similarly influenced by the temperature and pressure of the gas with which the solid phase is in contact. Mechanochemical synthesis is an attractive prospect for the large scale production of MOFs because the lack of solvents ensures the production of high phase purity materials. The dry grinding method also eliminates much of the solvent waste and handling involved which makes the process economically and environmentally better.

With regards to a solution phase synthesis, a crystal may be grown from the direct combination of salts in a solvent phase. To form a crystal in solution, the reagent salts are dissolved into a solvent in which they are both soluble. When these solvents are mixed together the ionic species can interact and form a molecule of the product crystal which is ideally not soluble in the solvent. The initial reaction to form an immiscible product is called the nucleation phase, during which molecules of the product crystal begin to aggregate together to form seed crystals. These seeds act as a substrate for other molecules to precipitate out of solution onto and so the crystal seeds grow larger. The nucleation step can often be slow and one way to speed up crystal growth is to add a few large seeds of the product crystal which act as ready-made substrates. This is very useful if getting a particular polymorph from direct nucleation is difficult. Instead of trying to grow the desired crystal structure from scratch, you can add an existing crystal of the desired polymorph into a supersaturated salt solution which will then induce it to grow larger. Solution-based synthesis is not ideal for growing large crystals with a high phase purity as required for X-ray experiments. If a crystal is grown quickly then the chances of defects and solvent incorporation within the structure are higher. Crystals newly formed from salt reactions may also contain by-product contamination. To further purify and grow a crystal the technique of recrystallisation can be used. This technique will be familiar to chemists, who regularly use recrystallisation to purify both organic and inorganic reaction products. The unrefined product crystal is dissolved in a solvent which is then slowly allowed to evaporate. As the solvent is lost the concentration of the ions and respective counter ions becomes larger until the solution becomes supersaturated with respect to the product crystal. The rate at which the solution is allowed to reach supersaturation is limited so that the recrystallisation of the solid product happens slowly. The surface of the growing crystal may therefore redissolve and solidify a number of times before the structure is locked in place by the gradually decreasing solubility. This ensures that the deposited material is lower in structural defects. The energetics of slow recrystallisation select for a phase pure structure and so the impurities are left behind in the solvent. Very large crystals can be grown at the laboratory scale, on the order of hundreds of microns to millimetres, using this approach. The speed of growth varies widely and it can take minutes in a benchtop purification procedure or it could take months if the crystal has to be exceptionally phase pure. Controlling the rate at which the solvent evaporates can be done by sealing the saturated solution in a vessel and changing the size of an aperture open to air. Alternatively, an antisolvent miscible with the first can be setup to diffuse by convection into the supersaturated solution. The dilution of the supersaturated solution with antisolvent will drop the miscibility of the product crystal and nucleate growth. However carefully this is done you may find that some solvent molecules remain trapped within the crystal lattice.

In this recipe we will make a simple bipyridine—cobalt adduct using the mechanochemical method. A 4,4'-bipyridyl ligand is reacted with $CoCl_2$ to form a MOF coordination polymer with long range order. The cobalt chloride salt readily attracts water, so in a sense this mechanism is solvent assisted, but the reaction can be performed using the anhydrous $CoCl_2$ salt with a small drop of ethanol present to catalyse the reaction instead. An alternative route to the same material using the hydrochloride salt of bipyridine and cobalt hydroxide will also be explored. Water is generated by the reaction but this can be removed by exposing the product to a vacuum. Octahedral cobalt complexes have a distinctive spectrochemical series running from pink to blue to green which makes identifying the differences between the starting reagent and product easy to determine by eye. There will be a colour difference between the MOF formed from the chloride (blue) and the MOF formed from the carbonate salt (white green or grey) even though the end products are structurally indistinguishable by powder X-ray diffraction, elemental analysis and infra red spectroscopy. This is intriguing as obviously there must be something different between the two products or else they would be the same colour. One possible explanation is that residual impurities from by-product salts in the shift the size of the crystal field splitting parameter as described by crystal and ligand field theory (see in box).

$$nCoCl_2 \cdot 6H_2O_{(s)} + nC_{10}H_8N_{2(s)} \rightarrow [CoCl_2(4,4'\text{-bipy})]_{n(s)} + (n \times 6)H_2O_{(l)}$$

$$n[4,4'\text{-}H_2C_{10}H_8N_2]Cl_2 + nCo(OH)_2 \rightarrow [CoCl_2(4,4'\text{-bipy})]_{n(s)}nH_2O_{(l)}$$

Crystal Field Theory

Transition metals have two or more possible oxidation states and this arises from the ease with which electrons can be removed or added to the *d*-orbitals. The shape of an orbital is determined by the probability of finding an electron at a given point around the atomic nucleus. *s*, *p* and *d*-orbitals are all described by a set of four quantum numbers. These are *n* representing the electron shell, *l* which is the angular momentum or subshell, m_l which is a specific orbital within the subshell and m_s which is the spin of an electron occupying a quantum state described by the other three values of n, l and m_l. *d*-orbitals represent the most probable positions about an atom for electrons when the angular momentum quantum number *l* is 3. There are 5 possible m_l values for *d*-orbitals which means there are 5 distinct *d*-orbital electronic configurations around a transition metal centre. Each of these orbitals is able to accommodate 2 electrons with opposing spin. The five orbitals are named $d_{xy}, d_{xz}, d_{yz}, d_x^2 - d_y^2$ and d_z^2 based on their spatial position relative to one another in an x,y,z coordinate system centred on the atomic nucleus. For a lone metal ion all of these orbitals are equal in energy, which is to say there is an equal chance of finding an electron in any one of the five different orbitals. When a set of ligands coordinates to a metal centre they will arrange themselves into a geometry that has the lowest repulsion between the formed bonds due to electronic repulsion. Commonly this will result in a square planar, tetrahedral or octahedral complex. If an x,y,z coordinate system is laid over the metal and ligands for an octahedral complex, it can be seen that the d_{xy}, d_{xz} and d_{yz} do not experience as much overlap with the direct bonds as the $d_x^2 - d_y^2$ and d_z^2 do. Because of this, the probability of finding an electron in the d_{xy}, d_{xz} and d_{yz} orbitals becomes higher than in the $d_x^2 - d_y^2$ and d_z^2. Looking at this another way, it means that the energy level for the $d_x^2 - d_y^2$ and d_z^2 orbitals is now greater than the d_{xy}, d_{xz} and d_{yz} orbitals. The lower energy orbitals in an octahedral complex are labelled e_g and the larger energy orbitals are called t_{2g} by convention. The splitting of energy levels in the presence of a ligand field is called crystal field theory and it explains a number of spectroscopic and magnetic effects seen in transition metal crystals (Fig. 4.78).

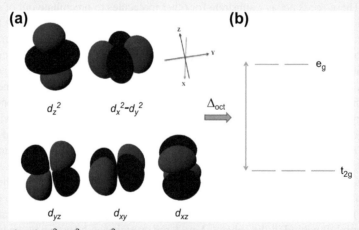

FIGURE 4.78   (a) Plots of the $d_{xy}, d_{xz}, d_{yz} d_x^2 - d_y^2$ and d_z^2 orbitals. (b) A diagram depicting how the energy levels of the five *d*-orbitals split when ligands are arranged around a metal centre as an octahedral complex. The gap between the lower and higher energy states is called the Δ_{oct}. This represents how much energy is required to promote an electron occupying a lower energy orbital into a higher one.

The gap between the split energy levels is defined as Δ_{oct} and is determined by the strength of electronic interactions of the coordinated ligand. In the presence of a 'strong' ligand like a cyanide anion the gap between the t_{2g} and e_g is large. If all the energy levels were the same then, as the orbitals are filled with electrons, they would fill each individual d orbital first before the electrons begin to pair up. In complexes with a sufficiently large Δ_{oct} the t_{2g} orbitals fill up to form paired electrons before any of the e_g orbitals are occupied and this is called a low spin configuration. In crystal fields where the Δ_{oct} value is low, unpaired electrons will fill both the t_{2g} and e_g orbitals before beginning to pair up. This situation is called a high spin configuration and happens in the presence of weak field ligands like I^- (Fig. 4.79).

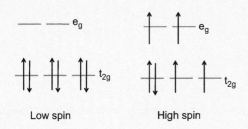

Low spin

High spin

FIGURE 4.79 Schematic of the low spin and high spin electronic d-orbital configurations for a metal centre having 6 electrons.

For 4 d electron and 7 d electron containing configurations having an octahedral geometry, the complexes can be switched between the low and high spin states. Small amounts of energy, like heat or light, can promote the low spin configuration into the high spin configuration. The Δ_{oct} for the transition in this near state is ~$K_b T$ where K_b is the Boltzman constant and T is the temperature in Kelvin. Often, switching between low spin and high spin states brings about a chromic response in the complex. This is because antibonding orbitals become filled during thermal or photonic excitation. Electrons in antibonding orbitals weaken the bond strength and so the bond is lengthened. This in turn alters the d-orbital electron energies which are responsible for the colour of the complex through metal to ligand charge transfer (MLCT) or ligand to metal charge transfer (LMCT) (Gütlich et al., 2000).

The exact energies and interactions giving rise to colour can be calculated using a Tanabe-Sugano diagram but these are specific to the complexes and geometries you wish to study. Alternatively, the concept of hard and soft ligands gives rise to an empirically derived spectrochemical series that can be used to predict the strength of the ligand field based on a samples colour. A transition metal solution will absorb certain parts of the spectrum and allow others to pass. A solution will therefore appear to be the opposite colour to the absorbed colour as seen on the colour wheel pictured below. A Co$^{(III)}$ complex coordinated to hard ligands will absorb in the violet and blue ends of the spectrum and will appear yellow or orange. When coordinated to soft ligands with low field strengths the absorbance will be in the red region and the solution will appear blue or green (Fig. 4.80).

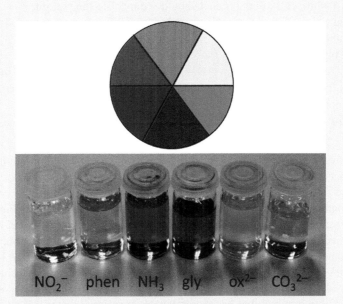

FIGURE 4.80 (a) A colour wheel shows the colour a compound appears opposite the colour the compound absorbs in the visible region. (b) A collection of cobalt compounds all having the same number of d electrons but with different ligand field strengths. The spectrochemical series runs strong to weak as $CN^- > NO_2^- >$ phenanthroline > ethylenediamine > NH_3 > glycine > H_2O > ox^{2-} > CO_3^{2-}.

To make [CoCl₂(4,4'-bipy)]ₙ you will need:

The bipyridyl ligand is toxic. Wear gloves and do not inhale or ingest it. Dispose of it in a marked solid waste container.

1. A mortar and pestle. A marble one will make it easier to collect the product. Avoid using porous materials such as porcelain. You will also need a spatula for scraping the product out with and a vial to keep it in.
2. A bell jar and vacuum pump for drying the product with. Ensure a cold trap is fitted between the jar and the pump.

The following reagents are required for the $CoCl_2$ based synthesis

1. 0.156 g (1×10^{-3} mol) of 4,4'-bipyridyl.
2. 0.242 g (1×10^{-3} mol) of $CoCl_2 \cdot 6H_2O$.

The following reagents are required for the $CoCO_3$ based synthesis

1. 4.68 g (0.03 mol) of 4,4'-bipyridyl.
2. 20 mL of 36% HCl in water in a 200 mL beaker. This is extremely corrosive and should be worked within a fume hood. Close the bottle when not in use.
3. An ice bath in which the 200 mL beaker can sit.
4. 500 mL of ice cold ethanol for precipitation and rinsing.
5. A Buchner funnel and filter with filter paper. This should be held by a clamp and used in a fume hood.
6. 0.119 (1×10^{-3} mol anhydrous) of $CaCO_3 \cdot xH_2O$. Because the salt is provided commercially with some level of hydration you may wish to try exposing the salt to a vacuum for a day or so try to dry it. The preparation should work with the hydrated form but you might see subtle differences in the product colour.

To make [CoCl₂(4,4'-bipy)]ₙ:
For the $CoCl_2$ based synthesis

1. Grind the pink $CoCl_2 \cdot 6H_2O$ and white 4,4'-bipyridyl together in the mortar and pestle until it forms a blue paste.
2. Scrape the blue paste into a sample vial. You now have a [CoCl₂(4,4'-bipy)]ₙ coordination polymer ready for analysis. For the $CoCO_3$ based synthesis.
3. Dissolve the 4.68 g of 4,4'-bipyridyl in the 20 mL of concentrated hydrochloric acid while cooling in the ice bath. This will form the hydrochloride salt of the bipyridyl ligand with the general formula [4,4'-H₂bipyridyl]Cl₂.
4. Precipitate the [4,4'-H₂bipyridyl]Cl₂ out of solution by adding 50 mL of ice cold ethanol to the beaker. If it does not precipitate out then add more ethanol. A white crystalline product should sediment to the bottom.
5. Setup the Buchner funnel and wet the filter paper with ethanol to stick it down in the funnel. Turn on the suction and collect the crystals. Rinse them thoroughly with cold ethanol.
6. Allow the collected [4,4'-H₂bipyridyl]Cl₂ to dry before further use.
7. Weigh out 0.247 g (1×10^{-3} g) of the [4,4'-H₂bipyridyl]Cl₂ and place it in the pestle and mortar.
8. Add the pink $CaCO_3$ salt to the mortar and grind the contents together until a white-pale green paste forms.
9. Use a spatula to scrape the contents into a vial.
10. Place both the vials containing the chloride and carbonate derived products into the bell jar and dry under vacuum for 24 h. After this time both samples are ready for further analysis.

The colour change is the most prominent feature that the reaction has gone to completion and can be monitored by UV-visible spectroscopy. Both products are stable under ambient conditions for a long time. Check the crystal structure by powder X-ray diffraction measuring the reflections between $2\theta = 10°-40°$ angles. You should observe that the patterns overlay one another and fit with an orthorhombic *Cmmm* space group. In the product the nitrogen in the bipyridyl ligands is directly bound to the metal centre. By acid treatment of the [CoCl₂(4,4'-bipy)]ₙ the nitrogen in the ligand can be reprotonated so that the ligands are held in position by hydrogen bonding through a metal chloride to the onium cation. The cobalt complex shifts from an octahedral to tetrahedral geometry in the blue acidified product. The shift can be reversed by heating the acidified MOF to 230 °C in a round bottomed flask for two hours during which time it will turn from a dark blue to light turquoise colour once more. Alternatively, the acidified complex can be ground with double the molar amount of KOH. The mechanochemical conversion is fast and happens within a minute but has the drawback of residual water and KCl mixed in as products. Because the cobalt MOF is poorly soluble the KCl can be extracted with a water rinse followed by exposure to vacuum to dry the solid (Fig. 4.81).

$$[CoCl_2(4,4'\text{-bipy})]_{n(s)} + HCl_{(g)} \rightarrow [4,4'\text{-H}_2C_{10}H_8N_2][CoCl_4]_{(s)}$$

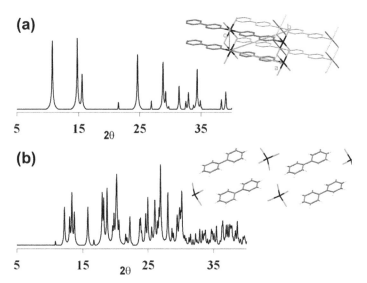

FIGURE 4.81 Powder X-ray diffraction patterns and the corresponding crystal structures for (a) $[CoCl_2(4,4'-bipy)]_n$ and after acid treatment to form (b) The hydrogen bonded coordination salt $[4,4'-H_2bipyridyl]Cl_2$.

PRUSSIAN BLUE ANALOGUES, A MOLECULAR MAGNET FRAMEWORK

It is easy enough to view the design of a metal organic framework as being mainly dependant on the structure of the organic tecton. Many tectons produce the same, or isomorphous, crystal structure even when a different transition metal centre is used to coordinate them together as a lattice. We should not forget that the choice of tecton, or metal, can also be used to impart functionality into the framework. In this section we will examine a basic metal organic framework that has some unique physical effects arising from spin transitions in metal centres of differing valances.

In 1704, a draper in Berlin boiled bull's blood in a basic medium and a bright blue compound was formed (Brown, 1724). At the time, no understanding of the chemistry involved was available but today this experiment is recognised as one of the first recorded synthesis of Prussian blue. The pigment is still used today in the colouration of ceramics and paints, and producing Prussian blue is for many a basic introduction to inorganic chemistry. It is formed easily by mixing aqueous solutions of hexacyanoferrate (II) and iron (III) salts.

$$3[Fe^{II}(CN)_6]^{4-}_{(aq)} + 4Fe^{3+}_{(aq)} + 14H_2O \rightarrow Fe^{III}_4[Fe^{II}(CN)_6]_3\eth_1 \cdot 14H_2O$$

The compound formed has a hexacoordinated high spin $Fe^{(III)}$ centre bound to the cyanide nitrogen and an $Fe^{(II)}$ low spin centre bound to the carbon. The $\eth$ represents the vacancies of $[Fe^{II}(CN)_6]^{4-}$ arising from a 4:3 reaction stoichiometry due to charge neutrality (Verdaguer and Girolami, 2004) (Fig. 4.82).

A large family of face centred cubic systems may be derived in this way with the general formula $M_nA_p[B(CN)_6]_q \cdot xH_2O$, where M is a monovalent cation (K^+, Na^+) located in tetrahedral spaces and free to move between sites to balance charge transfer between the metallic centres. Metal A occupies all of the centres and summits of the faces and metal B is arranged in octahedral sites. A simple Lewis acid−base equation represents the generic formula (Ludi and Hu, 1973):

$$X[B(CN)_6]^{z-}_{aq} (\text{Lewis base}) + z A^{x+} (\text{Lewis acid}) \rightarrow \{A[B(CN)_6]_x\}^0 \cdot xH_2O_{(s)}$$

Prussian blue analogues exhibit interesting optical and magnetic properties which arise from the interchange of paramagnetic and diamagnetic regions. Below a specific temperature (Curie temperature, T_c) some Prussian blue analogues become ferrimagnetic and may be attracted to a magnet. In addition, ferromagnets (Ni, Cu) and antiferromagnets ($V^{IV}O$) can be formulated (Ferlay et al., 1995).

In Prussian blue analogues the spin carriers responsible for magnetic behaviour are linked by a cyanide bridge between metal centres. This leads to double exchange interactions where delocalised electrons on the $Fe^{(III)}$ are polarised, which facilitate a rapid response to magnetic interactions over the entirety of its bulk. These manifest in the material as ferromagnetic behaviour. In order to interpret the magnetic behaviour of cobalt hexacyanoferrate an understanding of the electronic structure is required. A basic iron−iron Prussian blue consisting of two hetero-valent ions is a class II compound by Robin-Day classification. This means that

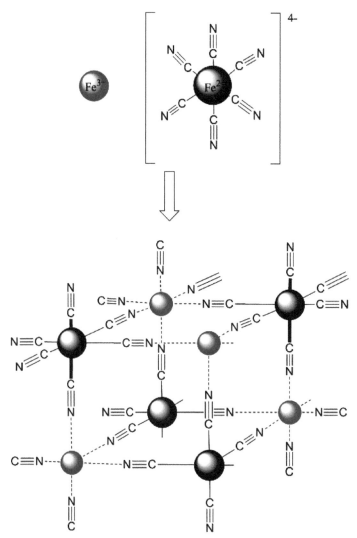

FIGURE 4.82 A diagram representing the structure of prussian blue. Red spheres represent an $Fe^{(III)}$ centre and the blue spheres represent an $Fe^{(II)}$ centre.

electronic exchange between the two metal centres is slow enough to consider them separately valent. Hence the transition to the high spin d^5, $(t_{2g})^3(e_g)^2$ Fe^{III} centre from the low spin diamagnetic d^6 $(t_{2g})^6$ Fe^{II} centre results in a blue colour. However, at low temperatures, for the Fe^{II}–Fe^{III} system 5.6 K, the separate magnetic dipole regions align and the solid is transformed in to a ferromagnet. In a cobalt–iron system T_c is reached at 16 K and ferrimagnetic behaviour is observed. In addition, photo-magnetic effects have been reported. It has been observed that upon further cooling to 5 K, irradiation with red light ($\lambda = 660$ nm) produced amplification of the magnetisation (by roughly double) and the increase of the T_c to 19 K. This was found to be reversible upon treatment with blue light. ($\lambda = 450$ nm).

These properties become clear, in part, when looking at the electronic structure of the cobalt-based Prussian blue analogue. Upon formation from the component salts the metals undergo a redox reaction that leaves them in a diamagnetic state. The iron centre exists as $Fe^{II}(t_{2g}^6)$ which is spin $= 0$ and the cobalt centre exists as $Co^{III}(t_{2g}^6)$ which is also spin $= 0$. By cooling the cobalt hexacyanoferrate, the electronic transfer rate across the cyanide bridge is increased and high spin Fe^{III} and Co^{II} centres are formed (Fig. 4.77). The cobalt hexacyanoferrate becomes paramagnetic and begins to exhibit magnetic behaviour (Fig. 4.83).

Current research with this type of MOF is to develop tuneable light/heat activated molecular based magnets for a diverse range of applications. For the bulk of materials investigated it has been extremely difficult to design a Prussian blue based molecular magnet which will alter states at room temperature. A spectacular exception to this has been reported in the development of $\{V_x^{(II)}V_{1-x}^{(III)}[Cr^{(III)}(CN_6)]_{0.86} \cdot 2.8H_2O\}$ (Ferlay et al., 1995). This material will undergo a transition between ferrimagnetic and paramagnetic states at a temperature of 42 °C and an ingenious device for energy production has been developed based on it. One problem with the vanadium system is that it is extremely sensitive to oxygen which makes synthesis difficult and prepared samples must be kept under an inert atmosphere to prevent oxidation. This has also limited its development as a component in memory,

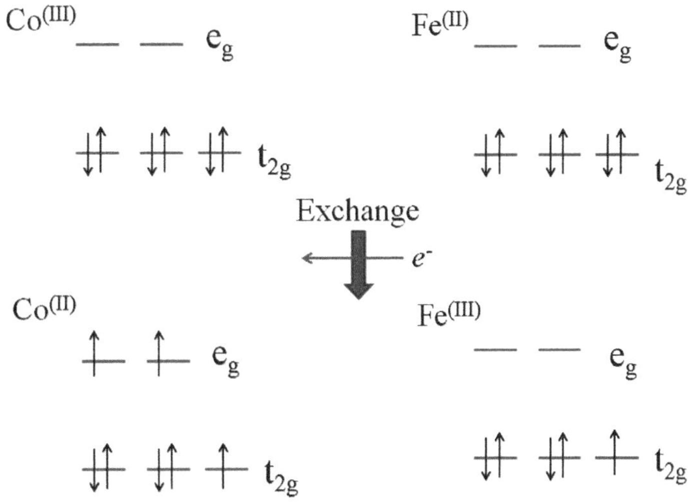

FIGURE 4.83 Electronic structure diagrams for metal centre exchange interactions in cobalt hexacyanoferrate.

computing and energy generating devices. Two routes for the development of molecular based magnets have been employed (i) trial and error through systematically testing various combinations which, although time consuming, may yield unexpected results and (ii) model predictions heavily influenced by the application of molecular orbital theory and the Neel formula for ferrimagnetic ordering.

MICROEMULSION PRODUCTION OF CUBIC PRUSSIAN BLUE PARTICLES

Adapted from Vaucher et al. (2000)

In this recipe we will use water droplets in a reverse microemulsion as vessels for the photolytic nucleation of nanocrystal lattices of an $Fe^{(II)}$, $Fe^{(III)}$ Prussian blue. Water in oil microemulsions stabilised by surfactant offer a unique approach to forming nanomaterials. Ionic reagents are dissolved into water droplets within a water in oil emulsion. Because the ions are immiscible in the continuous oil phase, crystals formed from the ions undergo nucleation and growth within the confines of a water droplet. Material exchange occurs through droplet interactions within the continuous medium allowing for ripening processes to homogenise the solids formed in the emulsion. Nanocrystals formed inside the water droplets are capped with a surfactant layer on the surface. This protects the nanocrystal product from aggregation and in some cases the hydrophobic interaction of the external aliphatic chains causes uniformly sized particles to pack together in a self assembled super structure. Sodium bis(2-ethylhexyl)sulfosuccinate, also known as sodium AoT is used as the surfactant for producing a microemulsion. The size of the water droplet is influenced by the ionic strength of the reagents within it and the ratio of water to surfactant. For formulation purposes reverse microemulsions are described by the value ω which is defined as the molar ratio of water to surfactant.

$$\omega = \frac{[H_2O]}{[AoT]}$$

As has been explored previously, the slower a MOF crystal forms the more defect free it will be. To produce good-quality nanocrystals with tightly defined size distributions the nucleation and growth steps must be slower than the rate of droplet interchange throughout the emulsion. Direct mixing of $[Fe^{(II)}(CN)_6]^{4-}$ and $[Fe^{(III)}(CN)_6]^{3-}$ salts would result in a Prussian blue lattice but it would grow too fast and result in ill defined crystals. Because of this an ammonium iron (III) oxalate is used which can be reduced photolytically to produce iron (II) cations which will then undergo reaction to form the Prussian blue structure. In order to remove potassium from the reaction solution the aqueous mixture containing the iron oxalate and cyanide salts is heated until a portion of insoluble $K_3[Fe(C_2O_4)_3]$ precipitates out. This is removed from the solution by filtration and the aqueous solution is then used to prepare a microemulsion. The microemulsion is exposed to light and the Prussian blue begins to nucleate within the confines of a water droplet in the emulsion.

To make cubic Prussian blue nanoparticles using an inverse microemulsion you will need:

1. 2.568 g (0.006 mol) of $(NH_4)_3[Fe^{(III)}(C_2O_4)_3] \cdot 3H_2O$
2. 0.978 g (0.003 mol) of $K_3[Fe(CN)_6]$. Grind the powder finely with a mortar and pestle before use.

3. A small vial wrapped in aluminium foil containing 3 mL of water and a small magnetic stir bar.
4. A heating mantle with an oil bath and magnetic stirring. You will need a clamp to hold the small vial within the bath.
5. A sonic bath.
6. A cupboard or dark box. Parts of this preparation will need to be conducted in the dark.
7. A syringe filter and syringe with a ~20 mL foil wrapped vial to collect the filtrate in.
8. Water for diluting the iron salt solution to the correct concentration with.
9. Volumetric pipettes in the 10–100 μL and 1–5 mL range and tips.
10. 4.456 g (0.001 mol) of sodium bis(2-ethylhexyl)sulfosuccinate (NaAoT) weighed into a 25 mL foil wrapped vial with a lid. This surfactant will readily absorb water from the atmosphere so it must be stored with a desiccant. If it has a gummy texture when you are weighing it out then it has water in it. Because the water content can affect the behaviour of the emulsion you will need to dry it out. To dry the surfactant expose it to vacuum in a bell jar for a few hours or longer as required.
11. 10 mL of 2,2,4-trimethylpentane (isooctane). *This is highly flammable and should be worked within a fume hood.*
12. A strong light for the photoreduction but placing the solution in strong sunlight will also work.

To make cubic Prussian blue nanoparticles using an inverse microemulsion

1. Set the temperature of the heating mantle to 50 °C and allow it to equilibrate.
2. Clamp the vial containing 3 mL of water in place in the oil bath. Because of the volume it should be at the correct temperature after 2 min.
3. Dissolve the $(NH_4)_3[Fe^{(III)}(C_2O_4)_3] \cdot 3H_2O$ salt in the water and add a stirrer bar.
4. Add the ground $K_3[Fe(CN)_6]$ to the solution with vigorous stirring. Place the lid on the vial to prevent evaporation and allow the solution to stir for 20 min.
5. Remove the sealed vial from the oil bath and switch off the heating mantle. Transfer the vial to a sonic bath and sonicate for two min.
6. Transfer the vial and solution into a dark cupboard or box and allow it to cool and stand overnight. Green crystals of $K_3[Fe(C_2O_4)_3]$ should sediment out of the solution.
7. **In the dark** swirl the vial around by hand and then draw the vial contents into the syringe. Fit the filter to the end and flush the solution into the 20 mL foil wrapped vial.
8. Make up the volume of the solution to 10 mL to give an equimolar stock solution of $[Fe^{(III)}(C_2O_4)_3]$ and $(NH_4)_3[Fe^{(III)}(CN)_6]$ with a concentration of 0.3 mol/L.
9. Add 6 mL of the isooctane to the vial containing the NaAoT to dissolve the surfactant. This may take a while and magnetic stirring can be used to speed the process up. Make up the volume of the isooctane and surfactant solution to 10 mL with more isooctane.
10. A value of $\omega = 15$ should result in uniform cubic particles. As there are 0.001 mol of AoT in the continuous isooctane phase we will need 0.027 mol of water. This equates to a volume of 0.27 mL. Add 0.27 mL of the iron salt stock solution to the isooctane and surfactant solution. Replace the lid and shake the vial vigorously to speed up the spontaneous formation of the microemulsion.
11. Remove the foil covering from the vial and expose it to sunlight. You should initially see a transparent yellow solution. If you see any opacity then the water droplets are too large and some phase separation is taking place. If this is the case you may want to try a more dilute iron solution or lower the amount of water. Allow the sample to sit for four days during which time you should see the emulsion turn blue from the formation of Prussian blue particles. The sample is now ready for analysis. Leaving the sample for two weeks will allow the solution to age and result in particles that should pack together upon drying.

UV-visible spectroscopy of the emulsion solution should show a peak absorbance at 680 nm consistent with a spin exchange transition from one iron centre to another in the MOF lattice. This can be represented as:

$$\{Fe^{(III)}[(t_{2g})^3(e_g)^2]Fe^{(II)}[(t_{2g})^6]\} \rightarrow \{Fe^{II}[(t_{2g})^4(e_g)^2]Fe^{(III)}[(t_{2g})^5]\}$$

Infra red analysis of the solid precipitate should also show a characteristic peak for the Fe–CN stretching mode at 2069 cm^{-1}. Transmission electron microscopy should reveal cubic particles with an average side length of around 16 nm. Grids can be prepared by directly drying a droplet of the microemulsion solution onto a carbon coated copper grid. If the particles are uniform in size then you will see them assembled into 2D arrays with long range order on the grid. These arrays can reach up to 200 nm across with the particles separated by layers of surfactant 2–3 nm thick. A similar approach is used to make cubic particle arrays of the cobalt iron Prussian blue analogue. Instead of using a photolytically reduced iron salt in the emulsion an emulsion containing only iron cyanate is mixed with an emulsion formed using a surfactant having cobalt as the surfactant head group counter ion. When the two aqueous phases meet the cobalt forms a cobalt hexacyanoferrate (Vaucher et al., 2002) (Fig. 4.84).

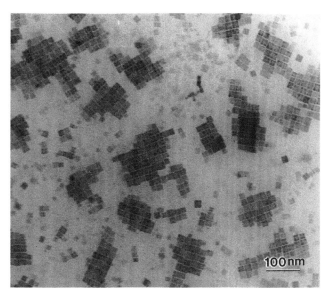

FIGURE 4.84   Transmission electron microscopy images of Prussian blue nanoparticles formed at a ω value of 15 and a reactant concentration of 0.01 mol/L. Scale bar is 100 nm. *(Image courtesy of Mei Li)*

MAKING A MOF WITH LARGE PORES

With thanks to Avinash Patil and Kamendra Sharma. Adapted from Bauer et al. (2008).

Research in metal organic frameworks has progressed to the point that it is now possible to use a method called reticular synthesis to design them (Yaghi et al., 2003). This approach is slightly different to those explored previously in this section in that two separate building phases are used to construct the MOF. Initially a short-range ordered unit is formed consisting of rigidly arranged octahedral or tetrahedral coordinated metal oxide or fluoride units bound together by shared bonds or a carbon oxygen (carboxylate) bridge. This basic building unit is then assembled further into a long range porous framework by organic ligand interactions. MOFs are often given a name to help classify their composition and the general method by which they are formed. In this recipe we will prepare a compound called MIL 101-NH$_2$. The MIL is an acronym for Material Insitut Lavoisier after the laboratory where this particular composition was developed. This convention is similar to the name MCM, Mobil Composition of Matter, given to porous surfactant templated silica. The suffix 101-NH$_2$ denotes that it is the hundred and first compound developed and has been modified to posses amine groups within the structure (Férey et al., 2004). The MIL series of compositions demonstrate extremely high porosities and surface areas. The largest pore size occurring in a natural mineral is around 30 Å, for cloverite. Standard approaches using ligand to metal binding to simulate these figures in a synthetic material often result in short-range order over a few nanometres. Where the pore diameter is expected to be tens of Ångstroms there is also the risk that the forming lattices will interpenetrate one another. By using large inflexible inorganic blocks instead of only metal centres these problems are minimised and large volumes and pore sizes are achievable. The MIL-101 material is made using a chromium or iron metal oxide arranged as a trimeric octahedral unit. These units are then bound together using terephthalic acid as a crosslinking agent to form a lattice with long range order. MIL-101 demonstrates a two tiered porosity and has micropores around 6.5 Å and mesopores around 25–30 Å in diameter. Internal surface areas of up to 5,900 m^2 g^{-1} are possible. These impressive figures make the MIL MOFs attractive for use in gas storage, as molecular sieves and as nanoscale reactors. Non-toxic metals and ligands mean that the MIL and its eventual breakdown products can be designed for biocompatibility. The large surface area means that, when the MOF is produced as a nanomaterial, it can be used as a carrier for poorly soluble and otherwise difficult to deliver drugs (Horcajada et al., 2009).

In this recipe we will make a biocompatible and biodegradable MIL 101-NH$_2$ (Bauer et al., 2008). An iron salt undergoes a hydrothermal treatment in the presence of hydrofluoric acid and the organic linker moiety. The basic inorganic units that form comprise three octahedral iron centres coordinated to four oxygen atoms contributed by bidentate carboxylates, one bridging oxygen between each iron centre and one terminal fluorine or water group. The bridging dicarboxylic acid then assembles these trimers into the larger porous metal organic framework. In this particular recipe 1,4-benzene dicarboxylate, also known as terephthalic acid, has been modified to posses an amine group. The reagents are all mixed together in a PTFE container sealed within a stainless steel hydrothermal bomb. Metal organic frameworks require high temperatures to drive the diffusion processes of formation but cannot be used with the organic components or else they will combust. By preparing the MOF under hydrothermal

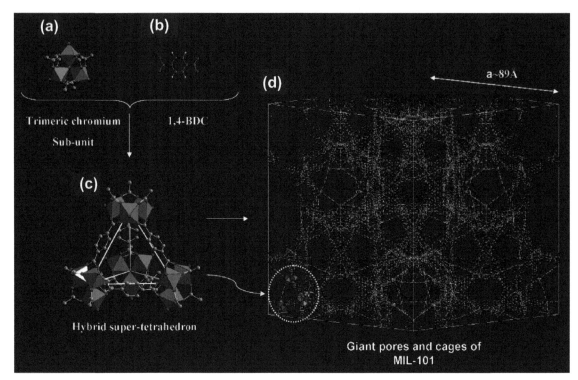

FIGURE 4.85   Schematic of MIL 101 formation from smaller units to form a porous network. *(Reprinted with Permission from Férey, G. et al., 2004)*

conditions the lattice rearrangements required to form the framework are made possible without degrading the organic parts (Fig. 4.85).

To make MIL 101-NH$_2$ by hydrothermal reaction you will need:

This procedure uses hydrofluoric acid! Before starting you must consult with whoever is responsible for laboratory health and safety. Special handling measures and full protective equipment is required when working with this strong acid. You will need calcium gluconate gel and a nearby safety shower and a plentiful water supply. Areas in which HF is being handled must be clearly marked along with any containers holding it in any concentration.

1. 0.240 g (1.34 mmol) of 2-amino-1,4-benzenedicarboxylate.
2. 2.484 mL of 0.2 mol/L hydroflouric acid. *This is extremely corrosive and hazardous. Depending on local guidelines you will need to take special measures when working with HF. You must work with this in a fume hood at all times and keep in a sealed well marked container when not in use. The container and receptacles in which the HF is used must not be glass as HF will dissolve them. Calcium gluconate gel must be kept on hand as a first aid treatment for skin contact with HF. A small amount of stock HF solution can be prepared by diluting 0.1 mL from a 48 wt% (~27.5 mol/L) HF stock solution up to a volume of 13.75 mL. Keep this in a well sealed and clearly marked thick screw top 20 mL PTFE bottle in the fume hood.*
3. 20 ml of water in a 25–30 mL vial with a lid.
4. 0.214 g (0.8 mmol) of FeCl$_3$·6H$_2$O.
5. A stainless steel autoclave bomb with a thick PTFE container housed within it. The container should accommodate a volume of around 25 mL.
6. An oven capable of reaching 150 °C. You will need this for a total of two days so check with other users before starting the experiment.
7. A centrifuge and centrifuge tubes large enough to accommodate around 25 mL of liquid. You will need scales for balancing the tubes before they are run.
8. Acetone for rinsing the product with and a waste solvent container assigned to handling hydrofluoric acid waste. Waste handling procedures may differ from laboratory to laboratory so make sure you follow the correct disposal guidelines.
9. Volumetric pipettes in the 10–100 μL and 1–5 mL ranges.
10. A magnetic stirrer and stir bar. You will also need a pair of PTFE tweezers to remove the magnetic bead from the reaction solution. You could use another magnet to get it out but this risks flicking some of the HF containing solution around and is best avoided.
11. Have a vial containing water and calcium carbonate on hand to dip the magnetic bead in once you have finished with it.

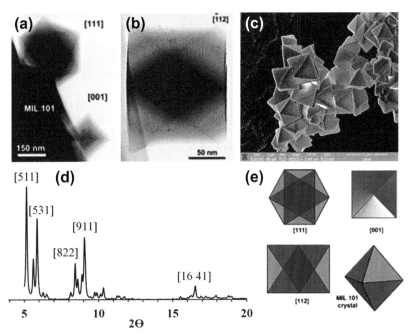

FIGURE 4.86 (a) and (b) Transmission electron micrograph showing the expected structure of hydrothermally prepared MIL-101 particles. (c) Scanning electron micrograph of MIL-101. Scale bar is 1 μ. (d) A calculated powder diffraction pattern for MIL-101 with the major reflections labelled. The chromium centred and iron centred forms are isomorphous. (e) How crystals of MIL-101 appear when viewed along certain lattice planes. *(Adapted from Ferey, G. et al., 2004 Chem. Mater. 2005, 17, 6525-6527)*

To make MIL 101-NH$_2$ by hydrothermal reaction:
To be performed in a fume hood with the sash down as far as possible and protective clothing while handling the HF.

1. Place the PTFE sleeve on top of the magnetic stirrer and add the 20 mL of water and a magnetic stir bar
2. Dissolve the iron chloride and 2-amino-1,4-benzenedicarboxylate into the water.
3. Carefully add the 2.484 mL of 0.2 M HF to the solution using a combination of the volumetric pipettes. Dispose of the tips in the calcium carbonate neutralisation solution.
4. Place the lid on the PTFE vessel and seal the whole thing within the stainless steel autoclave bomb.
5. Transfer this to the oven and set the temperature to 150 °C for 48 h.
6. Clean up or dispose of any items that have come into contact with HF by neutralisation and rinsing.
7. After the hydrothermal bomb has cooled to room temperature dismantle it and carefully remove the PTFE container.
8. Open the container and transfer the contents into a centrifuge tube. The MOF product may have settled to the bottom depending on the size of the granules formed. You may find using a pipette to agitate the sediment and drawing it up may help the transfer. Remember to prepare a counterweight for the sample.
9. Centrifuge the sample on full speed for 10 min or until all the precipitate has settled to the bottom of the centrifuge tube. Decant the top solution into a properly marked waste container.
10. Redisperse the pelleted product in acetone and centrifuge again. Decant the top solution into a waste disposal container. Repeat twice more.
11. Dry the product in the oven at 100 °C. You may need to transfer it to another container.

The product will be a very dark red/brown powder. The structure can be checked by powder X-ray diffraction. The calculated X-ray pattern for MIL-101 is provided below. Under transmission or scanning electron microscopy the formed particles will look cubo-octahedral in shape. BET surface area measurement should give an internal surface area in the thousands of square metres per gram (Fig. 4.86).

REFERENCES

Adams, C.J., et al., 2010. Towards polymorphism control in coordination networks and metallo-organic salts. CrystEngComm 12 (12), 4403−4409.
Aegerter, M., Leventis, N., Koebel, M. (Eds.), 2011. Aerogels Handbook. Springer.
Airoldi, C., Farias, R.F., 2004. Alkoxide as precursors in the synthesis of new materials through the sol−gel process. Quimica Nova 27 (1), 84−88.
Alexandrescu, R., et al., 2004. TiO$_2$ nanosized powders by TiCl$_4$ laser pyrolysis. Nanotechnology 15 (5), 537−545.
Alivisatos, A.P., et al., 1996. Organization of nanocrystal molecules' using DNA. Nature 382, 609−611.

Al-Salim, N.I., et al., 2000. Characterisation and activity of sol−gel-prepared TiO_2 photocatalysts modified with Ca, Sr or Ba ion additives. Journal of Materials Chemistry 10 (10), 2358−2363.

Anderson, T., 1855. Vorläufige notiz über die zersetzungen der platindoppelsalze der organischen basen. Justus Liebigs Annalen Der Chemie 96 (2), 199−205.

Andrews, R., et al., 1999. Continuous production of aligned carbon nanotubes: a step closer to commercial realization. Chemical Physics Letters 303 (5−6), 467−474.

Azad, A.-M., 2006. Fabrication of yttria-stabilized zirconia nanofibers by electrospinning. Materials Letters 60 (1), 67−72.

Barrera, C., Herrera, A.P., Rinaldi, C., 2009. Colloidal dispersions of monodisperse magnetite nanoparticles modified with poly(ethylene glycol). Journal of Colloid and Interface Science 329 (1), 107−113.

Batzill, M., Diebold, U., 2005. The surface and materials science of tin oxide. Progress in Surface Science 79 (2−4), 47−154. Available at: http://linkinghub. elsevier.com/retrieve/pii/S007968160500050X [Accessed June 14, 2011].

Bauer, S., et al., 2008. High-throughput assisted rationalization of the formation of metal organic frameworks in the iron (III) aminoterephthalate solvothermal system. Inorganic Chemistry 47 (17), 7568−7576.

Baughman, R.H., Zakhidov, A.A., De Heer, W.A., 2002. Carbon nanotubes−the route toward applications. Science 297 (5582), 787.

Baumann, T.F., et al., 2005. Synthesis of high-surface-area alumina aerogels without the use of alkoxide precursors. Chemistry of Materials 17 (2), 395−401.

Bawendi, M.G., Steigerwald, M.L., Brus, L.E., 1990. The Quantum mechanics of larger semiconductor clusters ('quantum dots'). Annual Review of Physical Chemistry 41 (1), 477−496.

Beck, J., et al., 1992. A new family of mesoporous molecular sieves prepared with liquid crystal templates. Journal of the American Chemical Society 114 (27), 10834−10843.

Berger, P., et al., 1999. Preparation and properties of an aqueous ferrofluid. Journal of Chemical Education 76 (7), 943.

Berry, C.C., Curtis, A.S.G., 2003. Functionalisation of magnetic nanoparticles for applications in biomedicine. Gene Therapy, 198.

Bhushan, B., Ruan, J., Gupta, B., 1993. A scanning tunnelling microscopy study of fullerene films. Journal of Physics D: Applied Physics 26, 1319.

Biju, V., et al., 2008. Semiconductor quantum dots and metal nanoparticles: syntheses, optical properties, and biological applications. Analytical and Bioanalytical Chemistry 391 (7), 2469−2495.

Brandhuber, D., et al., 2005. Glycol-modified silanes in the synthesis of mesoscopically organized silica monoliths with hierarchical porosity. Chemistry of Materials 17 (16), 4262−4271.

Briggs, J.B., Miller, G.P., 2006. [60] Fullerene-acene chemistry: a review. Comptes Rendus Chimie 9 (7−8), 916−927.

Brinker, C.J., Scherer, G.W., 1990. Sol−gel Science. Academic Press, London.

Brinker, C., et al., 1991. Fundamentals of sol−gel dip coating. Thin Solid Films 201 (1), 97−108.

Brown, J., 1724. Observations and experiments upon the foregoing preparation. By Mr. John Brown, Chymist, F. R. S. Philosophical Transactions of the Royal Society of London 33 (381−391), 17−24.

Brownlee, D., et al., 2006. Comet 81P/Wild 2 under a microscope. Science 314 (5806), 1711−1716. New York, N.Y.

Bruchez Jr., M., 1998. Semiconductor nanocrystals as fluorescent biological labels. Science 281 (5385), 2013−2016.

Brus, L., 1986. Electronic wave functions in semiconductor clusters: experiment and theory. The Journal of Physical Chemistry 90 (12), 2555−2560.

Brust, M., et al., 1994. Synthesis of thiol-derivatised gold nanoparticles in a two-phase Liquid-Liquid system. Journal of the Chemical Society, Chemical Communications (7), 801.

Caillard, R., et al., 2008. Fullerenes from aromatic precursors by. Nature 454 (August).

Cameron, N.S., Corbierre, M.K., Eisenberg, A., 1999. 1998 E. W. R. Steacie award lecture asymmetric amphiphilic block copolymers in solution: a morphological wonderland 1. Langmuir 1326, 1311−1326 (June 1998).

Cami, J., et al., 2010. Detection of C 60 and C 70 in a young planetary nebula. Science 70 (July).

Cao, W., Hunt, A.J., 1994. Improving the visible transparency of silica aerogels. Journal of Non-Crystalline Solids 176 (1), 18−25. Available at:

Chan, W.C., Nie, S., 1998. Quantum dot bioconjugates for ultrasensitive nonisotopic detection. Science 281 (5385), 2016−2018.

Chan, W.C.W., et al., 2002. Luminescent quantum dots for multiplexed biological detection and imaging. Current Opinion in Biotechnology 13 (1), 40−46.

Chatelon, J., et al., 1994. Morphology of SnO_2 thin films obtained by the sol−gel technique. Thin Solid Films 247 (2), 162−168.

Chatterjee, D., Mahata, A., 2001. Photoassisted detoxification of organic pollutants on the surface modified TiO_2 semiconductor particulate system. Catalysis Communications 2 (1), 1−3.

Chen, X., et al., 2011. Increasing solar absorption for photocatalysis with black hydrogenated titanium dioxide nanocrystals. Science 331 (6018), 746−750. New York, N.Y.

Cheung, C.L., et al., 2000. Growth and fabrication with single-walled carbon nanotube probe microscopy tips. Applied Physics Letters 76 (21), 3136−3138.

Choi, W., Termin, A., Hoffmann, M.R., 1994. The role of metal ion dopants in quantum-sized TiO_2: correlation between photoreactivity and charge carrier recombination dynamics. The Journal of Physical Chemistry 98 (51), 13669−13679.

Chu, Cw, et al., 1993. Superconductivity above 150 K in $HgBa_2Ca_2Cu_3O_8$ at high pressures. Nature 365, 323−325.

Collins, A.M., et al., 2009. Structure and properties of silicified purple membrane thin films. Biomacromolecules 10 (10), 2767−2771.

Conradt, R., 2008. Chemical durability of oxide glasses in aqueous solutions: a review. Journal of the American Ceramic Society 91 (3), 728−735.

Correa-duarte, M.A., Giersig, M., Liz-marzan, L.M., 1998. Stabilization of CdS semiconductor nanoparticles against photodegradation by a silica coating procedure. Chemical Physics Letters, 497−501 (April).

Costa, C.A.R., Leite, C.A.P., Galembeck, F., 2003. Size dependence of Stöber silica nanoparticle microchemistry. The Journal of Physical Chemistry B 107 (20), 4747−4755.

Culverwell, E., Wimbush, S.C., Hall, S.R., 2008. Biotemplated synthesis of an ordered macroporous superconductor with high critical current density using a cuttlebone template. Chemical communications 7345 (9), 1055−1057 (Cambridge, England).

Dabbousi, B.O., et al., 1997. (CdSe) ZnS core-shell quantum dots: synthesis and characterization of a size series of highly luminescent nanocrystallites. Journal of Physical Chemistry B 9463 (97), 9463−9475.

Dai, H., et al., 1996. Single-wall nanotubes produced by metal-catalyzed disproportionation of carbon monoxide. Chemical Physics Letters 260 (3−4), 471−475.

Daniel, M.C., Astruc, D., 2004. Gold nanoparticles: assembly, supramolecular chemistry, quantum-size-related properties, and applications toward biology, catalysis, and nanotechnology. Chemical Reviews 104 (1), 293−346.

Depla, A., Verheyen, E., Veyfeyken, A., Gobechiya, E., Hartmann, T., Schaefer, R., Martens, J.A., et al., 2011. Zeolites X and A crystallization compared by simultaneous UV/VIS-Raman and X-ray diffraction. Physical Chemistry Chemical Physics: PCCP 13 (30), 13730−13737. Available at: http://www.ncbi.nlm.nih.gov/pubmed/20715277 [Accessed August 2, 2011].

Díez, I., Jiang, H., Ras, R.H. a, 2010. Enhanced emission of silver nanoclusters through quantitative phase transfer. Chemphyschem?: a European journal of chemical physics and physical chemistry 11 (14), 3100−3104. Available at: http://www.ncbi.nlm.nih.gov/pubmed/20715277 [Accessed August 2, 2011].

Díez, I., et al., 2009. Color tunability and electrochemiluminescence of silver nanoclusters. Angewandte Chemie (International ed. in English) 48 (12), 2122−2125. Available at: http://www.ncbi.nlm.nih.gov/pubmed/19206134 [Accessed July 17, 2011].

Dobosz, A., Sobczyński, A., 2003. The influence of silver additives on titania photoactivity in the photooxidation of phenol. Water Research 37 (7), 1489−1496.

Dresselhaus, M., et al., 2002. Raman spectroscopy on isolated single wall carbon nanotubes. Carbon 40 (12), 2043−2061.

Ducati, C., et al., 2006. Full papers crystallographic order in multi-walled carbon nanotubes synthesized in the presence of nitrogen. Carbon Nanotubes, 774−784.

Dujardin, E., Mann, S., 2004. Morphosynthesis of molecular magnetic materials. Advanced Materials 16 (13), 1125−1129.

Eaton, P., Cole, T., 1964. An X-Ray diffraction study of nonplanar carbanion strictures. Journal of American Chemical Society 86 (15), 3157−3158.

Ebelmen, J.J., 1846. Justus Liebigs Annalen der Chemie 57, 331.

Englebienne, P., Weiland, M., 1996. Water-soluble conductive polymer homogeneous immunoassay (SOPHIA). A novel immunoassay capable of automation. Journal of Immunological Methods 191 (2), 159−170.

Englebienne, P., Hoonacker, A.V., Verhas, M., 2003. Surface plasmon resonance: principles, methods and applications in biomedical sciences. Spectroscopy 17, 255−273.

Ewijk, G.A.V., Vroege, G.J., Philipse, A.P., 1999. Convenient preparation methods for magnetic colloids. Langmuir 201, 31−33.

Fahlman, B.D., 2002. Chemical vapor deposition of carbon nanotubes an experiment in materials chemistry. Journal of Chemical Education 79 (2), 203−206.

Férey, G., et al., 2004. A hybrid solid with giant pores prepared by a combination of targeted chemistry, simulation, and powder diffraction. Angewandte Chemie 116 (46), 6456−6461.

Ferlay, S., et al., 1995. A room-temperature organometallic magnet based on Prussian blue. Nature 378 (6558), 701−703.

Fernandez-Bertran, J.F., 1999. Mechanochemistry: an overview. Pure and Applied Chemistry 71 (4), 581−586.

Filipponi, L., 2010. NANOYOU Teachers Training Kit−Experiment C: Colorimetric Gold Nanosensor, pp. 1−21.

Flemming, J.H., et al., 2006. A packed microcolumn approach to a cell-based biosensor. Sensors and Actuators B: Chemical 113 (1), 376−381.

Frank, S., Poncharal, P., Wang, Z., 1998. Carbon nanotube quantum resistors. Science 280, 1744−1746.

Frens, G., 1973. Controlled nucleation for the regulation of the particle size in monodisperse gold suspensions. Nature 241 (105), 20−22.

Fritzsche, W., 2001. DNA-gold conjugates for the detection of specific molecular interactions. Journal of Biotechnology 82 (1), 37−46.

Fuge, G.M., Holmes, T.M.S., Ashfold, M.N.R., 2009. Ultrathin aligned ZnO nanorod arrays grown by a novel diffusive pulsed laser deposition method. Chemical Physics Letters 479 (1−3), 125−127. Elsevier B.V.

Gao, G., Çagin, T., others, 1998. Energetics, structure, mechanical and vibrational properties of single-walled carbon nanotubes. Nanotechnology 9, 184.

Gao, X., et al., 2004. In vivo cancer targeting and imaging with semiconductor quantum dots. Nature Biotechnology 22 (8), 969−976.

Gao, P.X., Song, J., Liu, J., Wang, Z.L., 2007. Nanowire piezoelectric nanogenerators on plastic substrates as flexible power sources for nanodevices. Advanced Materials 19 (1), 67−72. Wiley Online Library.

Geffcken, W., Berger, E., 1939. German Patent 736, 411.

Geim, A.K., Novoselov, K.S., 2007. The rise of graphene. Nature Materials 6 (3), 183−191.

Gill, I., Ballesteros, A., 1998. Encapsulation of biologicals within silicate, siloxane, and hybrid sol−gel polymers: an efficient and generic approach. Journal of the American Chemical Society 120 (34), 8587−8598.

Gill, R., et al., 2006. Probing biocatalytic transformations with CdSe-ZnS QDs. Journal of the American Chemical Society 128 (48), 15376−15377.

Gill, I., 2001. Bio-doped nanocomposite polymers: sol−gel bioencapsulates. Chemistry of Materials 13 (10), 3404−3421.

Gole, A., Murphy, C.J., 2004. Seed-mediated synthesis of gold nanorods: role of the size and nature of the seed. Chemistry of Materials 16 (19), 3633−3640.

Govender, K., Boyle, D.S., Kenway, P.B., Brien, P.O., 2004. Understanding the factors that govern the deposition and morphology of thin films of ZnO from aqueous solution. Materials Science, 2575−2591.

Green, N., 1965. A spectrophotometric assay for avadin and biotin based on binding of dyes by avadin. Biochemical Journal 94, 23c−24c.

Greene, L.E., Law, M., Tan, D.H., Montano, M., Goldberger, J., 2005. General route to vertical ZnO nanowire arrays using textured ZnO seeds. Nano.

Grogan, M.D.W., et al., 2011. Structure of plasmonic aerogel and the breakdown of the effective medium approximation. Optics Letters 36 (3), 358−360.

Gütlich, P., Garcia, Y., Goodwin, H.A., 2000. Spin crossover phenomena in Fe (II) complexes. Chem. Soc. Rev. 29 (6), 419−427.

Hall, S.R., et al., 1999. Template-directed synthesis of bi-functionalized organo-MCM-41 and phenyl-MCM-48 silica mesophases. Chem. Commun. (2), 201−202.

Hall, S., et al., 2006. Fabrication of porous titania (brookite) microparticles with complex morphology by sol-gel replication of pollen grains. Chemistry of Materials 18 (3), 598−600.

Haneda, K., Morrish, A., 1977. Magnetite to maghemite transformation in ultrafine particles. Journal de Physique 4 (38), 321−323.

Hare, J.P. et al., 1991. The IR spectra of fullerene-60 and -70, 60, 412−413.

Harper, J.C., et al., 2011. Encapsulation of S. cerevisiae in poly (glycerol) silicate derived matrices: effect of matrix additives and cell metabolic phase on long-term viability and rate of gene expression. Chemistry of Materials, 2555−2564.

Hebbard, A., et al., 1991. Superconductivity at 18 K in potassium-doped C60. Nature 350, 600−601.

Hench, L., 1997. Sol−gel materials for bioceramic applications. Current Opinion in Solid State and Materials Science 2 (5), 604−610. Available at.

Hernandez, Y., et al., 2008. High-Yield Production of Graphene by Liquid-Phase Exfoliation of Graphite, pp. 563−568.

Hevesy, G. de, 1962. Adventures in Radioisotope Research. Pergamon Press, New York.

Horcajada, P., et al., 2009. Porous metal-organic-framework nanoscale carriers as a potential platform for drug delivery and imaging. Nature Materials 9 (2), 172−178.

Hornyak, G., et al., 2009. Introduction to Nanoscience and Nanotechnology. C. Press, Boca Raton.

Hostetler, M.J., et al., 2008. Alkanethiolate gold cluster molecules with core diameters from 1. 5 to 5. 2 nm: core and monolayer properties as a function of core size. Microscopy 7463 (c), 17−30.

Hu, F., et al., 2009. Ultrasmall, water-soluble magnetite nanoparticles with high relaxivity for magnetic resonance imaging. The Journal of Physical Chemistry C 113 (49), 20855−20860.

Huang, L., et al., 2006. Facile synthesis of gold nanoplates by citrate reduction of $AuCl_4^-$ at room temperature. Chinese Chemical Letters 17 (10), 1405−1408.

Hummers, W., Offeman, R., 1958. Preparation of graphitic oxide. Journal of the American Chemical Society 208, 1339.

Huo, Q., et al., 1994. Organization of organic molecules with inorganic molecular species into nanocomposite biphase arrays. Langmuir (8), 1176−1191.

Hurd, C.B., 1938. Chem. Rev., 403−422.

Huynh, W.U., Dittmer, J.J., Alivisatos, a P., 2002. Hybrid nanorod-polymer solar cells. Science (New York, N.Y.) 295 (5564), 2425−2427. Available at: http://www. ncbi.nlm.nih.gov/pubmed/11923531 [Accessed March 10, 2012].

Ito, A., et al., 2005. Medical application of functionalized magnetic nanoparticles. Journal of Bioscience and Bioengineering 100 (1), 1−11.

Jana, N.R., Gearheart, L., Murphy, C.J., 2001. Seed-mediated growth approach for shape-controlled synthesis of spheroidal and rod-like gold nanoparticles using a surfactant template. Advanced Materials 13 (18), 1389−1393.

Jansen, M., 2008. The chemistry of gold as an anion. Chemical Society Reviews 37 (9), 1826−1835.

Jiang, X., Herricks, T., Xia, Y., 2003. Monodispersed spherical colloids of titania: synthesis, characterization, and crystallization. Advanced Materials 15 (14), 1205−1209.

Johnson, C.J., et al., 2002. Growth and form of gold nanorods prepared by seed-mediated, surfactant-directed synthesis. Journal of Materials Chemistry 12 (6), 1765−1770.

Journet, C., et al., 1997. Letters to nature large-scale production of single-walled carbon nanotubes by the electric-arc technique. Nature 388, 20−22 (August).

Kareiva, A., Karppinen, M., Niinistö, L., 1994. Sol−gel synthesis of superconducting $YBa_2Cu_4O_8$ using acetate and tartrate precursors. Journal of Materials Chemistry 4 (8), 1267−1270.

Kim, Y.-Y., et al., 2009. Controlled nanoparticle formation by enzymatic deshelling of biopolymer-stabilized nanosuspensions. Small 5 (8), 913−918.

Kistler, S., 1931. Coherent expanded aerogels and jellies. Nature 127 (3211), 741.

Kobayashi, Y., et al., 2011. Control of shell thickness in silica-coating of Au nanoparticles and their X-ray imaging properties. Journal of Colloid and Interface Science 358 (2), 329−333.

Kresge, C., et al., 1992. Ordered mesoporous molecular sieves synthesized by a liquid-crystal template mechanism. Nature 359, 710−712.

Kroto, H., et al., 1985. C_{60}: Buckminsterfullerene. Nature 318, 162−163.

Lagow, R.J., et al., 1995. Synthesis of linear acetylenic carbon: the 'sp' carbon allotrope. Science 267 (5196), 362−367 (New York, N.Y.).

Lamouroux, E., Serp, P., Kalck, P., 2007. Catalytic routes towards single wall carbon nanotubes. Catalysis Reviews 49 (3), 341−405.

Law, M., Greene, L.E., Johnson, J.C., Saykally, R., Yang, P., 2005. Nanowire dye-sensitized solar cells. Nature materials 4 (6), 455−459. Nature Publishing Group.

Lee, S.Y., et al., 2011. Inorganic nanomaterial-based biocatalysts. BMB Reports 44 (2), 77−86.

Letchford, K., Burt, H., 2007. A review of the formation and classification of amphiphilic block copolymer nanoparticulate structures: micelles, nanospheres, nanocapsules and polymersomes. European Journal of Pharmaceutics and Biopharmaceutics: Official Journal of Arbeitsgemeinschaft für Pharmazeutische Verfahrenstechnik e.V 65 (3), 259−269.

Levene, L., Thomas, I.M., 1979. US Patent, 3,640,093.

Leventis, N., et al., 2009. Smelting in the age of nano: iron aerogels. Journal of Materials Chemistry 19 (1), 63.

Li, L.-shi, Hu, J., Yang, Weidong, 2001. Band gap variation of size-and shape-controlled colloidal CdSe quantum rods. Nano Letters 1 (7), 349−351.

Li, Y.-li, Kinloch, I.A., Windle, A.H., 2004. Direct spinning of carbon nanotube fibers from chemical vapor deposition synthesis. Science 304, 276−278 (April).

Li, C., et al., 2008. A facile polyol route to uniform gold octahedra with tailorable size and their optical properties. ACS Nano 2 (9), 1760−1769.

Lian Zhang, Erhan Ata, Stephen C. Minne, Paul Hough, 2007. Single-Walled Carbon Nanotube Probes for AFM Imaging.

Ling, L.I., et al., 2011. Coating multi-walled carbon nanotubes with uniform silica shells: independent of surface chemistry. Science and Technology 27 (2006), 181−184.

Liu, J., Lu, Y., 2006. Preparation of aptamer-linked gold nanoparticle purple aggregates for colorimetric sensing of analytes. Nature protocols 1 (1), 246−252. Available at: http://www.nature.com/nprot/journal/v1/n1/abs/nprot.2006.38.html [Accessed November 18, 2011].

Liz-Marzán, L.M., Giersig, M., Mulvaney, P., 1996. Synthesis of nanosized gold-silica core-shell particles. Langmuir 12 (18), 4329−4335. Available at: http://pubs. acs.org/doi/abs/10.1021/la9601871 [Accessed November 18, 2011].

Lu, X., et al., 1993. Thermal and electrical conductivity of monolithic carbon aerogels. Journal of Applied Physics 73 (2), 581.

Lu, Y., et al., 1997. Continuous formation of supported cubic and hexagonal mesoporous films by sol−gel dip-coating. Nature 389 (6649), 364−368.

Ludi, A., Hu, G., 1973. Structural chemistry of polynuclear transition metal cyanides. Structure and Bonding 14, 1−21.

Lynam, B.C., Moulton, S.E., Wallace, G.G., 2007. Carbon-Nanotube Biofibers, 2522, 1244−1248.

Lyon, L.A., Musick, M.D., Natan, M.J., 1998. Colloidal au-enhanced surface plasmon resonance immunosensing been realized using particle enhancement, with the theo. Science 70 (24), 5177−5183.

Ma, T., et al., 2005. High-efficiency dye-sensitized solar cell based on a nitrogen-doped nanostructured titania electrode. Nano letters 5 (12), 2543−2547.

Mann, S., Sparks, N.H.C., Board, R.G., 1990. Magnetotactic bacteria: microbiology biomineralization. Palaeomagnetism and biotechnology. Advances in Microbial Physiology 31 (1975), 27731.

Mann, S., 2008. Life as a nanoscale phenomenon. Angewandte Chemie (International ed. in English) 47 (29), 5306−5320.

Manna, Liberato, Scher, E.C., Alivisatos, A.P., 2000. Synthesis of soluble and processable rod-, arrow-, teardrop-, and tetrapod-shaped CdSe nanocrystals. Journal of the American Chemical Society 122 (27), 12700−12706.

May, P.W., 2000. Diamond thin films: a 21st-century material. Philosophical Transactions of the Royal Society of London. Series A: Mathematical, Physical and Engineering Sciences 358 (1766), 473−495.

Medintz, I.L., et al., 2005. Quantum dot bioconjugates for imaging, labelling and sensing. Nature Materials 4 (6), 435−446.

Meldrum, F., et al., 1993. Electron microscopy study of magnetosomes in two cultured vibrioid magnetotactic bacteria. Proclamations of the Royal Society 251 (1332), 237−242.

Meyer, M., Fischer, a., Hoffmann, H., 2002. Novel Ringing Silica Gels That Do Not Shrink. The Journal of Physical Chemistry B 106 (7), 1528−1533. Available at: http://pubs.acs.org/doi/abs/10.1021/jp013371q.

Miranda, A., et al., 2010. One-pot synthesis of triangular gold nanoplates allowing broad and fine tuning of edge length. Nanoscale 2 (10), 2209−2216.

Mitchell, G.P., Mirkin, C.A., Letsinger, R.L., 1999. Programmed assembly of DNA functionalized quantum dots. Journal of the American Chemical Society 121 (35), 8122−8123.

Miyaji, F., et al., 1999. Organic crystal templating of hollow silica fibers. Communications, 3021−3024.

Mohanan, J.L., Arachchige, I.U., Brock, S.L., 2005. Porous semiconductor chalcogenide aerogels. Science 307 (5708), 397−400.

Morales, A.M., Lieber, C.M., 1998. Semiconductor nanowires a laser ablation method for the synthesis of crystalline semiconductor nanowires. Science 279, 208.

Mucic, R.C., et al., 1998. DNA-directed synthesis of binary nanoparticle network materials. Journal of the American Chemical Society 120 (48), 12674−12675.

Mulik, S., Sotiriou-Leventis, C., Leventis, N., 2007. Time-efficient acid-catalyzed synthesis of resorcinol-formaldehyde aerogels. Chemistry of Materials 19 (25), 6138−6144.

Murray, C.B., Norris, D.J., Bawendi, M.G., 1993. Synthesis and characterisation of nearly monodisperse CdE (E=S, Se, Te) semiconductor nanocrystallites. Journal of the American Chemical Society 115, 8706−8715.

Naik, S.N., et al., 2010. Production of first and second generation biofuels: a comprehensive review. Renewable and Sustainable Energy Reviews 14 (2), 578−597.

Nakamura, H., Matsui, Y., 1995. Silica gel nanotubes obtained by the sol−gel method. Journal of the American Chemical Society 117 (9), 2651−2652.

Novoselov, K., et al., 2004. Electric field effect in atomically thin carbon films. Science 306 (5696), 666.

Novoselov, K.S., et al., 2005. Two-dimensional atomic crystals. PNAS 102 (30), 10451−10453.

O'Connell, M.J., et al., 2001. Reversible water-solubilization of single-walled carbon nanotubes by polymer wrapping. Chemical Physics Letters 342 (3−4), 265−271.

Ogasawara, W., et al., 2000. Template mineralization of ordered macroporous chitin-silica composites using a cuttlebone-derived organic matrix. Chemistry of Materials 12 (10), 2835−2837.

Oster, G., 1951. The isoelectric points of some strains of tobacco mosaic virus. Journal of Biological Chemistry 190 (1), 55.

Pacholski, C., Kornowski, A., Weller, H., 2002. Self-assembly of ZnO: from nanodots to nanorods. Angewandte Chemie (International ed. in English) 41 (7), 1188−1191.

Pal, M., et al., 2007. Size-controlled synthesis of spherical TiO2 nanoparticles: morphology, crystallization, and phase transition. Journal of Physical Chemistry C 111 (1), 96−102.

Papell, S., 1965. Low Viscosity Magnetic Fluid Obtained by the Colloidal Suspension of Magnetic Particles.

Park, S., Ruoff, R.S., 2009. Chemical methods for the production of graphenes. Nature Nanotechnology 4 (4), 217−224.

Patil, A.J., Vickery, J.L., Scott, T.B., Mann, S., 2009. Aqueous stabilization and self-assembly of graphene sheets into layered bio-nanocomposites using DNA. Advanced Materials 21 (31), 3159−3164.

Pekala, R.W., et al., 1992. Aerogels derived from multifunctional organic monomers. Journal of Non-Crystalline Solids 145, 90−98.

Peng, Z.A., Peng, Xiaogang, 2001. Mechanisms of the shape evolution of CdSe nanocrystals. Journal of the American Chemical Society 123 (7), 1389−1395.

Peng, X., et al., 2000. Shape control of CdSe nanocrystals. Nature 404 (6773), 59−61.

Poelz, G., Riethmuller, R., 1982. Preparation of silica aerogel for cherenkov counters. Nuclear Instruments and Methods in Physics Research 195 (3), 491−503.

Pong, B.-Kin, et al., 2007. New insights on the nanoparticle growth mechanism in the citrate reduction of gold (III) salt: formation of the Au nanowire intermediate and its nonlinear optical properties. Journal of Physical Chemistry C 111, 6281−6287.

Posthuma-Trumpie, GA., Korf, J., van Amerongen, A., 2009. Lateral flow (immuno)assay: its strengths, weaknesses, opportunities and threats. A literature survey. Analytical and Bioanalytical Chemistry 393 (2), 569−582.

Precker, J.W., 2007. Simple experimental verification of the relation between the band-gap energy and the energy of photons emitted by LEDs. European Journal of Physics 28 (3), 493−500.

Qin, Y., Wang, X., Wang, Z.L., 2008. Microfibre−nanowire hybrid structure for energy scavenging. Nature 451 (7180), 809−813. Nature Publishing Group.

Rao, C.N.R., Satishkumar, B.C., Govindaraj, A., 1997. Zirconia nanotubes. Chemical Communications 2 (16), 1581−1582.

Renzo, F.D., Cambon, H., Dutartre, R., 1997. A 28-year-old synthesis of micelle-templated mesoporous silica. Surface Science 10, 283−286.

Reynolds, R., Mirkin, C., Letsinger, R.L., 2000. Homogeneous, nanoparticle-based quantitative colorimetric detection of oligonucleotides. Journal of the American Chemical Society 122 (15), 3795−3796.

Rong, J., et al., 2009. Tobacco mosaic virus templated synthesis of one dimensional inorganic−polymer hybrid fibres. Journal of Materials Chemistry 19 (18), 2841.

Saito, R., Dresselhaus, M.S., Dresselhaus, G., 1998. Physical Properties of Carbon Nanotubes. Imperial College Press, London.

Satishkumar, B.C., et al., 2000. Synthesis of metal oxide nanorods using carbon nanotubes as templates. Journal of Materials Chemistry 10 (9), 2115−2119.

Schumacher, K., Ravikovitch, P., Chesne, A.D., 2000. Characterization of MCM-48 materials. Langmuir 16 (10), 4648−4654.

Schwarzenbach, G., Muehlebach, J., Mueller, K., 1970. Peroxo complexes of titanium. Inorganic Chemistry 9 (11), 2381−2390. Available at: http://pubs.acs.org/doi/abs/10.1021/ic50093a001.

Scott, C.D. et al., 2001. Growth Mechanisms for Single-Wall Carbon Nanotubes in a Laser-Ablation Process. 580, pp. 573−580.

Sen, R., Govindaraj, A., Rao, C., 1997. Carbon nanotubes by the metallocene route. Chemical Physics Letters 267 (3−4), 276−280.

Serpone, N., Lawless, D., Khairutdinov, R., 1995. Size effects on the photophysical properties of colloidal anatase TiO₂ particles: size quantization versus direct transitions in this indirect semiconductor? The Journal of Physical Chemistry 99 (45), 16646–16654.

Shchipunov, Y., 2004. A new precursor for the immobilization of enzymes inside sol–gel-derived hybrid silica nanocomposites containing polysaccharides. Journal of Biochemical and Biophysical Methods 58 (1), 25–38.

Shenton, W., et al., 1999. Inorganic–organic nanotube composites from template mineralization of tobacco mosaic virus. Advanced Materials 11 (3), 253–256.

Sherman, F., 1991. Getting started with yeast. Methods Enzymol 41 (2002), 3–41.

Sheu, T.S., et al., 1992. Cubic-to-tetragonal (t') transformation in zirconia-containing systems. Journal of the American Ceramic Society 75 (5), 1108–1116.

Simon, R., et al., 2006. Fabrication of porous titania (brookite) microparticles with complex morphology by sol–gel replication of pollen grains. Chemistry of Materials 18 (3), 598–600.

Singh, C., Shaffer, M.S.P., Windle, A.H., 2003a. Production of controlled architectures of aligned carbon nanotubes by an injection chemical vapour deposition method. Materials Science 41, 359–368.

Singh, C., Shaffer, M.S.P., Koziol, K.K.K., et al., 2003b. Towards the production of large-scale aligned carbon nanotubes. Chemical Physics Letters 372, 860–865.

Skomski, R., 2003. Nanomagnetics. Journal of Physics: Condensed Matter 15, 841–896.

Smith, D.M., Deshpande, R., Brinker, C.J., 1993. Ceramic Transactions 31 (Porous Materials), 71–80.

Sonawane, R.S., Hegde, S.G., Dongare, M.K., 2002. Preparation of titanium (IV) oxide thin film photocatalyst by sol–gel dip coating. Materials Chemistry and Physics 77, 744–750.

Stöber, W., 1968. Controlled growth of monodisperse silica spheres in the micron size range 1. Journal of Colloid and Interface Science 26 (1), 62–69.

Strawbridge, I., James, P., 1986. Thin silica films prepared by dip coating. Journal of Non-Crystalline Solids 82, 366–372.

Subramanian, V., Wolf, E., Kamat, P.V., 2001. Semiconductor–metal composite nanostructures. To what extent do metal nanoparticles improve the photocatalytic activity of TiO₂ films? The Journal of Physical Chemistry B 105 (46), 11439–11446.

Sun, Y., Xia, Y., 2002. Shape-controlled synthesis of gold and silver nanoparticles. Science 298 (5601), 2176–2179 (New York, N.Y.).

Sun, Y., et al., 2003. Polyol synthesis of uniform silver nanowires: A plausible growth mechanism and the supporting evidence. Nano Letters 3 (7), 955–960.

Sun, Y.-K., 1996. Preparation of ultrafine YBa₂Cu₃O₇-x superconductor powders by the poly (vinyl alcohol)-assisted sol–gel method. Society 5885 (95), 4296–4300.

Tanigaki, K., et al., 1991. Superconductivity at 33 K in CsxRbyC60. Nature 352 (6332), 222–223.

Tate, M.P., et al., 2010. How to dip-coat and spin-coat nanoporous double-gyroid silica films with EO19-PO43-EO19 surfactant (Pluronic P84) and know it using a powder X-ray diffractometer. Langmuir: The ACS Journal of Surfaces and Colloids 26 (6), 4357–4367.

Terriera, C., Rogera, J.A., 1995. Sb-doped SnO, transparent conducting oxide from the sol–gel dip-coating technique. Thin Solid Films 263, 37–41.

Thostenson, E.T., Ren, Z., Chou, T.-W., 2001. Advances in the science and technology of carbon nanotubes and their composites: a review. Composites Science and Technology 61, 1899–1912.

Tillotson, T., Hrubesh, L., 1992. Transparent ultralow-density silica aerogels prepared by a two-step sol–gel process. Journal of Non-Crystalline Solids 145, 44–50.

Vaucher, Â., Li, M., Mann, S., 2000. Synthesis of Prussian blue nanoparticles and. communications. Angewandte Chemie International Edition 39 (10), 1793–1796.

Vaucher, S., et al., 2002. Molecule-based magnetic nanoparticles: synthesis of cobalt hexacyanoferrate, cobalt pentacyanonitrosylferrate, and chromium hexacyanochromate coordination polymers in water-in-oil microemulsions. Nano Letters 2 (3), 225–229.

Vayssieres, L., 2003. Growth of arrayed nanorods and nanowires of ZnO from aqueous solutions. Advanced Materials 15 (5), 464–466. Wiley Online Library.

Verdaguer, M., Girolami, G., 2004. 9 Magnetic Prussian Blue Analogs Prussian Blue Analogs (PBA), Brief History, Synthesis.

Vijayaraghavan, A., Blatt, S., Weissenberger, D., 2007. Ultra-large-scale directed assembly of single-walled carbon nanotube devices. Nano Letters, 1–5.

Walsh, D., Wimbush, S.C., Hall, S.R., 2009. Reticulated superconducting YBCO materials of designed macromorphologies with enhanced structural stability through incorporation of lithium. Superconductor Science and Technology 22 (1), 015026.

Watanabe, S., et al., 2002. Enhanced optical sensing of anions with amide-functionalized gold nanoparticles A gold nanoparticle surface-modified with amide ligands. Society, 2866–2867.

Wei, A., Sun, X.W., Wang, J.X., Lei, Y., Cai, X.P., 2006. Enzymatic glucose biosensor based on ZnO nanorod array grown by hydrothermal decomposition. Applied Physics Letters 89, 123902.

Wilcheck, M., Bayer, E. (Eds.), 1990. Avidin–Biotin Technology. Academic Press, San Diego.

Wilson, W.L., Szajowski, P.F., Brus, L.E., 1993. Quantum confinement in size-selected, surface-oxidized silicon nanocrystals. Science (New York, N.Y.) 262 (5137), 1242–1244.

Winkler, L.D., et al., 2005. Quantum dots: an experiment for physical or materials chemistry. Journal of Chemical Education 82 (11), 1700.

Wohltjen, H., Snow, A.W., 1998. Colloidal metal-insulator-metal ensemble chemiresistor sensor. Analytical Chemistry 70 (14), 2856–2859.

Wu, X., et al., 2003. Immunofluorescent labeling of cancer marker Her2 and other cellular targets with semiconductor quantum dots. Nature Biotechnology 21 (1), 41–46.

Xu, J., Luan, Z., He, H., Zhou, W., Kevan, L., 1998. Chem. Mater. 10, 3690–3698.

Yaghi, O.M., et al., 2003. Reticular synthesis and the design of new materials. Nature 423, 705–714.

Yangping, S., Yang, J., Xinzheng, L., Chun, W., 2010. Mechanism and growth of flexible ZnO nanostructure arrays in a facile controlled way. Journal of Nanomaterials 2011.

Yi, G.-chul, Wang, C., Park, W.I., 2005. ZnO nanorods: synthesis, characterization and applications. Semiconductor Science and Technology 20, s22–s34.

Yu, D., Yam, V.W.W., 2004. Controlled synthesis of monodisperse silver nanocubes in water. Journal of the American Chemical Society 126 (41), 13200–13201.

Yu, W.W., et al., 2003. Experimental determination of the extinction coefficient of CdTe, CdSe, and CdS nanocrystals. Chemistry of Materials 15, 2854–2860.

Zhang, Q., et al., 2002. Thermal conductivity of multiwalled carbon nanotubes. Physical Review B 66 (16), 165440.

Zhang, Y., Yu, K., Jiang, D., Zhu, Z., 2005. Zinc oxide nanorod and nanowire for humidity sensor. Applied Surface Science 242, 212–217.

Chapter 5

Physical Techniques

CHEMICAL VAPOUR DEPOSITION

With thanks to Jason K Vohs. Adapted from Vohs et al. (1986)

Chemical vapour deposition, or CVD, is an invaluable tool in applying thin film coatings to surfaces. By surface modification it is possible to make films resistant to corrosion, impart specific electrical characteristics, make them conductive or even put an antireflective coating on them. CVD was first discovered by Mond in the nineteenth century after it was noticed that a nickel film would form upon the thermal decomposition of $Ni(CO)_4$ (Kumar and Ando, 2010). Since then the technique has been employed in the production of a diverse range of coatings. In this recipe, you build a CVD chamber and use it to coat the surface of a silicon substrate with a thin film of metal oxide. This is achieved by vaporising a volatile precursor in an inert gas and letting this pass over the surface of the substrate to be coated. If the surface of the substrate is hot, then molecules of the precursor chemisorbed to the surface can be made to react either by heat decomposition or by the addition of a second reactant. Any by-product from the surface reaction can then be carried away by the inert gas in an outlet so they will not foul the film. The majority of these processes may require specialist equipment but it is comparatively easy to make a setup capable of coating a substrate in aluminium oxide using easy-to-obtain materials. The first recipe deals with the manufacture of a simple CVD chamber that will be used to make an aluminium oxide thin film on a flat silicon substrate. After this, we will explore adaptations to the setup that will allow you to make silica and titania films on smooth substrates. Finally, we will explore the use of this technique to form replica coatings of butterfly wings, silks and other natural templates.

Normally to make an aluminium oxide film, a volatile small aluminium-centred molecule will be used as the precursor. $(Al(CH_3)_3)_2$, $AlCl_3$ and $Al(acac)_2$ are all routinely used but these require higher temperatures in order to work. These compounds have labile ligands that will easily undergo nucleophillic attack by water to eventually form an oxide. However, it also means that the precursor will react with moisture in the air easily and so you will need to work with the chemicals maintained under an inert atmosphere. In this case, the substrate will be a small piece of silicon wafer and the precursor will be trimethylaluminium, which can be vaporised easily in a hexane solution. This method is an adaption of the standard sol−gel mechanism of hydrolysis and condensation of an alkoxide to produce a metal oxide except that the experimental setup ensures that the condensation reaction occurs at the surface of the substrate being coated.

To use chemical vapour deposition to make an aluminium oxide film you will need:

1. A tank of argon gas as the gas supply fitted with a feed valve so that you may control the flow. Alternatively, you may be fortunate enough to have access to a fume cupboard fitted with an argon line, which should have controllable flow. *Make sure you are fully trained in the safe use of compressed gases before starting this experiment.*

2. Two gas flow meters, which ideally will have fittings to let you regulate the gas, flow through them. If these are unavailable, then you can use a flow indicator but this will lead to difficulty in reproducing the results.

3. Lots of gas piping and something to cut it with. 1−2 m might suffice but it will depend on how your setup is positioned. You will want stand clamps to hold everything in place and screw clips to hold the gas pipes onto the glass tubes and fittings.

4. Two 100-mL Schlenk flasks. Each should be fitted with a rubber bung having a glass tube stuck through it through it almost to the bottom of the flask where the argon will bubble through the aluminium precursor or water. The outlet neck on the side should also have a rubber bung with a glass tube fitting for attaching to a gas pipe.

5. A glass tube to act as the CVD chamber. A large glass test tube with the end removed will do (a glass workshop will be able to do this); at least a few cm in diameter is needed so you can get samples in and out. One end will have a rubber bung with one glass tube through it to act as the exhaust and the other will have a rubber bung with two glass tubes to act as the inlets for the water vapour and the isopropoxide vapour, respectively. The inlet tube for the water feed should be as close to the area to be coated as possible. This will help ensure that the hydrolysis reaction occurs at the substrate surface and not elsewhere in the chamber.

6. An oil bubbler for the exhaust to run through. This will help keep track of the gas flow through the chamber and prevent air from getting back in.

Nanotechnology Cookbook. DOI: 10.1016/B978-0-08-097172-8.00005-9

7. A way to heat the CVD chamber uniformly throughout. Electrical heating tape hooked up to a power supply is used in the source reference but this may not always be to have around. Use a temperature sensor to allow you to monitor the temperature of chamber. In this recipe, you will be heating the tube to between 150 and 200 °C, and maintaining tight control of this is important in controlling the deposition rate.

8. Two heating mantles with oil baths for heating the Schlenk flasks. This will allow you to more readily vaporise the reagents and can make the CVD proceed faster. They are not absolutely required in the case of aluminium oxide deposition but can help if you decide to use other precursors.

9. A silicon wafer that will fit inside your test tube for coating.

10. Assemble these items together in a fume cupboard as shown in the diagram. *You will not want to breathe in any vapour that makes it out of the reaction chamber so make sure the exhaust vents correctly!* Test the gas flow through the various fittings and ultimately the tube itself without the reagents in just to make sure nothing will pop off and that the flow rates are reasonable. By this I mean there should not be any vigorous escaping gas noises but rather a gentle flow. The actual gas flow rate will depend on what you find gives the best results. Typically, a feed rate of 3–6 cm^3 of argon per minute is suitable as measured at the exhaust.

11. A 2-M strength solution of $(Al(CH_3)_3)_2$ in hexane. The molecular weight of this compound is 144.18 g/mol and the density is 0.752 g/mL. Therefore, to make 10 mL of a solution of the correct strength you will need 2.88 g in 10 mL of hexane or rather 3.83 mL made up to 10 mL using hexane. This should be mixed under an argon atmosphere, which can be achieved by purging a Schlenk flask with argon and adding hexane by injection through a rubber septum cap. This can then be repeated using the $(Al(CH_3)_3)_2$. Once prepared, the flask should be kept separate from the CVD system until it has been purged of air.

12. Add ~20 mL of distilled water to the second Schlenk flask and fit it into the CVD setup. The water is in excess for reaction with the aluminium precursor (Fig. 5.1).

To use chemical vapour deposition to make an aluminium oxide film:

1. Thoroughly clean the silicon wafer. Rinse with pure water and acetone before allowing it to dry.

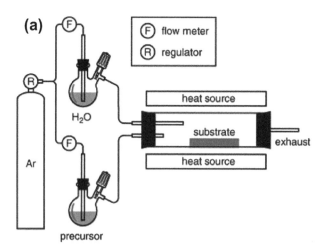

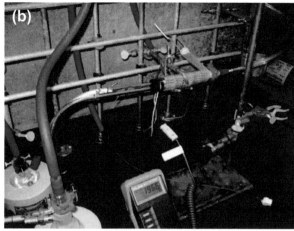

FIGURE 5.1   (a) A schematic of the experimental setup for chemical vapour deposition. (b) Photographs of the basic setup where heating tape is used to heat the deposition chamber. *(Reproduced with permission from Vohs et al. (2010). Copyright 2010 The American Chemical Society)*

2. Place the cleaned wafer into the glass tube so that the surface is facing up.
3. Purge the glass chamber and gas lines for ten minutes with argon to remove all the air. You will want all the air removed from the system. Bubbling argon through the distilled water will also help to degas it. The temperature in the chamber should be sufficient that you do not see any water condensation inside and certainly not at the surface of the substrate to be coated. The rate is variable but this will be somewhere in the region of $3-6$ cm^3/min leaving the exhaust.
4. Turn on the heating source for the chamber. Ideally, it will be 175 °C uniformly throughout the tube.
5. Make up the $(Al(CH_3)_3)_2$ solution in hexane while the CVD setup is purging. You can opt to use the pure precursor but it can be safer to handle it dissolved in hexane as it will fume less if accidentally exposed to moisture in the air.
6. After the system has been purged, add the Schlenk flask containing the aluminium trimethoxide and allow the gas to bubble through it. You may wish to sit this in a heating mantle as in the diagram but the coating will work well without heating.
7. Using the bubbles as a marker, adjust the gas flow in the water and the precursor so that they appear to be at the same rate.
8. A blue-coloured coating of aluminium oxide should form within thirty minutes to an hour. This will be quicker if the precursor and water solution are heated. Generally, if aluminium trimethoxide in hexane solution is used, room temperature will provide the best results.
9. You may stop the reaction at any time and do not have to let the entirety of the aluminium precursor solution evaporate. The deposition can be stopped by turning off the gas flow slowly and turning off the heating to the chamber. Once it has cooled to room temperature, you can analyse the film.

A white powder may form which is aluminium oxide that has not formed by growth on the surface. This can indicate that the CVD chamber temperature is too high. Films tend to be more homogeneous when they are grown slowly. Inconsistencies in the chamber temperature will also produce observable defects in the film and this can be solved by ensuring that heat is distributed evenly in the chamber. If there is too much water in the feed, then you will observe white fumes forming in the chamber also. This can be solved by lowering the rate of gas passing over or through the water. Experiment with changing the positions of the glass rods feeding into the chamber and precursors too as this will affect the local concentrations of water and $(Al(CH_3)_3)_2$. Films that adhere poorly to the silicon wafer could indicate that they are not clean enough.

At the visual level, you should see a thin-coloured layer on the surface. The tape test is also a very rapid assessment of how well the film is adhered to the surface. Applying some sticky tape on top of the coated substrate, you press down firmly for a few seconds before pulling it off. If nothing comes away, then the film is well adhered. This may seem like a quick and dirty test but it is in fact an industry-accepted method of assessment. In a similar vein, you can try drawing pencils across the surface of different hardness. If the pencil scratches the surface, then the film is said to be of the same hardness grade as that of the pencils (Fig. 5.2).

More in-depth analysis can be performed by atomic force microscopy (AFM) to examine the surface topology. If the coating produced is conductive, then you may wish to try conductive AFM techniques to investigate current flow in the coating surface. Larger areas can be examined rapidly by scanning electron microscopy (SEM) and this has the added advantage that most electron

FIGURE 5.2   A photograph of a silicon wafer with a thin film of aluminium oxide deposited on it. *(Reproduced with permission from Vohs et al. (2010). Copyright 2010 The American Chemical Society)*

microscopes allow you to perform energy dispersive X-ray (EDX) analysis, which will let you look at the elemental composition of the film. This is especially useful if you have doped the film to produce a semiconductor. For a more quantitative assessment of the elemental composition, X-ray photon spectroscopy (XPS) can be performed on the film.

Alternative precursors will behave differently due to having different vapour pressures and boiling points. Instead of using $(Al(CH_3)_3)_2$, you might wish to try aluminium tri-isopropoxide. This is an alkoxide precursor, which is not as pyrophoric as the trimethylaluminium, and can be handled in air without fuming (although this will spoil the chemical). You will find that it will need to be heated in order to be deposited in the CVD chamber somewhere near its boiling point (125 °C). By first experimenting with slow variations on the aluminium precursor, you can expand into making other metal oxide films. $Si(CH_3)_4$ and $Ti(CH_3)_4$ can also be used to form TiO_2 and SiO_2 films with little adaption to the CVD setup. Titanium, zirconium and silicon oxide films can be readily prepared by also using the respective elemental alkoxides. As with the aluminium isopropoxide, these may require heating to get the alkoxides to vaporise. More volatile precursors, such as $SiCl_4$ and $TiCl_4$, can also be attempted, though these are very reactive with water and are probably best attempted after you are experienced with the safe handling and use of them. As with most experiments, it will require much tweaking of parameters, such as the temperature, gas flow, tube positions and water to reagent ratios, to get a good film.

CVD USING SILICA OR TITANIA PRECURSORS

Adapted from Taylor and Timms (1997) and Cook et al. (2003)

In this recipe, hydrogen peroxide is used as an oxidiser to form a SiO_2 coating from $Si(CH_3)_4$ vapour on biological templates of interest. The original reference uses SiH_4 as a silicon source but this is more pyrophoric and therefore less safe to work with. The beauty of this technique is that the substrate is maintained at only 5 °C, meaning that biological samples of interest can be templated easily and with a high degree of replication. The precursor and oxidiser are flash evaporated into the coating chamber at the same time. The product of this reaction sprays across the surface of the template before the small molecules of forming glass have time to condense into large lumps — a bit like spray painting a wall and then letting it dry. Under the low vacuum, these monomeric units can easily penetrate into high aspect ratio features to form a uniform layer over even rough surfaces:

$$Si(CH_3)_{4(g)} + 4H_2O_{2(vapour)} \rightarrow SiO_{2(s)} + 2H_2O_{(l)} + 4CH_3OH_{(g)}$$

It is important to keep the template cool as this prevents the glass from condensing too quickly and solidifying before it has a chance to coat the substrate evenly. Furthermore, biological structures with protein or carbohydrate surfaces are rich in surface oxygen through which the silica can bind through Si—O bonds. This helps the formed film of glass adhere to the substrate ensuring that delicate samples do not crack upon removal from the chamber. Additionally, films formed by this method experience less shrinkage upon heating. 'Wet' coating methods of intricate samples, like a butterfly wing, with sol—gel precursors can experience catastrophic cracking after drying. During drying, a sol—gel glass becomes denser as more Si—O—Si bonds form and water is generated as a by-product. If the glass is heated (or sometimes even left at room temperature), the trapped water in the structure is forced out generating stresses, which result in cracks and the glass falling apart. Controlled and slow drying is one method to get around this problem in water-based sol—gel methods. By using CVD at low vacuum, the excess water is driven off before the glass is even fully formed over the surface of the butterfly wing. This means that if the butterfly wing substrate is removed by burning it at ~300 °C, the surface coating does not crack and you will be left with a good replica of the butterfly wing made of pure silica.

This recipe will require some vacuum line techniques, which are described in brief here. Some of the vacuum techniques used will depend on what equipment you have been able to get together. However you manage to cobble a vacuum line together suitable for the delivery of your precursors, the most important aspect of this system is the deposition chamber. You will require a way of cooling the substrate in the chamber and you will also need a flash evaporator. A flash evaporator is where a liquid is exposed very suddenly to a vacuum (sometimes with the aid of some heat) which causes it to instantly vaporise. Schematics are provided for a glass-based flash evaporator, though you are not limited to this material, and PTFE-lined aluminium can also be used for construction.

For the chemical vapour deposition of silica at low temperatures you will need:

1. A glassblowing workshop. This is not absolutely required but most chemistry departments should have an on-site workshop that can make things for you. Rough schematics are provided below for the main sections of the setup. The dimensions will change depending on what glassware and tubing are to hand.

   The precursor line: In order to deliver the volatile precursor into the flash evaporator and chamber, it will need to be stored in a 1-L glass bulb fitted with a tap valve. This presents us with a problem in that we want only the precursor to be present without any other gas. How do we get the precursor into the 1-L bulb before we attach it to the gas line? The vapour pressure of $Si(CH_3)_4$ at 20 °C is 80 Pa which means it will still be a liquid under atmospheric pressure. If the precursor is added to a bulb containing an inert gas such as argon and this is cooled in liquid nitrogen, it should be possible to evacuate the gas while the cold precursor remains liquid. The glass bulb containing the precursor is fitted to the line, which will already be under vacuum between 100 and 500 Pa and the tap slowly opened. A pressure increase will be observed in the fitted manometer (a device for sensing the pressure) proportional to the amount of precursor in the vapour phase. The precursor can be released into the deposition chamber through the needle valve at a controlled rate. This rate can be calculated by the speed of the pressure drop in the manometer. Glassware should have strong connections sealed with silicone-based vacuum grease. *Many of the precursors used in this system will burst into flames if exposed to air! Take every precaution possible to make sure this does not happen!* The flash evaporator: This is basically a heated borosilicate glass tube 35 mm in diameter into which the hydrogen peroxide is drawn and subsequently vaporised. The vapour will be ejected through a glass nozzle where it will mix with the silica precursor to form a spray pointed at the substrate. The schematic shown has two inlets for different types of precursors. For the volatile silane precursor discussed in this recipe, the inlet is through a polyethylene tube 1 mm in diameter at the outlet that is fitted inside the outlet of the flash-evaporated hydrogen peroxide to form two concentric nozzles. Less volatile precursors, such as those for forming titanium and aluminium oxides, can be added through a second intake on the side to which heat may be applied to aid vaporisation. You will also need a 75-W electrical heater to surround the tube and maintain it at 95 °C.

The deposition chamber: This will be a larger glass tube, which will stand vertically plugged on top of the flash evaporator. The chamber will have an outlet to the vacuum line that should run through the cold trap to take any products or reagents out of the flow and prevent them from damaging the pump. There should also be a connector fitted to a pressure gauge to let you monitor the pressure in the chamber. The main feature is the sample holder, which will hold the substrate between 3 and 6 cm from the spray nozzle. Depending on the setup this should be adjustable in height, which can be achieved by having an adjustable sliding seal as outlined in the diagram. At the end of the rod, you can use a small metal plate as the target and use either clips or a pin to hold samples in place upside down. In a pinch, even a paper clip can hold a sample in place. As mentioned previously, the coating is of a greater quality if substrates are cooled. Although this may not be a problem for the deposition of a silica film on a butterfly wing, it can be an issue with flat surfaces. You may wish to adapt the schematic provided to include a water-cooling coil within the glass holder rod and have this tipped with a copper plate to work as a heat sink (Fig. 5.3).

2. A vacuum line or pumps fitted with cold traps. The pumps can be oil diffusion pumps but they must use a silicone-based lubricant. Rotary vane pumps can also be used. They must be capable of evacuating to a pressure between 100 and 500 Pa. The cold traps can be cooled using a dry ice and ethanol mix. Install isolation valves on the cold traps so that you can remove them from the system without breaking vacuum if need be. *Remember to clean these out after the chamber is represssurised!*

3. A manometer to connect to the chamber and another connected to the precursor delivery setup. There are many different types available but a mercury-based column is suitable and reliable and it should be easy to find one that can be connected by quick-fit glassware. If you can afford, you may wish to use an electronic capacitance manometer.

4. A substrate for templating. A butterfly wing is a colourful demonstration of the coating and the changes in the optical interference patterns observed that make for a quick assessment of how the experiment has gone. Fly wings, lotus leaves, silk fibres and even stockings can all be mounted in the chamber for coating. Try anything that you think might be interesting.

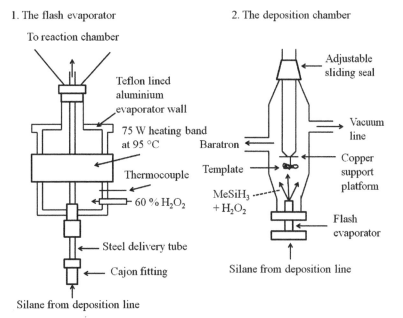

FIGURE 5.3 Schematics of the various components required for the chemical vapour deposition of silica.

5. A 60% solution of hydrogen peroxide in water. *Keep in the dark and keep it cold when not in use. This will degrade over time to produce H$_2$ gas; so never allow pressure to build up inside the containing vessel.* This will be placed into a graduated burette so that the rate at which it is used can be determined.

6. A silica precursor. In this recipe, we will use Si(CH$_3$)$_4$. *This chemical is very flammable and volatile. It boils at 26 °C and should be stored in a special container at 2−4 °C when not in use.*

7. At least 1.5 m of PTFE tubing having an internal diameter of 0.15 mm. This will be used for delivery of the hydrogen peroxide into the CVD chamber. You will also require polyethylene tubing having a 1-mm internal diameter in the construction of the flash evaporator.

8. Lots of vacuum tubing, fittings, silicone vacuum grease and taps.

9. A nitrogen or argon gas line for purging the equipment after use.

10. An oven capable of reaching up to 600 °C for removing organic substrates and a crucible for calcining them in.

For the chemical vapour deposition of silica at low temperatures:

Before you start, test the setup by placing it under vacuum and making sure nothing implodes and that there is no part that sucks in air. *Some volatile precursors will combust spontaneously with oxygen and water.* Although you are using a strong oxidising agent mixed with water for the reaction, you want this to take place in the nozzle and deposition chamber and nowhere else.

1. Evacuate the precursor line. This will feed into the flash evaporator by a needle valve used to control the flow of precursor. This should be closed as you evacuate the precursor line.

2. Attach the glass bulb containing the precursor to the precursor line. Open the valve to allow the precursor into the line ready to be dispensed. Make a note of the pressure on the manometer.

3. Affix the substrate to be coated to the glass rod and place it into the coating chamber. The substrate should be ~5 cm from the flash evaporator nozzle. This distance will vary depending on what you find gives the best results.

4. Warm up the flash evaporator using the electrical heater. Plug the coating chamber on top of it.

5. Evacuate the coating chamber slowly to make sure that the substrate is not deformed.

6. Connect the PTFE tube to the end of the burette full of hydrogen peroxide solution. This should be drawn through an adjustable tap so that the rate can be controlled. If the tap is open when you connect it to the vacuum, the solution will soak straight in! Make sure that this valve is closed before you start. Draw through the solution so that it fills the tube and there is no air in it.

7. Open the needle valve from the precursor line. You will observe a drop in pressure, which will correlate to how much silane is being drawn through. By adjusting the valve, you can adjust the flow rate.

8. As both solutions are drawn through by the pump evacuating the chamber, the substrate will begin to coat. The first few times you try this you alter the flow rates so that the silane to peroxide molar ratio is 1:6 per unit of time. The pressure in the reaction chamber while the reaction occurs should be somewhere around 200−250 Pa with the vacuum pump still on.

   From the literature, the ideal flow rate of hydrogen peroxide is 1 mmol/min (that does not include the water). The flow rate of silane precursor should be 0.16 mmol/min. A good coating will use up 1.5 mmol of hydrogen peroxide and 0.25 mmol of precursor. These rates should grow a silica film on the substrate at a rate of 50−200 nm/min. Hopefully, you will not see much happening in the chamber during the process but the reaction of the peroxide with the silane precursor can be *extremely vigorous!* If the flow of hydrogen peroxide is too high, you will see misting in the chamber and possibly even the precursor igniting. During initial runs you will want to turn up the hydrogen peroxide feed slowly to prevent this happening.

9. To end the coating reaction shut the valves supplying the hydrogen peroxide solution and the precursor, respectively. The precursor line should remain at vacuum and should be isolated from the deposition chamber and flash evaporator line.

10. Turn off the pump supplying vacuum to the deposition chamber and vent the chamber slowly to atmosphere. Remove the coated sample and replace the holder rod into the system.

11. Turn the vacuum back on and purge the flash evaporator and deposition chamber with N$_2$ or Ar gas. If you have used up the precursor, then the precursor line may also be purged.

12. Remember to clean out the cold traps. Isolate them from the system by valves so that you may maintain vacuum in other lines.

13. At this point, you should have a silica-coated butterfly wing. It should be stiffer than when it went in but still retain an iridescent sheen in light. The film of silica should have perfectly 'wetted' the surface and should not peel away from the template on handling. This can be analysed now or you can burn away the template by following the next steps.

14. Place the coated butterfly wing into a clean crucible for use at high temperature. The sample should be flat and not resting on an edge. With the organic component removed, the remaining replica will become brittle. If the sample is resting on an edge, this can cause breaks or fractures from its own weight.

15. Programme the oven (if you can) to raise the temperature by 1 °C/min to 500 °C in ambient atmosphere and hold it there for 12 h. This should ensure that the organic layer is burned away to gaseous products whilst leaving the inorganic matrix intact. The slow heating rate minimises damage to the replica during calcination. During heating, the glass will form more Si−O−Si bonds, which makes the silica contract slightly and generates water. If the heating rate is too fast, the stress generated by this process will fracture the replica.

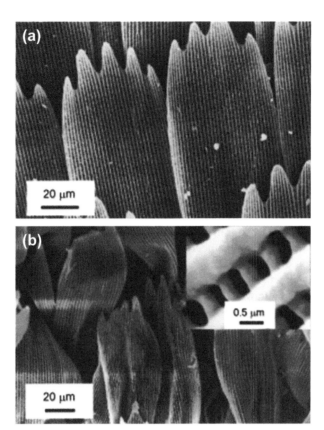

FIGURE 5.4   (a) Scanning electron micrographs showing the details of a peacock butterfly wing. (b) Calcined silica replica showing a slight degree of creasing as a consequence of heat treatment. Inset: high-magnification image. *Adapted with permission from Cook et al. (2003). Copyright 2003 Wiley-VCH*

It is important to remember that some metal oxides will change their crystal structure upon heating. A titania film deposited in this manner will be amorphous initially but will turn into the crystalline anatase phase if heated to 500 °C. For example, a titania film replica of a lotus leaf would have water-repellent properties. By heating this to 500 °C, it would convert to the photocatalytic anatase phase of titania and can now not only repel water but also destroy contaminants on the surface. This is the basis of a self-cleaning glass, though commercially these windows are not manufactured from templated butterfly wings!

16. Allow the oven to cool and remove the replica.

The fidelity and thickness of the produced structures can be performed by scanning electron microscopy. You will notice that calcined samples may have distorted slightly from densification but will hopefully not have cracked. If the template you use is a butterfly wing, you should still be able to see the iridescence. Samples can be analysed by SEM to examine the morphology. You can also run a solid-state diffuse reflectance spectrum to observe the optical characteristics. There will be a noticeable difference in wings that remain coated and once you have decided to calcine. Interestingly, the spectrum for a calcined glass replica will more closely resemble the spectrum for the native butterfly wing (Huang et al., 2006). This is because in the coated film the regular spacing, giving rise to light interference, has been slightly decreased by the thickness of the glass film. In the calcined replica, this periodicity is restored in the absence of the wing underneath. By further adaption of the setup, it is possible to introduce gas mixes into the precursor line in order to form mixed metal oxides. Boranes, phophanes and germane have all been used in this setup. Al_2O_3 films can be formed by using $(Al(CH_3)_3)_2$ as the feed precursor and adding it through the heated side nozzle in the flash evaporator. Heat must be applied to aid volatility of the $(Al(CH_3)_3)_2$ or else a 2-M mixture in hexane can be tried. Titania films can be formed using $Ti[N(CH_3)_2]_4$ and replacing the hydrogen peroxide with water. The yellow orange precursor will need to be heated above 50 °C to get it into the vapour phase and it is not always easy to get a smooth flow. Some fuming in the chamber will be observed during the reaction. Similarly, by replacing the hydrogen peroxide feed with ammonia, it is possible to deposit a titanium nitride film (Fig. 5.4).

ATOMIC LAYER DEPOSITION

Adapted from Higashi and Fleming (1989)

Atomic layer deposition (ALD) is a more refined form of CVD and analogous to the sol–gel method of sequential dip coating. It was first described as molecular beam epitaxy in the production of tantalum oxide thin films, though an earlier publication by Suntola had shown it could be used for forming zinc telluride films (Ahonen et al., 1980). The ALD technique is very important in the production of semiconductor thin films prepared from group II—VI elements for microwave and optoelectronic devices. ALD is also used in the formation of precisely controlled nitride films for corrosion resistance and increased toughness. One advantage of ALD over CVD is that laminar films composed of different materials may be deposited on top of one another to give rise to unique mechanical, optical or electronic properties. It is even possible to add long-chain organic molecules to the surface of a completed film in order to render it hydrophobic — a functional coating on top of a functional coating! Following on from the example used in the CVD recipe we shall continue with using a metal oxide as an example of a material deposited by ALD. A metal oxide precursor, such as $(AlMe_3)_2$, is allowed to vaporise into a chamber at low vacuum containing the substrate to be coated. The substrate, let us say a silicon wafer, has been chemically passivated before so that the surface is coated in hydroxyl groups and is maintained at a high temperature to make it reactive. A molecule of $(AlMe_3)_2$ dissociates into discrete $AlMe_3$ units and chemisorbs to the surface of the silicon. The surface hydroxyl group loses a hydrogen atom as the central Al atom undergoes nucleophilic attack by the lone pair on the oxygen of the surface-bound hydroxyl group. The displaced methyl group coordinates to the lost hydrogen to form a molecule of methane. This exposure to the $(AlMe_3)_2$ vapour is brief but timed correctly so that the entire surface of the silicon wafer is coated with $AlMe_2$ units. Since these will not react with $AlMe_3$ in the vapour phase, the coating on the surface is exactly one molecular layer thick. The unreacted $(AlMe_3)_2$ is then flushed from the system by a carrier gas or simply by removal with a vacuum pump and a pulse of water is allowed to vaporise into the chamber. This then hydrolyses the surface-bound $AlMe_2$ molecules to form Al—O—Al bridges and Al—OH units on the surface. The reactions for this process are given below. An excess of water present will not further react with the surface once it has been completely hydrolysed and so there is no further reaction within the chamber. The excess water and methane product are then drawn from the chamber.

The direct hydrolysis reaction of $AlMe_3$ in water is very exothermic — it will become hot. Remembering that $AlMe_3$ exists as a dimer

$$2Al(CH_3)_{3(g)} + 3H_2O_{(g)} \rightarrow Al_2O_{3(s)} + 6CH_{4(g)}$$

In the ALD reaction, the surface of the substrate is coated with a molecular layer of $AlMe_2$ bound by surface oxygen to the substrate:

A. $O-Al(CH_3)_{2(\text{surface bound})} + 2H_2O_{(g)} \rightarrow Al(OH)_{2(\text{surface bound})} + 2CH_{4(g)}$

B. $Al-OH_{(\text{surface bound})} + Al(CH_3)_{3(g)} \rightarrow Al-O-Al(CH_3)_2 + CH_{4(g)}$

It is important to remember that reactions within the chamber occur only at the surface of the substrate and nothing is formed in gas-phase reactions away from the surface as occurs with CVD. By repeating this binary reaction many times, a film of Al_2O_3 can be grown on the surface with atomic precision and with a high conformity to surface features. Growth rates are typically of the order of 1 Å per cycle, which may seem slow but films hundreds of nanometres thick can be grown in just a few hours as the cycles can be very fast. Reagents are allowed to vaporise for $1-2$ s before being purged after around 5 s. This is more than sufficient for molecularly thick coating to occur at the surfaces. In addition to the benefit of very precise control over the thickness of the films formed and the fidelity of feature replication, the low pressure allows the vaporised metal oxide vapour to penetrate into the tiniest of features that it might otherwise be excluded from by capillary forces under normal atmospheric conditions. This means that ALD can be used both to coat temperature-sensitive and complex templates, such as butterfly wings, and to form inverse opals of TiO_2 for photovoltaic applications or inverse opals made of tough WN (King et al., 2006; Rugge et al., 2003; King et al., 2006; Rugge et al., 2003). The technique has also been successfully applied to the growth of nanometre-thin Pt and Ir (Christensen and Elam, 2010). Given that this technique has angstrom-scale control, it can be applied successfully to nanomaterials and this has been demonstrated in the growth of Al_2O_3 on ZnO nanowires and also in the growth of zinc oxide on single-walled carbon nanotubes (Lin et al., 2011). It is also possible in most cases to form the metal oxide at relatively low temperatures. For Al_2O_3 formation, the activation energy is low and so the oxide will form as low as 100 °C. Most ALD reactions take place between 200 and 400 °C. If the reaction temperature is too high, the bonding of the volatile precursor at the surface cannot be sustained. If the temperature of the substrate is too low, then the activity of the chemisorption sites will decrease resulting in a reduction of the deposition rate.

The first synthesis of Al_2O_3 by ALD was performed by Higashi and Flemming at Bell Laboratories in 1989 (Higashi and Fleming, 1989). Their setup is described here, although it is not something easily accomplished with components found in most laboratories. If you wish to try ALD yourself, it can be easier to buy a readymade setup from a supplier who specialises in this equipment. However, if you are fortunate enough to have some of the components and some expertise to hand, why not have a go? This is more of an abbreviated overview — please see references for more in-depth explanations of how to go about making an ALD machine. It will be a lot of work:

1. A stainless-steel vessel that can be evacuated. It will need two inlets for the water and the precursor. It will also need to have a raised cavity on which the substrate for coating can sit on the inside of the chamber and on which an electric heating coil can

sit underneath on the exterior. It will need to have a vacuum gauge and temperature sensor fitted. If these can be linked to a computer, this is preferred. The smaller the chamber size, the better, although this will depend largely on what is to be coated. Remember, during each cycle there will be an excess of reagent in the chamber. A smaller volume will make it easier for this to be removed.

2. At least two solenoid valves with rapid opening and closing (on the order of milliseconds) that can be computer controlled. These will be used to open the vacuum chamber to the precursor solutions.

3. Suitable pipes and fittings for working with vacuum lines. All components find it favourable to operate in the 1×10^{-5} Torr range (roughly 0.0015 Pa).

4. Vacuum safe flasks into which you will be placing the precursor.

5. Ultrapure $(AlMe_3)_2$ and water for the reaction.

6. A vacuum pump that can monitor pressures. It should be fitted with filters to catch any oil vapour it might give off. Ideally, one that can be controlled by a computer using a feedback loop will give you a greater control over the deposition conditions.

7. An electric element heater with power source capable of reaching 500 °C to be placed externally under the substrate.

8. A computer that can be connected to the pump, the valves and any sensors on the setup. To correctly time the release of the precursors and evacuation cycles, you will need to make a program to control the valves and monitor the temperature and pressure. LabVIEW is a commonly used software package for use with laboratory equipment that will let you custom design your own program for controlling the deposition cycle.

9. A p-type Si-(100) silicon wafer. The (100) refers to the crystal plane presented at the surface on which the aluminium oxide will grow. The plane presented at the surface is important if you are going to etch the silicon surface as different crystal faces dissolve at different rates. In this particular recipe, it should not make much of a difference. The p-type Si means that it is a p-type doped silica and that it is a semiconductor. You may want to try other substrates but this is what was used in the original preparation.

10. The reagents and facilities to properly clean the silicon wafer to clean room standards. *See the section on cleaning with aqua-regia given elsewhere. Safety procedures apply* (Fig. 5.5).

To deposit the Al$_2$O$_3$ on silicon using atomic layer deposition:

1. Clean the silicon wafer as described elsewhere and place it into the ALD chamber.

2. Turn on the vacuum pump and allow the chamber to reach ~1×10^{-5} Torr. Then turn on the heater for the substrate and allow it to reach 450 °C. While the reaction is running, the temperature can be as low as 100 °C without affecting the film deposition. However, the best interfaces for electronic applications have been found to form when the substrate is maintained at 450 °C. The pump will remain on throughout the coating cycles.

3. The solenoid valve leading to the $(AlMe_3)_2$ reservoir is opened for 6 s. This will flood the vacuum chamber with vaporised precursor, which will form a monolayer at the surface. After 6 s the valve is closed and the chamber evacuated for a further 20 s to remove any remaining precursor not bound at the surface.

4. The solenoid valve leading to the water reservoir is opened for 0.25 s and then shut. This will flood the chamber with an excess of water vapour, which will react with the surface bound aluminium precursor. The pressure in the chamber will rise to ~10 mTorr and, depending on the pump strength, the pressure should then fall once more to 1×10^{-5} Torr. Depending on the size of the chamber, this will take 5–10 s to remove the water from the chamber. Although these are rough parameters, it is only important that all of the water (or aluminium precursor) has been flushed from the chamber before the next cycle.

5. The solenoid valve leading to the $(AlMe_3)_2$ reservoir is opened for 0.25 s to flood the chamber with precursor. As with the previous step, the chamber is once more allowed to evacuate.

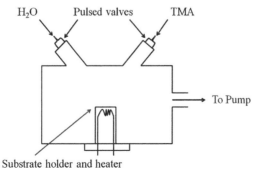

FIGURE 5.5   A schematic of a basic atomic layer deposition chamber.

6. Steps 4 and 5 are repeated as required. For each cycle of precursor and water used, the film will grow by 1 Å. This means that the film growth will occur at a rate of roughly 1 nm/min depending on the conditions.

7. Once the desired number of cycles has been performed, allow the final cycle to evacuate completely before turning off the heating element and allowing the chamber to cool while still under vacuum.

8. Bring the chamber back to atmosphere and remove the now-coated substrate.

Films can be analysed by performing capacitance voltage scans if the metal oxide has been grown on a conductive substrate and forms part of a p–n junction. This is an electrical technique whereby the coated silicon is placed on top of a metal contact. An electrode is then placed on the coated side of the film in an electrolyte solution. A voltage is applied across the coated silicon, which causes some of the electrons in the metal oxide film to be drawn away from the surface. In a semiconductor, the electrons rest in a valence band or a conduction band. The application of a voltage across a semiconductor film causes these bands to contract to form what is known as a 'depletion zone' near the surface of the film. By analysing the size of this depletion zone at various charges, the bandgap of the metal oxide film may be calculated. Additionally, it is possible to oscillate the applied voltage and observe the response times of the electronic structure to calculate doping parameters, Schottky defects and electrical permittivity using the Mott–Schottky equation.

Because of the homogeneity of the produced films and the uniform surface coating, it is possible to experiment with substrate replication from the nanometre to the micron scale on a range of substrates. In terms of further modification, this technique is not limited to the production of purely inorganic films; coatings using organic molecules can also be applied. This is invaluable in chemically tailoring a surface to produce organic/inorganic composites. This technique is called molecular layer deposition (MLD) and is slightly different from ALD in that the precursors generally tend to be homobifunctional molecules (Yoshimura et al., 2007). Early work demonstrated that polyamides and polyimides could be built up on a surface substrate by the sequential reaction of di-acid chlorides with diamines. In this manner, it has been possible to make films of nylon as a homogeneous surface coating (Du and George, 2007).

PHOTOLITHOGRAPHY PATTERNING

With thanks to James Vicary and Rebecca Boston

Photolithography is the method of using light to transfer a pattern onto a substrate, which can then be further etched into the material by other means. A substrate, normally a silicon wafer, is coated with a thin layer of a photoresist, which is a photosensitive polymer. The technique is widely used in the microelectronics industry to produce microprocessor chips. Microchips are composed of highly dense arrays of semiconductor transistors made by patterning doped silicon wafers using photolithography. As the demand for ever-smaller and more powerful microchips in computers and consumer electronics becomes greater, there is a pressure to produce arrays of single transistors in the nanodomain. The technique is not limited to the production of electronics and for a nanotechnologist it is a useful tool for fabricating nanoscale components using a top-down approach. It is used in the production of lab-on-a-chip devices and is the major production method for cantilever tips used in atomic force microscopy. Manufacturing an item by photolithography can take many tens of repeated cycles of lithography and etching; so, in this section, we will detail the major techniques of one etching cycle. This includes preparing a wafer, spin coating a photoresist onto it, exposing and developing the photoresist and plasma and chemical etching. The equipment used in a microfabrication laboratory is highly specialised; so it is unlikely you would be able to perform the procedures below in a normal laboratory. However, it is possible to outline some general practices so that the reader has a good idea of where to start if they wish to work with photolithography.

The examples given here are taken from various points in a 36-stage preparation method for manufacturing silicon nitride AFM cantilevers. In brief, a wafer of silicon with a thin coating of silicon nitride is cleaned, dried and then further coated in a thin layer of photoresist polymer. This is then patterned using a light source by shining it through a chrome mask patterned on a quartz wafer. These masks can be custom ordered from specialist suppliers and they are made using a range of methods depending on the size range of what you want to do. Chemical sputtering and etching can be used for making micron-scale features similar to making a circuit board for electronic components. Larger feature masks can be made quickly and cheaply by printing a pattern onto a gelatine film. This approach is often used for the microfabrication of lab-on-a-chip devices. If you plan to make a mask in the nanometre regime, then laser cutting, ion beam milling and electron beam lithography can be used, although the mask will become more expensive as the features become smaller. The mask is aligned over the substrate accurately using an alignment machine, which also holds it close to, or even on top of, the substrate. The first etching stage will be to put marks onto the substrate surface, allowing each subsequent step to be properly aligned. In a procedure with multiple etching steps, it is vital that the alignment between mask and sample is correct every time or you might end up dissolving away parts you wanted to keep. Strong light is then used to transfer the mask pattern onto the photoresist layer. The photoresist is developed chemically and unwanted photoresist is rinsed away to give exposed areas of substrate in the mask pattern for etching. The etching is done using reactive ion etching or chemical dissolution of the surface with a strong KOH solution. Both of these methods will be expanded on. Depending on what you are making, you may also need to evaporate metal layers onto

FIGURE 5.6 A researcher prepares to go into a clean room environment wearing coveralls and shoe covers.

the surface at various stages or flip the wafer over so that you can work with the front and back sides. For example, many AFM cantilevers have reflective gold coatings a few nanometres thick on the back so that the reflection of the laser is stronger. Putting something together by microfabrication can be a difficult and long process and careful consideration of how all the steps will influence each other is required for good results.

CLEAN ROOM PRACTICES FOR MICROFABRICATION

If you are using photolithography, you have to be in a clean room environment working under filtered lights that do not have a UV component. This ensures that the photoresist used does not spoil. For microfabrication procedures, it is important to maintain a dust-free environment. This means before you enter the laboratory you will have to put on a clean suit, hat and boots before you enter. Gloves should be worn at all times and for times when you will be up close to the wafers, you will need a facemask. High-grade laboratories will even have an airlock-type arrangement to prevent dust from getting in. The largest potential source of dust comes from shoes and so a procedure for putting on and taking off shoe covers should be put in place. This can involve a 'barrier' bench and a clean mat arrangement before you move through the laboratory door. As you are straddling the bench to get ready then a foot over the mat must be covered up by a protective layer. A foot on the other side in the unclean area should be a normal shoe. This avoids treading outside dirt into the clean room environment (Fig. 5.6).

CLEANING THE SUBSTRATE

Samples for photolithography must be flat and ideally surface roughness root mean square (rms) values on the subnanometre region are required. This ensures that the photoresist coats the surface evenly and that the etching occurs at identical rates for different points of the surface. Any dirt on the surface of the wafer will affect both photoresist coating and etching, so substrates must be extremely clean before use. This is accomplished by immersing the substrate wafer in sequential baths of methanol, acetone and isopropylalcohol and sonicating it each time. After solvent washing, the wafer is rinsed with water and blow-dried using a filtered stream of nitrogen. To ensure that the substrate is completely dry, the wafer is treated in an oven. Any moisture still remaining at the surface may influence the uniformity of the photoresist layer. The chrome and quartz mask will also need to be cleaned in a similar way, although one acetone rinse and oven dry will be sufficient.

To clean a silicon nitride wafer and mask you will need:
To be conducted in a fume hood.

1. The silicon/silicon nitride substrate and the chrome and quartz mask. PTFE tweezers for handling the wafers are recommended but make sure that they are wide nibbed or you could chip the wafer.
2. A clean glass beaker and a clean glass Petri dish that can sit inside it. Depending on the size of the wafer or mask, sitting the wafer in the Petri dish and then placing this in the beaker can make it easier to handle.
3. An oven maintained at 90 °C
4. An ultrasonic bath in which the beaker can sit.
5. Stock amounts of methanol, acetone and isopropanol. You will also need water for rinsing.
6. A dry nitrogen line and gun for drying the wafers.

To clean a silicon nitride wafer and mask:

1. Place the wafer into the glass Petri dish and then add some methanol. Fill the beaker halfway with methanol and place the Petri dish and wafer into it.
2. Sonicate the beaker and contents for 3 min.
3. Exchange the methanol for fresh methanol and sonicate again for 3 min. Repeat once more.
4. Repeat steps 1–3 with acetone instead of methanol.
5. Repeat steps 1–3 with isopropanol instead of acetone.
6. Rinse the wafer using water then dry using the nitrogen gun.
7. Dry the wafer in the oven for 10 min.
8. Place the mask plate in a beaker full of acetone and sonicate for 15 min.
9. Rinse the mask plate with water then dry with the nitrogen gun.
10. Dry the mask in the oven for 10 min.

When the mask and wafer are dry, you will be able to start using them in further processes.

RESOLUTION AND PATTERN TRANSFER ONTO A PHOTORESIST LAYER

In order to pattern a surface coated in a photoresist layer, strong ultraviolet light from a mercury or xenon lamp is projected through a patterning mask to create a light and shadow pattern on the substrate. Similar to light focussed onto photographic film in a camera, the pattern projected onto the photoresist layer causes a chemical change so that when the substrate is exposed to a developing solution the photoresist will be washed away in some areas and remain on the silicon surface in others. The photoresist layer that remains will protect the silicon from a subsequent chemical, plasma or ionic etching step that eats into the silicon surface. After the etching step the hardened photoresist layer can be chemically removed using a solvent to give a silicon substrate with the desired pattern engraved into it. To liken the process to photography once more, the photoresist layer can give a negative image with respect to the patterning mask or a positive image identical to the patterning mask. A negative photoresist means that areas exposed to light during patterning will become insoluble in the developing solution. A positive photoresist means that the areas exposed to light become soluble and will be rinsed away during developing (Fig. 5.7).

The resolution limit of the transferred pattern is determined by the wavelength of light, the dimensions of the mask through which it is projected and the optical aperture size. The current industry standard is to use light having a wavelength of 193 nm and because this is in the UV range it is ideal for inducing chemical reactions in the polymers used as a photoresist. However, the resolution limit on the best systems is down to 50 nm. The resolution limit on a photolithography setup is given by the following equation (Madou, 2002):

$$2b_{\min} = \frac{k_1 \lambda}{NA}$$

$$NA = 2\frac{r_0}{D}$$

where $b_{\min}$ is the minimum feature size obtainable, λ is the wavelength of light and NA is the numerical aperture, which is a number that represents the distance and angle of the projection lens. The numerical aperture is calculated from the radius of the lens, r_0, and the distance of the lens from the focal point, D. It can be seen by inspection that this is similar to the relationship of wavelength and numerical aperture for microscopy:

$$d_{\text{res}} = \frac{\lambda}{2NA}$$

The resolution can therefore be improved by using light of a shorter wavelength in a projection system with a large numerical aperture. Using higher-energy light gives rise to other issues as the masks can behave as diffraction gratings. In addition, different chemicals with sensitivity to high-energy light must be used as the photoresists. Many tricks can be employed in getting better resolution including engineering masks that incorporate the diffraction pattern into the design, changing the refractive index of the lithography medium and light phase-shifting masks. In the procedure listed below, the mask is actually placed in contact with the coated wafer so that the pattern will be a direct transfer.

Before the lithography can take place, the photoresist layer must be deposited onto the surface. To achieve this, the wafer is spin coated with a primer layer followed by the photoresist. The primer is normally composed of hexamethyldisilazane, which improves the interfacial adhesion of the inorganic substrate with the organic photoresist layer. The photoresist is then added on top and spun into a thin film. In this recipe, we will use a positive photoresist as these tend to give better resolution than

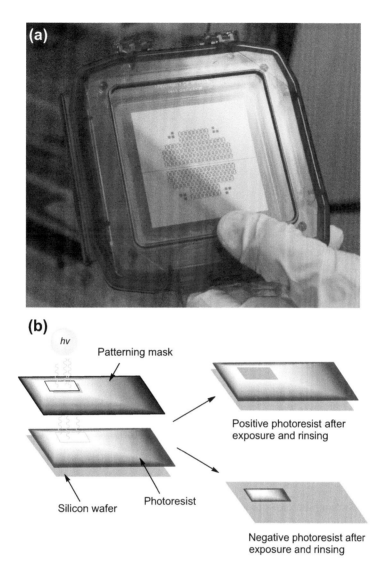

FIGURE 5.7   (a) A chromium mask printed on quartz. (b) A schematic of the photolithographic process using positive and negative photoresist layers.

negative ones. The major component is a phenol–formaldehyde polymer (Novolac) resin, which sets hard on the surface. The spin-coated wafer is 'soft' baked in an oven to remove solvent from the photoresist and to harden the resin ready for exposure. Full hardening of the resin layer is done after the developing stage so that the patterned photoresist area becomes resistant to the physical etching procedure. Dissolved into the resin is a diazonaphthoquinone that is normally insoluble until exposure to light induces a Wolff rearrangement to form a ketene and then a carboxylic acid. This makes the areas where this reaction has occurred many times more soluble in basic solution. When the film is then developed with a basic tetramethylammonium hydroxide solution, the UV-exposed parts of the photoresist are rinsed away. The resin layer is then fully set using a 'hard' bake oven treatment at a higher temperature, which drives the densification of the resin and makes it resistant to chemical degradation. In this step, we will spin coat the wafer with a primer and the commonly used positive S1813 photoresist to form a film around a micron thick. After this imprint, alignment marks will be patterned photolithographically onto the resist using the mask and a xenon UV lamp (Fig. 5.8).

To spin coat and pattern a photoresist layer onto a substrate you will need:

1. The clean silicon nitride wafer and chrome and quartz mask from the previous recipe.
2. A 5-mL syringe and a 0.45-μm filter. You will also need tissue. It is also handy to have a rack for the syringe and filter so that the assembly can be kept filter up.
3. A stock solution of S1813 photoresist. This will turn very dark purple.
4. A stock solution of primer in an eyedropper bottle. A specific primer will normally be recommended for the substrate and photoresist being used.

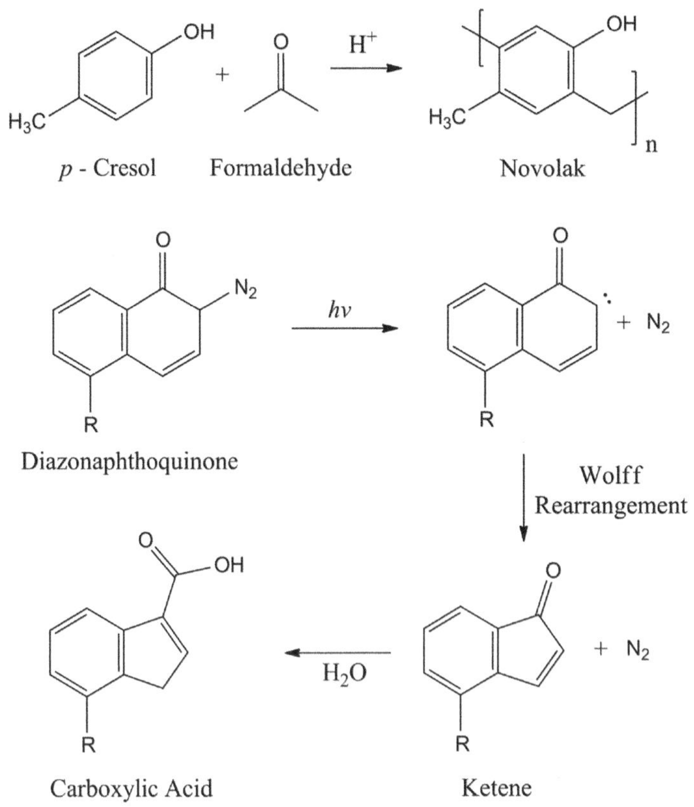

FIGURE 5.8 The chemical constituents of a Novolac positive photoresist. Exposure to light rearranges diazonaphthoquinone to a carboxylic acid and renders the Novolac polymer soluble in water.

5. A programmable spin coater.
6. An oven set to 90 °C for soft baking the cast photoresist layer.
7. A photolithography machine fitted with a xenon lamp light source.
8. A stock solution of MF-319 developer liquid. You will need a beaker to act as a bath for the developing solution.
9. Water for rinsing the developer away with.
10. A light microscope with a green light source for checking the pattern.
11. An oven set to 110 °C for hard baking the photoresist.

To spin coat and pattern a photoresist layer onto a substrate:

1. Place the 0.45-μm filter onto the syringe. Remove the plunger from the syringe and pour in the purple S1813 photoresist. Replace the plunger.
2. Have the filter pointing up and press in the plunger until the purple liquid starts to come through (Fig. 5.9).
3. Place the syringe with the filter pointing up into the rack or else rest it upright. There will still be air trapped in the liquid and you will need to leave the syringe for a few minutes for this to collect at the top by the filter.
4. Use the PTFE tweezers to place the wafer onto the spinner stage of the spin coater. Get it as central as you can before turning on the suction that will hold it in place. Some wafers may be very brittle and moving them about after you have turned on the suction can break them. Very often you may see the wafer deform where it comes in contact with the vacuum. Use this as a guide to see if the wafer is central or not.
5. Use an eyedropper bottle to add two or three drops of primer solution to the wafer and leave to stand for 5 s.
6. Run the spin coater for 10 s at 4000 rpm.
7. Take the syringe holding the photoresist out of the rack. Press the plunger gently so that the foamy portion of the photoresist containing air bubbles is pushed through. Wick this away using some tissue but be sure not to touch the tip of the syringe directly.
8. Add the photoresist though the filter onto the centre of the wafer until it covers the surface and allow this to stand for 5 s.
9. Run the spin coater for 30 s at 4000 rpm. This should form a film of the photoresist about a micron thick.
10. Turn off the suction on the spin-coater platform and use tweezers to transfer the coated wafer into the oven set at 90 °C. Leave this to 'soft' bake for 30 min before removing the wafer and allowing it to cool to room temperature.

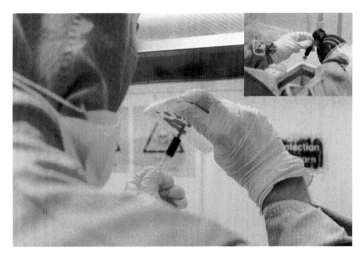

FIGURE 5.9   Photographs showing the syringe being filled with photoresist and dabbing the end of the syringe into tissue to remove air. Inset: the syringe is filled with photoresist by removing the plunger.

11. While the coated wafer is soft baking, turn on the photoresist machine. Many models will need at least 20 min for the lamp to warm up.
12. Fit the mask and substrate wafer into the photolithography machine and carefully align, then contact the mask to the wafer.
13. Expose the substrate to UV light for 60 s using a timer.
14. Remove the mask and remove the substrate from the photolithography machine. Transfer into a beaker of the MF-319 developer solution and leave to develop for 45 s. During this time, the portions of the photoresist mask exposed to light should solubilise.
15. Remove the wafer from the developer solution and rinse with water. Any soluble portions of the photoresist remaining should be removed by this. Dry the wafer off with the nitrogen gun. Check that the photoresist has fully developed by looking at it under an optical microscope with a green light source. You should be able to see the wafer substrate in the patterned areas; if not, then expose the wafer to the developing solution again for another 10–15 s.
16. Place the wafer in the 110 °C oven to 'hard' bake the photoresist layer for 30 min. The wafer will then be ready for the next etching stage. In this case reactive ion etching will be used to engrave the pattern into the silicon nitride.

REACTIVE ION ETCHING (RIE)

The process of reactive ion etching requires specialist equipment to perform. The basic technique is to create chemically reactive plasma that reacts with the substrate surface to evolve gaseous products that are then removed. The plasma is formed by a strong oscillating electromagnetic radiofrequency field inside the reaction chamber between two metal plates. The wafer to be etched sits on the bottom plate but is electrically isolated from it. A gas such as CF_4 or CHF_3 is passed into the chamber at low pressure and in the presence of the electric field becomes ionised to form a plasma. Simultaneously, the electrically isolated substrate wafer has its electrons stripped away, giving rise to a strong positive charge. The positive charge attracts highly energetic anions in the chamber, normally a fluoride, which then reacts with the silicon or silicon nitride to form gaseous silicon tetra fluoride and other product as in the empirical reactions outlined below (Pant and Tandon, 1999):

$$CF_x + Si_3N_4 \rightarrow SiF_4 + N_2, F_2, CFN_{(all\ gaseous)}$$

$$CF_x + SiO_2 \rightarrow SiF_4 + CO_2, CO + CF_y, COF_{2\ (all\ gaseous)}$$

The dominant reaction scheme when the gas becomes plasma is

$$CF_4 \rightarrow F^{\cdot} + CF_3$$

(where $F^{\cdot}$ is a highly energetic anion)

$$CF_4 + e \leftrightarrow CF_3^+ + F^{\cdot} + 2e$$

$$Si_{(s)} + 4F^{\cdot} \rightarrow SiF_{4(g)}$$

Although the plasma can be seen to aggressively degrade the exposed wafer, it does not degrade the photoresist resin so easily. Because of this only the exposed areas of the wafer not covered by the photoresist will be etched. Often, a mixture or combination of

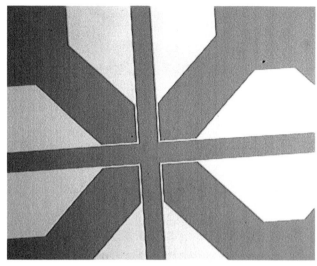

FIGURE 5.10   An example of a pattern etched into silicon nitride using RIE. The green areas are photoresist and the grey areas are exposed silicon nitride.

gases can be used to give the desired results. Once the RIE process is finished, the photoresist is removed by rinsing with acetone and can be further cleaned by O_2 plasma ashing to ensure that there are no more organic residues on the surface. O_2 plasma ashing is done in the RIE machine using pure oxygen as the feed gas at a slightly higher pressure. It is more aggressive to the organic photoresist layer and is useful for cleaning stages. The following procedure is very general and the times and pressures used will vary greatly depending on the machine in use. This example uses oxygen, CHF_3 and argon to form a mixed plasma. The method below should be used accordingly as a rough guideline to the process.

To etch the wafer using RIE to a depth of 100 nm you will need:

1. An RIE machine with argon, oxygen and CHF_3 gas feeds with adjustable pressures. Follow the manual instructions for use.
2. The wafer to be etched with the patterned photoresist layer coated on it.
3. Acetone in a beaker and an acetone wash bottle for rinsing away the photoresist.
4. Stock isopropanol and water for washing.
5. A nitrogen gun for drying.

To etch the wafer using RIE to a depth of 100 nm:

1. Place the wafer into the RIE machine and then follow the instructions for use as given in the manual.
2. Pump down the chamber and programme the machine to etch the wafer for 7 min and 30 s. This time may vary depending on the machine power and instructions. The following gases are used in combination with the listed partial pressures:
   O_2 at a pressure of 0.9 mTorr (0.12 Pa)
   CHF_3 at a pressure of 2.5 mTorr (0.33 Pa)
   Ar at a pressure of 10 mTorr (1.33 Pa)
3. Remove the wafer from the RIE machine and use a light microscope to check that the silicon nitride layer has been etched.
4. Remove the photoresist layer by dipping the wafer into an acetone bath. Remove the wafer and rinse it with more acetone.
5. Rinse the wafer with isopropanol thoroughly and then water.
6. Dry the wafer using the nitrogen gun.
7. Should you find there is still some photoresist on the surface then you can place the wafer back into the RIE machine and perform a plasma ashing step. This is done under an O_2 flow of 50 sccm at 20 mTorr (6.66 Pa) with an RIE power of 300 W for 6 min (Fig. 5.10).

CHEMICAL ETCHING WITH KOH SOLUTION

Adapted from Seidel et al. (1990) and Sato et al. (1998)

At this point, you will have a patterned layer of silicon nitride on the top of the silicon wafer core. This next step uses a strong hydroxide solution to chemically dissolve the underlying silicon wafer. It will not etch the silicon nitride on the surface. The rate at which the base will erode the silicon varies depending on the concentration of KOH and the temperature:

$$Si + 2OH^- + 4H_2O \rightarrow Si(OH)_2^{2+} + 2H_2 + 4OH^- \rightarrow Si(OH)_4 + 2OH^-$$

Below is a table listing the etching rates for various facets of crystalline silicon for a range of temperatures and concentrations of KOH solutions. This table was calculated using the following formula. For each facet, the calculated activation energy in eV for a silicon face to form a soluble hydroxyl species is given [etching rates reference]:

$$R = K_0[H_2O]^4[KOH]^{1/4}e^{-E_a/kT}$$

R = the etching rate in microns per hour.

For the [100] face of silicon the parameters for the rate equation are

K_0 = A constant of 2480 μm/h (mol/L)$^{-4.25}$.

$E_a = 0.595$ eV.

Silicon etch rates in microns per hour using the parameters above in the rate equation.

[100]	Temperature in °C								
% KOH	20	30	40	50	60	70	80	90	100
10	1.49	3.2	6.7	13.3	25.2	46	82	140	233
15	1.56	3.4	7.0	14.0	26.5	49	86	147	245
20	1.57	3.4	7.1	14.0	26.7	49	86	148	246
25	1.53	3.3	6.9	13.6	25.9	47	84	144	239
30	1.44	3.1	6.5	12.8	24.4	45	79	135	225
35	1.32	2.9	5.9	11.8	22.3	41	72	124	206
40	1.17	2.5	5.3	10.5	19.9	36	64	110	184
45	1.01	2.2	4.6	9.0	17.1	31	55	95	158
50	0.84	1.8	3.8	7.5	14.2	26	46	79	131
55	0.66	1.4	3.0	5.9	11.2	21	36	62	104
60	0.50	1.1	2.2	4.4	8.4	15	27	47	78

For the [110] face of silicon the parameters for the rate equation are:

K_0 = A constant of 4500 μm/h (mol/L)$^{-4.25}$.

$E_a = 0.6$ eV.

[110]	Temperature in °C								
% KOH	20	30	40	50	60	70	80	90	100
10	2.2	4.8	10.1	20.1	38	71	126	216	362
15	2.3	5.1	10.6	21.2	40	74	132	228	381
20	2.3	5.1	10.7	21.3	41	75	133	229	383
25	2.3	5.0	10.4	20.6	39	73	129	222	372
30	2.1	4.7	9.8	19.4	37	68	121	209	350
35	2.0	4.3	8.9	17.8	34	63	111	192	321
40	1.7	3.8	8.0	15.9	30	56	99	171	285
45	1.5	3.3	6.9	13.7	26	48	85	147	246
50	1.2	2.7	5.7	11.3	22	40	71	122	204
55	1.0	2.2	4.5	9.0	17	31	56	96	161
60	0.7	1.6	3.4	6.7	13	24	42	72	121

To etch a silicon wafer using KOH solution you will need:

You will be using an extremely strong solution of KOH, which will be highly corrosive. Take every precaution. Work within a fume hood with a water supply for rinsing spills.

1. 100 g (1.78 mol) of KOH.
2. 200 mL of water in a large plastic beaker.
3. The wafer to be etched.
4. A hot plate with magnetic stirring and temperature sensor. You will need a stir bar remover to get the stir bar out of the KOH solution.

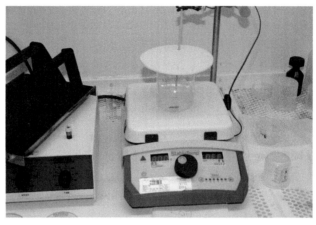

FIGURE 5.11   A photograph of a potassium hydroxide solution being heated in a beaker. Note the splash guard fitted over the top.

5. A splash guard that can fit over the beaker full of basic solution.
6. Water for rinsing the wafer and a nitrogen gun for drying.
7. PTFE tweezers for handling the wafer.

To etch a silicon wafer using KOH solution:

1. Add the KOH to the 200-mL water slowly under stirring and ensure that the solution does not begin to steam. After a few grams have dissolved, allow it to cool to room temperature and add a few grams more. Keep the splash guard on wherever possible. Once all the KOH has been added to the solution, remove the magnetic stir bar and rinse it off with water.
2. Place the KOH solution on the hot plate and place the temperature probe through the splash guard. Set the temperature to 80 °C (Fig. 5.11).
3. Once the solution is at the correct temperature add the wafer to the solution using the tweezers. Leave the wafer to etch from between one and one and a half hours.
4. Remove the wafer from the solution and rinse thoroughly with water. Dry the wafer off using a nitrogen gun.

You should now have a chemically etched pattern in the surface of the silicon. As before, check the results using a microscope. In the pictured example, the etched silicon surface appears as a rough reflective square surrounded by a dark border. The thin layer of unetched silicon nitride appears blue. The dark border is a result of the chemical etching occurring at a slower rate as it moves sideways through a different crystal face of silicon (Fig. 5.12).

FIGURE 5.12   A photograph of a wet-etched silicon nitride surface.

SOFT LITHOGRAPHY, PATTERN REPLICATION AND CONTACT PRINTING

With thanks to Alar Ainila and Rebecca Boston

Soft lithography is a rapid and useful technique for repeatedly patterning surfaces via a top-down approach. It holds many advantages over classical photolithography in that it is cheaper, can be applied to larger areas with a highly varied choice of substrate materials and can also be applied to curved or deformed surfaces to give interesting three-dimensional patterns. The technique has been extensively championed by the Whitesides research group and the reader is recommended to seek out the reviews the group has written (Xia et al., 1999; Sia and Whitesides, 2003; Xia and Whitesides, 1998). The basic concept is to use a material that is soft to replicate a master structure instead of a hard one to produce a template. An elastomer solution is mixed with a crosslinker and then poured over a positive master structure to be replicated. The elastomer, most commonly a polydimethylsiloxane (PDMS), then sets into a flexible silicone rubber negative of the original template. Because the negative copy is soft it can be easily separated from the template surface and can be used on malleable starting structures without deforming them. Templates formed in this way can be employed in a wide range of patterning and lithographic replication methods, although at a simple level this will involve either replication or contact printing. Direct replication involves using the negative PDMS template as a mould, which is filled in with another polymer material like polyurethane to give a replica of the master. The resolution of the copy is within 5 nm over surface areas of many millimetres and the flexible template is easily removed and reusable (Gates and Whitesides, 2003). Microcontact printing involves using the PDMS replica as a stamp to produce patterned self-assembled monolayers on metal surfaces (Wilbur et al., 1994; Wilbur et al., 1995). In this section, examples of positive replication and microcontact printing will be outlined as well as the use of PDMS in the fabrication of microfluidic devices.

CASTING A NEGATIVE MOULD USING PDMS

Poly(dimethylsiloxane) is the workhorse of soft lithographic techniques. It can be used in nonclean room environments and a template can be formed in hours, with resolutions from 30 nm to 500 μ. It has a low chemical activity and is biologically inert, which is useful when it is used in lab-on-a-chip devices with biomolecules or harsh chemicals (Ainla et al., 2010; Sia and Whitesides, 2003; Ainla et al., 2010). There are some drawbacks in that the PDMS will swell in certain solvents causing feature deformation. The softness of the PDMS can also lead to deformation in replicated patterns with high aspect ratio vertical features higher than 2. To some extent, these problems can be overcome by altering the composition of the siloxane or crosslinking agent to produce harder elastomers but this is often accompanied by a loss of optical transparency. Sylgard 184 is the formulation most often referenced in protocols and this particular compound is employed in lab-on–a-chip devices because it is transparent to light wavelengths down to around 300 nm. This makes it easy to track fluid in these systems using a microscope or by other spectroscopic means. Other formulations can be purchased which give a tougher elastomer at the cost of optical clarity or coloured dyes can be used in the crosslinker to ensure adequate mixing before a solution is cast. PDMS is made from a 3-component system involving a base, curing agent and a catalyst. The base is predominantly dimethylvinyl-terminated dimethylsiloxane. The curing agent is mostly comprised of dimethylhydrogen siloxane. Both of these will be mixed with additives depending on the required mechanical properties of the cured polymer. The third component, normally provided mixed in with the curing agent when a PDMS kit is purchased, is a metal-centred catalyst to promote crosslinking. This is a platinum complex that promotes a hydrosilylation reaction between the methylhydrogen siloxane units in the curing agent and the terminal vinyl groups in the base siloxane. The exact formulation is specific to the manufacturer but will not normally interfere with the chemistry of the end application as it is present in very small amounts (Fig. 5.13).

To make a silicone elastomer the base and curing agent are mixed together in a 10:1 by weight ratio. The curing agent needs to be mixed thoroughly so that the polymer will harden evenly. This often leads to air bubbles being trapped inside and these need to be removed before the siloxane sets. This is accomplished by degassing the sample repeatedly by exposing it to reduced pressure so that the air bubbles are drawn out. Once the mixture is degassed, it can be poured on top of the substrate to be replicated and then either left for a few days to set or placed in an oven between 60 and 70 °C to set in a few hours. If a thin film of the PDMS is required on a flat substrate spin coating can be used. For Sylgard 184 the following formula can be used to calculate the thickness in microns for a spin time of 60 s and an angular acceleration of 20 rpm/s (Zhang and Ferguson, 2004):

$$\text{Thickness (μm)} = \frac{KC^{\beta}[\eta]^{\gamma}}{\omega^{\alpha}}$$

where K is a constant of 6.2661

α, β and γ are exponential constants of 0.9450, 5.2707 and 6.2051, respectively; C is the concentration of polymer and is fixed at 0.90909; η is the intrinsic viscosity and is fixed at 4000 mPa/s; and ω is the spinning speed in rpm of the substrate.

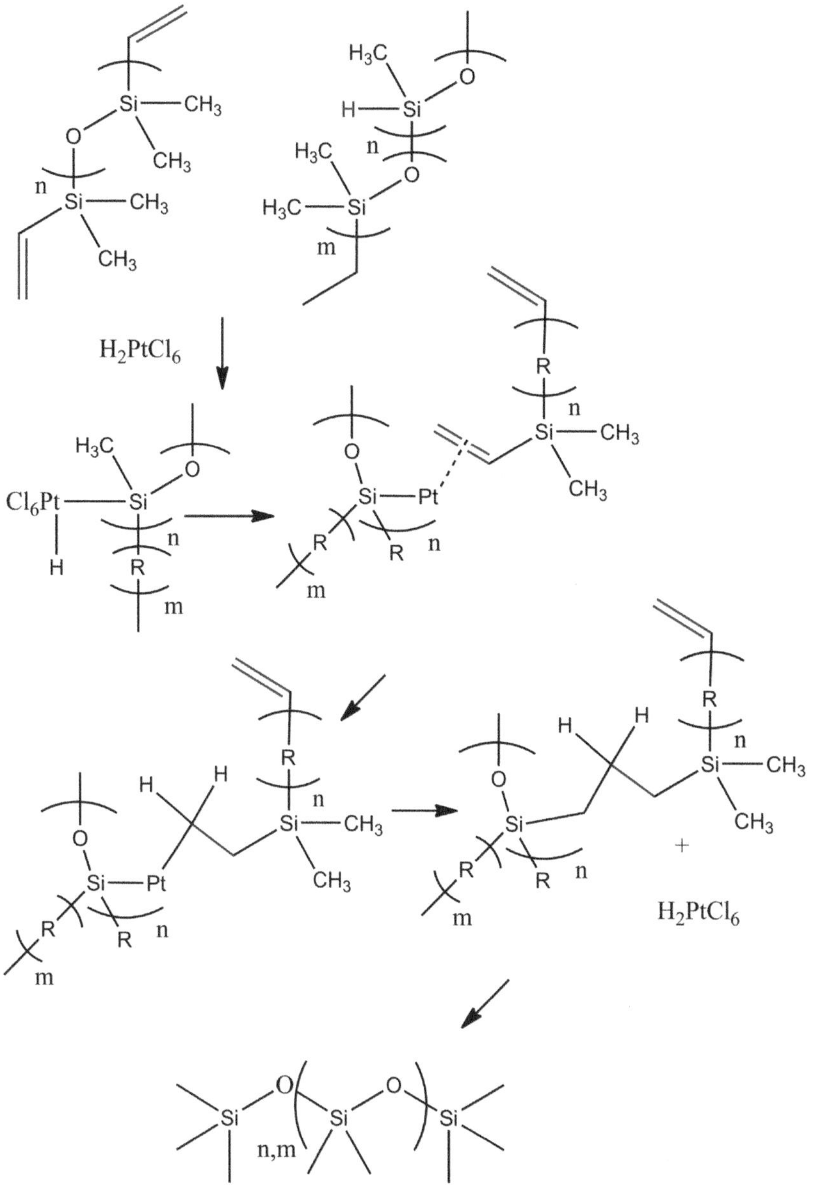

FIGURE 5.13   The chemical reaction of a dimethylvinyl-terminated dimethylsiloxane with a crosslinking agent to produce a poly(dimethyl siloxane).

Once the elastomer has formed, it can be cut or peeled from a substrate. The formed siloxane will exhibit 1% volumetric shrinkage from the liquid prepolymer state. Exposure to certain solvents will cause swelling as outlined in the table below and these should be avoided when printing or replicating, as their presence will distort the mould.

No Swelling	≤5% volume	<5% to 30%<	≥30%
Water, Glycol, Ethylene Glycol, DMSO, Nitromethane.	Dimethyl carbonate, Acetone, Pyridine, NMP, Acetonitrile, Methanol, Ethanol, Phenol, Propylene Alcohol.	Ethyl acetate, Benzene, Chloroform, 2-Butanone, Chlorobenzene, DCM, Dioxane, Propanol and *tert*-butyl alcohol.	Pentane, Hexane, Heptane, Triethyl amine, Ether, Cyclohexane, Xylene, Toluene, THF. Diisopropylamine will swell up to 2.1 times.

In the next recipe, we will manufacture a PDMS elastomer and use it for contact printing a self-assembled monolayer onto a gold surface.

PATTERNING A SELF-ASSEMBLED MONOLAYER ONTO A GOLD SUBSTRATE USING A PDMS STAMP

Adapted from Wilbur et al. (1994) and Xia et al. (1995)

Rubber stamps have long been used in combination with ink to make quickly replicated signatures, visa stamps and date markings. The basic concept is no different for PDMS elastomer stamps, which can be coated in a molecular 'ink' designed to stick to a contact surface when pressed into it. Here we will explore the production of self-assembled monolayers (SAMs) on thin gold films using an ink of alkane thiolates to produce a pattern in the gold metal with a PDMS stamp. The gold metal to be patterned is thermally evaporated onto a flat substrate that is 10−1000 nm thick. The source reference uses a [100] silicon wafer with a 5−50Å layer of titanium deposited on top, which will secure the gold in place, although a copper interfacial layer will also work. Both metals are deposited by electron beam or thermal evaporation, which requires special equipment, although the gold film can be deposited by other methods if they are available to you. An ink containing hexadecane thiolate dissolved in ethanol or diethyl ether and painted or dabbed onto the surface of the PDMS stamp. The elastomer swells slightly in these solvents so that, when the excess ink is blown off with air, some is still absorbed within the surface of the stamp. By applying light pressure to the back of the stamp when pressing it against a flat gold surface, some of the thiol will react with the surface by the following reaction:

$$CH_3(CH_2)_{15}SH + Au^0 \ \rightarrow \ CH_3(CH_2)_{15}S-Au + 0.5 \ H_2$$

If the gold were to be immersed in a 2×10^{-3} mol/L solution of the alkane thiolate, then the chemisorption of the thiols may take several minutes to form a monolayer. In contrast, the chemisorption from using a stamp occurs in seconds. For alkyl chain lengths of longer than 11 CH_2 units, the molecules pack tightly together due to van der Waals interactions. The alkyl chains lean over 30° horizontally as a result of the packing and form a molecular monolayer 1−2 nm thick. For shorter alkyl chain lengths of only 2−3 CH_2 units, long the van der Waals interactions are not strong enough for the molecules at the gold surface to assemble together. Under these circumstances, the thiols can be easily removed from the surface during the etching procedure. Because the PDMS stamp pattern will have areas that have not come into contact with the gold surface during the patterning you will be left with stamped areas covered in monolayer and clean gold areas. At this point, it is possible to coat the unpatterned gold areas with another alkane thiol terminated with a functional group so that you can alter the surface properties. In terms of a good visual demonstration, you can soak a patterned gold layer in a second ethanol solution containing thiols terminated with carboxylic acid groups. When removed from the secondary solution and allowed to dry, the gold surface will be completely passivated in a self-assembled monolayer patterned positively with hydrophobic methyl-terminated alkyl chains and negatively with hydrophilic-terminated alkyl chains. When water is condensed on the surface, the various attractive and repulsive forces at play on the surface will form a regular pattern that you should be able to see by eye. This works well when using a TEM grid as the original template.

The molecular monolayer is not thick enough to be used as a patterned resist for reactive ion or strong chemical etching as would be used in a photolithographic procedure etching but does passivate the gold surface against weak chemical oxidation. Cyanide anions will readily oxidise gold to form a gold cyanate complex by the following reaction:

$$4Au + 8CN^- + O_2 + 2H_2O \ \rightarrow \ 4[Au(CN)_2^-] + 4OH^-$$

By immersing the alkane thiol patterned gold into a 0.1 M KCN solution saturated with oxygen, it is possible to selectively etch away any gold not protected by the self-assembled monolayer. However, cyanide is extremely toxic and working with it presents large risks and specialist training. To get around this the much safer hexacyanoferrate anion is used. This iron-centred complex has been used for many years as an ingredient in the pigment known as Prussian blue. It has a low toxicity and is safe to work with on an open bench. The ferricyanide anion is a much weaker oxidant than the direct cyanide anion, so potassium thiosulphate is added to the solution that helps to lower the redox potential of the gold. Although weaker, the ferricyanide-based solution gives a better edge resolution and disturbs the monolayer less, which leads to lower defects in pattern replication compared to cyanide etching. It is also faster and can dissolve 20 nm of exposed gold in around 8 min as compared to 15 min for cyanide. Furthermore, the same etching solution can be used on silver and copper films patterned with alkane thiols, although the etching rate is much faster with a 20-nm silver film dissolving in around 15 s and a copper film dissolving in around 5 s. The overall reaction under basic conditions proceeds as (Burdinski and Blees, 2007)

$$Au^0 + 2S_2O_3^{2-} + [Fe^{(III)}(CN)_6]^{3-} \ \rightarrow \ [Au(S_2O_3)_2]^{3-} + [Fe^{(II)}(CN)_6]^{4-}$$

To pattern a gold substrate using microcontact printing of a self-assembled monolayer and a ferro/ferricyanide etchant you will need:

1. A pattern or substrate you wish to replicate using a self-assembled monolayer. The ideal features for a template will cover a large flat area and have feature aspect ratios of depth to width of between 0.5 and 2. A positive pattern master can be made

using positive photoresist on a silicone wafer following the procedures given in '*To spin coat and pattern a photoresist layer onto a substrate.*' For a quick pattern, you can try using a copper mesh TEM grid and placing it in contact with the photoresist before exposure to light.

2. A flat silicone wafer or glass substrate coated in a 0.5–5-nm-thick layer of copper or titanium then a 20-nm-thick layer of gold for patterning. Thinner gold films may result in defects. The wafer must fit inside the etching bath, which in this recipe will be a 400-mL beaker.

3. The base and primer solutions for making a silicone elastomer stamp. The most commonly used elastomer for this purpose is the Sylgard 184 kit.

4. A plastic cup and knife or fork for mixing the elastomer in along with tissue to place over the scale while you are weighing it out. You will also need a Petri dish for casting the stamp in.

5. A bell jar with a valve connected to a vacuum pump.

6. An oven set to 70 °C for curing the siloxane with.

7. A scalpel for cutting the stamp out of the bulk elastomer with should this be required.

8. 100 mL of a 2×10^{-3} mol/L 1-hexadecanethiol in ethanol solution to be used as the 'ink.' This can be prepared by adding 0.0615 mL (2×10^{-4} mol) to 95 mL of ethanol then making up to 100 mL. This solution should be prepared in a fume hood and kept in a sealed Scott bottle when not in use. Using a fresh solution is desirable over one that is a few days old. *Thiols smell very bad and can be an irritant.*

9. A cotton bud to be used as a swab to brush the ink over the surface of the stamp. Alternatively, you can use filter paper soaked in the ink and contact the stamp to it.

10. A nitrogen line for drying the stamp with.

11. A magnetic stirrer and a 2-cm-long stir bar.

12. 300 mL of gold etching solution in a 400-mL beaker to be used as the etching bath. 1 L of stock gold etching solution can be prepared by weighing the following amounts of each salt into 500 mL of water and making up to 1 L.

Salt	Weight/grams	Concentration in moles per litre
KOH	56.11	1.0
$K_2S_2O_3$	19.03 (anhydrous weight)	0.1
$K_3Fe^{(III)}(CN)_6$	3.29	0.01
$K_4Fe^{(II)}(CN)_6 \cdot 3H_2O$	0.42	0.001

The cyanide salts should not be exposed to strong acids or HCN gas will be evolved. Other than this, they are safe to work with.

13. Teflon tweezers for handling the etched substrate and copious amounts of water for rinsing away the etching solution.

14. A stopwatch for timing the exposure in the etching solution.

To pattern a gold substrate using microcontact printing of a self-assembled monolayer and a ferro/ferricyanide etchant:

1. Place tissue over the top of a set of scales and weigh out roughly 10 g of the silicone elastomer base into a plastic cup. It is very viscous and will take time to pour. Add 1 g of the corresponding curing agent.

2. Use a plastic knife to stir the base and curing agent together for at least 5 min.

3. Place the pattern master face up into a Petri dish. Pour the mixed elastomer into the Petri dish until it is a few millimetres from the top.

4. Place the Petri dish and elastomer into the bell jar sat on top of some tissue in case it spills over in the next step.

5. Turn on the vacuum pump and degas the elastomer. It will begin to bubble and foam as the air evacuates. If the bubbling becomes too much, return the sample to room pressure and repeat. After three or four cycles, you should see no more air bubbles evolved from the polymer.

6. Transfer the Petri dish into the oven and cure at 70 °C for two hours until the PDMS has solidified.

7. Remove the Petri dish from the oven and allow it to cool. Depending on the template either you will be able to pull the set siloxane out of the dish or you will have to cut it out with the scalpel. The original template should separate easily from the elastomer.

8. Taking care not to damage the pattern, cut the stamp into the desired shape. It should be thick enough that you will be able to hold the stamp in gloved hands by the sides. The PDMS stamp is now ready for use and can be kept in a clean Petri dish. It can be reused many times.

9. The following steps involve forming a SAM on the surface of the gold. Take the gold substrate and place it face up.

10. Dip a cotton bud into the thiol ink and swab it across the surface of the PDMS wetting it thoroughly. Alternatively, soak a piece of filter paper in the ink and press the stamp onto it briefly to wet it. In both cases, you must blow the excess liquid away using a nitrogen stream passed over the surface of the stamp for one minute.

11. Press the inked stamp with the pattern down onto the surface of the gold. Apply even hand pressure for around 5 s and then pull the stamp off. How you apply the stamp is important in getting good resolution. A peel-on and peel-off approach can work well, but results may vary. The best option is any motion that avoids smearing the stamp around on the surface. A self-assembled monolayer of alkane thiol should be adhered to the gold in the desired pattern. You may wish to examine the resolution and fidelity of the SAM by using atomic force microscopy at this point or proceed to the etching stage.

12. Sit the beaker containing the etching solution on top of the magnetic stirrer and turn on the stirring to 300 rpm.

13. Transfer the patterned wafer into the etching bath using the tweezers and start the stopwatch. Depending on the dimensions of the substrate, you can either rest the wafer against the wall of the beaker or sit it on the bottom. Contact with the stir bar must be prevented and you might like to try holding the sample in the bath with the tweezers and then clamping them in place. Any gold not protected by the patterned alkane thiol monolayer will be completely dissolved in the solution after around 8 min at room temperature.

14. After the 8 min is up remove the wafer from the etching bath and rinse with water or ethanol. Dry the patterned gold surface with a nitrogen stream. The sample is ready for analysis.

Scanning electron microscopy should reveal a high contrast between the etched gold pattern and the titanium or silica background. Gold will appear brighter than an inorganic background and you may see patches of brightness in areas where the gold has not been fully etched away. Elemental mapping should also reveal the difference clearly and allow you to determine if the brighter patches against darker areas are residual gold or not. Exposures of over 10 min will begin to appear like dark areas in the regions that are meant to be gold. This means the sample has been in the etching bath for too long. You may wish to either decrease the exposure time in the etching bath or try a concentration as high as 10×10^{-3} alkane thiol in the ink solution. Taking profiles of the etched gold after different etching exposure times will help you fine-tune the mask and etching procedure for whichever thickness of gold film you are working with.

Atomic force microscopy is a very useful tool both for examining the patterning on the gold layer with the thiol molecules and for profiling the eventual result post-etching. The formation of a self-assembled monolayer on the gold surface prior to etching can be observed by both tapping mode and contact mode techniques. If you are working in contact mode then ensure that you are scanning with the minimum tip force to avoid disrupting the monolayer. Lateral force contact mode imaging is particularly useful if you have patterned the surface with a mixture of hydrophobic and hydrophilic terminated thiols (Kumar et al., 1994). Lateral force imaging looks at the torsional response, the twist, in the cantilever as it moves backwards and forwards across the surface. The tip of an AFM cantilever being used in air is susceptible to capillary forces from ambient water molecules adhered to the surface of the sample or tip. Because the interactions of the hydrophobic methyl groups and hydrophilic groups, such as COOH, exert a different level of friction as the tip passes over them, it is possible to use the deflection signal from the tip twisting to build up and image. The relative friction experience by the tip increases with hydrophilic character so that the friction force and therefore the contrast in the image increases from $CH_3 < OH < COOH$ (Fig. 5.14).

Imaging the etched gold pattern by AFM should show a similar image as observed with the SEM. You should see the remaining gold standing proud of the substrate surface with well-defined and straight edges. As with the electron microscopy, you may see areas that look like pits in the gold if the sample has been etched for too long and the monolayer has been removed from the surface. Profile cross sections taken of these areas will appear as pits in what should be an otherwise flat surface. Taking profiles of the etched gold after different etching exposure times will help you fine-tune the mask and etching procedure for whichever thickness of gold film you are working with. You will need to image the gold substrate before treatment so that you can obtain a roughness value (normally quoted as the root mean square of the height over a given area) and will know how the etching bath affects this, if at all. Some metal deposition techniques, such as sputter coating, will produce a granular texture in the gold and this in turn will affect how well the self-assembled monolayer can pack together to form a chemically resistant layer. The flatter the gold or metal layer is to begin with, the better the results will be.

MAKING A MICROFLUIDIC DEVICE USING SOFT LITHOGRAPHIC TEMPLATING

With thanks to Andrew deMello and Alar Ainla

When designing a microfluidic device it is important to discuss mixing in these systems as, on the small scale, fluids behave very differently (Song et al., 2003; Microfluidics et al., 2006; Squires, 2005). As a simple example, let us say that we wish to make a device for reacting chemical A with chemical B and then analyse or isolate the product. We would like to use a microfluidic system as we have only a small amount of starting reagent (conversely, it may also be very expensive) and only need a small amount for analysis. It is not as easy as injecting the two fluids into a Y-shaped channel as they will not mix immediately together for reasons outlined below. You will also find that, when they do mix, the wave front of the product as it moves through the small channel will be

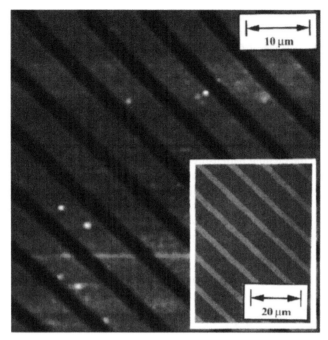

FIGURE 5.14 A lateral force microscopy image of a self-assembled monolayer terminated by different functional groups. The friction force experience by the cantilever as it moves over the surface changes with respect to the functional group at the surface producing contrast. *(Image adapted with permission from Kumar et al. (1994).Copyright 1994 The American Chemical Society)*

highly parabolic and therefore spread over a long distance in the capillary. Just as the water in the middle of a river moves quicker than the water by its banks, fluids in small capillaries experience more drag near the vessel wall, which spreads out the wave front like a cone. If you are hoping to use a spectroscopic method to look at the product, then you might want the product to be concentrated at one point as opposed to spread out over the length of a capillary. There are some unique and clever solutions to these problems that are outlined in this section, but in order to understand how they work we must first look at the problems more thoroughly.

The mixing and flow in a capillary are described by the Reynolds number (R_e). This is a dimensionless number that is used as a measure of how the inertial forces of a fluid, its fluidic turbulence and motion, compare to its viscosity. Expanding upon this, it gives a ratio of how a fluid of a given viscosity will behave when passed at a given speed through a tube of a known dimension as given by the formula below:

$$R_e = \frac{dv\rho}{\mu} = \frac{\text{Inertial flow}}{\text{Viscous flow}}$$

where d is the cross-sectional diameter through which the liquid is flowing in metres; v is the velocity of the fluid through the tube in metres per second; σ is the density of the fluid in kilograms per metre cubed; and μ is the viscosity of the fluid in kilograms per metre per second.

For situations where the Reynolds number is large (above 2000), the eddies and flows that occur in solutions will be the determining factor in how the fluid behaves. This behaviour is called turbulent flow and it can be seen by inspection of the Reynolds number formula that a large value of d, as you might see in a water pipe on the everyday scale, will give rise to this behaviour. If the water in a large pipe and flow rate is suddenly passed into a much smaller tube, it can be seen that the viscosity, μ, of the solution becomes dominant in providing a value for the Reynolds number and this will be very low compared to the situation for the water in a large pipe (Fig. 5.15).

At Reynolds number values below 2000, the dominant type of flow is called laminar and this describes a situation where the viscosity of the solution is relatively higher than the product of the diameter, flow rate and density such that turbulent flow is almost eliminated from the system. In a microfluidic system the micron-scale diameter of the capillary through which a fluid flows most often results in a low Reynolds number and exhibits laminar flow. This presents a problem in that, should we wish to mix two liquids together, for example, they will flow side by side as there is no turbulence within the small channel to mix them together. To describe how two fluids mix together another dimensionless figure, the Peclet (P_e) number, is used to describe the ratio of convection versus diffusion in a solution. The P_e is given by the following formula:

$$P_e = \frac{vd}{D} = \frac{\text{Convection}}{\text{Diffusion}}$$

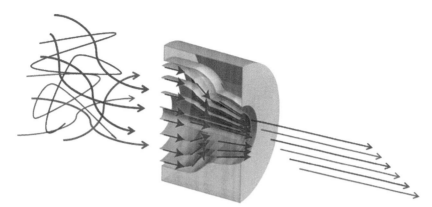

FIGURE 5.15   A schematic diagram of how flow changes depending on scale. Two different-coloured fluids are injected into a moving liquid flowing through (a) a pipe with a diameter of a few centimetres where turbulent flow is the dominant mixing method and (b) a pipe with a diameter of only a few microns where diffusion across a laminar flow is the dominant mixing method.

where the D, the diffusion constant, is defined roughly as

$$D \sim \frac{K_B T}{6\pi\eta a}$$

where K_B is the Boltzmann constant; T is the temperature in Kelvin; η is the shear viscosity in kilograms per metre second; and a is the diameter of a spherical molecule in the solution in metres.

For P_e values much lower than 1, as found in microfluidic systems, diffusion is the dominant mixing method. For values much larger than 1, the stretching and folding found in turbulent flow, known as Taylor dispersion, is the dominant mixing method. Of course, a simple solution for getting two fluids in a parallel laminar flow to mix is to change the values in the Reynolds and Peclet equations such that the dominant mixing method will be turbulent flow. It can be seen by inspection that increasing the flow rate, v, will achieve higher values for these dimensionless constants. In practice this is difficult to do with aqueous microfluidic systems as the flow rates have to be on the order of 1 mL/s and this uses up a lot of fluid very quickly. High flow rates also produce large pressures in the devices that can break them. One of the major attractions of using a microfluidic device to perform your chemistry with is that you only need to use small volumes; so increasing the flow rates by many orders of magnitude is undesirable. In a microfluidic device where two fluids are in a parallel laminar flow, mixing will occur slowly by diffusion over the interfacial surface where the fluids meet. The time it takes for mixing to occur is proportional to the square root of the distance a particle from one flow stream has to travel into another. This means that two laminar flows will mix together more quickly if there is a larger interfacial area between them. One way to achieve this is by passing the parallel flows through a very long capillary so that there is time for diffusion processes to take place between them. Another situation ideal for mixing is to have those two flows broken up into multiple parallel streams, which would also result in a large interfacial area between the two fluids. This process is called distributive mixing and can be designed into a microfluidic reactor so that two flows can be mixed together on the order of milliseconds (Bessoth and Manz, 1999). This is achieved by splitting a single reactant stream A into many thinner laminar flow channels and then blending them with a similarly segregated secondary flow B. The mixing can be further speeded up by passing the laminar flow around a series of bends so that one side of the separate flow streams has an increased surface area in comparison to the other. This is known as advection and is effectively folding the separate phases of liquid together like folding in dough. This is demonstrated easily by mixing two layers of coloured clay together then stretching and folding them repeatedly (Fig. 5.16).

The property of reagent fluids to move at different rates with respect to one another in adjacent laminar flows when directed around a curved capillary is useful for mixing purposes but gives rise to other problems should you wish to isolate or analyse products formed from reactions mixed in this way. A fluid moving through a narrow capillary will move slower nearer the walls of the capillary and faster in the centre, which gives rise to a parabolic profile when viewed side on. This is not a problem if you are mixing two reagents together to form product in a continuous stream as with the example of the mixing chamber just described. However, what if you have an application where a reagent is added into the flow at random times and you would like to analyse it? If you are adding another reagent at specific points to a microfluidic solution moving at a given rate, then you would hope to be able to predict where it will be in the device at a given time. However, from the point of reaction the reagent will be spread out across a length of capillary due to the difference in flow rates between the interior and exterior flow. Although a long capillary transit time helps two laminar flows to mix, it also results in the product being 'smeared' out along the length of the flow. Isolation of the product is therefore made difficult in applications where you would accurately like to gauge the concentration of the product or analyte in the system.

The solution to this problem is to introduce an immiscible phase into the device so that instead of a continuous laminar flow you have discrete bubbles of the reagents moved along in a carrier medium (Srisa-Art et al., 2007; Song et al., 2003). Because in this

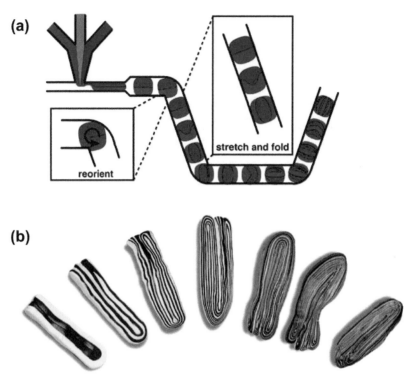

FIGURE 5.16   (a) A schematic representing advection between two fluids in a laminar flow moving around bends in the capillary. (b) This is similar to folding two different-coloured doughs together by stretching and compressing them together. (*Images used with the permission of Andrew deMello*)

arrangement you know the length of the capillary and the flow rate, you are able to determine how long the droplet components have been mixing for. This system therefore allows you to discretely mix and time reactions using droplets as the reaction vessels. Other microfluidic processes can be employed to sort the droplets independently of one another. If we look again at the problem of mixing two laminar flow fluids by this method, we still get a droplet composed of two halves. The contents of the droplet still need to be manipulated to promote mixing. As the droplet is forced along the capillary by the immiscible fluid, the respective contents will rotate inside the droplet by interaction with the walls. This circulation is still laminar such that the droplet is deformed by passing it through a series of tight bends in the capillary, which increases the interfacial surface of the laminar contents and promotes advective mixing just as it does in the distributive mixing-type chamber. Because these devices are manufactured predominantly from PDMS, it is vital to use an immiscible fluid that does cause swelling in the elastomer as the continuous phase. Perfluorocarbons such as perfluorodecaline are often used to fulfil this specification and have an advantage of being bioinert so that biomolecules can be used in the aqueous phase without fear of degradation or sequestration into the carrier medium. Fluorinated surfactants can also be added so that the continuous phase wets the inside of the capillary efficiently. Wetting in the continuous phase means that the droplets contained within the channel never touch the wall exterior and this prevents blockages in the device.

In this recipe, we will construct a very basic microfluidic device in which you should be able to observe laminar flow and microfluidic mixing. The channel pattern is cast as raised areas on a silicon wafer using a positive photoresist to form the Y shape. Silicone elastomer is then used to make a PDMS negative replica of the channel system. Holes are punched into the ends of the negative replica so that fluid can pass in and out of the device once the siloxane replica is adhered to a glass slide. The siloxane negative must adhere to the glass very strongly as the pressure in a microfluidic channel can be very high and you do not want liquid to seep out of the device. Glue and also unhardened prepolymer siloxane mix can be used to glue sections in place but this can yield poor results. The standard technique is to use an oxygen plasma cleaner to generate a reactive surface layer of SiOH on the siloxane surface so that when it is pressed against a similarly treated layer of glass Si–O–Si bonds it will form between the layers by condensation. This is far superior to using a glue, as the bonding is directly between the two substrates. Another advantage of the plasma treatment is that it renders the siloxane surface hydrophilic, which enhances capillary wetting in aqueous applications. However, prolonged exposure to the reactive plasma will remove the flexible organic components in the elastomer to leave behind only brittle SiO_2 (Fig. 5.17).

To make a positive template for a microfluidic channel device you will need:

Although the resolution of this template is not high enough to need preparation in a clean room environment, you will need to work in a laboratory setup for photolithography as you will be using a photoresist to pattern a silicon wafer.

1. All of the equipment and chemicals as outlined in the recipe detailing '*To spin coat and pattern a photoresist layer onto a substrate.*' The only difference is that the template you will be using as a lithographic pattern is a mask with your channel

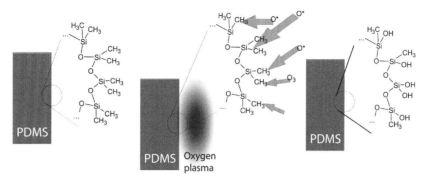

FIGURE 5.17   A schematic of the treatment of a PDMS stamp and a glass surface with oxygen plasma so that they can be bonded together. (*Image used with the permission of Alar Ainla*)

pattern on it. This can be printed onto a gelatine film using a high-resolution ink jet printer fairly cheaply. In this preparation, we will use a basic Y-shaped pattern with channels 100 μm wide.

2. A tin of Sylgard 184 silicone elastomer and the corresponding crosslinker solution. This will normally come as a kit. You will want to use a transparent product as this makes it easier to see air bubbles during processing and the substrate when the time comes to peel off or cut out the solidified stamp.
3. A plastic cup and plastic disposable knife for mixing the elastomer and crosslinker in.
4. A set of scales and tissue paper over the top of it.
5. A Petri dish for casting the elastomer in.
6. A vacuum pump and a bell jar. Line the inside of the bell jar with tissue.
7. A drying oven set to 70 °C.
8. A scalpel for cutting out a section of the silicone with.
9. A needle or sharp punch with the same diameter as the channels. A biopsy punch can be adapted for this purpose.
10. A clean glass slide. This will form the base of the microfluidic device.
11. Ethanol for cleaning and an air line for drying the components with.
12. A sheet of sand-blasted glass large enough to sit the slide and the cast silicone elastomer section on. It must be able to fit inside the plasma cleaner.
13. An oxygen plasma cleaner.
14. Tubing for connecting to the microfluidic device once it is assembled.
15. What goes in and out of the device is up to you but you will need a syringe and needle to deliver the fluid and an automated syringe pump to regulate the flow. You may have something completely different in mind for your device and these items are not strictly part of the device assembled in these instructions.

To make a positive template for a microfluidic channel device:

1. The first step is to prepare the channel template to be replicated. Here we are going to use a Y channel formed from a positive photoresist cast onto a silicon substrate. Follow the instructions for spin coating the photoresist onto the silicon and then exposing it to a Y-shaped pattern. After developing and rinsing the resist, you should have a section of photoresist standing proud of the flat substrate surface.
2. Place the silicon template into the Petri dish.
3. Place the plastic cup onto the scales and zero them. Open the lid of the elastomer and try to pour it into the cup. To fill a standard Petri dish will only require 10 g. It is extremely viscous but be patient, if you overtip the container then you might end up with lots coming out at once.
4. In the plastic cup add the crosslinker to 10% of the elastomer weight to within a gram or so. The more crosslinker there is, then the faster the silicone will set.
5. Stir the contents of the cup together using the plastic knife for at least 5 min. You must mix it thoroughly or you may end up with liquid voids in the set silicone.
6. Place the Petri dish into the bell jar and turn on the vacuum. You will see the silicone begin to bubble and foam as the trapped air in it expands. When it looks like it will spill over the sides of the Petri dish or is frothing up too much, then turn off the pump and let it settle back to ambient pressure. Repeat this cycle until no more air bubbles can be seen. This will take around 15 min though you will want to extend this to an hour if you are attempting to template features below a few hundred nanometres.
7. Place the Petri dish into the oven and let it polymerise at 70 °C for 2 h. After this time, the silicone should feel solid and rubbery. The silicone will polymerise at room temperature, although this will take a few days.
8. Cut out the patterned section and peel it from the positive template. You will now have a section of silicon with a negative Y channel in it.

9. Use the punch to drive holes through the silicone at the ends of the channels. You will not be able to do this when the section is stuck to the glass.

10. Rinse the silicone section and the glass slide in ethanol to clean it. Do not leave the silicone in the ethanol for longer than a minute or it will begin to swell and distort the channels.

11. Dry the glass and silicone thoroughly using an air or nitrogen line. Any remaining ethanol will make the plasma unstable.

12. Place the glass slide and the silicone section onto the blasted glass plate. Make sure the templated section of the siloxane is pointing face up.

13. Place the blasted glass plate into the plasma cleaner. Follow the machine-specific instructions and set the plasma cleaner for 2 min.

14. Once the plasma cleaner has finished you will have about a minute to assemble the microfluidic device before the reactive silica and siloxane surfaces become terminated by hydroxyl groups. Press the templated section of siloxane onto the silicon slide so that the template channels are next to the glass. Hold it in place firmly for a few seconds while the Si—O—Si bonds form and affix the two surfaces together. Baking the device at 70 °C for 30 min will improve the seal.

15. The next step is to cut up some suitable tubing and insert it into the holes you punched through the siloxane to act as inlets and outlets. The tubing should fit snugly but if you fear there may be leakage dab some of the unheated and unpolymerised siloxane around the join to form a seal. This will harden in a day or so.

16. Affix your inlet tube or tubes to the respective syringe pumps you will be using. You now have a very simple microfluidic device (Fig. 5.18).

Watching what happens in your device can be as simple as looking at it under a microscope. You should be able to focus on the channels through the glass slide easily. You can also focus on the channels through the siloxane if it is the optically transparent kind.

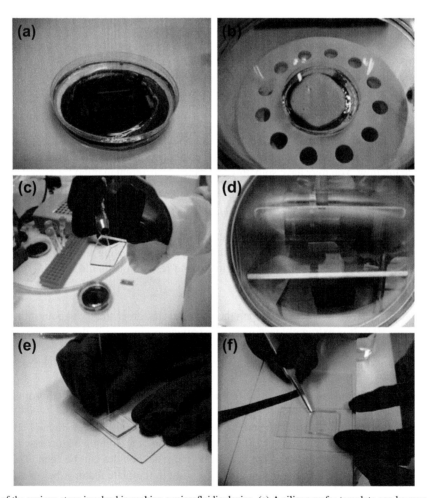

FIGURE 5.18 Photographs of the various steps involved in making a microfluidic device. (a) A silicon wafer template can be reused many times. (b) The mixed PDMS is degassed under a vacuum causing it to foam. (c) Once the PDMS has set it is cut out and peeled from the template. Holes are punched through to form inlets. (d) The PDMS component and a glass slide are plasma cleaned resting on sand-blasted glass. (e) The plasma-exposed faces are pressed together to form the microfluidic device. Inlet pipes are then pressed into the holes formed earlier.

Using plasma cleaning to make the surfaces reactive gives a very high binding strength of the siloxane to the glass, which is required for applications where high capillary pressures will be present. However, should you want to prepare a simple device without the use of a plasma cleaner you can try using a fine paint brush to paint the siloxane with the mixed prepolymer before pressing it into place on the glass. This can be very difficult as the still fluid PDMS has a tendency to spread out and block the channels in the device. You may also find the binding not as strong and the elastomeric section may come off under pressure. As a quick and fun demonstration you might like to put two different-coloured dyes in water solutions through each arm of the Y channel and see what happens when they meet. The liquids do not mix directly and will both form a separate flow next to one another that slowly blend together through diffusion further down the channel. Alternatively, you can try to form bead droplets in the main channel by passing coloured water through one and a fluorocarbon oil phase through the other.

MAKING WIRE TIPS FOR SCANNING TUNNELLING MICROSCOPY

As has been explored elsewhere, there are a variety of parameters that can affect the resolution of an STM or AFM image (Giessibl, 2003). Of all the parameters that determine the clarity of the gained image the sharpness of the tip used is the key to being able to visualise a surface at the atomic scale. Under ideal conditions, the very end of a tip will be sharp to the point of a single molecule or atom. For STM imaging this means the surface area through which electrons in close proximity to a sample surface will tunnel into the tip as a detectable current. Getting a sharp tip will require practice and patience and tips need to be used before you can tell if one is sharp enough. Standard imaging of a freshly prepared STM tip in an electron microscope can give you a rough idea of the geometry but you will not be able to see the 'business end' to the same resolution that an STM image will reveal. The recipes outlined in this section are concerned with the various methods for making platinum/iridium alloy and tungsten tips sharp enough for atomic-scale resolution. These tips should be capable of imaging highly oriented pyrolitic graphite (HOPG) which is commonly used as a calibration standard for STM scanners. The delocalised π electron system of the graphite layers acts like a pool of conductance, which should show up easily to an 'average' tip. The limits of STM and AFM resolution are still being explored and current ultra-high vacuum systems are capable of seeing individual atomic orbitals (Hembacher et al., 2004). This is accomplished by using the p orbital in a graphite surface as a probe in a combination of STM and force microscopy. The frontier p orbital of a carbon atom then becomes a point of interaction looking at a tungsten tip — like normal STM but with tip and sample reversed. This frontier orbital was used to image the four hybridised orbital lobes presented by a single tungsten atom. This is rather more advanced than the day-to-day imaging of conductive surfaces you may encounter in the laboratory. An advantageous quirk of STM tip preparation is that, while you should get a reasonable image with most of what you make, sometimes a tip will give an incredible image from being genuinely atomically sharp by chance. Only repeated practice and adjustment of your preparations will yield this occasional result; so always keep track of anything you did different during preparation variations. Surprisingly, it is relatively easy to prepare STM tips when compared to the complicated microfabrication and lithographic methods used for making AFM cantilever probes. In fact, a tip capable of seeing atoms can be made with a simple pair of pliers and a wire cutter (Fig. 5.19) !

MAKING A PLATINUM/IRIDIUM STM TIP

This recipe is by far the easiest in this book to do and for STM imaging performed under ambient conditions a 90% platinum 10% iridium alloy wire tip should be the first choice. This noble metal composition has an extremely low conductivity (0.000025 Ω cm^{-1}) and is highly resistant to oxidation. These properties have seen this alloy used as electrical contacts in pacemakers, catalysis, thermistors, high-end spark plugs and other applications where resistance to chemical reaction is important for the function of the device. Other metals, such as tungsten, must be used under vacuum because surface oxide layers impede tunnelling which renders a tip unusable. In open air, the platinum alloy does not suffer this limitation and a freshly made tip will last for many months or longer if kept free from contamination. It is better to use a platinum/iridium tip if attempting to image under an oil or liquid. This consideration is especially important for electrochemical STM applications where the tip must resist redox reactions. The most surprising aspect of this recipe is that, using a pair of pliers and a wire cutter, you will be able to get a tip capable of atomic resolution. One comparison for this is that it is like imaging a golf ball using a whole mountain turned upside down. This is due to the malleability of the alloy, which means it can be drawn to a fine point. The major limitation is that these tips are only suitable for imaging flat surfaces such as HOPG or conductive molecular layers assembled on an HOPG surface. The sheared metal beyond the very point end of the tip is too ragged for the clear imaging of samples standing proud of the scanned area. This can be overcome by etching platinum/iridium tips but this is comparatively more difficult (Fig. 5.20).

To make a 90% platinum 10% iridium STM tip you will need:

1. A spool of 90% platinum 10% iridium alloy wire. The diameter of the wire will normally be 0.25 mm for STM but check the diameter of wire that fits in your STM before you order. An 80% platinum and 20% iridium wire is also commonly available but be sure to not use this composition. Platinum alloys containing a higher percentage of iridium are slightly stiffer and will not draw as well.

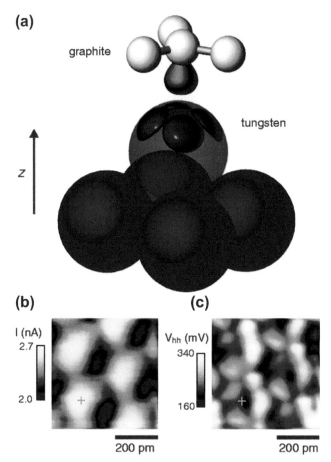

(a)

graphite

tungsten

z

(b)   **(c)**

I (nA)   V_hh (mV)
2.7   340

2.0   160

200 pm   200 pm

FIGURE 5.19   (a) A diagram showing the 'tip' orbital protruding from a graphite layer (white). The black spheres are tungsten atoms and the red lobes denote the distribution of tungsten electrons about the nucleus. (b) An STM image showing a constant current height taken of a tungsten surface and (c) the higher order harmonic image of force interactions during the scan, which map the distribution of electrons about the tungsten atom. *Adapted with permission from Hembacher et al. (2004). Copyright 2004 Nature Publishing Group.*

2. A pair of flat nose pliers. It can be better if the metal grips do not have cross-hatching in the grip, but it is not so important. You want to avoid squeezing the wire flat or deforming it as it might not fit in the STM holder.
3. A pair of clean and sharp wire cutters. You want the kind that chops edge to edge and not with overlapping blades like scissors have.
4. Ethanol for cleaning wire and any of the tools to be in contact with the tip.

FIGURE 5.20   An STM tip on the surface of an HOPG layer is capable of atomic resolution and this is comparable in scale to imaging a golf ball using a mountain. Not to scale.

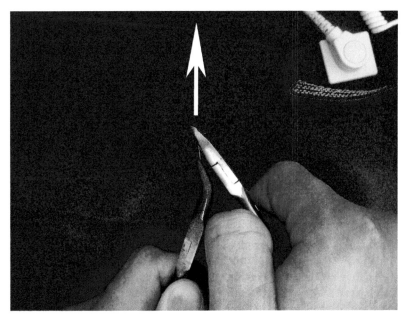

FIGURE 5.21   A photo showing a Pt/Ir wire being pulled into a tip directly ready for use in STM experiments.

5. Optional: A plasma cleaner with a metal bulldog clip for holding the wire tips upright with.
6. A tip box lined with polystyrene or foam that the blunt end of a tip can be poked into for storage. It should have a lid so that dust in the air cannot attach itself to the tips, which will ruin them.

To make a 90% platinum 10% iridium STM tip:

1. Cut a section of wire 2—3 cm long from the spool. The exact length of wire needed will depend on what your STM system is designed to hold. You will need the desired length of the tip plus another 0.5—10 mm, which will be torn from the end. You may find that the existing STM tip is long enough to be sheared again. Recycling a used tip is a very good way to keep costs down.
2. Clean your pliers and wire cutter with ethanol. Also clean the wire by dipping it in the ethanol then allowing it to dry. After this point, do not touch the wire with anything other than the pliers. The bulldog clip can also be cleaned at this stage if you are using a plasma cleaner.
3. Hold the straight piece of wire with the pliers gripping over half the wire length. You will need a firm grip but not so hard that the wire deforms.
4. Using the wire cutters grip the end of the wire diagonally 5—10 mm in from the free end. **Do not cut through the wire!**
5. Using the wire cutters pull hard so that the wire is torn rather than cut. You should hear a ping as the wire breaks. This tearing motion is what produces an ultra-sharp point and you may need to practice this a few times.
6. The tip should be handled with clean tweezers or pliers applied a few millimetres behind the sharp end. The tip can now be mounted in the STM or stored by poking the blunt end into supporting foam in a clean box and closing the lid. For optimal results, you may wish to use a plasma cleaner to ensure the tip is as clean as possible before use.
7. To plasma-clean the freshly prepared tips they need to be held in a bulldog clip or a similar metal clamp. Any support made of plastic will be degraded in the plasma cleaner. Make sure the clip is positioned on its side such that the tips never touch anything.
8. Place the clip and the tips in the plasma cleaner and run a standard cleaning program as given by the manufacturer.
9. The tips can then be used immediately or stored for later use (Fig. 5.21).

MAKING A TUNGSTEN STM TIP BY WET ELECTROCHEMICAL ETCHING

With thanks to Kane O'Donnell and Natalie Wood

Tungsten tips are almost exclusively used for ultra-high vacuum applications. Etched platinum—iridium tips require a tricky two-step procedure; so it is normally much easier to use tungsten, which is more amenable to etching and is a one-step process (Musselman et al., 1990). The downside is that tungsten tips must be placed into a vacuum for storage or else will form an oxide layer at the surface within 10 h. For imaging this means getting them into a vacuum quickly and imaging immediately. In an ultra-high vacuum STM you can begin to image and wait for the oxide layer to be worn off or use thermal techniques to remove the oxide. In this recipe, we will explore some popular methods for etching tungsten tips using direct current and alternating current in a basic medium. It is possible to purchase etching kits but here we will stick to using very basic equipment. The electrochemical etching of tungsten follows these basic reactions:

$$\text{Cathode: } 6H_2O + 6e^- \rightarrow 3H_{2(g)} + 6OH^-_{(aq)}$$

$$\text{Anode: } W_{(s)} + 8OH^- \rightarrow WO^{2-}_{4(aq)} + 4H_2O + 6e^-$$

$$\text{Overall: } W_{(s)} + 2OH^-_{(aq)} + 2H_2O \rightarrow WO^-_{4(aq)} + 3H_{2(g)}$$

It can be seen from this equation that the solid tungsten wire is oxidised to form a soluble tungstate anion in water. This action eats away the tungsten predominantly at the meniscus of the air/liquid interface where the wire enters the solution. Eventually this reaches a point where the wire is so thin it will snap and the bottom end will drop away. When this happens the current flow should be stopped as quickly as possible as any further etching will only blunt the tip. The section that has dropped away will normally be the sharper tip as current flow in the wire is ceased the moment it breaks away. In this way, electrochemical etching can provide you with two tips for every section of wire you use, although harvesting two tips is only really possible when using the 'loop' method explored further on. The section of wire acting as the electrode may be slightly blunted unless the current flow is stopped extremely rapidly after the wire has broken into two. Some direct current etching power supplies have differential analogue electronic circuits, which can detect the sharp drop in current as the wire gets thin and close to breaking. These are called dip etch controllers and are not easy to make from scratch unless you are good with electronics. Because these controllers can be expensive, the best compromise if you do not have one is to harvest the section of wire that falls away during etching.

The choice of alternating current or direct current etching depends on the shape of tip you would like. Direct current etching produces tips more suitable for ultra-high vacuum applications as they have a low aspect ratio. These short tips are relatively inflexible and comparatively strong, eliminating imaging errors arising from flex in the tip shaft. Alternating current-etched tips have very high aspect ratios, making them useful for rougher surfaces, and are better for wax coating if the tip needs to be used in electrochemical applications.

For either etching methods it is important to use tungsten wire with a smooth surface and a high level of uniformity about its radius. Surface roughness along the tip can lead to localised fluctuations in current density and will lead to noise in the STM signal. Similarly, defects in the conical shape at the apex will lead to differences in the reaction of the piezo stage during trace and retrace cycles as one side of the tip may be more susceptible to conduction than the other. Some of these issues are resolved by polishing a good portion of the tip length with a light etching process to smooth it before sharpening. The positioning of the counter-electrode is also influential in the geometry of the finished tip. The counter-electrode should be made from a gold or platinum wire formed into a ring so that the electric field surrounding the tip during etching is as uniform as possible. This produces an even etch rate around the wire circumference ensuring symmetry in the point. Because the dissolution of the tungsten is fastest at the meniscus surrounding the wire it is important that this remains uniform also. Hydrogen is generated at the counter-electrode during DC etching and this can disturb the surface of the liquid leading to a distorted meniscus. Even worse, the bubbles may accumulate around the tip disrupting the meniscus completely. Ultimately, this produces a tip with a serrated appearance and will result in a noisy image from uneven conduction. To get around this, the ring sits around the outside of a cylindrical PTFE or alumina shield sitting above and below the liquid layer. Bubbles formed on the cathode then break the surface but never reach the tip or disturb the meniscus.

ETCHING A TUNGSTEN TIP USING ALTERNATING CURRENT IN A BEAKER

With thanks to John Mitchels. Adapted from the Veeco user manual for STM.

The solution used as an etchant is sodium nitrite, which is used as a saturated solution. An attraction of this particular recipe is that you can etch more than one tip at a time. This is accomplished by sticking the tungsten wire into an integrated chip (IC) holder. This is a cheap electronic component used to hold chips onto prototyping chipboard (PCB). An IC holder can be slotted into a PCB board such that each wire is connected to the variac terminal in parallel. This means the AC current will flow through each tip allowing them to be etched as a batch.

To etch tungsten tips in a beaker using alternating current you will need:

1. A variac autotransformer. You will also need suitable wires and clips for making electrical contact with the respective platinum and tip electrodes. This is a device that will adjust the voltage of the applied alternating current. Keep it unplugged until it is required. **There is a risk of electrical shock working with this equipment. Take all necessary precautions. Do not touch any part of the circuit while in use.**

2. 40 mL of a 5 wt% solution of sodium nitrite ($NaNO_2$) in water. You will use around 40 mL for a batch of 10−12 tips at once. As the solution can be reused until it gets cloudy, you will find that 40 mL will be able to etch between 60 and 80 tips. Make this up fresh in a 50-mL beaker. Alternatively, you can use A 2 mol/L sodium or potassium hydroxide solution.

3. Ethanol for rinsing the wire and tips with. This should be kept in a 50-mL beaker.
4. A loop of platinum wire that will act as the counter-electrode. This should sit around the interior circumference but not touching the sides and have a centimetre of clearance from the bottom. If you are working with an STM, it is likely you will have some platinum/iridium wire to hand and this may be used alternatively. You will also need a crocodile clip and a wire for connection to the variac. Have a blob of blue tack to hand which will be used to keep the electrode in place.
5. A reel of tungsten wire 0.25 mm in diameter or whichever diameter is accepted by your STM tip holder. You will need wire clippers for cutting it into sections.
6. Optionally, 40 mL WD-40 in a 50-mL beaker can be used as an antioxidant to apply to the tips after etching and rinsing with ethanol. If you will not be placing the tips into a vacuum immediately then placing them in WD-40 will help to slow the oxidation.
7. A tip holder made from an IC chip holder and plugged into a PCB so that everything is connected electrically. If you use a large-enough piece of chipboard, then this will sit flush on the top of the beaker and hold the tips straight in the solution. You may find it necessary to adjust the height at which the tips sit in the solution. You can do this by holding the chip holder in a clamp or adding and removing solution as required.
8. A 20−100× magnification microscope. This will be used to make a rapid assessment of the tip quality. Any major defects can be spotted at this low magnification.

To etch tungsten tips in a beaker using alternating current:

1. The variac should be unplugged. Set the dial to 30 V.
2. Cut the tungsten wire into 1.25-cm sections with the wire clippers. Briefly check that they will fit into the STM holder. Sometimes wire cutters can flatten the end in a way that prevents it from being placed in tube-style tip holders.
3. Load the tips into the IC chip holder so that they are all sticking up straight. Attach a crocodile clip to the electrical contact of the IC chip holder. The other end of the wire is connected to one of the variac terminals.
4. Place the platinum ring electrode into the beaker containing the sodium nitrite solution. If possible, use the pouring lip of the beaker to feed the end of the platinum wire out of. Clip a crocodile clip onto this and affix it to the exterior of the beaker. The other end of the wire is connected to the remaining variac terminal.
5. With the tips pointing down, immerse the ends 2 mm below the surface of the sodium nitrite solution. Any deeper than this and there will be so many bubbles from the tip that the end result will be misshapen. However, there still needs to be enough weight on the end so that the wire breaks away during etching.
6. Plug in and turn on the variac. The tips should produce bubbles where they are immersed and may begin to glow. When the bubbling and sparking stop or you see the end of the wire fall off then stop the variac immediately. Over time you will begin to see the solution darken or a very small precipitate falling away from the wire. If there is too much of this, then use fresh solution.
7. Reduce the depth of the tip end to 1 mm. Turn on the variac for 15 s at 30 V to re-etch the tip then turn it off again. The longer you leave the tips in the solution while the variac is on, then the blunter the tip will be. Try adjusting the voltage or etching times to see what works best for you.
8. The very moment you decide the tips are etched then turn off the variac and remove them and rinse the ends of the tips in the beaker of ethanol. When you pull out the tips hold them upright so that solvent drains away from the point. You do not want to risk any potential contamination on the end.
9. Allow them to dry very briefly and then store in a vacuum or dip into the beaker of WD-40. It is a good idea to protect them from oxidation as fast as you can.
10. Check the tip shape and smoothness under the light microscope. Any tips appearing obviously bent or serrated will probably not be any good for use (Fig. 5.22).

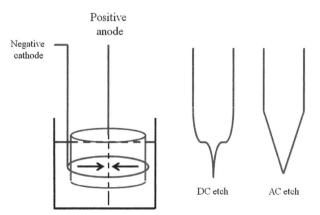

FIGURE 5.22 A picture of the setup for etching tips using an alternating current.

ETCHING TUNGSTEN TIPS USING DIRECT CURRENT

With thanks to Kane O'Donnell and Natalie Wood

Using a direct with thanks to Kane O'Donnel and Natalie Wood current will give tips that are shorter and stronger. The tip ends will appear concave under a high-magnification light microscope. You will need to have a tip etch controller to get good results from this particular method. The voltages and times should be played with depending on how you find the setup works. If you do not have a tip etch controller you can try using the loop method and this will be described further on.

To etch tungsten tips using direct current you will need:

1. A tip etch controller. This will apply a DC voltage to the system. Do not plug it in until it is to be used. *Take all the necessary precautions for working with electrical power supplies.* You will need wires and crocodile clips for making connections with the respective electrodes.
2. 40 mL of a 5-mol/L solution of NaOH in a 50-mL glass beaker. This can be prepared by dissolving 8 g of NaOH into 40 mL of water. *This solution is corrosive, handle with care!* Alternatively, you can use KOH instead at the same molarity. The concentration of the OH^- ions is the important factor.
3. A height-adjustable clamp that can be raised or lowered while holding a tip. In contrast to the AC method, you will be etching one length of wire at a time. The setup depicted uses a clamp to hold the wire and this is mounted on a thread that can be accurately raised and lowered. This is important for smoothing the surface of the wire with an initial polishing step.
   Because you will be raising and lowering the wire into the etching solution while the circuit is live, *make sure the parts of the system carrying current will be isolated from the parts you will be touching.* The etching is performed at 9 V, which should not be enough to give you a shock, but why risk finding out?
4. 40 mL of isopropanol in a 50-mL beaker and 40 mL of acetone in a 50-mL beaker. This will be used for cleaning the wire and tips.
5. A sonic bath for cleaning. This must be filled such that the beaker of isopropanol can sit in it. You will also need a clamp for holding the etched tips in the isopropanol during sonication. *Do not hold tips in an active sonic bath with your hands!*
6. Tungsten wire that fits with your STM and wire cutters for making the tips with. You can use a tungsten/rhenium alloy, as this will give better-quality tips. For ultra-high vacuum systems, the tips will normally be mounted onto a metal holder, which can serve as an electrical contact.
7. A platinum wire ring cathode. This should fit inside the beaker with about half a centimetre of clearance from the interior wall. Bubbles will form on this electrode during the etching process, so you will need a cylindrical shield that the wire can sit around to stop the formed bubbles from disturbing the meniscus. The shield should be made from PTFE or Al_2O_3. The cheapest option is to get a small piece of PTFE piping, which fits into the etching beaker with plenty of room, then wrap the platinum wire around it. The electrode and shield should be placed in the beaker according to the diagram below. Note the gap at the bottom that will allow charge carriers to move freely in solution. The top of the shield protrudes from the solution so that the liquid surface inside of the shield will remain calm.
8. A $20-100\times$ optical microscope for examining the tips.

To DC etch tungsten tips:

1. Cut the tungsten wire into 1.25-cm lengths or whatever length is required by your STM. Take a length of wire and clamp it in a holder.
2. Place the beaker containing acetone into the sonic bath. Dip the wire into the acetone and clamp it in place in the holder. Tips mounted in metal holders must be submerged into the acetone completely as they will need to be ultraclean. Sonicate the wire for 5 min to ensure it is clean.
3. Take the holder out of the acetone and let the wire dry. Clamp the holder in place over the NaOH solution so that the wire points directly down.
4. Connect the red positive terminal (anode) to the electrical contact for the wire. Connect the black negative terminal (cathode) to the platinum wire. Turn on the tip etch controller and set it to supply 9 V potential difference. Once you have done this, turn it off again.
5. Lower the wire into the NaOH solution to a depth of 5 mm. Turn on the tip etch controller at 9 V and slowly draw the wire out of the solution over 30 s. This step will remove the surface roughness on the tip exterior. If you do not do this, then you may end up with serrations in the tip. Turn off the power.
6. You might want to remove the wire from the holder and examine it at this point to check if the surface is smooth. Replace it in the holder for further etching.
7. Dip the end of the tip 2 mm below the surface. Turn on the tip etch controller at 9 V. During the etching, the current will be somewhere around 60 mA and you may see a small stream of sediment falling from the wire. If this sediment is undisturbed,

then this signifies a lack of turbulence, which generally leads to good tips. The controller will turn off the current the moment it detects a significant current drop as the bottom section of the wire drops off.

8. Even without the voltage applied there will be some etching still occurring in solution; so as soon as the power is off lift the tip out of the solution and rinse it in the isopropanol. Keep the tips pointing upwards so that the solvent drains away from the apex.
9. Check the tip under the optical microscope again. You may need to repeat step 7 two or three more times to get a really good tip.
10. After you are satisfied that the tip is done, let it dry briefly and store immediately under ultra-high vacuum. This level of vacuum requires specialist equipment, which should be available if you have an ultra-high vacuum system; otherwise, store the tip in WD-40 and use as soon as possible (Fig. 5.23).

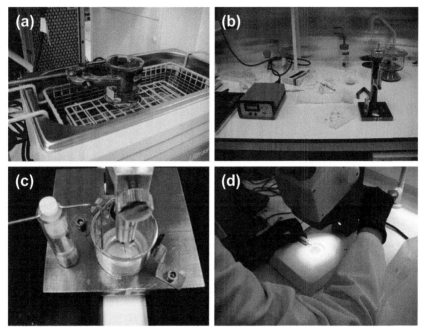

FIGURE 5.23 Photographs of the steps to etch a tip using a direct current. (a) The tips are cleaned by sonication. (b) The tips are mounted and placed into the holder. In this case, a specially constructed set. (c) The tip is lowered into the solution and etched. (d) The basic morphology is checked by light microscope.

DIRECT CURRENT ETCHING OF TUNGSTEN TIPS USING THE 'LOOP' METHOD

The loop method is a simpler variation that does not require a tip etch controller to give a really sharp tip. An added bonus is that you will yield two tips from one piece of wire as the section that falls away can be gently caught. The fallen tip will normally be the sharper of the two by a wide margin. The major difference is that instead of dipping wire into bulk solution the wire is pushed through an aqueous membrane held in a platinum wire loop electrode. The weight of the wire gently pulls it during the etching, leading to a very sharp high aspect ratio tip.

To DC etch a tungsten tip using the loop method you will need:

1. Tungsten wire and wire cutters.
2. A direct current power supply with wires and crocodile clips. Keep this unplugged until use.
3. 40 mL of a 2-mol/L solution of NaOH. This can be prepared by adding 3.2 g (0.08 mol) of NaOH to 40 mL of water. You will only be using a very small amount, so this can be kept as a stock for future use.
4. A platinum electrode made from wire and formed into a loop. Generally, most preparations use a wire 0.4–0.5 mm in diameter as this is strong enough not to bend when a water droplet is placed on it. Make a loop from the wire ~0.5 cm in diameter and twist the end so that you have a continuous hoop with a long piece of straight wire extending from it. The long wire will be used to make electrical contact to the power supply. The wire will be held in place with a clamp stand. Ensure that any component that will be electrically live is isolated from the clamp stand. Tape and/or putty can be used for this.
5. A crocodile clip or integrated chip holder for placing the wire in and holding it straight. This will be attached to a clamp stand; so again ensure that the circuit components are isolated from the clamp.
6. A can of shaving foam. This will be used to provide a soft landing for the tip that drops away. Alternatively, you can use a small glass tube shorter than the wire section that will drop off. In this case, the tip will fall into the tube without the end making

contact with anything. Shaving foam will give a better result as there is no risk of bending the tip. You will also need a Petri dish to sit the foam in.

7. Sonic bath and a beaker of acetone for cleaning the tip in. It is good practice to have a metal tip holder that can be sat in the acetone bath. Holding tips in with a pair of tweezers often leads to the tip falling in completely, which will render it useless (Fig. 5.24).

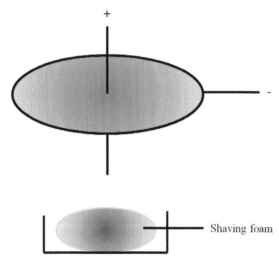

FIGURE 5.24   A schematic of the setup for etching with direct current using a loop.

To DC etch a tungsten tip using the loop method:

1. Cut a piece of wire roughly twice the length of what you will need to use as a tip. Do not use a length of greater then ~4 cm as this will apply too much load during the etch and the wire will snap. This will not result in a good tip.

2. Place the wire into the holder and clamp it in place. Connect the red positive (anode) output of the power supply. The bottom of the wire should only be a centimetre or so from the Petri dish, which should be positioned underneath it.

3. Dip the platinum hoop into the NaOH solution and draw it out slowly. This should leave a meniscus film of solution in the hoop.

4. Carefully move the loop so that the tungsten wire goes through the centre of the meniscus until it is halfway up. Clamp the platinum loop in position and connect it to the black negative (cathode) output of the power supply.

5. Squirt a small amount of shaving foam into the Petri dish under the wire.

6. Set the output to between 3 and 5 V then turn it on. Below 3 V, the etching will take an impractically long time; above 5 V, you may see oxygen being evolved and the tip blackening from the formation of WO_3. You may find that the optimum voltage varies depending on your setup.

7. The etching will take a few tens of minutes but this can vary depending on the thickness of the tungsten wire. The etching will be over when the bottom tip falls into the shaving foam. Hopefully, the newly sharpened end will not come into contact with the foam or anything else. Turn off the power supply immediately and remove the top tip from the loop. If you use a normal DC power supply, then this may already be too blunt for use no matter how quick you are.

8. Using tweezers to hold the tip briefly rinse away the foam with acetone. Mount the tip in a metal holder and then sonicate the fresh tip for five minutes fully immersed in the beaker of acetone. When removing the tip from the bath keep it pointing upright.

9. Allow the tip to dry briefly, then store under vacuum. It is now ready for use but will need to be thermally annealed in an STM to remove the oxide layer.

Scanning electron microscopy of the top and bottom sections should reveal the bottom tip to be much sharper and longer. You should be able to image the very tip, which will look almost spherical. The radius of this sphere is often a good indicator of the sharpness with a low radius end being more desirable. However, the final indicator of quality is how well the as-produced tip images in the STM.

STICKING A PARTICLE ONTO A SCANNING PROBE MICROSCOPY TIP

With thanks to Esben Thormann

Much of the focus in making tips for scanning probe microscopy (SPM) is on the sharpness and symmetry. Although these aspects are definitely important to get a good picture, they are not always the right tool for force or friction measurements. For example, if you wanted to perform a force curve on a cell stuck to a polycarbonate plate, then you might find a sharp tip penetrating right through the cell instead of squashing it under load. For applications like this, you need to stick a particle with a similar radius of curvature to the end of your cantilever so that the applied load is spread over a larger area. This might seem trivial in concept but it is impossible to simply pick up a micron-sized sphere with tweezers, dip it in glue and affix it to the end of a tip by hand. Taking a top-down approach to affixing a particle requires micromanipulation techniques to get right as the human eye and hand, however steady, are limited in control at the small scale. In this recipe, we will use a micromanipulator setup to attach micron-sized silica particles to the end of an SPM cantilever with a heat setting glue. The method described here uses a commercially available micromanipulator system mounted on an inverted light microscope with optics capable of a large depth of field. A large depth of field means that the microscope can focus at a given magnification over a long vertical distance. This is very important as you will be positioning components above one another and need to bring them in and out of focus as you line them up. At its most basic, the micromanipulator is a sharp stick that can be moved around slowly and accurately on the micron scale up and down, left and right, backwards and forwards. The tip of the manipulator does not have any kind of 'gripping' system and relies on capillary forces to stick a single silica sphere to the end. When the sphere encounters a surface that is more attractive then it will detach from the end of the tip; in this case, the surface tension of glue at the end of a cantilever will pull it in. If you do not have a micromanipulator, then it is possible to work around the problem by using the motors and optics of an SPM in a similar way. Notes are provided on this method, although you might be limited by the movement range and optical range of the microscope.

To make an SPM probe functionalised with a colloidal tip you will need:

1. An inverted overhead light microscope with a range of lenses having long focal distances. In order to coat tips you will be working at 20× magnification with the tips and 200× magnification for the silica particles. You may need even higher magnification depending on how small your silica particles are.
2. Tipless cantilever probes and tip tweezers for moving them around with. You can order these in boxes and it is up to you if you want to use triangular or rectangular probes. Rectangular probes are better for lateral measurement applications because the torsional deflection can be more easily calibrated.
3. A thermosetting glue or epoxy resin. This is going to form the bond between the colloidal tip and the end of the cantilever. If you decide to use a two-component epoxy resin, be aware that it will begin to set from the moment you start mixing and this limits the amount of time you have for manipulation. In this recipe, we will be using thermosetting glue.
4. A particle for attaching to the end of the tip. Here, we will be using uncoated colloidal silica spheres 7 μm in diameter suspended in water.
5. A heating block capable of reaching the melting point of the glue you are using, which can be mounted under the microscope. In this recipe, the melting point of the glue is 110 °C, which can easily burn you. *Make sure you do not touch the block when it is hot!* If you are attempting this procedure with an SPM, then you might have access to a heated sample stage. This item can also serve as a hot plate under the microscope.
6. Glass slides. These are going to be your working surface on the hot plate.
7. tungsten wire 0.25 mm in diameter, though this can vary depending on your setup. You will need at least two 2−3-cm-long sections for making the tips. These sections will be etched into sharp points.
8. A direct current power supply set to between 10 and 20 V. *Risk of electric shock!* Keep this turned off when not in use. You will need the terminals attached by wires to crocodile clips. The negative terminal (black) can be clipped to a metal spatula, which will act as a counter-electrode. The crocodile clip connected to the positive (red) terminal should be covered in rubber, so you can hold it.
9. 250 mL of a 1 mol/L solution of KOH in a 300-mL beaker. *Corrosive − do not get any on your skin.* This can be prepared by weighing out 14.2 g of KOH and dissolving it in 250 mL of water. The beaker should be clamped in place or held inside another glass pot with a stable base.
10. A beaker full of water for rinsing the KOH off the tungsten tips with.
11. A micromanipulator with a tip holder on it attached to the microscope. Tungsten tips will be attached to the end of a holder using putty.

To make an SPM probe functionalised with a colloidal tip:

1. Take the positive terminal connected to the spatula and place it in the KOH solution.
2. Take a section of tungsten wire and grip it with the crocodile clip connected to the positive terminal. Hold the clip by the rubber part and wear nitrile or rubber glove to ensure that you do not get an electric shock.
3. Turn on the power supply and dip the end of the tungsten wire into the KOH solution for a few seconds at a roughly 45° angle. You should hear a hissing sound and see hydrogen evolving on the spatula. With a smooth movement, dip the wire in and out of the solution 5 or 6 times. This will sharpen the tip and also curve it slightly, which helps during the manipulation stage to keep the rest of the tip away from the surface (Fig. 5.25).

FIGURE 5.25   A photograph of a tungsten wire being sharpened by direct current etching in a KOH solution.

4. Turn off the power supply. After you have sharpened the wire, dip it in the beaker of water to rinse the KOH from the end. Remove it from the clip and place it on a glass slide so that the end is kept clean. Repeat the process with three to four more pieces of wire so that you have enough to work with.
5. With all the sections of wire on the glass slide, check them for quality under the microscope at $20\times$ magnification. Reject any that have visible lumps on the end. The tip should taper to the point that you cannot focus on the end.
6. Take a tip and affix it with putty to the arm of the micromanipulator so that there is a slight downwards curvature (Fig. 5.26).

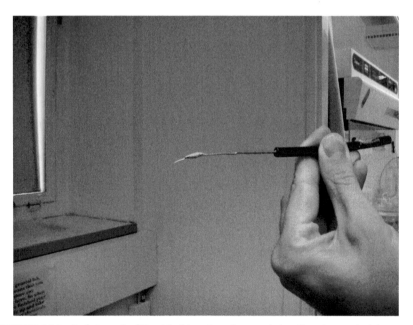

FIGURE 5.26   A photograph of the etched tungsten wire mounted on the micromanipulator arm.

7. Using tweezers place the SPM probes onto a glass slide. They should have been positioned in the box with the bottom surface of the probe pointing up. Keep them in this orientation with the cantilever pointing towards the tungsten tip.
8. Place a few grains of the thermosetting glue onto another glass slide and place it on the heating block between the cantilevers and the tungsten tip.

9. On another glass slide add a drop or two of the colloidal dispersion and allow it to dry. You will want the colloid to be dilute enough that the particles will not clump together when they dry down. You are aiming to get lone particles sitting on the glass.
10. Turn on the heating block and let it get to the temperature required to melt the glue. You should be able to watch this under the microscope as the white crystalline solid melts (Fig. 5.27).

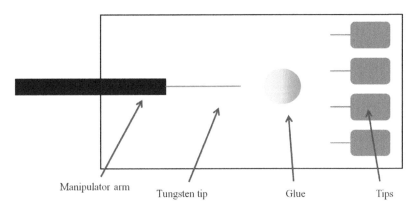

Manipulator arm   Tungsten tip   Glue   Tips

FIGURE 5.27   A schematic of how the manipulator arm, glue, dried colloidal spheres and tips to be functionalised are arranged in relation to one another.

11. Using the microscope to keep track of the relative positions, move the tungsten tip over the glue. Bring the tip into focus and then focus on a point just below it. Raise the stage until the glue comes into focus. You will now be able to lower the tip into the glue then pull it back out.
12. Move to a clean section of glass and dip the tip up and down a few times on different areas. You should see globules of glue left behind. This step will allow you to get rid of excess glue. If you put too much on the end of the cantilever, then the entire particle may be immersed, making the probe useless.
13. Raise the manipulator arm and move the stage so that the arm is positioned above the end of a cantilever. As before, focus on the tungsten tip then focus below it and raise the stage. When the tungsten tip is just above the centre and end of a cantilever lower the arm to apply a small blob of glue. Raise the arm away and then move to another cantilever and repeat the process. You may be able to apply glue to 3 or 4 cantilevers before needing to pick up more glue.
14. As the cantilevers are resting on the stage, they will remain hot enough that the glue will not set. Raise the manipulator arm away from the stage and remove the arm from the micromanipulator. You must use a fresh tip for picking up the silica or else residual glue will not let it go when you attempt to attach it to the cantilever. Give a fresh tungsten tip a quick rinse in water and then affix it to the manipulator arm.
15. Reattach the manipulator arm in the microscope setup and position it above the silica colloid dried on the glass slide. As before, focus on the tungsten tip and then move the focus to below it. Change to a $200\times$ or other suitable magnification so that you will be able to see individual particles. Raise the stage until you see the dried colloid come into focus.
16. Lower the manipulator arm onto one particle. You should see the particle jump onto the end. This action is driven by capillary forces from residual moisture in the air. Lower the stage and move it until the captured silica sphere is above a blob of glue on the cantilever.
17. Use the manipulator to lower the silica particle into the glue. It should be drawn from the tungsten tip and into the glue.
18. Repeat the placement of silica spheres until all of the cantilevers are terminated with a silica particle.
19. Turn off the heating block and allow the cantilevers to cool to room temperature. The glue should set holding the silica particle in place.
20. For quality-control purposes, it is a good idea to take photos of the tips if you have a camera attached to the microscope. This will let you accurately measure the dimensions of the cantilever, which is important if you need to calculate the torsional or normal force constants of the tips.
21. Use the tip tweezers to move the colloid tipped probes back into a safe storage box. They are now ready for use.

If you do not have a micromanipulator you may be able to perform these steps using a stationary tungsten tip held centrally over a sample in an SPM. Most systems have a microscope that is normally used for positioning the tip over a specific point. You should also be able to raise and lower the stage as well as move it in the X and Y planes. By placing a cantilever on a heated SPM stage, you can manipulate its position such that the steps outlined here can be performed to get the glue and the particles onto it. This is generally more laborious but will get the job done. For SPM systems that are not heated you can try using a two-component epoxy like Araldite. However, this type of glue will normally be too viscous to work with after about 10 min, so be as quick as you can (Fig. 5.28).

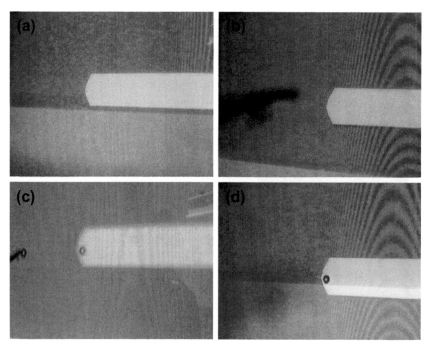

FIGURE 5.28 Images of an SPM cantilever being functionalised with silica spheres. (a) A tipless cantilever under the microscope. (b) A sharpened tungsten wire with glue on the end is positioned above the end of the tip. (c) The cantilever is raised into the tip and back down to leave a dot of glue on the end. (d) A fresh tip is used to pick up a silica sphere with surface tension and this will be deposited in the glue to produce a functionalised cantilever.

ELECTROSPINNING FIBRES FROM POLYMERIC SOLUTIONS

With thanks to Joe Harris. Adapted from Fletcher et al. (2011)

Electrospinning is the technique by which a fibre is spun from a polymeric solution by extrusion through a needle onto a target. The target and the needle are maintained at a potential difference and a solution under pressure is forced from the needle and attracted to a collection plate electrode. The process is similar to the technique of electrospraying where an aerosol solution is deposited onto a charged surface, whereas electrospinning is focussed on producing homogeneous unbroken fibres and not a surface coating. The process of electrospinning can be used to create fibres ranging from as narrow as 50 nm to those with a diameter greater than 1 µm (Vasita and Direndra, 2006). Fibres are spun from a solution of polymer dissolved in a solvent, which readily evaporates in air. The electrospinning apparatus consists of a syringe containing the polymeric solution, a flat-tipped needle or spinneret (which is maintained at a positive or negative charge by being attached to a direct current power supply) and a grounded collector plate, which is placed directly opposite the tip of the needle at a given working distance (Fig. 5.29).

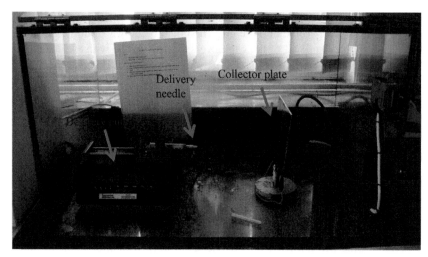

FIGURE 5.29 A photograph of the experimental setup for electrospinning a polymeric solution to form a fibre mat.

During the process of electrospinning, the polymeric solution is pushed from the syringe through the needle at a constant rate either by gravity or, more commonly, by use of an advancement pump. The polymeric solution forms a droplet at the end of the needle as a result of surface tension and, without a voltage, the droplet would simply drop to the ground under gravity. The application of an electrical potential provides the solution with a charge, which in turn causes mutual charge repulsion between the polymer molecules, a force which is directly opposite to that of the surface tension. At first an increase in the potential causes the surface of the liquid at the end of the needle to extend into a cone (called the Taylor cone after professor Sir Geoffrey Ingram Taylor who provided a theoretical understanding of the electrostatic forces involved) (Taylor, 1969). Once the potential is increased above the critical voltage, the charge repulsion forces of the liquid overcome those of the surface tension and a jet is emitted from the apex of the Taylor cone. Once airborne, the jet is drawn to the grounded collector plate. In flight, the jet whips as a result of electrostatic repulsions between like charges. Whipping causes the jet to elongate and therefore thin, whilst the extended time of flight due to the chaotic pathway taken allows for greater solvent evaporation. The polymer solution dries in flight and is deposited on the plate as a continuous randomly arranged nanofibre mat. There are a wide range of parameters that must be finely tuned to produce a fibrous mat from electrospinning and these are tabulated below. If any of these aspects of fibre production are not correct, then the most likely result is that the polymer solution will be sprayed or else the fibre will have significant beading along its length (Table 5.1).

TABLE 5.1 Parameters that Influence the Product of the Electrospinning Process

Substrate and solution-related parameters	Process-related parameters	Environment-related parameters
Molecular weight of polymer	Hydrostatic pressure in capillary tube	Humidity
Molecular weight of polymer	Voltage	Temperature
Molecular weight distribution	Gap distance	Air velocity
Viscosity	Flow rate of solution	
Elasticity	Needle aperture diameter	
Conductivity		
Surface tension		
Chemical interaction between molecules		

Electrospinning is used in the production of materials for a variety of applications such as filtration, sutures, textile fabrication and as scaffolds in tissue engineering (Burger et al., 2006). Until the mid-1990s it had been mainly artificial polymers which had been electrospun; however, a drive towards the production of biomaterials has seen an increase in the use of naturally occurring polymers as this technique is used for biomedical applications. Electrospun nanofibrous mats are an area of great interest as potential scaffolds for tissue engineering as their diameter can be altered to mimic the extracellular matrix of natural tissues by adjustment of the process parameters. Furthermore, the 3-D structure, large surface area, porosity and the ability to dope the fibrous mats with bioactive agents make them ideal structures to be seeded with cells. In this recipe, we will use the electrospinning technique to produce a fibrous polymer mat loaded with amorphous calcium phosphate, which can be used to repair tooth enamel. This will demonstrate that a polymer fibre can be formed in addition to seeding it with a material of choice.

Teeth are subject to some of the harshest conditions in the human body and as a consequence they are composed of the hardest material in the body. The primary role of the tooth is mastication (chewing of food); during this process, teeth are placed under high compressive forces ~700 N. In order to withstand these forces and repetitive acid attack the tooth has to be both hard and tough. Consequently, the tooth is a composite material with an enamel layer covering the underlying layer of dentin. Enamel is the hardest material in the body, it is 96% mineral and comprised of bundles of ion-substituted hydroxyapatite rods intermeshed together. These are embedded in the more resilient dentin, which has a lower mineral content and is therefore tougher. Although extremely hard, the enamel is constantly undergoing a process of dissolution and remineralisation. During normal enamel formation, an amorphous calcium phosphate initially forms at the surface of the tooth which then converts to the crystalline polymorph hydroxyapatite, which is more stable in the mouth environment at neutral pH. Calcium and phosphate salts within the saliva provide the material source for the mineralisation process. Exposure to acidic drinks or dietary acids shifts the equilibrium so that the tooth dissolves faster than new mineral can be deposited from saliva. This leads to cavities where hydroxyapatite crystals have dissolved to the point that the softer dentine underneath is exposed, leading to sensitive teeth. As the cells responsible for the formation of enamel are lost after the tooth is formed, enamel repair is impossible by natural means; therefore, a different route is sought to combat dentin hypersensitivity. Amorphous calcium phosphate cannot be directly added into pH neutral toothpaste as it would convert to the hydroxyapatite polymorph before it has a chance to bind to the surface of a tooth. To get around this, the amorphous calcium phosphate has to be delivered in a nonaqueous medium to the site where we would like mineralisation to occur. This is done by blending the mineral into a fibrous mat, which can be adhered to a damaged tooth. The porous fibre mat allows saliva to flow through and forms a gel on the tooth, releasing amorphous calcium phosphate directly onto the tooth surface. The calcium phosphate then transforms to apatite on the existing apatite crystal repairing the damage done by decay. In this preparation, an amorphous calcium phosphate precipitate is prepared and then mixed with poly(vinylpyrrolidone) in ethanol solution. The calcium phosphate particles will be 20−200 nm in diameter. This mixture will be placed into a syringe and then electrospun onto an aluminium foil target to produce a dry fibre mat composed of a few hundred nanometres to microns thick fibres, depending on the spinning parameters. This can be used for a wide range of polymers, though you will have to play about with the setup to get it working with the desired results.

There is a large risk of electrical shock from this procedure as you will be using large potential differences generated from a power supply. Electrical arcing and sparks from the equipment are common. Before you begin, consult someone with electrical expertise about safe practice and isolation of the experiment!

To make an electrospun mat from poly(vinylpyrrolidone) and amorphous calcium phosphate you will need:

1. A transparent insulated box or cage large enough to hold the electrospinning apparatus with a door on the front. *It is highly recommended to have a safety switch that kills the power to the system if the door is opened.*
2. A metal plate about 10 cm in diameter and aluminium foil. Use a clamp as a stand with an insulated base. The aluminium foil is used to coat the target so that samples can be easily lifted off.
3. An automated syringe pump with adjustable speed.
4. A 10-mL syringe fitted with a 19-gauge (1.1-mm diameter) flat-ended needle.
5. A ruler or tape measure for keeping track of the distance from the needle tip to the target.
6. A high voltage power supply capable of reaching 30 kV or possibly even higher. The terminals should be fitted with crocodile clips. *Ensure that you are fully competent with its use and also ensure that any electrical contacts are isolated in such a way that no one can come into contact with them.*
7. 17.76 g (2.2×10^{-3}) of poly(ethylene glycol) (molecular weight 8000 g/mol). This decreases the solubility of the amorphous calcium phosphate when it forms so that it may be precipitated out of solution in large quantities for use in the experiment. It also serves to slow down the transformation of amorphous calcium phosphate to more crystalline polymorphs.
8. 400 mL of a 0.1 mol/L solution of calcium chloride in water in a 600-mL to 1-L beaker with a stir bar. This can be prepared by dissolving 4.44 g of $CaCl_2$ into 400 mL of water.
9. 200 mL of a 0.13 mol/L solution of sodium phosphate in water. This can be prepared by dissolving 4.36 g of Na_3PO_4 into 200 mL of water.
10. A magnetic stirrer with a large ice bath on top that can accommodate the beaker containing the calcium chloride solution.
11. A large Buchner funnel and filter paper large enough to collect ~10 g of precipitate. You will also need large amounts of water and bench ethanol for rinsing the precipitate.
12. A sonic bath and a float or rack to stand a vial in the bath with.
13. 5 mL of bench ethanol in a sample vial with a lid. This will be used to dissolve the poly(vinylpyrrolidone) in so that during spinning it will evaporate rapidly leaving only the polymer fibre behind.
14. 0.66 g of poly(vinylpyrrolidone) (molecular weight 1,300,000 g/mol)
15. A set of scales and weighing boats.

To make an electrospun mat from poly(vinylpyrrolidone) and amorphous calcium phosphate

1. Setup the apparatus for the electrospinning as outlined in the schematic below. **Do not turn the power on yet.** Ensure that everything is inside the box for safety. The tip of the needle must be 8 cm from the plate with the needle pointing towards the middle of the target. The distance and the positioning may change depending on what results you get. Gravity tends to cause the fibres to aggregate lower than where they were aimed. Place the aluminium foil on the target. Attach an electrode to the bottom of the plate, ensuring that it comes into contact with the foil.
2. Dissolve the poly(ethylene glycol) into the calcium chloride solution under stirring in the ice bath. The temperature of the solution should be maintained around 5 °C.
3. Add the sodium phosphate solution to the calcium phosphate solution. A white precipitate of amorphous calcium phosphate will form immediately. This should be transferred to a Buchner funnel.
4. Collect the precipitate in the Buchner funnel. Rinse the sample with water followed immediately by ethanol to remove salts and quench the reaction. Exposure to water for long periods will induce transformation to a more stable polymorph. Allow the amorphous calcium phosphate to dry in air or vacuum dry.

5. Weigh out 0.5 g of the amorphous calcium phosphate powder on the scales.
6. Add this to the 5 mL of ethanol and cap the vial. Sit the sample in the sonic bath and disperse the amorphous calcium phosphate in the sonic bath for an hour to form a suspension.
7. Add the poly(vinylpyrrolidone) to the ethanol and calcium phosphate suspension. Continue to sonicate the solution until the polymer has fully dissolved to form a viscous solution.
8. Draw the entirety of the viscous solution into the syringe that will be used for the electrospinning. Do not have the needle fitted to the syringe when you do this, as the solution will be too thick to draw through.
9. Affix the needle to the filled syringe and clamp it into the automated dispenser.
10. Attach the live wire terminal to the base of the needle so that it will not get in the way of the solution jet.
11. Ensure that the needle tip is 8 cm from the target plate. You may find other distances work better for you.
12. Turn on your automated syringe pump and set the desired program. A flow rate of around 50 µL/min should give good results. Turn it on and immediately shut the box. The flow should be slow enough so that you will have time to turn everything on before the solution makes it to the end of the needle.
13. Ensure that the closed box is electrically isolated. Wait until you see a droplet forming at the end of the needle and then turn on the power supply. The droplet at the end of the needle should then suddenly leap towards the aluminium target dragging a thin filament along behind it. This will be very difficult to see unless it is against a dark background. If the syringe pump setup is not properly earthed, then during the spinning you may see arcing and you will hear a sharp crack when this happens. This should not be a problem and will not affect the filament formation.
14. Allow the electrospinning to continue until you feel the formed fibre mat is large enough. Turn off the power supply and open the box. Turn off the automated plunger. The mat should remain stuck to the surface of the foil and can be lifted or cleaved off easily using a razor blade. The fibre mat is now ready for analysis.

Scanning electron microscopy of the fibres is a quick method for assessing the morphology of the fibres and the density of the mat. The fibres should be uniform in thickness along their length. Energy dispersive X-ray analysis of the elemental composition should show the presence of calcium and phosphate from the embedded particles. Smaller fibres can be delicately placed across a grid for transmission electron microscopy, though they will readily melt in an intense electron beam. You should be able to see darkened areas where the relatively electron dense calcium phosphate provides high contrast. Diffraction patterns of the particles should not demonstrate any peaks as the mineral phase is amorphous at this point but often the energy of the beam itself will be strong enough to promote a phase change in the crystal structure.

Exposing the fibres to water will cause them to swell and form a hydrogel. In this state, and depending upon the pH, the amorphous calcium phosphate will begin to crystallise into hydroxyapatite crystals. In the source reference, the fibre mats were placed onto sections of teeth and soaked with a synthetic saliva composition for an hour. Conversion to hydroxyapatite will also occur in other high-moisture environments. The fibre mats provided a source of calcium phosphate, which then mineralised at the tooth surface and filled in channels of dentine. The recipe for synthetic saliva is included as this is important for simulating remineralisation processes in the mouth (Fig. 5.30).

To make 1 L of synthetic saliva add the following to water:
    0.040 g (2×10^{-4} mol) of $MgCl_2 \cdot 6H_2O$
    0.147 g (0.001 mol) of $CaCl_2 \cdot 2H_2O$
    0.544 g (0.004 mol) of KH_2PO_4
    1.192 g (0.016 mol) of KCl
    0.240 g (0.0045 mol) of NH_4Cl
    4.766 g (0.02 mol) 4-(2-hydroxyethyl)piperazine-1-ethanesulphonic acid, N-(2-hydroxyethyl)piperazine-N'-(2-ethanesulphonic acid) or HEPES buffer
    Adjust this to pH 7 using 1 mol/L NaOH solution. This should be stored in the fridge and you may see sediment form after a few days as calcium phosphate precipitates out of solution.

MAKING A SOLAR CELL USING TITANIUM DIOXIDE NANOPARTICLES AND CONDUCTIVE GLASS ELECTRODES

Adapted from Smestad and Gratzel (1998)

Photovoltaic devices are a ubiquitous technology employed wherever there is sunlight to be harvested. The attraction of using a device that can run from the light of the sun alone is obvious and having a small portable power source is useful and applicable to a wide range of applications. In this recipe, you will make a basic Grätzel cell capable of powering a small liquid crystal display or digital watch. You may wish to adapt the size of the panel to get more power out of it by making the solar cell larger. The major

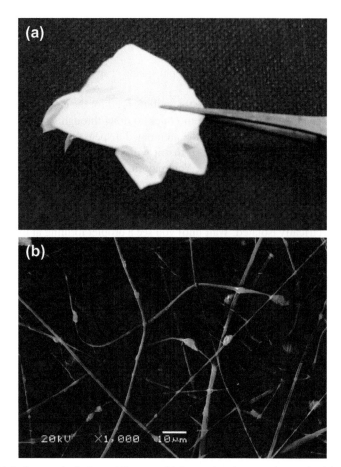

FIGURE 5.30   (a) A photograph of a formed fibre mat. (b) A scanning electron micrograph of the electrospun fibres.

component of a Grätzel cell is the use of nanoscale titanium dioxide (TiO_2) or 'titania' as the light-driven electron source. Titania is a wide-bandgap semiconductor, which can absorb light in the ultraviolet part of the spectrum when formed in a specific crystal arrangement known as anatase. Upon absorbing a photon of the correct energy an electron is promoted into the conduction band of the titania and makes its way to the surface. This may be siphoned away for use in an electrical device when the titania forms part of an electrical circuit as in a solar panel.

Anatase-phase titanium dioxide is an n-type semiconductor (bandgap ~3.2 eV) and is utilised in solid-state electronic devices. The bandgap structure arises from electrons in the $3d$ band of the titanium atom occupying the oxygen $2p$ band when bonded. As the $3d$ band is full the compound is dielectric. An alteration of ratios gives holes in the $3d$ and $2p$ bands leading to a controllable level of semiconductance. The low bandgap of titanium dioxide has also led to its incorporation into photoelectric solar cells, or Grätzel cells, as named after their inventor (Grätzel, 2001).

For a solar cell to be capable of harvesting light in the visible range the titania must be stained with a dye (Fig. 5.31). The dye should complex with the surface of the titania and upon illumination there will be a charge transfer from the dye into the titania. The staining effectively lowers the bandgap energy required to transfer an electron into the conductive band. This can be achieved with a ruthenium bipyridyl dye for a long-lasting cell or for demonstration purposes raspberry and blackberry juice. These juices and some teas contain flavanoid and cyanin compounds, which absorb in the visible spectrum, giving the juice a red colour, and can complex with titania at the surface of the nanoparticles. Not all fruit juices will do this and the cyanins must be rich in $=O$ and $-OH$ groups to allow for complex formation with a Ti^{4+} ion. Recent work in this field has focussed on the sensitisation of titania with a variety of dyes to bring the absorbance region into the visible frequency. Strong UV radiation ($\lambda < 400$ nm) is required to stimulate electronic emission in anatase and solar illumination consists of a higher proportion of visible frequencies. A chemical dye is bound to the surface of the titania through a carboxylic acid functionalised ligand in a visibly active transition metal complex. Stimulation by a light then injects the excited electron from the metal complex directly in to the conduction band of the anatase. This may be incorporated into a circuit.

One problem with dye staining is that the organic components, such as the bipyridyl ligand bridge, are themselves degraded by the titania, so cells of this type only work for short periods. Porphyrin and phthalocyanine dyes have also been adsorbed, which act as p–n heterojunctions at the surface to increase electron transfer efficiency. Currently, the maximum conversion efficiency

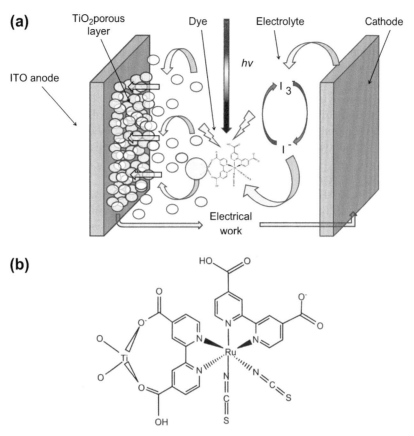

FIGURE 5.31 (a) A schematic of a Grätzel cell. A dye molecule (S) is adhered through a carboxylic acid group to the surface of the titania. The dye is photoactive in the visible region and electrons promoted by visible light excitation are of the correct energy to pass through the conduction band of the n-type semiconductor film. This is then directed through the circuit by a layer of transparent conducting oxide, which is normally tin oxide. (b) Di-tetrabutylammonium *cis*-bis(isothiocyanato) bis(2,2′-bipyridyl-4,4′-dicarboxylato)ruthenium(II) is an example of a sensitising agent. The organic components are broken down over time by photo-oxidation, which shortens their usable life.

achieved in dye-sensitised cells has been in the region of 11%. The light to electrical energy conversion encompasses only 1% of the total visible light available. To overcome this problem, research is continually involved in the direct addition of additives to the titania structure to lower the bandgap. To this end, the introduction of crystal defects using nitration and hydrogenation has shifted the absorbance of titania nanocrystals into the visible region. Other routes towards introducing defects into the anatase crystal structure are explored elsewhere in the 'How to make black titania' recipe.

Anatase-phase titania is also used in wastewater reclamation as a photocatalyst. A nanoparticle of titania dispersed within water has a surface layer of hydroxyl groups. When illuminated with light the electron generated in the conduction band can move to the surface of the particle and generates a hydroxyl free radical group. These free radicals are highly reactive with organic components in the water and eventually they are reduced to carbon dioxide. Remaining free radicals are scavenged by the positive holes in the titania particle and the process can continue again until the illumination stops. This is the same mechanism that is employed in self-cleaning windows. A surface coating of titania nanoparticles is applied to a pane of glass. The titania particles are too small to be seen in the visible range and so the windows are still transparent but now dirt on the windows will be broken up by free radicals generated by the UV portion sunlight passing through them. Ultimately, this type of technology can be used to split water into hydrogen and oxygen for use in fuel cells, meaning that titania is important in the development of both free electrical energy and free chemical energy sources.

This recipe requires a tin oxide conductive glass for use in the cells. Conductive glass is very useful for making optically transparent electrodes or for electrodeposition applications where the colour of the deposited sample needs to be assessed. A method for making conductive glass can be found after this recipe.

To make a Grätzel cell you will need:

1. Conductive tin oxide-coated glass with a resistance of between 10 and 30 Ω/cm. This needs to be antimony or indium doped and the conductive coating on commercially available glass will be normally on just one side. You will need two pieces to form the front and back of the solar cell and in this recipe we will use conductive glass 2.5×2.5 cm^2 and only a few millimetres thick.

2. 12 g of nanoparticle anatase-phase titania. Degussa P25 is a commercially available industry-standard titania powder. Depending on how large or how many solar cells you are making, you may wish to scale this down. For a 2.5 cm by 2.5 cm electrode, only milligram amounts will be used to cover the surface of an electrode, so this can go a long way.

3. A dry (no water) ethylene glycol solution containing both 0.5 mol/L potassium iodide and 0.05 mol/L iodine. The iodine will undergo redox switching between I^- and I_3^- during illumination of the cell and carries an electron from the carbon electrode back to the titania to fill the positive holes generated by light.

4. An oven or, if unavailable, a paint-stripping heat gun. These should both be capable of reaching temperatures up to 500 °C

5. 20 mL of nitric or acetic acid solutions in water of between pH 3 and 4 to be used in making a titania paste.

6. Cellotape and some clamps or bulldog clips to hold the cell together. A resin-type glue can be used to seal up the cell for long-term use.

7. A glass rod and a large-enough mortar and pestle to hold greater than 30 mL.

8. A staining agent for the titania. Tris(2,2′ bipyridyl) ruthenium (II) chloride in water solution will last significantly longer than the fruit juice preparation. Only a few milligrams in 10 mL of water will be required to stain the titania properly. *This solution is toxic and should be handled with care and disposed of in a heavy metal solvent waste container if not to be stored.*

   Alternatively, and perhaps more fun, you can use a punnet of raspberries to make a staining solution. A juice stain will degrade over time while the cell is in use; so this is not a good option if you would like to use the solar cell for prolonged periods. Crush the raspberries and strain the juice through some filter paper. Add 1 mL of water for every 5 mL of juice you extract. You can store the juice long term in the dark in a fridge.

9. Ethanol or isopropanol for rinsing.

10. A carbon rod or soft graphite pencil for making a contact layer with the iodide electrolyte.

11. An electrical multimeter capable of measuring current, voltage and conductivity.

12. Crocodile clips and wires for connecting the solar cell to something.

To make a Grätzel cell:

1. Use the multimeter to determine which sides of your glass electrodes are conductive. It is very important that the conductive layer is the one that gets coated in titania. Lay the nonconductive side down with the conductive side up on a bench. Do this for both your electrodes. One is going to be coated in titania and the other in carbon.

2. For the titania electrode use tape to mask off 3 mm of the surface on three sides. Then mask off 5 mm using tape on the fourth side. The fourth side will later be used to make a contact.

3. Place the titania powder into the mortar and pestle and begin to add the acid solution 1 mL at a time using a pipette. Grind the powder until it is smooth and add more of the acid solution. Repeat this until a smooth titania paste has formed.

4. Add the titania paste to the conductive side of the taped glass electrode. Add around 5 μL of the titania paste per cm² of exposed surface area. Use the glass rod to roll over the paste and make it continuous and even within the taped off area. Now allow it to air dry or use heat to drive the excess moisture off. Once dry, carefully remove the tape from the edges (Fig. 5.32).

5. Using a carbon rod or a pencil coat the entire conductive surface of the second electrode. Sometimes the pencil might not take to the glass. This can be remedied by holding the conductive side in a candle flame to form a carbon layer over the surface. This layer will aid the regeneration of tri-iodide to iodide in the electrolyte solution during use. Once covered, the electrode is ready for use but can be made to last longer by heating along with the titania electrode in the next step for only a few minutes.

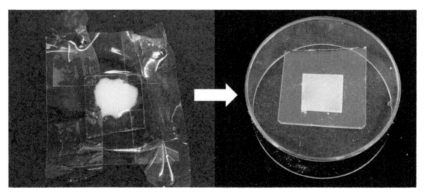

FIGURE 5.32 Areas of the electrode you do not wish to be coated in titania are covered using tape. Once the titania solution is dry the tape is removed and the sample is ready to be placed in an oven.

6. Place the titania-coated conductive glass into an oven and heat it to ~450 °C for 30 min. This sintering step anneals the titania to the surface of the tin oxide layer and forms links between the nanoparticles that make the titania layer more like a rigid foam. Without this step the titania would come away from the surface of the tin oxide layer and the circuit would not work. Allow both electrodes to cool to room temperature before use.

7. Soak the titania-coated electrode in the staining agent solution for about ten minutes. If either side of the electrode looks white, then leave the electrode in the stain for longer. If you are using the ruthenium dye, then ensure that you are wearing gloves before handling. The stained electrodes are first rinsed with water and then with ethanol or isopropanol before being blotted dry on a tissue. Ruthenium waste should be disposed of in the correct manner.

   If you are staining using the raspberry juice but are not going to use the electrode within the next hour or so, then store it in pH 4 acetic acid in water solution (you can use vinegar in water) in the dark.

8. Place the titania-coated electrode face up on the bench and place the carbon-coated electrode face down on top of it. The carbon electrode should be offset so that the uncoated 5-mm section of the titania-coated electrode is sticking out on the bottom. This will also leave an overhang of the carbon-coated electrode at the opposite side. Use bulldog clips to hold these together.

9. Using a teat pipette add the iodine—tri-iodide electrolyte solution to the interfaces of the electrodes. You should hopefully see capillary action drawing the solution in between the two layers. If it does not seem to be going in, you can pull the electrodes apart a little bit and add some more. Use a tissue to wipe away the excess.

10. Attach the crocodile clips to the exposed overhanging sections of the electrodes.

11. The solar cell is now complete. You may wish to seal it with glue so that you can remove the clamps or even solder permanent electrical contacts onto the cell. The electrodes may be reused indefinitely but juice-based dye will degrade from use and if reassembled the cell will need fresh electrolyte.

12. Using the multimeter you should see a voltage across the wires when the solar cell is exposed to sunlight! Depending on how strong the generated current is you can try to power something with it. A digital watch can be a good place to start (Fig. 5.33).

FIGURE 5.33 A completed solar cell ready to use.

MAKING A CONDUCTIVE GLASS ELECTRODE

Adapted from Tanaka and Suib (1984)

Optically transparent electrodes can be used in many different applications. For example, switchable windows are made by pressing a liquid crystal and particle dispersion between two glass electrodes. By applying an electric field between them, the liquid crystal suspended particles can align and let light through (Rauh, 1999). Conductive glass electrodes are also vital in solar cell applications where light must pass through into the solar cell.

To make an antimony-doped tin oxide-coated glass electrode you will need:

1. A furnace capable of reaching at least 600 °C. In addition, you will need oven tongs and a small thermal brick on which the slides can be rested. You will also require an insulated mat outside of the oven on which a hot brick can rest.
2. Soda lime glass to be coated, depending on size. The quantities here should be enough to cover many tens of centimetres of glass. Microscope slides work well.
3. 25 g (0.0713 mol) of $SnCl_4 \cdot 5H_2O$ dissolved in 6 mL of methanol.
4. 1 g (0.0034 mol) of Sb_2O_3 dissolved in 3 mL of concentrated HCl (32%). To this is added 17 mL of methanol. *The acid must be handled in a fume hood.*
5. An atomiser or spray gun. Only small volumes will be used; so it may prove necessary to experiment with the atomiser setup first. The finer the mist, the better the results. *The spraying must be done in a fume hood.*

To make an antimony-doped tin oxide-coated glass electrode:

1. Add 1 mL of the Sb_2O_3 solution to 3 mL of the $SnCl_4$ solution. Place this in the atomiser ready for use.
2. Place the glass to be coated onto the thermal brick inside the oven and heat it to 600 °C.
3. Remove the brick from the oven and place it on a mat in a fume hood. Immediately spray the antimony-doped tin chloride mist over the slides for only a few seconds. It is important the slides remain hot so that the tin chloride will oxidise to tin oxide. Place the slides on the brick back into an oven for about 2 min before removing and spraying again. This may be repeated in the event the coating is not conductive enough. *Do not breathe in the mist. Do not be tempted to spray into the oven, as the mist will blow directly back out.*
4. After the required number of coatings has been performed, you may take the slides out of the oven and allow them to cool. They should have a colourful iridescence to them from the doped tin oxide coating. The thickness of the film will be between 100 nm and 1 μ and the colour of reflected light from the surface can be used as a rough indicator of the film thickness. White or yellows and reds will be 100−300 nm in thickness while the appearance of blues and greens indicate that the films is over 500 nm in thickness. Use a resistance meter to measure the conductivity.

REFERENCES

Ahonen, M., Pessa, M., Suntola, T., 1980. A study of ZnTe films grown on glass substrates using an atomic layer evaporation method. Thin Solid Films 65 (3), 301−307.

Ainla, A., et al., 2010. A microfluidic pipette for single-cell pharmacology. Analytical Chemistry 82 (11), 4529−4536.

Bessoth, F., Manz, A., 1999. Microstructure for efficient continuous flow mixing. Analytical Communications 36, 213−215.

Burdinski, D., Blees, M., 2007. Thiosulfate-and thiosulfonate-based etchants for the patterning of gold using microcontact printing. Chemistry of Materials 19 (2), 3933−3944.

Burger, C., Hsiao, B.S., Chu, B., 2006. Nanofibrous materials and their applications. Annual Review of Materials Research 36, 333−368.

Christensen, S.T., Elam, J.W., 2010. Atomic layer deposition of Ir−Pt alloy films. Chemistry of Materials 22 (8), 2517−2525.

Cook, G., Timms, P.L., Göltner-Spickermann, C., 2003. Exact replication of biological structures by chemical vapor deposition of silica. Angewandte Chemie. (International ed. in English) 42 (5), 557−559.

Du, Y., George, S.M., 2007. Molecular layer deposition of nylon 66 films examined using in situ FTIR spectroscopy. Journal of Physical Chemistry C 111 (24), 8509−8517.

Fletcher, J., et al., 2011. Electrospun mats of PVP/ACP nanofibres for remineralization of enamel tooth surfaces. CrystEngComm 13 (11).

Gates, B.D., Whitesides, G.M., 2003. Replication of Vertical Features Smaller than 2 nm by Soft Lithography, pp. 14986−14987.

Giessibl, F.J., 2003. Advances in atomic force microscopy. Reviews of Modern Physics 75 (3), 949.

Grätzel, M., 2001. Photoelectrochemical cells. Nature 414, 338−344 (November).

Hembacher, S., Giessibl, F.J., Mannhart, J., 2004. Force microscopy with light-atom probes. Science 305 (5682), 380−383.

Higashi, G., Fleming, C., 1989. Sequential surface chemical reaction limited growth of high quality Al_2O_3 dielectrics. Applied Physics Letters 55 (19), 1963−1965.

Huang, J., Wang, X., Wang, Z.L., 2006. Controlled replication of butterfly wings for achieving tunable photonic properties. Nano letters 6 (10), 2325−2331.

King, J.S., et al., 2006. Conformally back-filled, non-close-packed inverse-opal photonic crystals. Advanced Materials 18 (8), 1063−1067.

Kumar, M., Ando, Y., 2010. Chemical vapor deposition of carbon nanotubes: a review on growth mechanism and mass production. Journal of Nanoscience and Nanotechnology 10 (6), 3739−3758.

Kumar, A., Biebuyck, H.A., Whitesides, G.M., 1994. Patterning self-assembled monolayers: applications in materials science. Langmuir (10), 1498−1511.

Lin, Y.-H., et al., 2011. Atomic layer deposition of zinc oxide on multiwalled carbon nanotubes for UV photodetector applications. Journal of The Electrochemical Society 158 (2), K24.

Madou, M., 2002. Fundamentals of Microfabrication. CRC Press.

Microfluidics, D.-based, et al., 2006. Reactions in droplets in microfluidic channels. Angewandte Chemie International Edition, 7336−7356.

Musselman, I., Peterson, P.A., Russell, P.E., 1990. Fabrication of tips with controlled geometry for scanning tunnelling microscopy. Precision Engineering 12 (1), 3−6.

Pant, B.D., Tandon, U.S., 1999. Etching of silicon nitride in CCl_2F_2, CHF_3, SiF_4, and SF_6 reactive plasma: a comparative study. Plasma Chemistry and Plasma Processing 19 (4), 545−563.

Rauh, R.D., 1999. Electrochromic windows: an overview. Electrochimica Acta 44, 3165−3176 (September 1998).

Rugge, A., et al., 2003. Tungsten nitride inverse opals by atomic layer deposition. Nano Letters 3 (9), 1293−1297.

Sato, K., et al., 1998. Characterization of orientation-dependent etching properties of single-crystal silicon: effects of KOH concentration. Sensors and Actuators A: Physical 64 (1), 87−93.

Seidel, H., et al., 1990. Anisotropic etching of crystalline silicon in alkaline solutions. Journal of the Electrochemical Society 137 (11), 3612−3625.

Sia, S., Whitesides, G.M., 2003. Microfluidic devices fabricated in poly (dimethylsiloxane) for biological studies. Electrophoresis 24, 3563−3576.

Smestad, G.P., Gratzel, M., 1998. Demonstrating electron transfer and nanotechnology: a natural dye-sensitized nanocrystalline energy converter. Journal of Chemical Education 75 (6), 752−756.

Song, H., Tice, J.D., Ismagilov, R.F., 2003. A microfluidic system for controlling reaction networks in time. Angewandte Chemie 115 (7), 792−796.

Squires, T.M., 2005. Microfluidics: Fluid physics at the nanoliter scale, 77(July).

Srisa-Art, M., deMello, A., Edel, J., 2007. High-throughput DNA droplet assays using picoliter reactor volumes. Analytical Chemistry 79 (17), 6682−6689.

Tanaka, J., Suib, S.L., 1984. Surface conductive glass. Journal of Chemical Education 61 (12), 1104.

Taylor, M.P., Timms, P.L., 1997. Studies on the reaction between silane and hydrogen peroxide vapour; surface formation of planarized silica layers. Journal of the Chemical Society, Dalton Transactions (6), 1049−1054.

Taylor, G., 1969. Electrically driven jets. Proceedings of the Royal Society of London. A. Mathematical and Physical Sciences 313 (1515), 453.

Vasita, R., Direndra, S., 2006. Nanofibers and their applications in tissue engineering. International Journal of Nanomedicine 1 (1), 15−30.

Vohs, J.K., et al., 2010. Chemical vapor deposition of aluminum oxide thin films. Journal of Chemical Education 87 (10), 1102−1104.

Wilbur, J.L., et al., 1994. Microfabrication by microcontact printing of self-assembled monolayers. Advanced Materials 6 (7−8), 600−604.

Wilbur, James L., et al., 1995. Scanning force microscopies can image self-assembled monolayers. Langmuir 11 (3), 825−831.

Xia, Y., Whitesides, G.M., 1998. Soft lithography. Annual Review of Materials Science 28 (1), 153−184.

Xia, Y., et al., 1995. A selective etching solution for use with patterned self-assembled monolayers of alkanethiolates on gold. Scanning Electron Microscopy (15), 2332−2337.

Xia, Y., et al., 1999. Unconventional methods for fabricating and patterning nanostructures. Chemical Reviews 99 (7), 1823−1848.

Yoshimura, T., et al., 2007. Orientation-controlled molecule-by-molecule polymer wire growth by the carrier-gas-type organic chemical vapor deposition and the molecular layer deposition. Applied Physics Letters 91 (3), 033103.

Zhang, W., Ferguson, G., 2004. Elastomer-supported cold welding for room temperature wafer-level bonding. Micro Electro Mechanical 2 (c), 741−744.

Chapter 6

Biological Nanotechnology

Biological science has examined ever deeper and more closely into the composition of life, and in the last century it has become possible through biotechnology to not only examine, but fundamentally alter, the working mechanisms of living things. As a macroscopic focus moves from the classification of animals down to the microscopic cellular and ultimately chemical make up of living creatures, the defining line between alive and chemical reaction becomes blurred. The length scale on which a chemical process becomes a spontaneously ordered and self-assembled dynamic system is the nanoscale and a practical understanding of the engineering of active biological species coupled to their interaction with materials on the same scale is vital to the integration of nanotechnology with medicine, industry and everyday life.

In this chapter some common techniques for genetic manipulation and basic nanobiotechnology will be explored. Genetic transformation in bacteria and yeasts is a useful tool for splicing genes into plasmid sequences and for turning those sequences into useful proteins. This section provides a complete work-up for the production of the green fluorescent protein intended to provide the reader with a basic template for the techniques involved in genetic modification. Later on we will look at analysis techniques and DNA amplification using PCR and using DNA as a versatile construction material at the nanoscale.

CLONING AND GENE EXPRESSION TO PRODUCE PROTEINS IN *ESCHERICHIA COLI*

With thanks to Ross Anderson, David Green, Thom Sharp and Nigel Savery

This section provides the basis for understanding recombinant DNA technology and a number of techniques routinely used in molecular biology. It describes a method by which you can grow large amounts of a particular DNA sequence, such as the coding for a particular gene, and then insert this gene into an *E. coli* cell in a process called transformation. The transformed *E. coli* can then be multiplied in a culture before the relevant cellular machinery is turned on so that the inserted sequence will be expressed as a protein. It is not as easy as placing the lone strand of DNA into a cell and first the gene must be grafted into a vector capable of moving into the cytoplasm and being read by the transcription mechanisms. This is done by inserting the gene into a vector, normally a plasmid, and sticking it back together with the gene in place — essentially recombining the DNA together which is why it is called recombinant DNA. This technique is highly successful for the production of large amounts of protein that you might otherwise have to harvest directly from a creature of interest. A great example of this is in the production of the hormone insulin which diabetics need to regulate sugar levels in their blood. Early insulin therapies used a variant of insulin harvested from pigs which was a lengthy and impractical procedure yielding disappointing amounts of the vital hormone. By isolating the DNA sequence, or gene, that produced human insulin it was possible to splice this DNA into a ring plasmid and insert it into a bacterial culture. Bacteria transformed to contain this insulin gene could then be grown into vast quantities and then induced to express large amounts of insulin. Similarly, Bacillus strains are used to produce antibiotics as part of their fermentation process (Madigan et al., 2000). Herein a complete multistage procedure for the production of green fluorescent protein (GFP) by expression in *E. coli* is provided as an example of recombinant DNA technology.

The preparation has been broken into stages which can be performed by themselves as the starting point for other experiments depending on how you adapt them. In brief, a gene of interest will be excised from a plasmid, placed into another vector plasmid designed to controllably express the protein we want and then inserted this into a competent *E. coli* cell. This *E. coli* carrying the desired genes will then be cultivated before the expression of the protein is triggered so that it can be produced in large quantities. The last step will be to purify and analyse the protein.

In this case we will be using *E. coli* strain BL21(DE3) which is routinely used in the laboratory and is not very pathogenic. That is to say, it should not make you very ill if you are accidentally exposed to it. Many microorganisms can be pathogenic to the human body so *all the following procedures must be performed using biohazard-level one safety procedures in a laboratory setup for handling the specific microorganism used.*

INSERTING A GENE INTO A PLASMID VECTOR

Genes are strings of DNA sequenced such that, when it is read by cellular machinery, it produces messenger RNA which is then translated further into proteins. This important mechanism is sometimes referred to as 'The central dogma of molecular biology' so called after a publication by Francis Crick who coined the term (Crick, 1970). DNA therefore provides the instructions for manufacturing the components required for life to function. In research you may wish to focus on one particular gene sequence of DNA that encodes for a particular protein. It may be that the DNA itself is of interest or it may be that you wish to further use that DNA to produce a large quantity of protein. In this section, we look at the basic manipulation techniques for transferring a gene from a native stretch of chromosomal or plasmid DNA and grafting it into a form of DNA we can work with in the laboratory so that it can be further multiplied. This process is gene cloning and involves cutting out (excising) the gene of interest using suitable restriction enzymes and then inserting it (ligation) into a plasmid vector which can be inserted into a bacteria where it will be replicated. In biochemistry you will most commonly find that the interesting genes of complex organisms will have been sequenced by someone and can be purchased commercially. You may also have a particular sequence in mind and a custom gene sequence can be prepared from scratch to order. When obtaining a gene it will most likely be sent to you as part of a plasmid to work with.

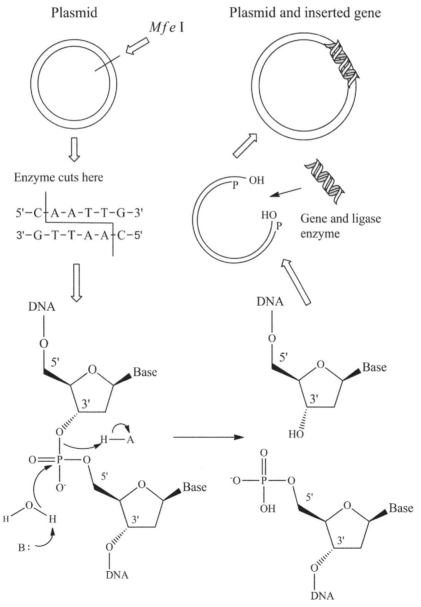

FIGURE 6.1 A schematic of a plasmid being cleaved using restriction enzymes. The break can occur at different points in the interior and exterior DNA strands. A new gene sequence can be stuck onto these overhanging ends using a ligation enzyme. A plasmid ring is reformed carrying the inserted gene.

Plasmid DNA consists of two identical strands of DNA arranged top to tail with one another and joined together to form a ring. This ring can be further twisted up in what is called a supercoiled plasmid structure. You may find that the plasmid that the gene is supplied in is not the one that will work well with your bacterial system and you may need to transfer the gene into another plasmid that has the function you want. One example may be that the supplied plasmid is resistant to the antibiotic kanomycin but you want to work with a plasmid resistance to amoxicillin instead. You might also want to place the gene into a plasmid that will express the protein with a histidine tag on the end of the protein so that it can be easily purified from solution.

There are a number of methods for preparing recombinant plasmid DNA, and two of the most common methods are outlined in this section. The first, perhaps most common method, is an enzymatic digestion of a plasmid containing a gene of interest followed by the enzymatic ligation of this gene into a suitable plasmid vector ready for transformation. This method is technically more involved and is provided as an exploration of the underlying techniques rather than a quick preparation.

The second plasmid preparation involves cutting the gene of interest from plasmid DNA quickly using oligomeric primers instead of restriction enzymes and running a PCR reaction to pull the plasmid DNA apart and replicate the gene in large quantities. The major advantage of this method is that it generates a large amount of the pure gene with only very trace amounts of the initial plasmid. This is advantageous if the plasmid you are working with is hard to come by.

Genes have a specific nomenclature depending on if they are by themselves or spliced into a plasmid during the PCR process.
If the genetic sequence coding for a particular protein is by itself, then it is referred to as a 'gene.'
If the gene resides within a plasmid, then it is referred to as a 'GENE'.

Both of these methods will produce plasmids that can be used in the transformation reaction with competent cells weakened chemically so that they will take up the exogenous DNA.

PLASMID MODIFICATION – USING RESTRICTION ENZYMES

Removing a particular gene of interest from a section of plasmid is done using enzymes called restriction endonucleases. These enzymes cut the ring DNA at specific points of a given 4-, 5- or 6-base-pair-long sequence by hydrolysing the phosphate diester linkage in the DNA backbone.

When the restriction enzyme latches on to a particular sequence on the outside strand of the plasmid, it hydrolyses the sequence and breaks the DNA. This reaction forms one end with a terminal phosphate group in the $5'$-position on the organic ring and a terminal $3'$-hydroxyl group at the opposite end. The same occurs for the identical sequence on the inside of the plasmid running in the opposite direction resulting in a linear section of double-stranded DNA. This break on the interior strand may not be adjacent to the break on the outside sequence and so the excised gene will have a staggered cut in it giving an overhanging single strand of DNA at each end. This is very useful as it provides a point where either end of a gene can be anchored into a new plasmid vector. Similarly, a ring plasmid vector of choice can be opened up which will have complementary ends to which the gene can join to form a ring of double-stranded DNA once more. A ligase enzyme is used to promote condensation of the DNA backbone and reform phosphodiester bonds between the end of the inserted gene and the plasmid vector (Fig. 6.1).

Restriction Enzymes and Names
The nomenclature for naming restriction enzymes is based on a simple system from the bacterial strain from which they were first isolated. In nature these enzymes have evolved so that foreign DNA within the cell will be broken down before it has a chance to interfere with anything. Because there is a high chance that some of the sequences in the host DNA will be the same as the foreign DNA, the host organism produced methylases corresponding to the same sequences that the endonucleases cut and block them with methyl groups. The name of an endonuclease does not relate specifically to the sequence where it breaks the DNA bond. Instead, it is named after the source organism from which it was isolated. The first letter is the genus, the second two letters are the species name. If there is a single letter after the name, then this represents the strain of the organism and finally there will be roman numerals denoting the order in which it was discovered. The enzymes used in this experiment for example are:

Mfe I – Mycoplasma fermentas
This enzyme cleaves the sequence:
$5'$...C$^{\blacktriangledown}$AATTG...$3'$
$3'$...GTTAA$_{\blacktriangle}$C...$5'$
Xho I – Xanthomonas holcicola
This enzyme cleaves the sequence:
$5'$...C$^{\blacktriangledown}$TCGAG...$3'$
$3'$...GAGCT$_{\blacktriangle}$C...$5'$

Restriction enzymes are very sensitive to the temperature and concentration of salt so make sure the restriction buffer you use is the right one. Most enzymes will be provided with a sheet specifying the correct conditions but a table of general buffers for use with various restriction enzymes is provided below. Adapted from Sambrook and Russell (2001).

Enzyme	Common Isoschizomers	Useful salt concentration	Incubation Temperature	Recognition Sequence	Compatible Cohesive Ends
Ava I		Med	37 °C	G▼PyCGPuG	Sal I, Xho I,
Bam H I		Med	37 °C	G▼GATCC	Bcl I, Bgl II, Mbo I, Sau 3A
Bgl II		Low	37 °C	A▼GATCT	
Bst E II		Med	60 °C	G▼GATCC	
Eco R I		High	37 °C	G▼AATTC	
Eco B			37 °C	TGA(N)$_8$TGCT	
Eco K			37 °C	AAC(N)$_6$GTGC	
Eco R I			37 °C	▼AATT	
Hae III		Med	37 °C	GG▼CC	Blunt
Hin d II		Med	37 °C	GTPy▼PuAC	Blunt
Hin d III		Med	37–55 °C	A▼AGCTT	
Kpn I		Low	37 °C	GGTAC▼C	
Mbo I	Sau 3A	High	37 °C	▼GATC	Bam H I, Bcl I, Bgl II, Xho I
Pst I		Med	21–37 °C	CTGCA▼G	
Pvu II		Med	37 °C	CAG▼CTG	Blunt
Sau 3A	Mbo I	Med	37 °C	▼GATC	Bam H I, Bcl I, Bgl II, Xho I
Sma I	Xma I		37 °C	CCC▼GGG	Blunt
Taq I		Low	65 °C	T▼CGA	Acc I, Acy I, Asu II, Cla I, Hpa II
Xba I		High	37 °C	T▼CTAGA	
Xma I	Sma I	Low	37 °C	C▼CCGG	Ava I

Many different plasmid vectors are available which can be used to insert exogenous DNA into a host organism. You must use a plasmid that will provide the correct reading frame for the transcription of the inserted gene. This is determined loosely by the position of the 'Shine–Delgano box' which is the starting point for the binding of the ribosome which will be reading off the DNA and turning it into protein. At a given space up to 20 base pairs from the Shine–Delgano box, there will be an ATG sequence which the ribosome will begin transcription from by reading three letters from the sequence at once. It is therefore important to make sure that the reading frame for the gene you ultimately wish to express is correct. After the ATG sequence anything read from the DNA will result in a continuous protein chain being produced. Most plasmids will have a 6xHis code inserted after the ATG sequence such that the N-terminus of the produced protein is capped with a histidine tag so that the protein binds to nickel and can be isolated easily. Many plasmid vectors also have multiple cloning sites positioned after an ATG code so that more than one gene can be expressed or else to produce a modified protein after expression. This is controlled by having multiple points at which the plasmid can be cleaved by a restriction enzyme and a new gene inserted. If the genetic sequence read off by the ribosome does not include the sequence that denotes the 'end' of replication, then a hybrid protein is formed from reading one gene after the first without a break. After the insertion point there will be a stop codon telling the ribosome to stop reading and producing protein. There are three different stop codons occurring naturally. These are TAG, TAA and TGA also colourfully labelled amber, ochre and opal (or umber), respectively (Fig. 6.2).

PLASMID MODIFICATION FOR GENE EXCISION – PRIMERS AND PCR

PCR is a method of amplifying DNA (Mullis and Faloona, 1987). This is to say that a small amount of DNA can be replicated into larger amounts. In this case we will be working with a plasmid which is two strands of complimentary DNA that form into a ring. The PCR reaction works by heating up the DNA and causing the two strands that form the DNA to come apart or unzip from one another. The heating generates thermal motion in the DNA that overcomes the hydrogen bonding holding the two strands together between roughly 45 and 60 °C. The actual temperature at which the strands come apart is determined by the ratios of GC nucleotide pairings to AT nucleotide pairings. The GC interaction has three hydrogen bonds holding the pair together, whereas the AT pair has only two. It therefore requires more thermal energy to disrupt the GC couple. This means that the more GC pairings occur in a sequence of DNA; the higher the temperature required to split, the double stranded DNA will be (Fig. 6.3).

For replication of the unzipped DNA to occur in a normal PCR reaction, a short piece of tailor-made DNA is needed that will stick to each end of the DNA strands after they have been split apart to form two separate rings. This is called a primer and is typically 10–25 base pairs long so that there is a good overlap with the ends of the DNA sequence to be copied. The primer is

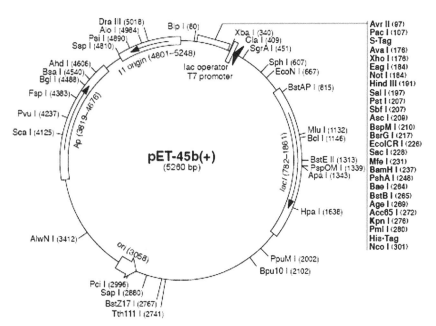

FIGURE 6.2   A map of the pET45 plasmid. A position of a gene or particular location is given by a number noting the number of base pairs from the 12 O′clock position on the map. Arrows denote the direction in which the sequence is read: clockwise for outer ring 'sense' DNA and anticlockwise for inner 'antisense.' A sequence is included for antibiotic resistance (marked AP) and a multiple cloning site is present after the ATG into which a gene can be inserted (marked lac operon) after cleaving the plasmid open with restriction enzymes and subsequent ligation. *Adapted from the Novagen product specification sheet.*

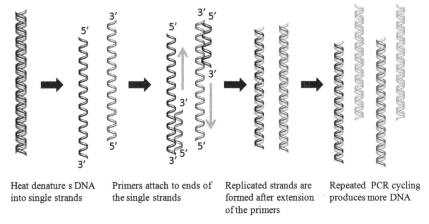

| Heat denatures DNA into single strands | Primers attach to ends of the single strands | Replicated strands are formed after extension of the primers | Repeated PCR cycling produces more DNA |

FIGURE 6.3   A schematic of the polymerase chain reaction using primers.

needed so that another protein called DNA polymerase can stick onto the DNA and start building a complimentary DNA strand. The primer is designed so that it only sticks to the start and end of the gene we wish to amplify. When the plasmid is heated initially in the PCR reaction, it will split into the inside and outside rings. The primers are present in an excess concentration so that, when the reaction is cooled down and the DNA recombines through hydrogen bonding, the primers stick to the single strands at the desired point instead of the complementary inside or outside ring of the plasmid DNA.

A DNA polymerase is present in the solution that then begins to read and replicate the DNA sequence starting from the primer and moving along the gene sequence. Much like a starting block for a runner, the polymerase requires this primer on the end to get going. In a sense we are tricking the polymerase into thinking it is working to replicate a much longer piece of DNA like it might in a natural cell. The target ends of the gene DNA strands are called the three prime (3′) and five prime (5′) ends. The primers are complimentary to the 3′-end of sense and antisense DNA (or the inside and outside strands in a plasmid). You will need a primer complimentary to both the inside and the outside strands of your plasmid (or DNA). The polymerase will always start working on the DNA strand from the primer paired 5′-end and work its way along to the 'naked' 3′-end. The primer that binds to the inside or 'bottom' strand is known as the forward primer. The primer that sticks to the outside or 'top' strand is known as the reverse primer. During this process the DNA polymerase will use up a feedstock of deoxynucleotide triphosphates to make the amplified DNA. The DNA polymerase enzyme used in the PCR reaction is specially cultivated from thermophillic extremophile sources as it needs to retain functionality after the elevated temperatures (up to 95 °C) used to split up the DNA strands. There are two general strains commonly available and both have advantages and disadvantages.

Taq polymerase was originally isolated from the bacterium *Thermus aquaticus* which lives in hydrothermal deep-sea vents. It works extremely fast to replicate DNA but has a high error rate. It can read 100 base pairs a second and will read up to 10,000 before disengaging from the DNA strand (Lawyer et al., 1993). Because we wish to obtain high-quality cloned genes that can produce the protein we want, we will not be using Taq polymerase.

Vent polymerase was similarly isolated from the bacterium *Thermococcus literalis* which is also a deep-sea thermophile (van den Burg, 2003). This polymerase and its genetically modified variants like Phusion polymerase have a much higher accuracy in duplicating the source DNA. Because of this it is the polymerase of choice for applications where high fidelity of the cloned gene is important.

Here we will be using the PCR reaction and primers to replicate many copies of the GFP gene and then using an enzyme to ligase the cloned genes into a new plasmid vector. In this reaction the primer sequence is designed to stick to the ends of the opened plasmid DNA and insert cleavage sites into the amplified DNA. The chunk of DNA produced at the end of the PCR reaction will have restriction enzyme sites at the correct positions so that it can be ligated into the correct plasmid vector (containing antibiotic resistance) for transformation. It is important to remember that the rest of the plasmid we start with is not replicated and only the gene selected by the primers is amplified. The source plasmid sequence, in this case pEGFP-C1, will be known beforehand and this will allow you to get primers made up by a specialist company. For this experiment, the pEGFP-C1 does not have cleavage sites in the position we would like at the beginning and the end of the gene. We are going to use a tailor-made primer sequence for each of the forward and reverse DNA sections in the pEGFP-C1 plasmid so that after the PCR reaction, the MfeI and XhoI cleavage sites will be in the correct position on the cloned GFP gene. Both of these cleavage sites are already present in the pET-45b vector as they are part of the multiple cloning sequence site where genes are inserted and ligased.

Once the gene has been amplified, it will need to be purified before ligation. This can be done by gel electrophoresis or by using a specific kit for purifying PCR-generated DNA. The kit is a quick column-based centrifuge preparation but the procedure for using gel electrophoresis is included as this is also a routine method for checking purity of the product and can be used for gene isolation (Fig. 6.4).

To subclone restriction sites into pEGFP-C1 and amplify the GFP gene using primers and PCR, you will need:

1. A stock of the pEGFP-C1 plasmid. This was originally made available publically by a company called Clontech into the biochemistry community but is available elsewhere. It is commonly available in the biochemistry community and, in general, people will hopefully be kind enough to send you some. You will only need 100 nanograms to a microgram in 1 mL of water or buffer.

2. A stock of the pET-45b vector plasmid which is available commercially.

3. Custom-ordered oligomeric primer DNA. For this particular experiment, you will need the following sequences for the PCR reaction with the pEGFP-C1 plasmid:
   Forward primer: 5′-GATCC<u>CAATTGG</u>**ATG**GTGAGCAAGGGCGAGG-3′
   Reverse primer: 5′-GTAG<u>CTCG</u>AGATCTGAGTCC-3′
   The underlined sequences are the cleavage points for the restriction enzymes. You will start from the ATG preceding the his-6 site. The ATG in the primer simply ensures that the gene will be expressed when inserted near a Shine—Delgano box in case the gene is to be inserted after another protein sequence in the multiple cloning site of the vector. You will require a stock solution of each at a concentration of 10 μmol/L. The primer is available as a powder along with instructions for the correct dilution in water for use. You will be using 2.5 μL of each of the stock primer solutions.

4. A DNA polymerase kit including a 10× DNA polymerase buffer. Many different kits and types of DNA polymerase are available but the source preparation uses the Vent$_r$ DNA polymerase kit from New England Biolabs for this particular PCR reaction. The buffer is 10× the working concentration and is diluted to the correct concentration by the addition of the other reagents. This will contain Mg^{2+} which is vital to the reaction and is a cofactor for the DNA polymerase. Typically, the concentration of magnesium is adjusted between 2 and 6 mmol/L for optimal results. The following reagents form 1 L of the 10× buffer which should be at pH 8.8 at 25 °C:
   20 mmol/L (3.152 g) Tris(hydroxymethyl)aminomethane hydrochloride
   10 mmol/L (1.321 g) $(NH_4)_2SO_4$
   10 mmol/L (0.745 g) KCl
   2 mmol/L (0.240 g) $MgSO_4$
   0.1% Triton X-100 nonionic surfactant
   The kit will also include a 10 mmol/L solution of deoxynucleotide solution which acts as a feedstock for the PCR reaction. The final component in the kit will be the DNA polymerase which is added in small amounts to the PCR tube. The polymerase is often quoted in units and you will need one unit for this reaction which will equate to 0.5 μL of polymerase stock solution as outlined in the kit. For the DNA polymerase, this means 1 unit will extend DNA at a rate of ~1 kilobase pair per minute.

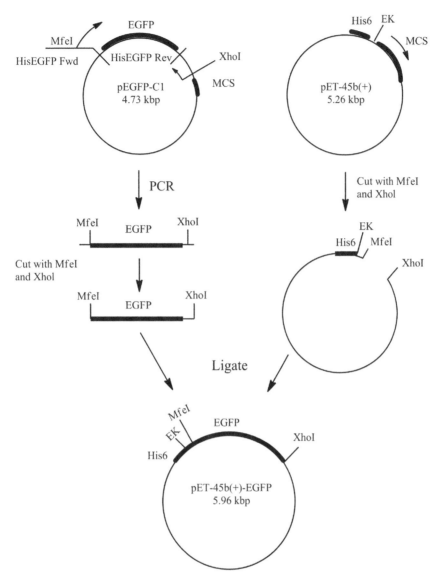

FIGURE 6.4 Schematic for the production of pET45b(+)-EGFP plasmid encoding a 6× His tagged fusion protein. PCR is performed using primers that encode for EGFP in both the forward and reverse reading positions. Both vector and inserts are digested using MfeI and XhoI. EK stands for enterokinase protease site.

5. PCR machine and PCR tubes 0.5 mL in volume. They are small and designed to transfer heat efficiently. Additionally, DNA will not stick to the tube interior.

6. A benchtop microcentrifuge.

7. A plasmid purification 'miniprep' kit. These are commercially available and are routinely used in the laboratory for DNA purification. A more involved procedure for the isolation and purification of plasmid and gene DNA is provided elsewhere.

To subclone restriction sites into pEGFP-C1 and amplify the GFP gene using primers and PCR:

1. **Wear gloves throughout the experiment**. First and foremost you do not want any contamination. Whilst this is true of everything else in this book, DNA and PCR can be very sensitive. Human hands are covered in nonspecific nucleases. This means that they will cleave the DNA you are trying to make at random places and will make the DNA you have amplified unusable.

2. Take all your polymerase enzymes and primer solutions out of the freezer and allow them to defrost on a bed of ice. Keeping them cool will protect them. Be very careful not to expose the contents of the vials to melted water.

3. Take the PCR tubes. Place them into a rack for easy pipetting.

4. Use a vortex to mix up the primers as supplied. Before you use them the primers may have settled so it is good practice to ensure you are working with a homogenised solution.

5. To the PCR tubes add the following in this order. It can be easy to add small droplets to the walls of the PCR tube and then mix by tapping it down and giving the tubes a quick centrifuge.

a) 5 μL of the 10× polymerase reaction buffer. By adding the buffer, first the DNA and proteins are being placed into a stable environment and should experience minimal damage.

b) 1 μL of the deoxynucleotide phosphate solution. The final working concentration will be 200 μmol/L.

c) 2.5 μL of the forward primer and 2.5 μL of the reverse primer. This will give a final working concentration of 0.5 μmol/L for each primer.

d) 100 ng of the pEGFP-C1 plasmid in 1 μL of water.

e) 0.5 μL of the DNA polymerase containing one unit of the active enzyme.

f) Make up the total volume of the PCR reaction solution to 50 μL with sterile water.

6. Tap the PCR tube down so that nothing sticks to the wall and briefly centrifuge them in a microcentrifuge. Flick the tube to get the reactants to the bottom of the tube and vortex them to ensure adequate mixing. A few seconds will be more than enough. Excessive stirring could damage your sample.

7. Put all your enzyme and DNA stocks back in the freezer. The longer they are out, the more they might degrade and lose activity. Biological products tend to be very expensive so you do not want to lose them.

8. Take the closed PCR vials and place them into a PCR machine. The samples should sit in a hot block which will seal when you close the machine. This will ensure that the temperature is evenly distributed in the samples. If it does not have a hot block, then you can add some mineral oil to the top of the solution which will seal it to evaporation during the heating cycle. Using oil does require some careful pipetting afterwards to remove it. Most PCR machines can be setup to run a program so as to find a user manual. You may get prompted to run a 'hot start.' Unless you are working with a polymerase that specifically requires this, do not use it. Here is the PCR heating cycle:

a) 94 °C for 5 min (first melt).

b) 30 cycles of: 1 min at 94 °C to melt, 45 s at 65 °C for the primers to attach and 1 min at 72 °C for extension. For DNA the extension step allows 1 min for every kilobase pair of DNA.

c) 72 °C for 5 min it make sure all the polymerase fills in the DNA ends.

d) Maintain at 4 °C. This can be held indefinitely to store the DNA.

9. The DNA may be stored at 4 °C for long periods at this stage. For the sake of ease, use a commercially available plasmid purification kit to isolate the DNA from any contaminants.

Before further use the cloned gene must be checked to ensure that the correct restriction sites have been included. Normally the plasmid is sent to a laboratory for sequencing which is cheap and quick. You may wish to check the plasmid yourself using gel electrophoresis with a 1% agarose gel and comparing the expected size of the cloned plasmid and the starting plasmid with a known DNA 'ladder.' The procedure for this is outlined in the next section. The concentration of the DNA can be determined by UV-vis spectroscopy of the plasmid in water over the 260–320 nm range. A reference sample of a known concentration is measured and used as a standard. The concentration of DNA in your final product is given by the formula, keeping in mind that you may need to dilute the DNA further to get the absorption under 1 unit:

$$\text{DNA conc in moles per litre} = \text{Absorbance at 260 nm} \times 50 \times (\text{dilution factor if used})$$

The quality of the DNA can be checked by comparing the ratio of the absorbance at 260 to the absorbance at 280. A ratio of between 1.5 and 2 indicates pure DNA.

If the sequence is correct, then the subcloned purified GFP gene is ready to be ligated into the pET-45b(+).

Determining the Concentration of a Primer in Solution

The concentration of the primer can be calculated directly from a known sequence using the Beer–Lambert law. Take the sequence of the DNA you have ordered and total up how many of each C T G and A there are in the sequence. For example, a primer might contain 7 C, 5 T, 8 G and 3 A.

The extinction coefficient formula is

$$\varepsilon = C(7.05) \times G(12.01) \times A(15.02) \times T(8.4) = \text{mol/cm}$$

or substituting the example numbers in:

$$\varepsilon = 7(7.05) \times 8(12.01) \times 3(15.02) \times 5(8.4) = 232.49 \text{ mol/cm}$$

Next take a known volume of the primer from the stock solution as provided and dilute it in a known amount of water. Place this solution in a 1 mL path length cuvette and run a UV-visible spectrum by making note of the absorbance of the dilute primer solution at 260 nm. The absorbance needs to be below 1 to be of use so if you find that your absorbance is above this, then dilute it further.

$$\text{Concentration of stock in moles (mol/L)} = \text{Optical absorbance} \times \text{dilution factor/extinction coefficient.}$$

CUTTING THE GFP GENE FROM THE SUBCLONED PLASMID AND TRANSFERRING IT TO THE PET-45B PLASMID VECTOR USING RESTRICTION ENZYMES

In this section, we will be taking the cloned and amplified gene for the expression of the green fluorescent protein and inserting it into the plasmid vector pET45b. Once the DNA sequence of choice has been inserted, the vector is renamed pET45b(+) by convention to indicate that it is carrying the recombinant gene. The digestion sites on the gene will be cleaved using restriction enzymes in one tube to get rid of the dangling primer ends while the same enzymes will open up the plasmid vector to form linear DNA in another, making both components ready to splice together using a ligation enzyme. Gel electrophoresis is then used to separate out the digested components by size fractionation. The components will be stained with a UV fluorescent dye so you can see them in the gel. By comparing the digest components to a known gene ruler run in the same gel, you will be able to pick out both the gene you want from one channel and the linearised transformation vector from the other. The genetic parts you want are cut directly from the gel using a scalpel and then purified to get rid of the agar gel and dye to leave behind only the DNA. Fine silica is used to separate out the DNA which binds to silica in the presence of a strong chaotropic salt such as sodium iodide at high concentrations. The silica with the DNA stuck on it is centrifuged and remains in the tube whilst the rest of the solution is rinsed away. When the salt concentration is lowered, the DNA becomes water soluble once more and can be extracted from the silica with water. The linear vector and the GFP gene are then mixed in a ligation buffer containing the bacteriophage T4 ligase enzyme needed to covalently bond the plasmid back into a ring ready for use in transformation.

Both the gel step to isolate the plasmids and the purification step using silica are performed in a laboratory using commercially available kits. The inclusion of the more involved procedures here should give the reader a better understanding of the process but may prove slightly more difficult than the 'minipreps' custom made for specific functions.

To ligate the gene for GFP into another plasmid vector you will need:

Gloves must be used and standard biohazard level 1 protocols should be followed throughout this experiment!

1. The cloned GFP gene from the previous section and a plasmid vector for insertion (pET-45b). You will need 10 μL each of a 25 ng/μLe solution of plasmid DNA in water. The concentration can be checked by UV-vis spectroscopy. Each separate plasmid should be in a microcentrifuge tube.

2. A stock of restriction buffer containing 20 mM Tris acetate, 10 mM $MgCl_2$, 50 mM potassium acetate and 1 mM dithiothreitol at pH = 7.9. The restriction enzymes are sensitive to temperature and salt concentration so the composition will vary depending on what enzymes you intend to use. This can be ordered premade from the same place you have purchased your enzymes.

3. MfeI and XhoI stock restriction enzymes. This will be provided from a supplier in a 50% glycerol buffer solution which is quite thick and special care will be needed when measuring this out with a volumetric pipette. The concentration will be 10,000 units/mL where 1 unit will digest 1 μg of DNA in 1 h at 37 °C in a reaction volume of 50 μL. The enzymes will degrade rapidly at room temperature and must be kept at −20 °C in a freezer. *Do not handle without gloves on! Nonspecific nucleases on your fingers can completely ruin these enzymes.*

4. The equipment for running agarose gel electrophoresis on DNA as outlined elsewhere. The gel will be composed of 1% agar and 90 mM Tris−borate buffer and 2 mM EDTA adjusted to pH = 8.3. The same buffer without the agar is also to be used as the running buffer. A UV gel imaging setup so you can see the DNA in the gel.

5. A gene ruler for running with the digests on gel electrophoresis. You will need a 10-μL aliquot of the ruler ready in a microcentrifuge tube.

6. A clean scalpel. Rinse with ethanol and pass through a flame after each use. Make sure the blade is large enough that the parts of gel will not get trapped where the blade meets the handle.

7. A microcentrifuge and microcentrifuge tubes for performing reactions in.

8. An incubator that can be set to 37 °C and 60 °C with a stand for microcentrifuge tubes. Most biochemistry laboratories will have a range of incubators set to different temperatures.

9. A stock solution of calf intestinal phosphatase. This is commercially available and is used to keep the vector plasmid ring open until the ligation reaction. Because the linearised vector DNA will have 5′-phosphate groups remaining at the ends of the DNA after the digestion, there is a chance that these could undergo condensation with the 3′-hydroxyl bonds and close off the plasmid before insertion. The phosphatase removes the phosphate groups so that the loose ends cannot undergo this reaction.

10. A stock solution of agarose gel sample buffer containing SBYR-green I, 20 mmol/L EDTA, 3 g/L bromophenol blue and 50 g/L Ficoll. This solution is added to the digest to stain the DNA for gel electrophoresis. The solution is dark blue in colour and without this you will not be able to see the DNA in the gel. *SBYR Green intercalates with DNA and is potentially toxic. Wear gloves and avoid contact with your hands!*

11. A stock solution of sodium iodide. This can be prepared by dissolving 90.8 g (0.605 mol) of NaI and 1.5 g of (0.011 mol) Na_2SO_3 in 100 mL of water. This is a colourless solution but will stain yellow upon drying; so, avoid spills.

12. A suspension of 200 mg of diatomaceous earth in 1 mL of water. This is fine powdered silica which the DNA will stick to under the correct conditions and is used for separating out the DNA.

13. A stock solution of wash buffer. This can be prepared by adding 50 mL of ethanol to 40 mL of water. Add 0.157 g (0.001 mol) Tris—HCl, 0.584 g (0.01 mol) NaCl, 0.146 g (0.5 mmol) EDTA and adjust to pH 7.5. Make up to 100 mL.

14. Bacteriophage T4 ligation buffer which will normally be provided directly by the supplier. The buffer concentrations will normally be 500 mmol/L Tris—HCl, 100 mmol/L $MgCl_2$, 100 mmol/L dithiothreitol, 10 mmol/L adenosine triphosphate and 0.1 μg/mL bovine serum albumen at pH = 7.5. T4 ligase is another enzyme which catalyses the formation of the phosphodiester bonds. To work properly, it requires magnesium ions and ATP which are cofactors for the reaction in linking a 5′-phosphate end of DNA to a 3′-hydroxyl end. This enzyme is very sensitive to the concentration of NaCl in solution, and salt concentrations above 70 mmol/L should be avoided as this will impair the ligation reaction.

15. A fridge at 4 °C for incubation of the ligase reaction.

16. Volumetric pipettes and disposable tips in the 1—20-μL and 200—1000-μL range.

17. A marker pen for keeping track of the microcentrifuge tubes and what they contain.

18. A vortex for separating out the diatomaceous earth.

To ligate the gene for GFP into another plasmid vector:
Never use the same pipette tip twice. Use a fresh tip every single time to avoid contamination or transfer of material! (**Fig. 6.5**).

1. Add 1.3 μL of the restriction buffer to the 10 μL of the GFP gene DNA in the microcentrifuge tube.

2. Add another 1.3 μL of restriction buffer to the microcentrifuge tube containing the 10 μL of pET-45b plasmid vector.

3. To each tube add 1 μL of each of the restriction enzyme stock solutions. The enzyme solution should be on ice when not in the freezer. Because the amount is so small, it can be difficult to measure out without getting an excess of the enzyme on the outside of the pipette tip. Let the end of the tip penetrate to only just below the surface of the solution before you draw the enzyme solution in.

4. Close the tube lids and flick quickly so the small volumes drop to the bottom of the tube. Spin them very briefly in a microfuge to mix. Do not use a benchtop vortex to mix as this can damage the DNA through shear forces.

5. Place the plasmid and DNA digests into an incubator at 37 °C for 1 h. After 30 min you will need to add 1 μL of calf intestinal alkaline phosphatase to the tube containing the plasmid vector (pET45b). After adding this give it a flick to mix the solution properly and a brief spin on the microfuge and then place back in the incubator for the remaining 30 min. This should only be added to the plasmid vector that you want to linearise and insert the gene into. It is *not* to be added to the plasmid solution you are excising the GFP gene from.

6. Prepare a solution of 1% agarose and running buffer and cast a gel in a tank with taped sides. Remember to insert a comb to produce running wells and let it set.

7. Once the gel is set and cool, remove the tape and place it into the gel electrophoresis tank. Pour running buffer over the gel to a depth of 5 mm. Remove the comb gently to leave wells in the gel.

8. After an hour, remove the digestions from the incubator and add 2 μL of the blue dye-containing agarose gel sample buffer to each tube and to the 10 μL aliquot of gene ruler.

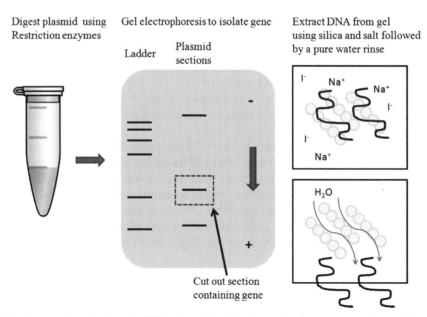

FIGURE 6.5   An experimental schematic of digesting a plasmid and extracting the target gene using gel electrophoresis.

9. Use a volumetric pipette to transfer 12 µL of the gene ruler into the well furthest left in the gel. Skip over the next well and then add 12 µL of the plasmid vector digest to another well. Once more skip a well and then add 12 µL of the GFP gene digest into another well. You may wish to place another aliquot of gene ruler in the final well. The spacing is so that there is a minimal risk genetic material from one track will contaminate another.

10. Place the lid onto the gel tank and run the gel at 150 V for 1 h. Remember that DNA is negatively charged and will move towards the positive terminal. This will separate out the products of the digestion by size.

11. Remove the gel tray with the gel in it from the tank and transfer it to the UV illuminator. Take special care not to let the gel drip on anything.

12. Inspect the gel and compare the sizes of the digest components against the gene ruler. The linearised plasmid vector should appear as a single band at 5.26 kb pairs. The GFP gene is 0.7 kb pairs.

13. Use a clean scalpel to cut the section containing the GFP gene and the lineraised plasmid vector from the gel. *Make sure the scalpel is cleaned between cutting sections out.* Place the gel slices into separate and marked microcentrifuge tubes so you know which is which.

14. Add 1 mL of sodium iodide solution to each tube.

15. Shake the suspension of diatomaceous earth to resuspend it and immediately transfer 25 µL of the suspension into the tubes containing the gel slices. Use a vortex to mix everything together.

16. Place the tubes in an incubator set to 60 °C for 10 min. The agarose should dissolve, but if it does not then leave the sample in until it has. You will need to shake the samples every few minutes to keep the earth suspended. The high concentration of the chaotropic sodium iodide will cause the DNA to stick to the surface of the silica which can be separated from the agarose solution in the next step.

17. Centrifuge the tubes using the microfuge for 60 s so that the diatomaceous earth sediments to the bottom. Use a pipette to remove the top solution carefully so that none of the silica is disturbed.

18. Add 1 mL of the ethanol containing wash buffer to the tube and vortex it to resuspend the silica. The ethanol present in the wash buffer forces the DNA to precipitate on the surface of the silica so that the agar and other unwanted reactants can be removed.

19. Centrifuge again for 60 s to form a pellet and remove the top solution with a pipette. Repeat steps 18 and 19 twice.

20. Use a small volumetric pipette (with fresh tips) to remove as much of the remaining wash buffer from the top of each silica pellet in the tubes as possible. Replace the tubes with the lids open back into the incubator at 60 °C so any remaining ethanol will evaporate away. You do not want ethanol present for the next step as it will disrupt the ligation reaction. Prolonged heating will damage the DNA so let the samples dry for a maximum of two minutes.

21. Add 50 µL of sterile water to each tube and vortex for a few seconds to resuspend the silica. Let the solution stand for two minutes during which the DNA will dissolve in the water, allowing the silica to be separated.

22. Centrifuge the tubes for 1 min so all the silica sediments to the bottom. Very gently transfer the tubes into a stand so that none of the earth is disturbed. You will now have a tube containing the GFP gene sequence and a tube containing the linearised plasmid vector into which you will be inserting the gene. The next step is to perform a ligation reaction to bind these two sections together to form a ring plasmid.

23. Stand an empty microcentrifuge tube in the stand. This is to be the vessel for the ligation reaction. Transfer 10 µL from the plasmid vector containing solution into the ligation tube. Next, transfer 5 µL of the GFP gene solution into the ligation tube. Take these aliquots from the top solution so that none of the earth is transferred across. Add 2 µL of water, 2 µL of ligase buffer and finally 1 µL of the stock T4 ligase enzyme. The T4 enzyme should be placed back into the freezer immediately after use.

24. Mix the ligation solution by flicking the bottom of the tube. Place it into a stand and leave it to incubate in a fridge at 4 °C overnight to allow the ligation reaction to occur.

This solution should now contain a plasmid vector with the GFP gene spliced into it and can be used directly in the transformation reaction.

TRANSFORMATION AND CLONING

With thanks to David Green and Nigel Savery

For research purposes, having lots of a given gene is useful if you want to analyse it. It is not as simple as extracting DNA from the organism of interest and cutting the required section from it using enzymes. The gene you might wish to look at will be a tiny percentage of the total DNA so that it becomes impractical to cut a gene from the bulk. Instead, it is better to cut the gene from the DNA and place it into a plasmid which has very little genetic information of its own so that the gene makes up a large percentage of its mass. Because the foreign gene has been grafted into a plasmid, it can now be inserted into a bacterium where duplicate plasmids will be generated as part of the natural process of the bacteria. Getting the recombinant DNA plasmid into a bacterium is known as a transformation procedure. The duplicate plasmids are genetic clones of the original recombinant plasmid vector which is why the term cloning is used. As the host bacterium replicates so to does the amount of plasmid within it so that you are eventually left with a large number of cloned plasmids that you can then isolate to extract DNA from.

In this section, we transform a colony of *E. coli* by inserting the plasmid vector containing the GFP gene prepared in the previous ligation step. This is done by making the *E. coli* receptive to the uptake of the foreign DNA by making them 'competent.' This is essentially a procedure that weakens the *E. coli* and makes its outer membrane permeable enough for plasmids to enter. There are a number of ways to do this but we will use a simple method involving exposure to a strong calcium salt solution before heat shocking the bacteria at 42 °C for a brief time which opens pores up in the cell wall. Another common method used in the laboratory is electroporation which involves passing a current through a colony and renders the cell wall receptive to DNA uptake. However, this can be more technically difficult to execute and so will not be covered here. The transformation reaction is not very efficient, and only roughly 1 in every 10,000 *E. coli* cells will successfully uptake the recombinant plasmid. This means that you will be left with a large number of live cells that will not produce the gene you want. The plasmid vectors used in transformation and cloning have an additional active gene in them which confers antibiotic resistance. To select only the *E. coli* that has undergone transformation, you will need to spread a colony of the transformed suspension onto a selection plate which contains an agar gel growth medium and an antibiotic. Because only the transformed *E. coli* will contain a gene for antibiotic resistance, any untransformed cells will die. In this way, you can grow only successful transformants.

One thing to keep in mind with a transformation procedure is that you are trying to amplify a functional gene but not express that gene as a protein until you need to. As the bacteria replicate, they will be generating identical cells composed of proteins and enzymes read from the plasmid coding. Why does this cellular mechanism not also express the inserted gene? Most of the time it will but for some applications, where the protein produced will be toxic to the bacteria for example, you need to grow enough of the bacteria so that you can generate a large amount of protein all at once. For a situation like this, the plasmid vectors into which the excised genes are spliced include a sequence called a *lac operon* which behaves as a switch for turning the expression of the protein from the plasmid DNA on and off. The lac operon sequence is composed of three genes strung together that are masked by a protein in the absence of lactose. This protein is called the lac repressor and it prevents the transcription mechanisms from working until a lactose molecule binds preferentially to the lac repressor which exposes the lac operon. Once the lac operon is unbound then a ribosome can lock onto it and set to work forming RNA and eventually proteins. When this sequence is present in a vector plasmid, the production of the protein coded after it is suppressed meaning that although the plasmid will be copied, the proteins it codes for will not be produced. Once you have grown a large enough amount of bacteria, a chemical trigger such as isopropyl β-D-1-thiogalactopyranoside (IPTG) is added which behaves similarly to lactose in the bacterial cell except it cannot be broken down as a metabolite in the same way. Large amounts of the recombinant gene protein will then be generated which can be harvested for use.

To transform E. coli with pET-45b(+)EGFP, you will need:

1. An autoclave and sterile glassware including Schott bottles and plates. All glassware must be autoclaved before use. Any solutions to be used can be sterilised by filling a Schott bottle with them half way and then screwing the lid on so that it is finger loose before autoclaving it. Both growth mediums and water used in the experiment must be treated in this way before use.

2. A gas flame for keeping the surrounding area sterile. The working area should be wiped down with ethanol beforehand but *ensure that no ethanol remains before the flame is ignited!*

3. A stock of super optimal broth (SOC) for cultivating the cells with. This broth contains a number of ingredients that will strengthen the weaken *E. coli* after the heat shock step. This can be prepared by dissolving the following reagents in 1 L of water:
    20 g Bacto-tryptone
    5 g Bacto-yeast extract
    0.6 g NaCl
    0.19 g KCl
    2.04 g $MgCl_2$
    2.48 g $MgSO_4 \cdot 7H_2O$
    3.60 g Glucose
    The pH should not need adjustment, but if it deviates significantly from pH = 7 then adjust using a 1 mol/L solution of NaOH or HCl. This solution should be sterilised before use in a bottle in an autoclave.

4. A colony of *E. coli* cells. It is possible to purchase an *E. coli* strain made competent for expression directly and this is the normal choice for laboratory research. These can either be designed to express genes or simply to amplify the DNA with no further expression. You will want to use a strain that is capable of gene expression to produce protein. In this experiment we will use the strain BL21(DE3) which can be obtained commercially.

A general method for making any *E. coli* cell competent is provided although you should work only with safe strains that are commonplace in a biochemistry laboratory. You will start the transformation experiment with an aliquot of 100 μL of competent cells in a sterile microcentrifuge tube. *The E. coli should remain on ice.*

Making E. coli *Cells Competent*

Just before you begin take 1 mL aliquots from an *E. coli* colony in broth while it is in the exponential growth phase and place them into a 1.5 mL microcentrifuge tube. The logarithmic phase can be estimated by measuring the optical density of the broth containing the *E. coli*. Take an aliquot when the optical density is between 0.3 and 0.6.

You will also need a stock solution of 0.1 mol/L calcium chloride solution. This can be prepared by dissolving 11 g of anhydrous $CaCl_2$ in 1 L of water.

1. Centrifuge the 1 mL of *E. coli* culture in a microcentrifuge for 30 s. Get rid of the top solution and sit the formed cell pellet on ice in the bucket.
2. Transfer 1 mL of the calcium chloride solution into an empty microcentrifuge tube and sit this in the ice bucket.
3. Add 200 μL of the cold calcium chloride solution to the tube containing the cell pellet. Use the pipette to gently mix the contents by drawing liquid in and out slowly. The *E. coli* cells will be delicate. Seal the tube and leave to stand on ice for 10 min.
4. Centrifuge the *E. coli* in calcium chloride dispersion for 30 s to form a pellet again and get rid of the top solution. Sit the *E. coli* pellet in the sealed tube on ice.
5. Resuspend the pellet once more in 100 μL of cold $CaCl_2$ and flick the tube to disperse. The cells will be weakened so any vigorous motion like vortexing could damage them. Allow the sample to sit for another 10 min on ice although they can remain at this temperature for hours if need be.

These cells will now be competent and can be used in the transformation reaction. However, it is easier to use a colony purchased specifically for transformation.

5. The pET-45b(+)EGFP plasmid vector containing the GFP gene as prepared in the previous section.
6. A large ice bucket for keeping all the biological material cold.
7. A microcentrifuge and centrifuge tubes.
8. A stopwatch or countdown timer set to 2 min. This is very important for timing the heat shocking correctly.
9. Sterile Petri dishes with lids for making plate colonies with. You will need a marker pen for labelling the samples as you will be making many plates.
10. Agar for making the agar selection plates with dissolved in lysogeny broth (LB) medium. Dissolve 1 g of agar into 100 mL of LB medium boiled in a bottle on a hot plate. Unless you are going to pour the selection plates immediately the bottles can be screwed shut and stored at 60 °C to prevent the agar from setting. The LB medium is prepared as a stock by dissolving the following into 1 L of water:
    10 g Bactotryptone
    5 g Yeast extract
    5 g NaCl
    1 g Glucose.
    This should be mixed in a bottle and then autoclaved to sterilise it.
11. 1 mL of a 100 mg/mL solution of carbenicillin in sterile water. This antibiotic hydrolyses slower than ampicillin so that the selection plates made using it will last longer. The ampicillin resistance gene also provides resistance for carbenicillin. 100 μL of the antibiotic will be added to the agar solution so that the agar gel in the plate will only be colonised with bacteria with an ampicillin resistance from transformation.
12. An incubator set to 37 °C with a rocker in it.
13. A water bath maintained at 42 °C for performing the heat shocking reaction. Use a thermometer to ensure that the temperature is correct.
14. A glass spreader rod with a stand and a beaker containing bench ethanol for sterilising it.

To transform a plasmid vector into E. coli:

Wherever samples containing biological materials are open to air, it is important to have them near the flame so that airborne contaminants have little chance of getting in Fig. 6.6.

1. Allow a bottle containing 100 mL of the 1% agar in LB medium to cool until it can be held in a gloved hand. You do not want it to cool so much that it sets. To this add 100 μL of the carbenicillin solution and swirl to mix thoroughly.
2. Decant the mixed agar and carbenicillin in LB medium into sterile Petri dishes sat near to the flame. You will make duplicates of the cultures to get the best chance of a successful transformation. You should have enough solution for three to four dishes. Fill the dish so it is half full as you will need a gap between the lid and the gel for the selection incubation. Leave the lid of the dish slightly ajar for five minutes to let moisture evaporate away and then place it on fully. The gels should set fully within 30 min.
3. Take the microcentrifuge tube containing the ligated plasmid vector containing the GFP prepared earlier from the fridge. Centrifuge it briefly to ensure that the contents are in the bottom of the tube and place it on ice in the bucket with the lid clear of the ice to avoid contamination.

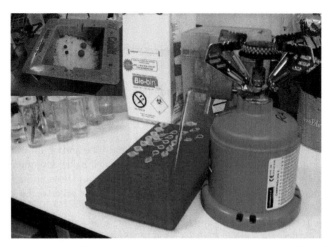

FIGURE 6.6   An example of a transformation experiment bench setup working near a flame. An ice bucket is maintained nearby for holding the biological components in. Any vessels are opened in the vicinity of the burner.

4. Transfer 1 μL of the plasmid vector (about 10 ng of DNA) into a 100 μL of a competent *E. coli* colony. Tap the tube gently with a (gloved) nail a few times. A commercially obtained BL21 colony will contain ~100 nanograms of bacteria per microlitre. You must not shake or flick the competent cells. Handle only the top of the tube as any rough treatment at this stage will kill the *E. coli*. Let it sit in the ice for 30 min.

5. Get the countdown timer or stopwatch ready and place it next to the water bath at 42 °C. Holding the tube by the closed lid, immerse the tube containing the *E. coli* and plasmid mixture into the water so the contents are below the water line and simultaneously start the stopwatch. Hold the sample in place for 45 s during which time the heat shock will cause the membranes of the competent *E. coli* cells to become permeable and allow the plasmid DNA to enter. After the time is up remove the sample immediately.

6. Add 950 μL of the SOC medium to the tube so that cell growth can begin. Place this in an incubator at 37 °C for 1 h with shaking at 200−250 rpm.

7. Mark the Petri dishes containing the set agar selection gel on the part actually containing the gel as the separate lids can sometimes get accidentally mixed up. This is a problem if you are running more than one transformation. It is good practice to mark it with the sample contents, the date and your name. Keep these near the flame any time they are opened to air.

8. If you have used your own competent cells, then there will still be high levels of calcium salts in the transformation mixture so this needs to be exchanged for fresh growth medium. Centrifuge the transformed cells in a microcentrifuge to form a pellet and use a pipette to remove and discard the top solution. Add 200 μL of fresh SOC medium. You do not need to do this for a commercially obtained BL21 colony.

9. The transformed cells now need to be spread onto the agar selection gels where the ampicillin will kill off any *E. coli* that has not got the gene for ampicillin resistance. To do this we will use a glass spreader that needs to be sterilised with each use. Dip the glass spreader into an ethanol solution and allow the excess to drain off. Wave the ethanol coated glass spreader through the flame but do not hold it directly in the flame. The ethanol should burn off in a few seconds and this will leave the glass rod sterilised and ready for use. You do not want the glass rod to become significantly hot or else the cells could be damaged. One sign of this would be the agar gel hissing when you deposit the rod upon it. Place the glass spreader in a stand with the spreading end close to the flame if you are not using it immediately. *Take care not to set the ethanol solution on fire and keep this a respectable distance from the flame! Do not place the spreader back into the ethanol solution until you are sure it is not still burning* (Fig. 6.7).

10. Keeping the lids closed as long as possible, transfer 50 μL of the transformed *E. coli* solution to the centre of a selection gel. Using the freshly sterilised spreader quickly, but with minimal pressure, smear the drop over the surface of the selection gel. Replace the lid and seal using some sterile parafilm. Invert the gel so that the gel is uppermost in the Petri dish. This will ensure that any condensation will drop into the lid and not roll about on the surface where it could mix the growing colonies. The aim is to cultivate single bacterial colonies which we can isolate. Repeat this process on the other selection gels.

11. Transfer the upside down selection gels into a stationary incubator set to 37 °C for 16 h. During this time *E. coli* cells that have been successfully transformed will begin to grow. Ones that have not been transformed will be killed off by the carbenicillin. It is important to remember that the gene for conferring antibiotic resistance is expressed in these cells but the gene for expressing the GFP is not. This means that large quantities of the *E. coli* can be grown before you turn on the expression of the GFP using IPTG in the next section.

FIGURE 6.7   A photograph depicting *E. coli* being spread across a selection gel in the vicinity of the burner. At the bottom of the picture the glass spreader rod is stored upright next to the burner after an ethanol rinse and burn.

FIGURE 6.8   A photograph of a selection plate with transformed *E. coli* colonies growing on it. This example has a slightly high density which may make it hard to select a single dot of colony from the gel.

Because the number of transformed cells is very low compared to the overall number of *E. coli*, the transformants should be distant from one another when they start to multiply. This means that you should see many single dots in the Petri dish where single cells have divided into many. Where the dots overlap there is a chance that some genetic information may have been swapped between the two different bacteria and there is an increased chance of mutation. You do not want to take samples from any colonies that have grown together in such a manner (Fig. 6.8).

The next stage is to isolate and cultivate a single transformed strain of the *E. coli*. To determine the success of the cloning procedure you may need to check the DNA using a 'miniprep' to digest the cloned plasmids and check the excised GFP gene is of the correct size. If the plasmid sequence is correct, then the transformed *E. coli* can be further cultivated in large quantities and then triggered to express the GFP protein.

GROWING UP A COLONY AND TRIGGERING EXPRESSION OF THE GFP PROTEIN

To grow up a colony of transformed E. coli and express the GFP protein you will need:

1. Much of the general equipment required for protein expression has already been listed in the transformation step. To recap briefly, you will need an incubator at 37 °C fitted with an orbital shaker that can run at 200–250 rpm. In this preparation you will need to be able to shake a sample at 25 °C overnight so ask other users before you alter the incubator shaker temperature to avoid arguments! You will need the flame burner setup as before. You will also need an optical density reader sampling

at 600 nm for determining the rough concentration of the bacterial colony. You will need volumetric pipettes capable of dispensing in 50−100 μL and 1−5 mL volumes. Make sure the disposable tips are autoclaved before use.

2. An inoculation loop. This is a long piece of metal wire with a hoop at the end that is used to scrape up bacteria and then deposit it on a gel or in a broth.

3. 1 mL of a stock solution of 34 mg/mL of carbenicillin in sterile water. This antibiotic solution is slightly more dilute than the stock used in the transformation. Make it up fresh and pass it through a 0.22 μ syringe filter. Ensure the filter is sterile before use.

4. A stock solution of sterile LB medium. You will need 1 L and 50 mL for this preparation and the solution is prepared as outlined in the previous section item number 6. Prepare the required volumes in a 100 mL and 2-L Schott bottle, respectively. Autoclave the solutions, screw down the lids and allow them to cool to room temperature.

5. 0.238 g (0.001 mol) of isopropyl β-D-1-thiogalactopyranoside (IPTG) dissolved in 10 mL of sterile water. You will need a 0.22-μ filter and 10-mL sterile syringe for the addition of this stock solution to the medium. Prepare the solution immediately before use. This will be added to a large culture to initiate the expression of the protein.

6. A large centrifuge that can handle large volumes of liquid. Most laboratories will have a centrifuge that can accommodate many 50-mL falcon tubes.

7. 1 L of a stock solution of lysis buffer which will be used to break apart the cells for the extraction of protein. The buffer contains imidazole which is chemically similar to the histidine ring and it will help in the purification stage explained further on.

Lysis Buffer (1 L):
6.90 g (50 mmol) $NaH_2PO_4 \cdot H_2O$
17.54 g (300 mmol) NaCl
0.68 g (10 mmol) imidazole
Adjust pH to 8.0 using NaOH

To grow up a colony of transformed E. coli and express the GFP protein:

1. Turn on the burner flame and hold the inoculation loop in it until the metal becomes red hot. Place the inoculation loop upright and let it cool near the flame for about one minute. If the gel sizzles when you try to extract a colony, then it is still too hot. Should this happen then repeat the process (Fig. 6.9).

2. Open the smaller 100-mL Schott bottle containing the LB medium and add 50 μL of the 34 mg/mL carbenicillin solution. Hold the bottle lid in your hand as you do so to prevent it coming into contact with any potential contamination sources.

3. Select a single dot on the agar gel which represents a colony grown from a single cell. Scrape the loop across the surface through the dot two or three times which should be enough for some of the bacteria to cling on to the loop.

4. Open the 100-mL Schott bottle and dap the loop into the 50 mL of LB medium and give it a gentle swirl. Do not touch the sides of the bottle. If necessary, tilt the bottle on its side so that the loop can reach the liquid. Replace the lid. Enough *E. coli* should have been transferred that a colony will grow in the medium.

5. Place the 50-mL colony into the incubator at 37 °C and mount it securely on a shaker. Set the shaker to 200 rpm and leave the colony to grow overnight.

6. After the colony has been allowed to grow overnight, open the larger Schott bottle and add 100 μL of the carbenicillin solution to the 1 L of LB medium.

7. Remove the 50-mL colony from the orbital shaker and transfer the contents by sterile pipette into the 1 L of LB broth to form an even larger colony. Replace the lid on the 1-L colony.

FIGURE 6.9   A photograph of a metal selection loop being sterilised in the burner flame.

FIGURE 6.10 Photograph of a solution of transformed *E. coli* colony that has been induced to express green fluorescent protein. You can see that the medium has a visible green shade. On the right is another photograph showing the *E. coli* cells, full of GFP collected in the bottom of the centrifuge tube.

8. Secure the 1-L colony on the orbital shaker and incubate at 37 °C with the shaker set to 200−250 rpm. During this time, every 1−2 h, turn off the shaker and remove a 1-mL aliquot of the colony. Make sure you maintain the sterility of the colony by not letting the pipette touch the walls of the bottle or putting the lid down on anything. Use the optical density reader to monitor the optical density at 600 nm. Once the density reaches between 0.6 and 0.8, the colony has reached its mid-exponential growth phase and the expression of the protein is ready to be induced.

9. Lower the temperature of the incubator to 25 °C and prepare the stock IPTG solution in sterile water. Before you induce expression, take a 1-mL aliquot from the colony and store it for later testing in a sterile centrifuge tube. Use the syringe and filter to drip the IPTG solution into the 1 L of culture medium. The lac repressor protein will be removed from the lac operon and the transcription of the gene to produce GFP will begin. Replace the lid on the bottle.

10. Set the orbital shaker to 200−250 rpm and allow the colony to incubate at 25 °C for around 18 h. During this time the medium should become visibly green denoting the production of the GFP. After this time take another 1 mL aliquot and store in a sterile centrifuge tube for testing (Fig. 6.10).

11. Remove the colony from the incubator and transfer the green solution into centrifuge tubes. The protein is still inside the bacteria so the next step will be to break open the cells and isolate the protein. Sediment the cells by centrifuging at 8000 rpm for ten minutes. You should see a pellet form in the bottom of the tube with no green colour remaining in the top solution.

12. Decant away the top solution into a suitable biowaste solvent container. At this point, the isolated cells can be stored until required at −20 °C. Resuspend the cell pellets from each centrifuge tube in a total of 100−200 mL of lysis buffer. Ideally, you want all of your harvested cells in the same tube. This can be achieved by redispersing the pelleted cells in small portions of the lysis buffer and directly decanting the contents into one tube. In order to determine the amount of nickel−agarose beads you will need for isolating the protein, weigh the tube in which the cells are to be collected before you use it.

13. Centrifuge the cells once more at 8000 rpm for 10 min to form a cell pellet. You can now proceed to the protein isolation stage or freeze the cell pellet at −80 °C for storage. Weigh the tube containing the pellet so you have a rough idea of the wet weight of the contents.

PURIFICATION OF HIS₆ TAGGED PROTEIN

By this point you should have a visibly green pellet of *E. coli* cells in the bottom of a centrifuge tube where the GFP has been expressed but is still stuck inside the bulk of the cell. The purification procedure involves breaking open the *E. coli* cells so that the contents are set loose in the lysis buffer. The solution of cellular debris and product formed is termed the lysate. This solution is mixed with agarose beads containing Ni^{2+} ions that will preferentially bind the GFP through the histidine tag attached to the N-terminus of the protein. Other proteins present in the solution also contain histidine residues that can coordinate to the nickel and therefore there is a risk of unwanted impurities in the final product. Nonspecific binding is avoided by the inclusion of imidazole in

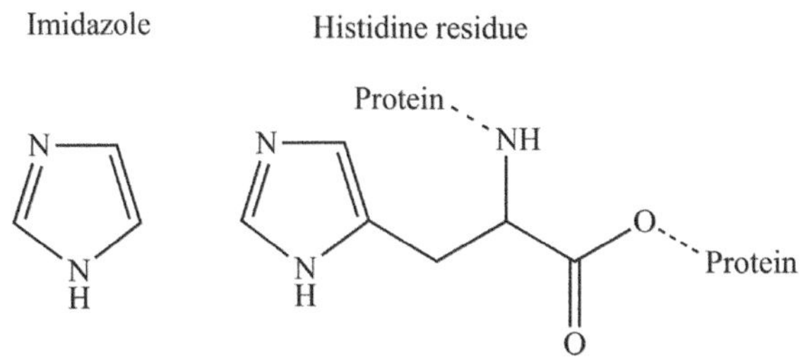

Imidazole   Histidine residue

FIGURE 6.11 The chemical formula of histidine and how it compares to imidazole. Histidine will coordinate to a Ni^{2+} anchored on an agarose bead. High concentrations of imidazole will displace the histidine and release the protein it is bound to.

the wash buffer and by having a sufficient salt concentration charge shield the metal ion. Imidazole is chemically similar to the histidine unit but binds only weakly such that it provides a chemical mask over the nickel ions until displaced by the very strongly binding histidine unit. This is more of a problem in recombinant procedures where the exogenous protein is not expressed at high levels in comparison to the endogenous ones. In this recipe there should be a large excess of the GFP in comparison to the debris of the *E. coli* (Fig. 6.11).

Although the competent cells were delicate and easy to kill, breaking the physical body of *E. coli* apart is surprisingly hard and requires some manual coercion in addition to an enzymatic treatment. Throughout the practical, great lengths have been taken in earlier steps to maintain the integrity of the weakened competent cells. By this point, they have served their purpose and we can now treat the sample much more roughly. The enzyme lysozyme is added which acts to break down the peptidoglycan bonds that make up the cell wall so that it is weak enough to rupture. The physical destruction of the cell wall can then be performed using a variety of methods. The lysis buffer and cell solution can be passed repeatedly under pressure through a narrow gauge needle or aperture using a syringe where the pressure and shear forces tear the cell apart. Similarly, the solution can be placed into a tube and capped with a tightly fitting PTFE bung at the end of a plunger. This is then pressed into the tube so that the solution shears past as the plunder moves down (Fig. 6.12).

FIGURE 6.12 A photograph of a hand-operated homogeniser being used to break up *E. coli* containing GFP. The operation is performed on ice to preserve the lysozyme function.

The cell and lysis mixture is very viscous to start with and then will thin rapidly as the cell contents are released and broken up. In some samples a large amount of DNA is released which makes the lysate solution formed very viscous and difficult to work with. In this case, DNA nucleases, enzymes that break down DNA, are also added to the lysate to thin it out. The solution needs to be fluid enough to be passed through a column of the nickel—agarose beads or mixed in a bed of them. Because the syringe method of breaking cells apart can be laborious and uses small amounts, a probe sonicator is commonly used instead. The sonicator used a piezoelectric oscillator to deliver high-amplitude (loud) sound waves into a solution. If tuned correctly, the sound waves cause cavitation, areas of very high and low pressure oscillating as a standing wave, which behaves like a shear force in the cells and cracks them open. Despite the rough treatment, the temperature must still be kept low so that the lysozyme retains functionality but the GFP itself could be denatured for the nickel extraction step if this made things easier. Denaturing a protein is done by using a strong chaotropic agent such as 8 mol/L urea or 6 mol/L guanidinium chloride which act to break up hydrogen and van der Waals bonding between adjacent protein chains. Chaotropic agents have a very specific behaviour and the same effect will not be seen by simply increasing the concentration of salts which are already in the buffer. The NaCl concentration of buffers used in the protein isolation step should be between 0.2 and 2 mol/L and serves only to prevent some of the nonspecific binding when the lysate is exposed to the nickel—agarose beads. One situation where denaturation might help is in working around the presence of 'inclusion bodies' which arise from proteins clumping together in solution. This should not be the case with the GFP as it is a fully water-soluble protein and dispersed easily in aqueous conditions. Most proteins can be rather hardy and can be isolated in their natural state or can be denatured and then refolded after extraction. Denaturing a protein prior to isolation may be required when the histidine tag grafted to it is inconveniently located in a recess or channel in the structure. By unfolding the entire protein chain, the tag becomes accessible to the Ni^{2+} and can bind to it. This will not be required for the GFP extraction as the binding tag stands proud of the protein surface. Because this procedure should produce large, tens of milligrams, amounts of the GFP, the lysate will be added as a batch to a bed of nickel—agarose beads and placed on an orbital shaker. This will maximise the absorbance of the beads which will then be packed into a large column so that washing and finally elution steps can be performed. With the gel beads packed into a column, the washing and elution buffers can be drawn through the gel by gravity. The washing step will flush unbound and unwanted cell material and protein off of the beads so that only the nickel-complexed target protein remains. The elution step will disengage the target protein from the nickel—agarose gel beads so that the pure protein can be flushed out and isolated. Control of the protein binding to the gel is done by regulating either the pH or the imidazole concentration. Both of these parameters greatly influence the strength of the histidine binding to the nickel centres and the choice of what to use depends on which procedure will maintain the integrity of your protein better. For this particular stage, we will be using a higher concentration of imidazole to extract the protein from the gel.

For pH-controlled protein release from the gel: The pH of the washing buffer is maintained at 6.3 which is low enough that endogenous proteins weakly bound by histidine groups to the nickel—agarose gel will be protonated, dissociated from the gel and subsequently flushed out of the column. After repeated washing you should be left with only the protein coordinated by the hitidine tag to the gel beads. The histidine tag is strongly complexed to the Ni^{2+}, but, just as the more weakly bound histidine groups were, it can be protonated by increasing the acidity further in order to disengage the target protein from the beads. The elution buffer is adjusted to between pH 4.5 and 5.5 and then rinsed through the gel to elute the target protein from the column.

For chemically controlled protein release from the nickel—agarose gel: The concentration of imidazole in the wash buffer is increased so that the equilibrium between binding of the endogenous weakly bound proteins and the imidazole molecules to the nickel ion is shifted towards the molecular species. Essentially the imidazole is 'crowding out' the weakly bound protein. The unbound protein can then be rinsed through. The concentration of imidazole for washing is between 0.01 and 0.05 mol/L. Similarly, to flush out the target protein with the elution buffer, the imidazole concentration is increased further to 0.1—0.25 mol/L.

The final step of the entire process is to dialyse the eluted protein into the buffer of choice and determine the product purity using SDS-PAGE analysis. As with many steps in this process, there are kits available to streamline procedures. For protein purification, you can purchase ready-packed columns for use with automated purification and elution systems. Most biochemistry laboratories will also have a cell homogeniser into which you can place your *E. coli* cells.

To extract and purify the GFP from the E. coli pellet you will need:

1. An ice bucket.
2. The cell pellet from the previous step. Allow it to defrost if it was frozen and keep it on ice throughout. You will need to know the rough wet weight in order to use the correct amount of nickel—agarose gel. This recipe assumes you are starting with your cell pellet in a large plastic centrifuge tube with a screw top lid.
3. A stock lysis buffer as prepared in the previous section containing 0.01 mol/L of imidazole at pH 8.0
4. 50 mg of lysozyme enzyme. This will come as a lyophilised powder and should be kept frozen until use. Keep on ice when it is not in the freezer. You will need 1 mg for every mL of lysate you produce. In this preparation we will prepare around 50 mL of lysate solution so you will need about 50 mg of the enzyme.
5. A dewar flask containing liquid nitrogen and a wide dewar basin for bowl which you can use as a bath for freezing samples in. If liquid nitrogen is unavailable, you will be able to achieve the same flash freezing effect using a dry ice and ethanol mixture. Tongs for handling frozen samples are also useful although a sample can be held in the basin by the lid with gloves on. *There*

will be a risk of cryogenic burns with handling these solutions. Follow all relevant handling procedures. Do not work with liquid nitrogen in an enclosed space.

6. A 200—300-W probe sonicator. If one is unavailable then you can replace this with a syringe and a small gauge needle through which the solution can be repeatedly drawn. You will need to sit the sample in an ice bath to keep it cool during sonication and will also need a clamp to hold the sample vial in place. *Ensure the probe tip is clean before use as this will be placed into your solution.*

7. A centrifuge and tubes large enough to handle 10—50-mL volumes. It is possible to use smaller centrifuge tubes but this will require you to transfer material about.

8. Microcentrifuge tubes for storing small aliquots of sample for running SDS-PAGE gel electrophoresis on.

9. 10 mL of a 50% suspension of nickel—agarose gel beads for protein purification in a sterilised 100-mL Schott bottle. These will be provided as a mixture in solution and must be kept wet. The beads are expensive and can be reused in other purifications if cleaned and kept wet.

10. An empty and clean glass column with a silica frit in the bottom with 20-μ pores. You will need a cap that fits over one end so that you can stop the flow from gravity if required. You will also need a clamp to hold it upright with and some space to place vials or tubes underneath for collecting fractions rinsed through the column. This will be packed with the gel beads after the target protein has bound to them so that they can be washed and the protein eluted.

11. An orbital shaker in an incubator at 4 °C.

12. Stock solutions of wash and elution buffers. These are both prepared exactly the same way as the lysis buffer except the imidazole concentrations are as follows:
    Wash buffer: 1.36 g (20 mmol/L) imidazole 1.36 g imidazole
    Elution buffer: 17.00 g (250 mmol/L) imidazole

13. Dialysis tubing with a molecular weight cutoff of 3.5—5 kDa and clips. You will also need a stock buffer for getting rid of the excess imidazole after purification. The choice of buffer depends on how you would like to store your protein. You will need a 1—2-L beaker for dialysing in.

To extract and purify the GFP from the E. coli pellet:

1. Resuspend the cell pellet in 50 mL of lysis buffer in a centrifuge tube.

2. Dissolve the 50 mg of lysozyme into the lysis suspension and mix thoroughly by swirling the tube. Seal it and allow the suspension to sit on ice for half an hour.

3. Decant some liquid nitrogen into the dewar basin. Remove the incubated suspension from the ice and plunge it into the liquid nitrogen to flash freeze it. *Be careful not to splash the liquid nitrogen about.* Allow the sample to thaw and then freeze it again. This process will soften up the cell walls making them easier to break apart.

4. After the sample has thawed, open the lid of the tube. Place the probe sonicator into the suspension with the tip at a depth of 1 cm but without the probe touching the sides. Sit the sample tube in an ice bath and clamp it into position.

5. Sonicate the suspension with three ten-second bursts and allow ten seconds of rest in-between. If you are not using a probe sonicator, then you would be using the syringe and needle setup at this point to break up the cells.

6. Centrifuge the sample at 13,200 rpm for 30 min. Afterwards you should have about 50 mL of a bright green solution containing mainly water-soluble GFP and a pellet of cell debris at the bottom of the tube. Decant the green top solution into the clean 100-mL Schott bottle containing the nickel—agarose beads. Store a 50-μL aliquot for SDS-PAGE analysis. Do not throw the pellet away as this may still contain some protein. If you wanted to harvest this as well, then you would have to follow a denaturing procedure for extraction. This is not outlined here but can be found in the references for this section.

7. Secure the Schott bottle in the orbital shaker within the incubator at 4 °C. Set it to 200 rpm and allow the solution and beads to incubate for 30 min. During this time, the histidine tags on the GFP molecules will bind to the nickel ions in the beads and they should become visibly green as they become loaded. If you find that the free solution remains very green, then add some more nickel—agarose beads. If this does not correct the problem, then you may need to lower the concentration of imidazole so that the relative binding affinity of the tag is increased. This can be done by dilution with imidazole-free lysis buffer.

8. Make sure the cap is on the glass column and then transfer the protein-loaded beads into it. Place a receptacle beneath the column and collect the first runoff after the cap is removed as this may still contain some of your protein.

9. Once the column has drained, replace the end cap and fill it with the wash buffer. Any unwanted proteins not having the histidine tag will be displaced by the imidazole and move into solution. Remove the cap and let it drain through the gel. Collect the washings so that these can be checked by gel electrophoresis. Repeat the washing process three more times. You do not want the wash solution to be very green although some of your product might leach out. If you feel the protein is being removed from the column, then reduce the concentration of imidazole in the wash buffer by dilution as before.

10. After washing the gel there should now be only the GFP bound to the nickel. Place a fresh and clean container underneath the column. With the cap on the end of the column elution buffer and then take the cap off and allow gravity to drain the solution. This time the solution should be green and will contain the purified GFP. You may find you need to perform more than one

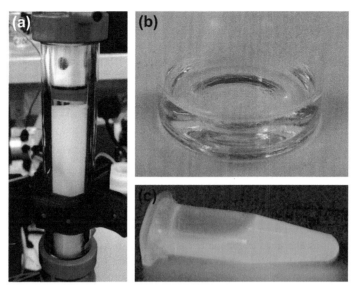

FIGURE 6.13 (a) A photograph of GFP being purified in an automated nickel–agarose gel column, (b) a vial of the purified GFP product under normal light and (c) glowing under ultraviolet light.

elution step and using repeated small increments of elution buffer will be more efficient and result in a higher concentration of GFP in the solution.

11. Soak suitable lengths of dialysis tubing in the large beaker. These should be long enough to fold and double clip each end.
12. Transfer the GFP solution into the dialysis tubing and dialyse against the buffer of your choice. Three changes over the course of a day will get rid of the imidazole present from elution (Fig. 6.13).

You should now have a solution containing green fluorescent protein made using some basic genetic engineering procedures. The most rapid way to assess the quality of the protein is by exposing it to UV light whereupon it should emit a green glow. More quantitatively, the purified solution should be analysed by SDS-PAGE gel electrophoresis to ensure that it is the correct size and that you have not generated an unwanted mutation.

SODIUM DODECYL SULPHATE – POLY(ACRYLAMIDE) GEL ELECTROPHOERESIS (SDS-PAGE) FOR SEPARATING PROTEINS

With thanks to Stuart Bellamy

The SDS-PAGE technique is used to assess the purity and quality of a protein. Small aliquots of a protein solution are denatured using a surfactant which also gives them a homogeneous surface charge. This solution is then placed into wells in a poly(acrylamide) gel and an electric field is applied which draws the proteins through the gel at a rate determined by their relative molecular weight. By mixing in a stain with the gel containing denatured protein solution, the result can be imaged in a light box with a camera. This is similar to the practice of separating DNA fractions in an agarose gel by applying an electric field gradient across the gel which drags the negatively charged strands of DNA through the gel. However, in contrast to DNA, proteins come in a wide range of shapes and sizes and often possess a range of charges over their surface. Sodium dodecyl sulphate (SDS) is used to denature the protein and gives it a negative surface charge. This treatment will often damage the protein so if you need to reclaim it from the gel, then you can try nondenaturing PAGE, without SDS. Instructions on how to do this will be provided at the end of the section. Poly(acrylamide) is used due to its resistance to high voltages and chemical stability. The polymer has a desirable porosity suitable for electrophoresis and was initially used as a biological fixative for microscopy. The density of the gel can be modified by altering the ratios in which the acrylamide monomer and the crosslinking agent such as poly(acrylamide) gels can be fine-tuned to give the best results for the protein you are working with. Gels are mounted vertically inside two plates with samples running from the top of the gel to the bottom. For the SDS-PAGE, the gel is split into two portions on the bottom and the top called the resolving and the stacking gels, respectively. The resolving gel forms the main body of the gel and is where the separation of the various components will occur. On top of this is the stacking gel which is comparatively low in density and used to form the wells into which the sample is loaded. The protein solution is mixed with a running buffer containing dithiothreitol (DTT) which prevents the denatured proteins from forming disulphide bonds with one another. Otherwise there is the potential that they would form dimers and move much more slowly through the gel than they should. The running buffer also contains bromophenol blue which acts as a marker to let you know when the gel is finished. When the dye reaches the bottom of the gel, it is time to stop.

Any laboratory working with biological materials will almost certainly have the equipment to run SDS-PAGE. In this recipe, we will be using a Bio-Rad® gel electrophoresis kit but the designs of other suppliers should not differ significantly. The gels can be purchased readymade but it is good practice to use freshly prepared gels you have made yourself. The example protein used here is a 1 mg/mL solution of recombinant green fluorescent protein as prepared by the method outlined elsewhere. A range of dilutions of the solution will be made prior to running them in the gel. This way you can compare different concentrations of the protein. This will help to spot impurities that otherwise might get masked by strong bands from high concentration components.

To run SDS-PAGE you will need:

1. An SDS-PAGE gel electrophoresis setup in a dedicated and marked area. This will include a power supply, a gel tank, gel holders and plates and a comb for making wells with. Depending on the design you will be able to run more than one gel in the tank at a time.
   When you start this preparation, the gel tank and plates will be in separate parts. Because acrylamide is used with these components, the whole lot should be on top of a layer of tissue on a dedicated bench area to mop up any spillages.
2. A protein solution for testing. A concentration of 1 mg/mL is a good starting point but this can vary depending on the protein. In this example, we will use a solution of green fluorescent protein prepared by methods outlined elsewhere in this chapter.
3. A number of microcentrifuge tubes and a stand for them.
4. A hot block set to 105 °C. This will be used for denaturing the protein in the SDS solution. If you do not have a hot block, you can use boiling water instead although you must take care not to get burned or get water in the sample.
5. Volumetric pipettes in the 10−100-μL and 1−5-mL range.
6. A benchtop centrifuge or spinner.
7. A plastic container with a lid large enough for a gel to sit in.
8. A seesaw rocker.
9. A gel-imaging system. This is not always required as you will be able to see the stained protein bands in the gel by eye.
10. A protein ladder standard for comparing the test samples. These are composed of proteins of a known size and are often stained with different colours to make comparison and band identification easier.
11. 10 mL of a stock double strength (2×) loading buffer. This can be prepared from other solutions as given in the table below. Depending on the temperature of your laboratory, you may encounter some difficulty getting the SDS to dissolve. This can be remedied by heating the solution and gently swirling. The 10% SDS solution will also be used in the running buffer. Glycerol can be purchased as an 85wt% solution commercially and can be diluted down.

Stock Solution and Strength (mol/L)	Weight of Component Required in 100 mL Volume to Make Stock	mL of Stock Required to Make Loading Buffer	Final Concentration in Loading Buffer
1 M Tris−HCl at pH 6.8	15.57 g (adjust with NaOH/HCl)	1	0.1 M
0.0346 M (10wt%) Sodium dodecyl sulphonate	10 g	4	0.013 M (4wt%)
0.54 M (50wt%) glycerol	50 g	2	0.216 M (20wt%)
Bromophenol blue	n/a	0.01 g	0.1wt%

12. 1 mL of a stock of dithiothreitol (DTT). This should be kept frozen until use as, if left out, it will spoil. When required, thaw the DTT and add to the loading buffer. You will need 20 μL of DTT for every 80 μL of loading buffer. Each aliquot of DTT should be discarded after use.
13. A stock solution of five-fold (5×) strength running buffer. This will be diluted to 1× or strength before use but it is good to have a stock. This can be prepared by dissolving the following components into 800 mL of water and then adjusting the volume to 1 L. You will need a large 2-L measuring cylinder for diluting the buffer with.

Component	Weight or Volume
Tris−HCl	15.1 g
Glycine	94 g
10wt% SDS in water	50 mL

14. Gel mixtures for both the resolving gel and stacking gel. The main component is a 30wt% solution of monomer having a 38:1 weight ratio of acrylamide to N′,N′-methylenbisacrylamide. This can be purchased directly from a commercial supplier pre-mixed. The bisacrylamide is a crosslinker and without it the gel formed would be an unusable thick viscous sludge. *This*

solution is harmful and should be used in a fume hood. The solution must be stored in a marked fridge when not in use. Once this has been polymerised, it is not as harmful and may be handled outside of a fume hood. A solution can be prepared following the amounts below.

30 g of acrylamide

0.8 g of N′,N′-methylenbisacrylamide (store in a fridge).

Make up to 100 mL and filter. Keep the solution in a darkened bottle in a fridge.

An ammonium persulphate free radical initiator is used in conjunction with a tetramethylethylenediamine (TEMED) catalyst to polymerise the gel. These components should not be added until you are ready to pour the gel as outlined in the preparation steps. The table in the following lists the various mixtures required to make resolving and stacking PAGE gel.

Component	Resolving Gel (%)				5% Stacking Gel
	7	10	12	15	
30wt% 38:1 acrylamide: bisacrylamide solution	2.33 mL	3.33 mL	4 mL	5 mL	833 μL
Water	3.77 mL	2.77 mL	2.1 mL	1.1 mL	3.47
0.375 mol/L Tris−HCl at pH 8.8	3.75 mL				0
0.125 mol/L Tris−HCl at pH 6.8		0			625 μL
10wt% SDS in water		100 μL			50 μL
10wt% ammonium persulphate (0.1 g) in water (1 mL)		50 μL			25 μL
Tetramethylethylenediamine		10 μL			5 μL

You will need two 25 mL vials with lids for mixing the resolving and loading gels with before adding them to the gel mould.

15. 50−100 mL of 0.2% coomassie brilliant blue staining solution. You will need enough to immerse the gel in the plastic container with the lid. Although a premixed solution is commercially available, you can make your own stock solution using ingredients as outlined below.

0.5 g coomassie brilliant blue R-250

530 mL water

400 mL methanol

70 mL acetic acid

16. Ethanol, water and tissue for cleaning with.

To run SDS-PAGE:

Wear gloves and appropriate safety equipment including safety spectacles. The acrylamide solution is hazardous and can squirt out at various points of this process. Work with it in a fume hood.

1. Clean the plates that will form the gel mould using ethanol and rinse with water.
2. Assemble the plates to form a gel mould as outlined in the instructions for the equipment. They will form a space about 1.5 mm thick and ultimately this will be the dimensions of the gel. Sometimes a tissue wedge can be used to exert pressure on the plates to the rubber seal. This will prevent the gel solution leaking out before it has set.
3. Remove the acrylamide solution from the fridge. Mix the components for the resolving and stacking gels as outlined in the table in separate vials taking care not to shake the gel mixtures. Because they contain surfactant, it would be easy to produce foaming and bubbles in the mixture which will affect the quality of the gel. You may wish to degas the solution by exposing it to a vacuum. Because this can release some of the acrylamide into the air, we have opted not to do so here.
4. Add the TEMED and ammonium persulphate initiator to the resolving gel only. Mix the contents of the vial but, as before, do not shake. Use a pipette to transfer the solution into the space between the plates. Fill the mould to three-quarters full. There should be some of the prepolymer solution left in the vial and seeing how long this takes to set will help you gauge when the gel in the mould has set.
5. Add a layer of water over the top of the gel in the mould using a pipette. This will prevent atmospheric oxygen from quenching the reaction. It will allow the gel to settle properly as it sets and make a good interface for the addition of the stacking gel (Fig. 6.14).
6. While you wait for the gel to set, turn on the hot block and set it to reach 105 °C. Remove the DTT from the freezer and allow it to thaw.
7. Arrange 3 microcentrifuge tubes on a rack. Add 10, 5 and 2 μL of the test protein solution to them. Make the volumes of the 2 and 5-μL samples up to 10 μL with water.

Add 5 μL of the protein ladder to a separate microcentrifuge tube and make the volume up to 10 μL using water.

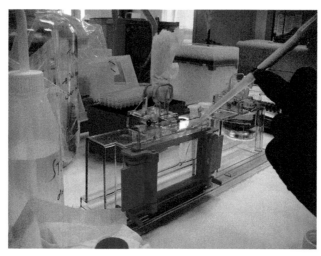

FIGURE 6.14   Adding acrylamide gel mixture to the mould using a pipette.

8. By this point the resolving gel should have set and you can check this by examining the traces in the vial you mixed it up in. Carefully hold the mould and pour away the water on top of the gel. Add the TEMED and ammonium persulphate initiator to the stacking gel solution and then use a pipette to transfer the solution into the mould so that it sits on top of the resolving gel.
9. Add the well comb to the gel. Place it in gently starting at an angle and easing it in. *Plunging the comb in directly will often cause acrylamide solution to squirt out!* As the gel sets, you may be able to make out fringes formed around the comb in the gel. Leave the comb in place.
10. Once the stacking gel has set, transfer the gel within the plates to the running tank. Depending on the design, you may need to put a dummy plate in the back of the cell (Fig. 6.15).

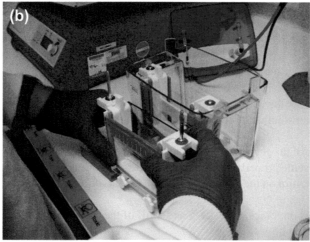

FIGURE 6.15   (a) A setting gel in a mould with the well comb inserted. (b) The gel after it has been transferred into the electrophoresis setup. A dummy gel has been placed near the front.

11. Dilute 200 mL of the 5× running buffer up to 1 L in the large measuring cylinder.
12. Fill the running tank up to a fill line that should be marked out on the tank with the running buffer. You only need enough of the buffer to cover the bottom electrode. Fill the cell containing the gel. If the plates or tank is not sealed fully, you will see that the running buffer from the cell go down and the fill level in the tank come up. This is not too much of a problem but you do not want to have to open up the tank and add more buffer to the cell while it runs. Often you may wish to re-bed the gel and plates in the tank although there is a risk of damaging the gel if you do so. Do not attempt to reassemble the gel until after the test solutions have made it from the stacking gel into the resolving gel.
13. Gently remove the comb from the stacking gel to leave behind sample wells.
14. Make up the loading buffer and transfer 80 μL of this solution into another microcentrifuge tube. To this add 20 μL of the thawed DTT. Replace the DTT back in the freezer.
15. Add 10 μL of the freshly prepared DTT containing loading buffer to all four of the test samples. Each of the test samples and the protein ladder control should now have a volume of 20 μL.
16. Place all the samples to be run in the gel into the hot block and boil for three minutes to denature the proteins. They should remain sealed but if they pop open simply replace the cap immediately.
17. Give the heated samples a quick spin in the benchtop centrifuge so that any condensation on the tube walls collects in the bottom. This maintains the concentration of the protein solution.
18. Transfer each of the test solutions and the ladder control into separate wells in the gel using a pipette (Fig. 6.16).
19. Place the lid on the tank and switch on the power. The gel should be run at 120–150 V for around 2 h or until the blue dye reaches the bottom of the gel. You will be able to see bubbles forming on the bottom electrode within the tank.
20. When the gel is ready, turn off the power and remove the gel and plates from the tank. Lay the gel flat on a tissue-covered surface and gently cleave the uppermost plate away. Trim the wells from the top using a plastic square and then slide this under the gel and gently pull this away from the bottom plate.
21. Fill the plastic container half way with water and then sit the gel in it flat on the bottom. Seal the container and place on the see-saw rocker for 5 min. This will wash out the surfactant and salts. Replace the water and repeat twice. At this point, if you have used a coloured ladder control, you should be able to see the various colours corresponding to particular molecular weights. It is worth making a note of the positions of these bands before you run the stain as they may not always retain their various colours after treatment (Fig. 6.17).
22. Pour away the water and add coomassie blue stain to the container. Seal the container and leave on the see-saw shaker for 1 h.
23. Pour away the stain and add water again. Allow the gel to soak overnight before changing the water.
24. Blue bands should now be visible by eye everywhere proteins are present. The gel can be stored in water or imaged using a light box and camera. For longer-term storage, the gel can be pressed between two sheets of plastic to prevent cracking and then allowed to dry.

An example gel is pictured below. The ladder for comparison is on the left-hand side while the three different concentrations run left to right in order of increasing dilution. In this example, it can be seen that the intensity of the stain drops with decreasing concentration. Although the main band expected for GFP is located in the correct position when compared to the ladder, some impurities can also be seen. Above the main band, traces of an impurity with a larger molecular weight than the GFP are observed. The faint traces below the main band are likely to be break down products from the GFP (Fig. 6.18).

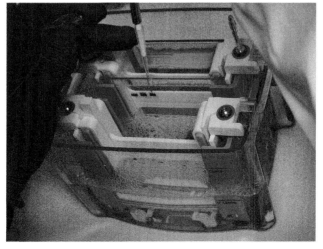

FIGURE 6.16 The lower half of the tank is filled with running buffer and also added to the top of the gel. The test samples are loaded into the well using a pipette.

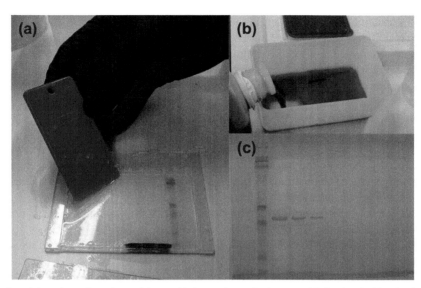

FIGURE 6.17 (a) Separating the gel from the surface of one of the mould plates using a plastic square. Notice that the blue dye is at the bottom of the gel and the coloured protein ladder can clearly be seen. (b) The gel is placed into a plastic container for rinsing in water on a rocker and then stain is added. (c) After soaking in the coomassie blue stain, the protein bands in the test samples can be seen by eye.

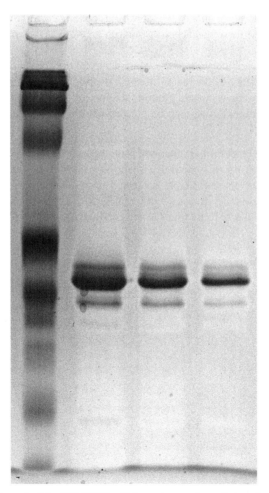

FIGURE 6.18 A photograph of the SDS-PAGE gel taken using a gel-specific imaging camera and light box.

TRANSFORMATION OF YEAST

With thanks to Katja Fisch. Adapted from Gietz and Woods (2002)

The taxonomic name for the common yeast used in baking and fermenting is *Saccharomyces cerevisiae*. It is a single-celled organism useful in biochemistry because it is a simple eukaryotic cell which means that the DNA containing nucleus of the yeast cell is compartmentalised within a membrane (Schiestl and Gietz, 1989; Sherman, 1991). The DNA within the nucleus is stored as a tightly wound chromosome which constitutes 85% of the genetic material within the cell. The other genetic material is 10% mitochondrial DNA and 5% plasmid DNA. This contrasts with bacteria which are known as prokaryotes meaning they store their genetic information floating freely within the cellular cytoplasm as plasmid rings. Bacteria and yeast cells are useful as tools for bioproduction processes and both can be modified to express proteins of choice. Yeast was the first eukaryotic organism to be completely sequenced and is well understood as a model mammalian cell. As it is nonpathogenic and easy to manipulate genetically, yeast is indispensible in the laboratory as a model in genetic disease research and is generally useful for ligation. Yeast has efficient DNA repair mechanisms which can ligate large 20 kb pair and larger genes into plasmid vectors with a high efficiency. This is not as easy to accomplish in bacterial transformations. Should you wish to insert a large gene into a plasmid and then express it as a protein, the ligation is easily done using yeast to stitch a linear plasmid and the gene together using primer sequences. The transformed plasmid is then isolated from the yeast and transferred into a bacterium where it can be expressed into a protein.

In this recipe we will use a transformation procedure to insert a plasmid of choice into a yeast cell so that a culture can be grown. Because of the biological differences, the transformation procedure for yeast is slightly different from that outlined for bacteria. A culture of yeast in its exponential growth phase is isolated and then placed into a transformation mixture of lithium acetate, poly(ethylene glycol), single-stranded DNA and a plasmid or DNA for insertion. The cell wall in a eukaryote is thick and the DNA has to cross both this and the nucleus membrane for a successful transformation. Because of this, the heat shock step at 42 °C is on the order of minutes rather than seconds as with *E. coli* so that the DNA has time to penetrate through the various membrane barriers. The long single-stranded DNA present acts as a carrier for the plasmid across the yeast membrane and boosts the efficiency of the transformation. The carrier DNA is made from denatured salmon sperm DNA which has been boiled but remains long enough to 'catch' inside the yeast cell and be dragged in along with the plasmid. Once inside the cell repair mechanisms are activated by the single-stranded DNA and begin to replicate the inserted vector.

Minor variation of this recipe will allow you to perform a ligation reaction in the yeast. The ligation of large exogenous DNA sequences into a vector plasmid can be accomplished by adding the DNA, primer oligonucleotides and a linear plasmid into transformation mix together. Once all the components have permeated into the yeast cell, they will self-assemble to form a ring plasmid containing the large inserted DNA fragment which can be amplified further by culturing the yeast. For yeast-based ligation, it is important that all the genetic components are present in the same concentration. This is easily achieved by running gel electrophoresis using known amounts of your primer, single-stranded DNA and gene sequences against a 'ladder.' In addition to the various sizes, a ladder can be purchased containing genetic components at varying known concentrations. When the gel is developed, the bands in the ladder will differ in intensity relative to their concentration. You can compare the intensity of these control bands to the intensity of the sequences you have run to get a rough concentration of the DNA you are using. You can then adjust the concentration by dilution or variation of the amount added into the transformation mixture so that your DNA components are all roughly equal before ligation. At the same time as you run the ligation, you should run a control transformation using only the linear plasmid. This way you will be able to determine if the gene insertion has been successful using gel electrophoresis to see the difference in weight between the plasmid having the inserted gene and the plasmid without it.

For the rapid transformation of yeast, you will need:

Yeast is classed as a level 1 biohazard and this transformation must be performed in a biohazard level 1 laboratory. You will need a laminar flow cabinet with a burner and ethanol for maintaining sterility. You will also need biohazard waste bins and decontamination tablets for the broth or else dispose of broth and solids by autoclaving prior to placing in a bin or down the sink.

1. An autoclave for sterilising equipment and reagents. Everything, save for the yeast, should be autoclaved before use.
2. Waterbaths maintained at 30 °C and 42 °C for heating and boiling solutions. One with temperature control is desirable so that you can set the correct temperature for heat shocking the yeast. For the heat shocking setup, you will need a large beaker of water and something to float a microcentrifuge tube in while it is heating. A piece of foam with a hole in will do.
3. A stock solution of 1 mol/L lithium acetate. This can be prepared by dissolving 5.1 g of lithium acetate dihydrate in 50 mL of water. Sterlise the solution in the autoclave for 15 min before use or else run the solution through syringe filter.
4. A 50wt% solution of poly(ethylene glycol) (molecular weight = 3350) in water. This can be made by dissolving 50 g of the polymer in 50 mL of water under stirring on a hot plate and then making the solution up to 100 mL. Transfer the mixed solution to a screw top bottle and autoclave before use. The polymer acts as a protectant for the yeast cells and carries DNA through the lipid double layer of the yeast membrane.
5. A 2 mg/mL stock solution of salmon sperm DNA in a Tris—HCL and Na_2EDTA buffer at pH 8. This can be prepared by dissolving 0.1576 g of tris(hydroxymethyl)aminomethane hydrochloride and 0.0372 g of disodium ethylenediaminetetraacetate

dehydrate into 80 mL of water and adjusting the pH before making up to 90 mL. Dissolve 200 mg of salmon sperm DNA in the solution under stirring for 2–3 h in an ice bath to keep it cool. Once the DNA has dissolved, make the solution up to 100 mL. The DNA needs to be denatured before use in order to act as a carrier for the plasmid.

On a hot plate with stirring, heat the DNA solution to boiling point for 5 min. Cool it in an ice bath afterwards. You can also denature the DNA quickly by bringing the solution to boiling point using a microwave. If you are reheating it in a microcentrifuge tube, be prepared for the top to pop open. Close it and continue to heat. Separate the solution into 1 mL aliquots and freeze for storage keeping a sample on ice ready for use in the experiment if you will be running it on the same day. Before use in the procedure, use a benchtop vortex to ensure that the single-stranded DNA is thoroughly mixed.

6. An incubator with a shaker inside set to 30 °C.
7. A microcentrifuge and microcentrifuge tubes.
8. A benchtop vortex.
9. Volumetric pipettes and tips capable of dispensing in the 10–100-μL and 1–5-mL range.
10. 1 μg of the plasmid you wish to transform into the yeast. Procedures for isolating and purifying plasmids are outlined elsewhere.
11. A stock solution of (2× YPAD) broth. The recipe is given elsewhere.
12. A colony of yeast cells. This procedure should work for many different strains but normal bakers' yeast can be used. A colony can be grown by inoculating a conical flask containing YP broth with yeast at the end of a sterilised toothpick. The solution should be cloudy before use.
13. Sterile water.
14. A stock of synthetic complete selection medium. This will allow you to grow up your transformed yeast.
15. Synthetic complete selection medium agar gel plates. These can be prepared from the synthetic complete medium and agar as outlined elsewhere. Keep plates upside down and in the fridge when not in use.
16. A beaker one-third full of ethanol with glass spreaders sat in it.
17. A conical flask containing further medium for growing the transformed yeast with. You may also wish to have a yeast plasmid isolation kit so that you can extract and check the quality of the transformation.

To perform the rapid transformation of yeast:

Make sure the burner is turned off and spray down your working area in the laminar flow hood using ethanol before you begin.

1. Grow up a colony of yeast cells in 5 mL of the 2× YPAD broth. Put some yeast into the broth in a sealable vial and incubate at 30 °C on a shaker plate at 200 rpm. Leave this overnight and they should be in the log phase of growth by the time you use them. See Fig. 6.19a

2. Use the water bath at 42 °C or else use the hot plate to heat a large beaker full of water to 42 °C. *Do not forget it is on and let this dry out!*

3. Using a volumetric pipette, remove 2 mL of the yeast culture before it has time to sediment. Transfer this into a microcentrifuge tube.

4. Centrifuge the cells at 13,000 rpm for 30 s to form a pellet in the bottom of the tube. Discard the top solution. This will look similar to Fig. 6.19b.

5. To the tube containing the pellet, add the following in the order listed to form the transformation mix.
    240 μL of the poly(ethylene glycol) solution
    36 μL of the lithium acetate solution
    50 μL of denatured salmon sperm DNA solution.

6. Dissolve the plasmid you wish to insert in water. This will most likely be in the bottom of a centrifuge tube, depending on how you have purified it. You will need 0.1–10 μg of the plasmid dissolved in 34 μL of water. Pipette the contents up and down three times to ensure that the plasmid is solubilised and then deliver the entire volume into the tube containing the transformation mixture. Seal the lid of the tube.

7. Use a benchtop vortex to resuspend the yeast into the transformation mixture.

8. Sit the sealed tube into a piece of foam and float it in the water bath at 30 °C for half an hour to adapt the cells. Then transfer the mixture to the water bath at 42 °C. This will heat shock the yeast making the cell wall permeable to the carrier DNA. Leave the tube in the water bath for 15–30 min. Depending on the strain, you may find that a longer time is needed.

9. Centrifuge the tube at 13,000 rpm for 30 s to form a pellet. Discard the top solution and add 1 mL of sterile water and redisperse the pellet by drawing the solution back and forth in a pipette.

10. Light the burner and pass the ethanol soaked glass spreaders through the flame to sterilise them. Sit them on an agar gel to cool shortly before use.

11. Use a volumetric pipette to transfer 100 μL of the yeast dispersion onto the synthetic complete medium selection plates. Use the spreader to smear the transformed yeast droplet over the surface. Incubate them at 30 °C for 3–4 days.

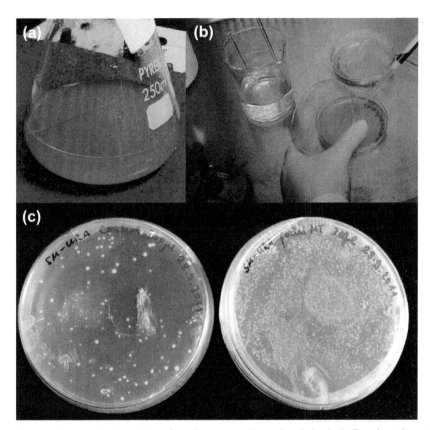

FIGURE 6.19 Photos of the various steps involved in the yeast transformation process. (a) A colony in broth. (b) Transformed yeast cells are spread onto selection plates using a glass spreader. The spreader is sterilised first using ethanol and the open flame. (c) Transformed yeast cells growing as single colonies on a selection plate. On the left is a control yeast and on the right is the transformed yeast.

12. You should now be able to see transformed yeast cells growing as colonies on the selection plates (see Fig. 6.19c). Select a single colony from the plate using a sterile toothpick. This can then be used to inoculate a conical flask-containing medium to grow up the amount of plasmid further. Alternatively you can use a plasmid isolation miniprep to purify the plasmid so that you can perform gel electrophoresis.

For protein expression, you would then further grow the yeast up as a large colony. Protein expression could then be triggered in a similar manner to *E. coli* expression. If the yeast has been used to ligate DNA fragments together to produce a large plasmid, then this can be isolated using a kit preparation. The isolated plasmid can then be transformed into *E. coli* for expression in large amounts.

DNA ORIGAMI

Adapted from Castro et al. (2011) with thanks to Hendrik Dietz

DNA origami involves using a single-stranded DNA scaffold up to several thousand base pairs long and stapling sections together with short complimentary oligonucleotide sequences (Rothemund, 2006). This self-assembles into close-packed arrangements of β-form helical DNA of a predetermined shape. In addition to binding the DNA together in specific configurations, the staple oligonucleotides can be functionalised to act as anchor points for proteins, nanoparticles and fluorescent dye molecules. Because of the exquisite control in the nanometre region and the large patterning areas possible, this type of templating is extremely useful for nanoscale surface design with large implications for the realisation of ultra-small scale electronics and lab-on-a-chip applications (Schreiber et al., 2011). DNA structures can also be prepared in 3 dimensions and can be modified such that they are mechano-chemically responsive (Dietz et al., 2009). One great example of this is in the production of a DNA origami box with a lid which can be opened and closed by a chemical trigger (Andersen et al., 2009). Each box was assembled from one strand of single-stranded M13-phase DNA stapled together with oligonucleotides. Because of the way in which the box was designed to fold together, the last section to go into place was much like the action of a lid closing. Two of the staple strands holding this lid closed were designed with overhanging DNA strands, like straps hanging out from under the lid. To open the box, two short complimentary oligomers were added to the solution that bound to the staple overhangs. This has the effect of weakening the staples enough that the lid of the box opens, as it would if you were to pull on straps trapped under a lid. The opening mechanism of the box was

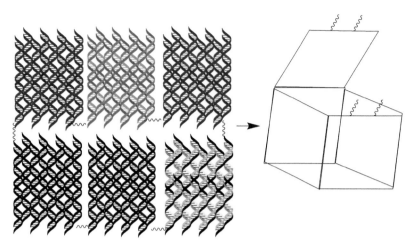

FIGURE 6.20 The unfolded DNA origami box laid flat. The star and circle mark where primer overhangs are placed. The DNA self-assembles as a box which can be opened by adding oligomeric DNA keys (orange and blue lines).

demonstrated quite elegantly by encapsulating two dye molecules inside the box. When in close proximity, certain fluorophore dyes experience a resonant energy transfer between one another (termed FRET, fluorescence resonance energy transfer) which can be monitored spectroscopically. When the box was opened and the dyes were released, the change in absorbance and emission of the dyes changed indicating that the dye molecules were not wedged together inside the DNA box (Fig. 6.20).

It is actually relatively simple to make an object with DNA origami as most of the design and sequencing is handled by software packages (Douglas et al., 2009). The actual practical work involved is pipetting mixtures into PCR tubes and purification with gel electrophoresis although you may also wish to isolate your own scaffold DNA for use. The DNA origami has two main components. Firstly, a known sequence of single-stranded DNA of a given length is used as a scaffold. The second component is a set of complimentary single-stranded DNA 'staples' which will bind to the scaffold DNA to form a double helix. The staples are manufactured sequences that will only bind to certain points in the scaffold DNA where we wish to put in a join or fold in the final structure. Two separate strands of double helix DNA are connected by designing the staple such that it forms what is called a Holliday junction (McCarthy, 2004). This is where a single strand of DNA 'jumps' out of the double helix and binds to another double helix. This will only occur at certain points along the double helix when the offshoot DNA sequence is matched to the joining DNA sequence and only if the helical angles of the DNA backbones are adjacent to one another. There are some selection rules that apply when designing DNA origami. The selection rules where staples can bind a double helical strand together are determined by the twist angle of the double helix. Imagine you could look down the length of a section of double-stranded DNA and pick a base pair to start from. Set this to the 12 O'clock position. You would see that the helical twist will not be in the same 12 O'clock position until you look 10.5 base pairs further down the length of the strand. Because a staple cannot interlink the helices at half a base pair, we look at double that distance to 21 base pairs and again find the backbone of the helices is passing through our 12 O'clock position. Every 7 base pairs along a 21-base-pair sequence, there are three points where crossovers can be engineered. As you look down the length of the 21-base-pair sequence, these possible crossover points appear at 0°, 240° and 480° around the centre spaced 7 base pairs apart. Crossovers can be designed to insert in other sections but this can introduce strain into the DNA which will cause it to twist. In some situations, this is desirable as it can engineer curvature into the final shape (Fig. 6.21).

By extending a DNA double helix lengthwise and placing it adjacent to other similar strands, the three crossover sites for each 21-base-pair sequence arranges the helices into a hexagonally packed array. This honeycomb structure may not be the ideal packing type for the DNA origami that you wish to design but is the easiest conformation granted by the natural geometry of DNA. It is possible to get around this natural limitation so that, should you wish, you can pack the adjacent double-stranded helices together in a square lattice. In this type of design, 4 points are required along a given base pair length at 90° angles to one another. This is accomplished by cheating a little bit and using a 32-base-pair basic length with crossover points located 8 base pairs apart rotated 270° from the last. This gives a figure of 10.67 base pairs per 360° rotation of the helices which is not quite the 10.5 base pairs observed naturally. The result of this is that torsional stress can warp the final origami structure you are trying to make. One way around this is to have some of the linkages not located at the 8-base-pair-spaced positions which relieve some of the strain. For end applications that will be two-dimensional, considering potential deformation in the square-packed DNA is not too important. Surface-binding interactions, for example adherence of the construct to a mica surface for AFM analysis, will pull the assembly flat.

After the design stage the methodology is surprisingly easy. The scaffold DNA and synthesised primer oligonucleotides are mixed together in a folding buffer containing Mg^{2+} ions which helps the assembly. The solution is heated to melt the DNA and then allowed to cool slowly so that the origami takes shape. This is done in a PCR machine so that the heating and cooling rates are

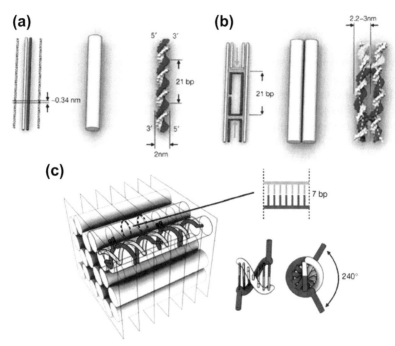

FIGURE 6.21 Examples of computer designed DNA origami shapes and how they appear under an electron microscope. Images courtesy of Hendrik Dietz.

tightly controlled. Flat shapes form quickly, in hours, and in a higher yield than more complex three-dimensional structures which may take weeks to self-assemble. After this the sample is purified by gel electrophoresis and finally analysed using a suitable microscopy to ensure the quality of the folding reaction.

The final size of the DNA origami object depends on how much DNA goes into making it. As a general rule, the length of a double helical domain is the number of base pairs times 0.34 nm. The width contribution to the cross section from each helices making up the shape will be in the range of 2.1–2.4 nm

DESIGNING A DNA ORIGAMI SHAPE USING A PROGRAM

Adapted from Douglas et al. (2009) with thanks to Shawn Douglas

Come up with a shape. If it is to be a nonplanar structure, then you will have to consider how the different units you design will assemble to one another. An advantage of the modular approach is that sections of the final design can be modified separately with anchor points or functional groups.

The design and layout of the origami scaffold is performed using software that is setup to automate much of the design process. You must decide if using a square lattice packing DNA or hexagonal packing DNA will best suit your needs. Most programs come loaded with a selection of DNA scaffold sequences and can automatically select the location and sequence of staples. In this recipe, we will discuss the design of the DNA origami using the caDNAno software. This program will let you model twisted components to render three-dimensional models and will let you move staples around when you wish to have loose end strands of DNA available for functionalisation.

In the caDNAno program, there are three windows which can be collapsed depending on which view you would like to look at. If you start up the program and select a known sequence of DNA as a scaffold, then you can begin to design your desired shape. The M13 sequence is most commonly used. On the left-hand side, the window represents the side view as you look down the length of the DNA. This window is filled with circles arranged in a hexagonal- or square-packed lattice. Each circle is a double helix DNA strand viewed end on. Clicking on one of the circles activates a DNA helix. In the middle window, you will see the length of the strand pop up. For the purposes of this description, we will discuss a basic design using the hexagonal-packed DNA design program. If you select six circles in the left-hand side view window, then you will see they become numbered in a particular order. This is to help you keep track of which strand is which when they pop up as an expanded length and height view in the middle window. The activated strand in the middle window is arranged as a sequence of boxes running from the $5'$- to $3'$-ends of a scaffold strand of DNA in the top row and back the other way in the bottom row to form a helices. For example, the blue DNA scaffold sequence will run along the top boxes in a row and the corresponding staples will run along the bottom when you add them. There is a marking every 7 base pairs denoting where the staples can be placed to form linkages between adjacent helices. The row of squares is by default 42 base pairs long and a coloured slice bar runs through the centre with blue arrows located along it. The blue bar represents the cross-sectional slice, or width view, you are looking at in the left-hand window. The blue lines in the running forwards and backwards along the centred in the cross-section bar are terminated by a rectangle and an arrow. The rectangles are where $5'$-break points are placed and the triangle end is where the $3'$-break points are placed. By dragging the square and the triangle break points, you can change the lengths of the scaffold DNA along that helical component. If you need to make the shape larger than the default size area in the middle window, then you can do this by dragging the slice bar in the middle window to the very end of the scaffold and clicking on an arrow which will extend the length. The overall shape to be formed can then be observed in the 3D representation in the far right-hand window (Fig. 6.22).

The next step is to calculate a route for the scaffold DNA and then generate the corresponding staple strands. As you click on a helices in the middle window numbers will appear by the sides representing where there are potential staple points between adjacent helices as viewed in the left-hand width window. If you click on the numbers that pop up in the middle, length and height windows the program will automatically install links between helices at these points. You may find that some loose ends are formed depending on where the break points in your DNA are located. Once this route is determined, then the program can calculate a complimentary set of staple sequences that will hold the scaffold in place to form your desired shape. Click on the autostaple option and the program will calculate a default set of sequences to give the shape you have just designed. These will appear in the empty top or bottom row boxes adjacent to the blue scaffold route. Some might appear thicker than others and this highlights staples that are shorter or longer than the 18–50-nucleotide sequence limits or else form a loop. To get around this, you can select a break tool from the tool bar and insert break points in the problem staple sequences so that all of them become around 30–40 bases in length. Those that are too short can be joined into other sequences and then broken elsewhere if they are too long. Once you have done this, the lines representing the staples will appear thin and sit complimentary to the given design coloured blue. The final design step is to select a 5'-end of the scaffold and then use the add sequence option so that the program will calculate the required set of oligomeric sequences to produce the structure. You will be presented with set of options for which scaffold DNA you would like the oligomer sequences to be used with. Remember, the oligomer list the program gives you will only work for this specific single-stranded DNA sequence. If the generated list of sequences contains some with question marks instead of base letters, then it means that the total scaffold path is discontinuous. You will need to go back into the design and make sure everything is joined together. The oligomer staple list can then be transferred directly into a spreadsheet and used for custom ordering from a commercial source. For some experiments, such as those requiring separate components to be assembled first before mixing with others, you can use the software to colour-code-specific sequences which can help you to group them together for your experiment. It is also possible to export graphics to other programs and sequence files that can be used for modelling. One particular program of interest is the CANdo program which can take a folding file generated in caDNAno and show you graphically what the torsional stresses and resulting deformations will be in a 3D image. More advanced information and some very helpful tutorials can be found on the caDNAno website listed with the reference (Douglas et al., 2009) (Fig. 6.23).

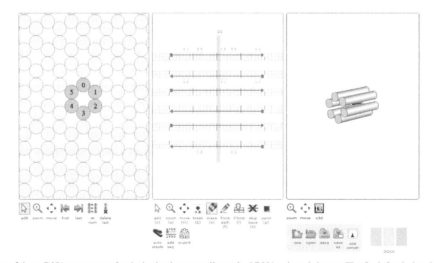

FIGURE 6.22 A screen shot of the caDNAno program for designing hexagonally packed DNA origami shapes. The far left window is a view down the plane of the helices. The middle window is the main work area showing the length of the helices to be formed. In this picture the blue lines are the lengths of scaffold DNA we would like. A 3D representation of the final DNA shape is provided in the right-hand window.

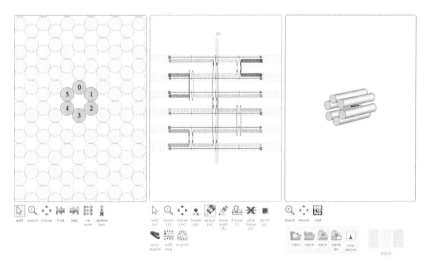

FIGURE 6.23   A screen shot of the caDNAno program. Here the autostaple button has been used to automatically put in complimentary oligomeric DNA. These are highlighted in different colours. The thickness of the staple line indicates they are within a defined length range suitable for synthesis. You may have to alter the positions and lengths of the staples manually so that when the add sequence button is pressed you do not get any question marks come up in the list of oligomer sequences required.

After the design stage, it is time to physically put together the DNA components. Getting hold of the oligomeric staples is most easily accomplished by ordering them from a custom supplier. DNA origami objects normally need a few hundred different staples so it is a good idea to use a company that will synthesise the staples for you supplied in 96-well plates which make it easy to pipette the contents. The staple molecules are on average 40 nucleotides long but normally range between 18 and 50 nucleotides. Staple sequences shorter than 18 bases mean that they will not be stable at room temperature in the final product which results in unfolded material. Conversely, there is a jump in the cost of producing pure staples above 50 bases in length and this is one of the reasons why the caDNAno program is setup to highlight any potential staples that are over this length.

There are many sources of single-stranded DNA which can be used as a scaffold and this too can be ordered from various bioscience companies. The source reference highlights the reproducibility and utility of using 8,064-base-pair-long genomic DNA isolated from M13mp18 bacteriophage. It is also possible to use an enzymatically opened plasmid ring as well as other custom sources of DNA, depending on what you would like to try. Herein the procedure for the isolation of M13 bacteriophage DNA is provided in case you would like to make your own.

MAKING A DNA SCAFFOLD FROM M13 BACTERIOPHAGE

For use as a scaffold in DNA origami, the DNA sequence must be unpaired with a complimentary strand. Single-stranded DNA occurs naturally in bacteriophages which are a form of virus. A bacteriophage is composed of a protein cage which acts as a delivery vehicle for a single-stranded string of DNA housed inside. The M13 phage carries this string into an *E. coli* cell where the existing cellular equipment forms a complimentary DNA strand. This forms a plasmid which codes for the replication of more M13 phage. Because the virus does not kill the host cell by lysis, large amounts of the phage can be grown by allowing it to replicate inside a phage competent *E. coli*. Heat shock is used to get the phage DNA inside the XL-1 blue *E. coli* strain where it will begin to replicate more phage. The transformed *E. coli* is then colonised out onto selection plates which are setup for blue/white screening. This is a technique that makes it easy to identify successful transformants from those that have taken in antibiotic resistance but are not expressing the desired gene. The selection plates contain IPTG to initiate gene expression along with bromo-chloro-indolyl-galactopyranoside, also abbreviated to X-gal which is initially transparent. X-gal is a modified sugar that breaks down into a coloured dye in the presence of β-galactosidase. The inserted vector plasmid DNA has a subunit in the multiple cloning site which can inhibit the production of the β-galactosidase when the desired gene is expressed and read. This means that, although all of the surviving bacteria colonies on a selection plate have taken in at least enough genetic material for them to be resistant to the antibiotic, only those that are expressing the genetic material we want will appear white. Any other bacteria will express the enzyme so that the X-gal is broken down into a blue dye. In the presence of a transformant carrying the desired gene, the X-gal is converted to produce a white compound which denotes the success of the transformation. In transformants where the *E. coli* contain a vector without the desired M13 gene sequence, the production of the enzyme β-galactosidase is switched on. This turns the X-gal blue. In this way, the phage plaques growing by *E. coli* colonies can be identified and harvested from the selection plate to be inoculated into a growth medium. A growth medium containing another *E. coli* colony is inoculated with the phage plaque and incubated. The colony is centrifuged to remove the bacteria, and the top solution is transferred into a vessel containing poly(ethylene glycol) and salt which causes the phage free in solution to precipitate out. This in turn is centrifuged to form a pellet which is treated with a base and detergent-based lysis buffer to crack open the protein coat on the virus. This is called alkaline lysis and is very important in forming the single-stranded DNA. The sodium dodecyl sulphonate breaks up the cell wall and denatures any proteins present. The sodium hydroxide acts to denature the DNA by breaking up the hydrogen bonding between the base pairs resulting in the single-strand structure. The single-stranded DNA within is then precipitated out of solution using ethanol and then purified and isolated.

To produce the DNA scaffold from the M13 bacteriophage you will need:

Before beginning this procedure ensure that all biohazard safety protocols are observed. Sterilise surfaces and utensils either by autoclaving or wiping with ethanol. Glassware and utensils should be autoclaved before use.

1. A colony of the competent XL-1 blue *E. coli* strain. This can be purchased from a bioscience supplier. This should be in 50-μL aliquots and should be stored in a −80 °C freezer for long-term storage.
2. About 1 L of a stock of LB medium sterilised in an autoclave. A recipe for this is given elsewhere.
3. A gas burner and metal loop for plating out bacteria.
4. Pre-prepared agar selection plates containing 12.5 μg/mL tetracycline, 40 μg/mL isopropyl β-D-1-thiogalactopyraniside (IPTG) and 100 μg/mL X-gal. These are cast from lysogeny broth (LB medium) with 1−2% agar which is heated until the agar dissolves. As the mixture cools, but before it sets, the requisite chemicals are added to the medium and it is poured into sterilised Petri dishes. These are always to be stored with the lids on and upside down so that condensation does not drip on the gel. Label them clearly.
5. A solution of 50 nmol/L M13mp18 phage plasmid single-stranded DNA in water.
6. A solution of 50 nmol/L wild-type M13 p7249 single-stranded DNA in water to be used as a positive control.
7. Volumetric pipettes in the 1−10-μL, 100−500-μL and 1−5-mL range with sterile tips.
8. An ice bucket for placing the samples into when not in use.
9. A water bath maintained at 42 °C for heat shocking the bacteria. Have a stopwatch nearby so that the heat shock step can be accurately timed.
10. 10 mL of LB medium containing 12.5 μg/mL tetracycline in a 125 mL sterilised conical glass flask with a stopper. This should be prepared freshly after you have isolated plaques from the selection plates.
11. 250 mL of double strength YT medium containing 12.5 μg/mL tetracycline and 5×10^{-3} mol/L magnesium chloride. A stock of this solution can be prepared as outlined in the box. Add the tetracycline only to the aliquots of solution to be used. This should be placed in a large 2-L conical flask with a lid. The volume will allow the solution to adequately aerate during incubation and shaking.

> *Double strength YT medium:*
> Dissolve the following into 900 mL of water under stirring.
> 16 g of bacto-tryptone
> 10 g of bacto-yeast extract
> 5 g of NaCl
> 0.254 g of $MgCl_2 \cdot 6H_2O$
> Adjust the pH to 7.0 and then make the volume up to 1 L. Sterilise the solution in an autoclave.

12. 10 g of anhydrous poly(ethylene glycol) molecular weight 8000 and 7.5 g (0.128 mol) of NaCl. These will be added to a solution containing the phage in order to precipitate it.
13. An orbital incubator set to 37 °C that can rotate at a speed of 250 revolutions per minute.
14. A temperature-controlled centrifuge at 4 °C that can accommodate 250 mL of volumes or larger. You will need at least at least 4 large centrifuge tubes/bottles that can take around 300 mL of liquid volume. Two are used in the process and two will either be for other samples or as counterweights.
    You will also need a cooled benchtop centrifuge that can take microcentrifuge tubes and 50 mL falcon tubes.
15. An ice bath large enough to accommodate one of the large centrifuge tubes mounted atop a magnetic stirrer. You may also need a clamp to hold the centrifuge tube in place.
16. A stock of 0.01 mol/L tris(hydroxymethyl)aminomethane buffer adjusted to pH 8.5.
17. A method of reading optical density, using either a spectrometer or a plate well reader.
18. A stock of lysis buffer consisting of 0.2 mol/L sodium hydroxide and 1% by weight sodium dodecyl sulphonate. This can be prepared by slowly dissolving 8 g of NaOH in 900 mL of water under stirring. Dissolve 10 g (0.0346 mol) of sodium dodecyl sulphonate into the solution and make up to 1 L. Sterilise by autoclaving.
19. A stock solution of neutralisation buffer containing 3 mol/L potassium acetate at pH 5.5. This can be prepared by dissolving 295 g of potassium acetate into 500 mL of water. Adjust the pH to 5.5 using glacial acetic acid. Make up the volume to 1 L and sterilise by autoclaving.
20. Stocks of pure ethanol and 75% ethanol in water solution. This will be used for precipitating out the single-stranded DNA. Keep this ice cold until used.

To produce the DNA scaffold from the M13 bacteriophage:

1. Take a 50-μL aliquot of the competent XL-1 *E. coli* out of the freezer and allow it to defrost sitting in the ice bucket. You will need more than one if you are running the positive control sample alongside. This procedure will describe the transformation and isolation of DNA from one *E. coli* aliquot for clarity.

2. Pipette a 5-μL aliquot of the single-stranded phage plasmid DNA to be cloned into the *E. coli* suspension. Allow the mixture to stand in the ice for 10 min.

3. Ensure that the vial or tube containing the cells is closed and heat shock the sample in the water bath at 42 °C for 30 s. Use a stopwatch to ensure this is timed correctly.

4. Immediately after the heat shock treatment, sit the tube containing the cells back into the ice bucket and leave for 5 min.

5. Add 200 μL of LB medium from the stock (without tetracycline in it).

6. Incubate the suspension at 37 °C for 30 min with shaking. Also place the selection plates into the incubator so that they are at physiological temperature prior to plating out the bacteria.

7. Heat the spreading loop until it is red hot using the flame and allow it to cool back to room temperature. Dip the end in the tube containing the *E. coli*. If there is a hiss when it comes into contact with the broth, then it is still too hot.

8. Use the loop to draw lines across the agar gel in a number of selection plates to colonise them. The more plates you colonise, the greater the chances of success. Four is normally sufficient. The antibiotic will kill off any bacteria that have not taken in the plasmid as this codes for antibiotic resistance in successful transformants.

9. Place the lids on the gels and keep them upside down in the fridge at 4 °C for 12−18 h. Ensure they are labelled correctly.

10. When you open up the plates, you should see colonies have grown in spots all over the surface. Each of these dots represents a colony grown from an individual *E. coli* seeded on the gel. Some may appear blue and others will appear white. We want to use only those colonies that appear white or clear as these are successful transformants. The blue colonies are *E. coli* that have taken in the DNA but are not expressing the gene we want. The wild-type positive control will also appear blue as this does not have the clone sequence for the DNA we are going to use.

    The plates can be kept for a day in the fridge at 4 °C until they are needed.

11. The next step is to isolate the transformants carrying the phage and grow a large culture of them. In this way, the cloned phage will infect and breed in other *E. coli* cells. For this, a fresh aliquot of XL-1 *E. coli* from the −80 °C freezer can be used. Scrape the surface of a frozen 50-μL aliquot using a pipette tip or metal loop. Swirl the loop in the 10 mL of LB medium contained in the 125 mL conical flask.

12. Place the lid on the inoculated flask and incubate at 37 °C overnight with shaking.

13. Transfer 3 mL of the incubated colony into the 250 mL of double strength YT medium in the 2-L flask. Place the lid on and incubate at 37 °C shaking at a rate of 250 rpm.

14. Every 20 min take a measurement of the optical density using a spectrometer or plate well reader. In around one and a half to two hours, the optical density should reach 0.4, indicating that the cells have reached a logarithmic growth phase. Infecting them with phage now will yield the largest amount of product.

15. Take the selection plates holding the transformed colonies out of the fridge. Use the sterilised metal loop to scrape a line of white colonies from the gel. All should be carriers of the desired phage. Swirl the metal loop into the 2-L flask with the double strength broth.

16. Replace the lid on the flask and incubate with swirly for a further 4 h.

17. The *E. coli* carrying the phage must now be isolated from solution before further processing can begin. Transfer the growth medium from the 2-L flask into a large centrifuge tube or bottle for centrifugation. Ensure the centrifuge is balanced. Centrifuge the sample at 4000 rpm until the solution is clear and all cells have sedimented to the bottom of the tubes.

18. Remove the top solution without disturbing the sediment and transfer it to a clean and sterile large centrifuge bottle. The solution should contain an abundance of viral phage but you will not be able to see it.

19. To the top solution add the 10 g of poly(ethylene glycol) and the 7.5 g of NaCl. Add a magnetic stir bar.

20. Clamp the bottle in place above a magnetic stirrer sitting in an ice bath. Turn on the stirring and leave for 30 min. During this time, the solution should become cloudy as the phage free in solution should precipitate. If it does not, then consider restarting the whole process again.

21. Centrifuge the solution at 4000 rpm for 15 min or until the precipitated phage sediments at the bottom of the tube. This should leave a smear in the tube as it packs down. Ensure the centrifuge is balanced.

22. The phage will only be loosely packed so any motion in the liquid will disturb it. Use a pipette to remove the solution gently from the tube, leaving the pellet as undisturbed as possible.

23. Add 2.5 mL of the 0.01 mol/L Tris pH 8.5 buffer to the pellet. Use the pipette to redisperse the sediment into the buffer. The solution should be transparent but some clumps of *E. coli* may still be present.

24. Transfer the 2.5 mL dispersion of phage in buffer into a smaller centrifuge tube and centrifuge at 15,000 rpm for 15 min at 4 °C so that any residual *E. coli* cells sediment out.

25. Decant the top solution which contains the phage into a roughly 10 mL centrifuge tube and seal. The phage solution can now be stored in a freezer below −20 °C until required for the next step.

26. Add 5 mL of lysis buffer to the phage solution. Replace the lid and gently mix the sample by repeated inversion.

27. Add 3.75 mL of neutralisation buffer and mix once more by inversion.

28. Sit the tube in an ice bath for 15 min.

29. Centrifuge the contents at 16,000 rpm for 10 min at 4 °C.
30. Transfer the top solution into a 50 mL falcon tube using a pipette. Stand this in an ice bucket.
31. Remove your ice cold ethanol solutions from the fridge and sit them in an ice bucket so that they remain cold.
32. Add 8 mL of pure ethanol to the falcon tube sat in the ice. Leave it to stand for 30 min with the lid on. This should precipitate out the single-stranded DNA.
33. Centrifuge the tube at 16,000 rpm for 15 min at 4 °C. Remove the top solution using a pipette and keep the sediment in the bottom of the tube.
34. Add 1.5 mL of the ice cold 75% ethanol solution to the tube and gently mix the contents together. Sit the closed tube in ice for further 10 min.
35. Centrifuge again for 15 min at 4 °C so a pellet forms in the bottom of the tube. Use a pipette to gently remove the top solution and get rid of it.
36. Resuspend the pellet in 1mL of the 0.01 mol/L Tris buffer at pH 8.5. You should now have around 1 mg of single-stranded DNA within the buffer solution.

Roughly check the concentration of the DNA by measuring the solution absorbance in a spectrometer. It should be worked out to be around 350 nmol/L. Check the quality of the DNA by running gel electrophoresis. Compare the DNA to be used with the positive DNA control on a 2% agarose gel containing 11 mmol/L $MgCl_2$ and 0.5 mg/mL ethidium bromide stain. Use a half strength Tris, borate and EDTA (TBE) running buffer. The single-stranded DNA solution can be frozen at -20 °C for storage until use.

RUNNING THE SELF-ASSEMBLY USING THE SINGLE-STRANDED DNA AND THE STAPLES

This is a relatively simple step in that all you are going to do is mix things together. One of the most important factors in getting the DNA components to fold together properly is the concentration of magnesium ions present in the solution. The first time a folding reaction is performed, it should be attempted in a range of magnesium concentrations to see which one gives the best results. The oligomers to be used may have been supplied in different concentrations so you will need to make up stock solutions of a known amount so that the final working concentration can be calculated. There needs to be a 10-M excess of the oligomers compared to the scaffold DNA so each oligomer must have a final concentration of 500 nmol for this protocol. This means that you will have to transfer the correct amount of oligomer from each well plate and pool it together in one solution. This can then be diluted to give the final working concentration. When doing this, it is extremely helpful to have a pipetting plan made up. Especially for products that are made from smaller folded components. The actual reaction occurs within a PCR tube as these are specifically designed for the thermal cycling required to get the DNA to denature and then renature into the target shape. The products will also not stick to the inside of the tubes. The PCR tubes are then placed into a PCR machine which is then programmed to cycle the temperature over the course of hours for small or flat assemblies or days to weeks for large multicomponent assemblies.

The shape the DNA assembles into should represent the lowest energy configuration according to the interactions of the Watson—Crick base pair interactions. However, there will be some differences to the design as electrostatic and solvent interactions will give rise to torsional strain which may deform the overall shape. The self-assembly is driven by heating and cooling which breaks up the DNA into single strands and then lets it reassemble into the desired morphology. By cycling, or annealing, the solution, a flat structure can be formed in hours. The presence of magnesium helps to shield the electrostatic repulsion the closely packed helices experience near one another and so minimises the packing stress. Larger multi-layer objects can take longer to assemble as there are multiple pathways through which the individual components can combine. In some cases, undesired conformations can form which are stable against further reconfiguration. This is known as a kinetic trap and half-folded products stuck like this need to be reheated and allowed to refold.

The procedure below is very general and provides rough guidelines suitable for most DNA origami assemblies. The protocol assumes that you will be running this reaction for the first time and so the quantities are geared towards running multiple samples at 8 different concentrations of $MgCl_2$. You can choose to run varying concentrations of scaffold DNA between 10 and 20 nmol/L and in a volume of 50 or 100 µL.

To assemble the DNA origami you will need:

1. At least 160 µL of the scaffold DNA at a concentration of 100 nmol/L.
2. The 96-well plates containing the oligonucleotide staple sequences.
3. Many PCR tubes for making oligonucleotide stocks and various other solutions in.
4. A 10 times working strength stock of folding buffer. This can be prepared by dissolving 0.292 g of NaCl, 0.605 g of tris(hydroxymethyl)aminomethane (Tris) and 0.292 g of ethylenediaminetetraacetic acid (EDTA) into 50 mL of water. Adjust the pH of the solution to 8 using HCl and NaOH as required. Make the volume up to 100 mL with water.
5. A volumetric pipette in the 10−100 µL range with tips.

6. A stock solution of 0.05 mol/L magnesium chloride solution. This will be further diluted to give a range of working concentrations for each iteration of the DNA folding samples. This can be prepared by dissolving 0.101 g of $MgCl_2$ in 10 mL of water.

7. A PCR machine with programmable thermal annealing cycles. It should fit with the PCR tubes used in the experiment. If you are going to make something composed of a number of different shapes, then the annealing program will take around a week to complete. Ask around the laboratory to make sure it is OK to use the equipment for that long!

To assemble the DNA origami:

1. Following the pipetting plan, transfer all the oligonucleotides required for each modular component of the final structure into one PCR tube from the 96-well plates.

2. Make up the volume of the stock oligomer solutions so that the final concentration for *each oligomer staple present* is at a concentration of 500 nmol/L. There may be many tens of or even hundreds of oligonucleotide sequences in the stock pool and all of them are at a concentration of 500 nmol/L.

3. Make up a range of different magnesium chloride concentrations on water at ten times the strength of what their final working concentration will be in the folding reaction. 10% of the final reaction volume of the folding reaction will be composed of the magnesium solution so that the working magnesium concentration is correct. To do this, line up 9 PCR tubes and add the listed amounts of the stock 0.05 mol/L magnesium chloride solution to the given amounts of water:

μL Mg^{2+} from stock	20	24	28	32	36	40	44	48	52
μL water	80	76	72	68	64	60	56	52	48
Effective millimolar working concentration in final reaction.	10	12	14	16	18	20	22	24	26

4. There are four possible configurations for running the folding reaction and these are tabulated below. You may find you need to alter them depending on results but, in general, these are a good place to start. Select a volume and DNA concentration and pipette the listed amounts of the component solutions into PCR tubes. You will end up with 9 samples, one for each different concentration of magnesium. You will have four times this if you decide to run all the variations.

For a DNA concentration of 20 nmol/L, add the following to a PCR tube:

Solution	Volume for a 100 μL total	Volume for a 50 μL total
100 mM DNA scaffold solution	20	10
500 mM Oligonucleotide staple solution	40	20
10× Strength EDTA and Tris buffer	10	5
Water	20	10
Given strength of $MgCl_2$	10	5
Total volume	100	50

For a DNA concentration of 10 nnmol/L, add the following to a PCR tube:

Solution	Volume for a 100 μL total	Volume for a 50 μL total
100 mM DNA scaffold solution	10	5
500 mM Oligonucleotide staple solution	20	10
10× Strength EDTA and Tris buffer	10	5
Water	50	25
Given strength of $MgCl_2$	10	5
Total volume	100	50

5. The samples now need to be thermally annealed using a PCR machine. No matter what the size and shape of the DNA object is to be, the first step is to heat the sample up to 80 °C so that all the DNA melts into single strands. The rate at which the sample is allowed to cool depends on the objects size. If you are assembling a component that consist of a single layer or unit, then, after the initial heating, the sample can be allowed to cool to room temperature and will fold on the order of hours. If you have designed a large structure composed of more than one differently shaped module shape then getting those larger units to assemble requires a much longer cooling period to avoid kinetic traps. In this case, heat the sample to 80 °C and allow it to

cool to 60 °C at a rate of 5 °C/min. Once it has reached 60 °C, allow it to cool further to 25 °C at a rate of 300 minutes per degree. This will take several days.

When the program has finally finished, then you should have a PCR tube containing your folded DNA object. The next step is to purify and analyse it.

PURIFICATION AND ANALYSIS OF DNA PRODUCTS

The contents of the PCR tube are run through a 1−2% agarose gel using electrophoresis in a running buffer containing magnesium salts. The species that migrate the fastest through the gel, with the exception of remaining unused staple sequences, are generally the folded objects with the lowest defect rate. The desired bands are cut out of the gel and the agarose is separated out by centrifugation with a 'freeze 'n' squeeze' DNA purification kit. This kit consists of a special microcentrifuge tube with a sharp or 'dolphin-nosed' 2 mL tip into which the DNA product will pool. A chunk of agarose gel containing the DNA is broken up and centrifuged in one of these tubes through a cup that sits inside. The base of the cup is a 450-nm cellulose acetate filter through which the DNA product can pass to collect in the 'nose' of the microcentrifuge tube. Normally, the sample would be placed in a freezer for 5 min but this is omitted from the protocol when using DNA origami in case the freezing process damages the DNA shape. The extraction process described herein is also slightly modified from the protocol provided with the tubes. Cryogenic transmission electron microscopy and atomic force microscopy can then be used to assess the product morphology directly. Here is an outline for running the gel electrophoresis isolation of the DNA origami products. There are only some very minor variations from the recipe for running a standard gel electrophoresis of DNA PCR products outlined elsewhere.

To run agarose gel electrophoresis on DNA origami objects you will need:

This technique involves the use of ethidium bromide which is very toxic. Gels of this type should be run in specially sectioned-off areas with the correct disposal equipment.

1. 2.5 g of agarose in a 200 mL beaker.
2. 1 L of half strength Tris, borate and EDTA running buffer. This can be prepared by dilution of a TBE buffer prepared as outlined in the recipe for TAE but replacing acetic acid with an equimolar amount of boric acid.
3. A microwave for dissolving the agar. You will need some water for making the volume backup in the gel due to evaporation.
4. A 10 mg/mL stock solution of ethidium bromide. *Toxic. Wear gloves and protective equipment at every stage of handling.*
5. 1 mL of a 1.375 mol/L solution of $MgCl_2$ in water. This can be prepared by dissolving 2.795 g of $MgCl_2 \cdot 6H_2O$ in 8 mL of water and making up the volume to 10 mL.
6. Volumetric pipettes in the 1−10-µL, 10−50-µL and 1−5-mL ranges.
7. A gel electrophoresis setup including tank, tape and comb. This preparation assumes a gel size of 125 mL. Ensure the equipment has been specifically tasked for working with ethidium bromide.
8. A stock of six times the working strength of loading dye in a 30% glycerol solution.
9. A stock of a 1 kb premixed ladder for running with the samples.
10. A large ice bath into which the running gel tray can be placed to keep it cool while the electrophoresis is running.
11. A UV illumination lamp and box along with a camera or visualisation equipment. *Remember the UV lamp will be powerful so wear protective goggles and long sleeves and gloves when handling the gel under the light.*
12. A scalpel for cutting sections containing the DNA out of the gel with.
13. A glass rod and a microcentrifuge tube. This will be used for breaking up the agar gel.
14. A microcentrifuge.
15. A 'freeze 'n' squeeze' DNA gel extraction spin column. You will need one for each sample you will be looking at.
16. A disposal bucket for ethidium bromide containing gels and charcoal 'teabags' for destaining ethidium bromide solutions. *You should follow your established laboratory waste disposal procedures.*
17. The various solutions containing the DNA origami objects you wish to isolate from the last section.
18. A 100-nmol/L solution of the single-stranded phage DNA you used as a scaffold in the same concentration as the reaction solutions.

To run agarose gel electrophoresis on DNA origami objects:

1. Add the 125 mL of the half strength TBE buffer to the agar and stir. Place the solution into the microwave and heat at full power for a minute. Take the beaker out, stir it, and place back into the microwave for another 1−2 min or until all the agar has dissolved. It will probably be boiling so some water will be lost.
2. Add water to make the volume back up to 125 mL. Allow the solution to cool until the beaker feels lukewarm to the touch.
3. Add 1mL of the 1.375 mol/L magnesium chloride solution and stir. This will give a final working concentration of 0.011 mol/L Mg^{2+}.

4. Add 7 μL of the 10 mg/mL ethidium bromide solution. Stir the solution gently to mix but avoid trapping air bubbles in it.
5. Tape the sides of the gel mould and pour the agar solution into the tank to set. Be careful that there are no air bubbles in the gel. Place the comb in to form the sample channels.
6. When the gel has set, remove the tape from the sides of the mould and lower it into the running tank. Sit the assembly in the ice bath.
7. Add 4 mL of the magnesium chloride solution to 496 mL of the half strength TBE buffer. Pour this over the top of the gel to fill the tank. Use more if required. When the tank is full, remove the comb to form the sample loading channels. Take care not to tear the edges of the gel as you do so.
8. Transfer 12 μL of your DNA origami solution into a microcentrifuge tube containing 3 μL of loading dye solution. Repeat this for each sample you wish to run.
9. Pipette the DNA origami and dye mixture carefully into one of the channels. Keep the tip deep into the well and add slowly. Remember to leave the channels at either side of the gel empty for loading the DNA ladder and single-stranded phage DNA as a reference.
10. In a small microcentrifuge tube, pipette 1.2 μL of the single-stranded DNA control into 10.8 μL of water. To this, add 3 μL of the loading dye and then add into a channel.
11. Use a pipette to add 6 μL of the 1-kb ladder to one of the channels.
12. With the gel tank still sitting in the ice bath, place the lid on the tank and attach the black and red wires to the power supply. *The DNA will flow from black (negative) to red (positive)* so as to make sure you have the loaded wells near the black end and the rest of the gel pointing towards the red end. Set the voltage to 70 V. The samples should be suitably separated for isolation after 3–4 h.
13. When you feel the gel is ready, turn off the power supply. Transfer the whole gel very carefully to the UV viewing box with it supported on the mould and taking care that it does not slide off onto the floor.
14. Under UV viewing, identify the areas of interest where the DNA product shows up as a band. In general, the objects that have folded correctly will be the fastest to move through the agar gel, although this should not be confused with the remaining unused staples. At the far end of the gel from the well, you will see a diffuse band consisting of the unbound oligonucleotides. The DNA origami object will be closer to the well than the staples and should appear as a solid band under the UV light.
15. Use a scalpel to cut out the section of the gel containing the product. You will want to try and trim any gel not fluorescing under UV light as this will not contain any product. Put the DNA origami-containing agarose gel block into a microcentrifuge tube.
16. Use a small, clean glass rod to break up the gel chunk in the tube. Place the sample in a microcentrifuge and spin at full speed until the agarose sediments to the bottom after a few minutes.
17. Use the scalpel to cut the end of the tube containing the agarose debris. Place this upside down, so the contents can drain out through the bottom of the small filter cup, in the freeze 'n' squeeze spin column.
18. Centrifuge the column at full speed for 10 min. During this time, the DNA origami objects should pass through the filter and collect in the 'dolphin nose' end of the tube. The cup containing the agar debris can now be removed and you will have your DNA origami product ready for further imaging. The liquid should contain around 2–5 nmol of the target origami shape although this will change depending on the size of the item.

The final stage is to now look at what you have produced either by transmission electron microscopy or by atomic force microscopy. If it is the first time you have run a specific shape, then you will need to assess which concentration of magnesium salt gave the highest yield of correctly folded objects. In general, the most reliable concentration of Mg^{2+} for folding is around 20 mmol/L. DNA is not an electron dense material and so you will need to use a heavy metal stain in order to see it under the electron microscope. A protocol for negative staining with uranyl actetate is given in the methods section of this book. To prepare a grid for analysis, use a pipette to add 3 μL of the isolated solution containing the DNA object to the carbon film on top of a copper TEM grid. Allow it for 3–4 min to adsorb to the film and then use filter paper to wick away the excess solution. Extend this if you fail to see anything upon imaging. However, leaving a grid to dry out with the DNA solution on could result in remaining salts forming into crystals. Alternatively, the grid can be pretreated with a droplet of 0.5 mol/L solution $MgCl_2$ which can be wicked away before applying the DNA solution. This will result in different orientations on the carbon film. Apply a staining procedure so that you can see the material. In brief, this will involve placing a droplet of uranium salt solution onto the grid for a minute or so and then wicking it away.

Atomic force microscopy is desirable as the DNA object can be imaged under solution in varying environments. For basic imaging under liquid, use a mica substrate and a solution of the folding buffer as the liquid medium. Because the structures are delicate, you will need to use a tapping/intermittent contact mode so that the DNA is disrupted as little as possible. The interaction of the tip may cause the samples to lift off the surface, making it hard to image them. To compensate for this, add nickel acetate or chloride to the folding buffer at a concentration of 5–10 mmol/L. This will improve the binding of the DNA to the mica surface through the positively charged nickel cations. For smaller or flat objects, which will more readily deform in solution, you may find imaging the samples cast onto a dry surface works better. Allow a droplet of the solution containing the folded object to sit on a freshly cleaved piece of mica for ten minutes. Drain off the solution by tilting the mica on an angle. Add a droplet of water and then tip this off the surface also so that any residual salts are taken with it. Samples imaged in air will experience a loss of resolution along with some damaged or torn structures.

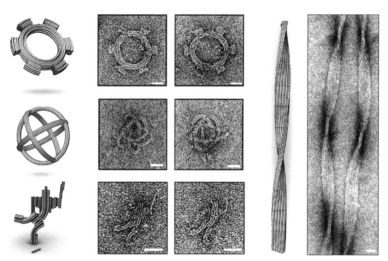

FIGURE 6.24   Examples of computer designed DNA origami shapes and how they appear under an electron microscope. *(Images courtesy of Hendrik Dietz)*

The DNA origami objects should be stable at physiological temperatures in a range of buffers and cell culture mediums but will start to come apart at around 55 °C. This is around the melting point of the DNA. DNAase enzymes will degrade the structures but at a greatly reduced rate compared to unfolded control DNA. This is likely due to the relatively greater density in the folded DNA structures. Other endonucleases will not have the same effect and, in some cases, you will see little change in the structure.

You may find that the product appears to be of the right shape but that it has warped or bent in some way. This is a result of strains within the DNA that put twisting forces into the structure upon folding. This type of distortion can be predicted from the caDNAno structure design template using another program called CANdo. This program models the mechanical interactions of the DNA based on your file to produce a 3D model showing where the likely twists and turns will appear in the final product. Designs can be uploaded as detailed in the references (Fig. 6.24).

AGAROSE GEL ELECTROPHORESIS FOR CHECKING THE DNA PRODUCT FROM A POLYMERASE CHAIN REACTION (PCR)

With thanks to Stuart Bellamy

Gel electrophoresis a staple of the biochemistry laboratory If you are getting into biochemistry and biology, then this is a commonplace technique that you will need to use at some point. This technique is used for separating out lengths of DNA in an agarose gel. It should not be confused with SDS-PAGE which uses a polyacrylamide gel and is used for separating proteins. This technique provides a method of separating out DNA of different molecular weights in an agarose gel. The DNA is placed into a well formed in the gel and then a charge is applied. A negatively charged molecule like DNA will be attracted to the positive end of the gel. Smaller fragments of DNA will move faster through the gel while heavier ones will trail slowly behind. This separates the various weights of DNA into bands which can be stained and imaged. A control solution containing DNA broken into known lengths is run alongside the test samples to allow for comparison. This is known as a 'ladder.' In this recipe, we will focus on analysing a DNA product from amplified DNA, that is to say DNA that has been replicated using the polymerase chain reaction (PCR). In this example, the DNA is from a plasmid that has been broken into fragments using a digestion enzyme. Separating the components out on a gel and comparing them to a known ladder will let us determine if the plasmid has been cut in the correct positions. The test DNA is added to the gel with a blue dye of a low molecular weight that will pass quickly through the gel under charge. This provides a visual indicator of the progress of the gel. When the blue dye reaches the opposite side of the well, then the process is complete. The agarose gel is permeated with a running buffer composed of Tris base, acetic acid and EDTA given the acronym TAE. This provides an electrolyte for the current and maintains the integrity of the DNA in the gel. A UV-visible fluorescent dye is used to stain the DNA in the agarose gel so that it can be imaged under a UV light source with a camera. The dye is ethidium bromide and this fluorescent chemical inserts itself between base pairs when used as a stain. This has a drawback in that it is extremely toxic. The protocols for handling this substance are provided and caution should always be used with this chemical.

To run an agarose gel electrophoresis, you will need:

This procedure will most likely be performed in a biolab so standard biosafety procedures should apply. This includes laboratory coats, safety glasses, gloves and alcohol rinses for surfaces.

Ethidium bromide is used and requires special handling protocols as outlined in the recipe. You must work with this with equipment and glassware dedicated to purpose in an area taped off for use and clearly marked.

1. Agarose for making the gel with. The amount depends on the size of the gel you are going to make and 1 g will be more than enough to produce a 100-mL volume gel. This is a polysaccharide extract also used for making gelatinous desserts. You will also need a conical flask (large enough for a few 100 mL) some saran wrap and a microwave (or water bath) for mixing the agarose with water.

2. The DNA you wish to analyse. In this example, it is a plasmid that has been amplified by PCR and cut into pieces of different lengths using restriction enzymes.

3. Stock TAE buffer (50× the strength you will use). You will need enough for casting the gel and for filling the gel electrophoresis tank. This recipe will use around 2 L.

4. A loading buffer. When you put the DNA into the gel and run it, you will not be able to see the DNA. By adding the DNA to be tested in with a dye and some glycerol, the DNA will sit in the gel properly and you will be able to see the dye moving through the gel. This will let you know how far the sample is moving across the gel. It is important to make sure you use a dye which will move faster through the gel than your DNA will.

> To make 1 L of 50× TAE buffer:
> 242 g (2 mol) of Trizma base
> 57.1 mL (1 mol) of glacial acetic acid
> 100 mL of 0.5 mol/L disodium EDTA
> Make up to 1 L and adjust it to pH 8

> Loading buffer dyes and the molecular weights of DNA with which they can be used:
> Xylene cyanol – 10,000–4000 base pairs
> Cresol red – 2000–1000 base pairs
> Bromophenol blue – 1000–400 base pairs
> Orange G – less than 100 base pairs
> Tartrazine – less than 20 base pairs

In this recipe, we are going to be running a gel on plasmid DNA which has around 1100 base pairs. Some of the DNA in the example has been cut by a digestion enzyme to make it smaller. This means using a sucrose and bromophenol blue loading buffer is a good choice of stain:

4 g of sucrose
25 mg of bromophenol blue
Make up to 10 mL using distilled water
If sterile and kept in a fridge, this can last for years.

5. A gel electrophoresis setup.
6. Autoclave tape or similar for 'taping the tank.'
7. A UV light and camera box for looking at the gels.
8. Ethidium bromide. This substance is *extremely toxic*. Ideally you will have the area where gel electrophoresis is to be taped off and kept separate from the laboratory. Have a large tray for use with solutions of ethidium bromide so that any spillages can be contained. It is bright red in colour and so should be fairly easy to spot drops on the benchtop. If available, have a different colour of nitrile glove to wear over the top of your normal laboratory gloves. This will remind you that you are working with something extra hazardous and are not to touch any nondedicated equipment.
   Have separate and dedicated tips and pipettes for the handling of ethidium bromide. The chemical should be stored, well marked, in the dark in a fridge. You will also need ethidium bromide destaining bags for removal of the waste from water.
9. A bucket specifically marked for the disposal of gels and waste (cleaning towels/gloves) from the experiments. Waste products from these experiments should be incinerated by specialist handlers.
10. A 'ladder' set of DNA. This is a mixture of known lengths of DNA to be run in the gel alongside the test DNA. It acts as a form of calibration standard in the gel. Every gel you run might be different so it is important to have a control in each one you run. We used a 1000-base-pair ladder purchased from New England Biolabs.

To run an agarose gel electrophoresis:

1. Much like making jelly, a gel has to be poured into a mould and allowed to set. The mould in this case is the gel tank. Using autoclave tape seal over the sides of the gel electrophoresis section where the gel sits. It does not have sides so that the formed gels can be slid out easily without risking damage. The mould does not need to be sterile. Make sure the tape is in place as liquid is going to be poured into it (Fig. 6.25).

2. Make the agarose solution. The density of the gel used determines how fast the DNA fragments will move through it. Large DNA fragments will move slowly so better resolution can be obtained by using a less dense gel which allows the different sized pieces to spread out further. For the plasmid DNA used in this recipe, we can use 0.8–0.5% agarose by weight in water.

FIGURE 6.25   Taping the sides of the gel tank.

Weigh out the agarose in a conical flask. The volume you need to make is determined by the size of the gel mould. You need enough to fill it. Add water to the agarose and swirl by hand. Place some saran wrap over the top of the conical flask and micro-wave for a minute. The aim is to dissolve the agarose and you do not want the water to boil off or the density of the gel will change. Keep swirling or heating as required until there are no air bubbles in the solution or no obvious lumps of agarose remaining. The solution should be completely homogeneous so that the passage of DNA through the gel will be continuous (Fig. 6.26). Let the solution cool slowly until it is warm to the touch. If you leave it to cool for too long, it will set and you will have to redissolve it.

3. Place the warm agarose solution in the conical flask into the tray for handling ethidium bromide. Using a pipette, add 5 μL of ethidium bromide to the agarose solution and swirl to mix (Fig. 6.27).

Add the solution to the taped tank and place the comb into the gel. The comb will be removed after the gel has set and forms wells into which the test DNA in the loading buffer can be added. The comb can be moved through the gel a few times before being placed to make sure there are no lumps in the solution. Lumps in the gel could skew the result. Any air bubbles present can be destroyed by poking them with a pipette tip. The gel can now be left to set which will take about an hour (Fig. 6.28).

4. While the gel is setting, the DNA can be prepared with the loading buffer. First make up the ladder solution in a small 1 mL vial:

    1 μL of 500 ng/μL solution of ladder DNA

    15 μL of water

    10 μL of the sucrose/bromophenol blue loading buffer.

This will make 30 μL of solution containing 500 ng of DNA. If the concentration of the DNA to be tested is unknown, then you can run a UV-vis spectrum to get the absorbance at 260 nm and use the formula outlined in the GFP ligation experiment to determine it.

FIGURE 6.26   The water is microwaved to dissolve the agarose and then the solution is swirled to mix.

FIGURE 6.27   A tray used for handling ethidium bromide in. The tank is lined with tissue so that any spills will show up clearly.

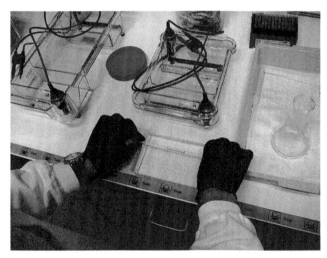

FIGURE 6.28   A comb is placed into the gel before it sets so that wells will form. DNA in loading buffer will be placed into these.

For test samples add an aliquot containing ~500 ng to a 1 mL vial and make up to 20 μL with water. Then add 10 μL of loading buffer. Do not vortex these solutions to mix them as this can tear the DNA fragments and skew your results.

5. Once the gel has set, remove the tape and place the gel into the gel electrophoresis tank still in the mould. Very gently, remove the comb so that there is a line of wells in the gel.

6. Add 50 μL of ethidium bromide to 1 L of the TAE buffer and mix. This is to be poured into the gel electrophoresis tank until the gel is submerged.

7. Starting with the ladder DNA on the outside of the gel, add the 30 μL aliquots of test DNA to the wells using a pipette. Add the samples gently and as deep into the well as possible.

8. Place the lid on top of the gel electrophoresis tank and connect the wires to the electrodes. These should be coloured red and black and have connection of the same colours to avoid confusion. In the setup pictured, the black connections are negative and the red connections are positive. The DNA should move away from the negative end and towards the positive end (Fig. 6.29a,b).

9. Turn on the gel electrophoresis set. The controls may vary but generally set the voltage to 75−100 V. The higher the voltage, the quicker the gel will run. In this recipe, the voltage was set to 75 V and the gel was allowed to run for 2 h. When the machine is turned on, there will be bubbling in the buffer reservoirs and some steaming may build up on the lid. Unless this is particularly vigorous, it is not something to worry about.

The progress of the gel can be monitored by watching the dye moving towards the electrode. It should be stopped before the dye appears to disappear off the end of the gel (Fig. 6.30a,b).

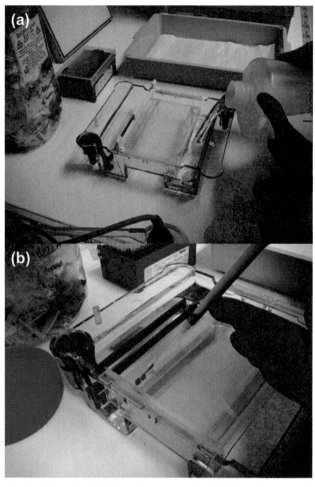

FIGURE 6.29   (a) Adding TAE running buffer into the tank so that the gel is submerged. (b) Adding the test samples of DNA in loading buffer to wells in the agarose gel.

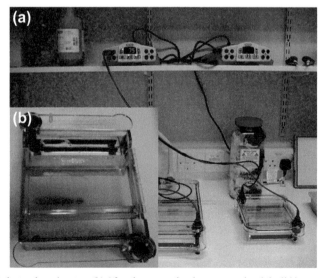

FIGURE 6.30   (a) A running agarose gel electrophoresis setup. (b) After the current has been stopped and the lid is removed, you should be able to see the blue dye about to move off the end of the gel.

10. Once the gel has been run, remove the gel and mould and slide the gel gently out of the side. Rest the gel on something flat (the lid of the gel electrophoresis bath for example) while the excess buffer is disposed of properly in a marked solvent waste bucket containing the ethidium bromide destaining bags for at least 24 h.

11. Place the gel into the UV lamp camera box. These tend to have very powerful UV lights which can damage your eyes so make sure it is closed off before you turn the lamps on. Once the lamps are on, the ethidium bromide can bleach out quickly so make sure your imaging system is aligned before you turn these on. Using a camera in the box, you should be able to see the blue dye remaining in the bottom of the wells and at the end of the gel and these can be used for visual alignment prior to UV illumination. Sometimes you may get background interference from the gel which will swamp out thinner bands you wish to see. If you find that the contrast in the gel is low, then you can destain the gel by soaking it in an excess of water for an hour.

12. Once you are happy that you have an image, you can get rid of the gel in a separate marked bucket for ethidium bromide waste. This will be disposed of by incineration.

13. You will now (hopefully) have a gel electrophoresis image. Using the DNA ladders at the side, you can tell how large the segments of DNA in your gel are. The largest fragments will be at the top while the smallest fragments will be at the bottom.

A common problem can be that the lines in the gel appear to have breaks in them, like a wave in the sea moving around a rock. This can be caused by the gel being thicker in places so good gel preparation is important. You may also see smearing like the line has a long tail behind it. This can be the result of damaged DNA where sections have been randomly cut from mechanical or chemical damage. This might suggest that nonspecific DNA cutting nucleases have made their way into the system somewhere or that the DNA has degraded. If this happens you may need to repeat the gel taking extra care to avoid contamination of the DNA. Even worse — the DNA preparation may have gone wrong and will have to be repeated (Fig. 6.31).

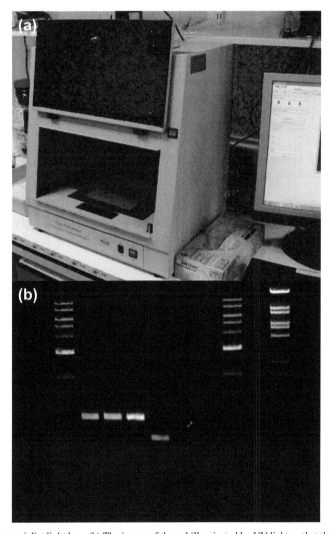

FIGURE 6.31 (a) The gel is imaged in a specialist light box. (b) The image of the gel illuminated by UV light so that the ethidium bromide dye intercalated in the DNA fluoresces.

A quick way to check the molecular weight of the DNA within the gel is comparing it with the ladder. Compare the band of interest with the nearest band in the ladder to obtain an approximation of the molecular weight of the DNA in it. The scale from the bottom of the gel to the top is not linear so making a direct measurement will be inaccurate. For more accurate estimates of the DNA length, a graph can be plotted comparing the molecular weight of the ladder bands against distance travelled in the gel. The bands of interest can then be plotted on this graph from their positions in the gel. It is important to remember that two separate gels cannot be compared directly as the running conditions like voltage or time might be variable and even the thickness of the gel will cause differences in the distance travelled by the DNA fragments. This problem can be solved by calibrating two different gels by using the relative distances the ladders have travelled. If you take the band that has travelled furthest in the first gel and define that as 1, then you can define the other bands in the same gel as a fraction of this distance. Doing the same to a second gel will allow you to directly compare two separate gels that may have been run under different conditions.

KEEPING BACTERIA LONG TERM IN A GLYCEROL STOCK

If you have a bacterial colony, it may be necessary to store them long term to be revived later on for more experiments. There are a few methods of achieving this, but for long-term storage with good reliability the glycerol stock method is recommended. In this recipe, we will mix a colony of bacteria grown in medium with glycerol and then freeze it. The glycerol acts as a preservative for the bacteria.

For storing the bacteria, you will need:

1. Liquid nitrogen and a dewar flask for using it. The setup will be a shallow bowl containing the liquid nitrogen into which you can easily place and retrieve vials for freezing. You will also require a device for fishing your vials out with. A small sieve in the end of a stick or a plastic spoon works well. *Risk of cryogenic burns. Observe the correct safety procedures for handling liquid nitrogen.*
2. A freezer maintained between −70 and −80 °C.
3. A propane benchtop burner for sterility control.
4. A colony of the bacteria you wish to store in a glass vial containing a few mL of lysogeny broth (LB) medium. It should be cloudy to the eye. You do not need to know the exact concentration of bacteria in the stock.
5. Fifty per cent glycerol in distilled water. This should be autoclaved before you begin so that it is sterile.
6. Two × 1000 μL (or 1 mL depending on how they are labelled in your laboratory) pipettes. One should be set to dispense 300 μL (or 0.3 mL) and the other set to dispense 700 μL (or 0.7 mL).
7. Plastic vials just a bit larger than 1 mL that can be sealed and a marker pen you can write on them with. These will be frozen so do not use anything that is likely to shatter during freezing. Have a rack you can stand them up in.
8. Bench ethanol and wipes for sterilising surfaces.

For storing the bacteria:

1. Before you begin you should be wearing a laboratory coat, safety glasses and gloves. Wipe down all the surfaces you will-be in contact with using the bench ethanol. The surfaces should be dry before you begin. Give your gloves a wipe with ethanol also.
2. Light the burner. This will be used to flame the edges of the vials storing your bacteria which should keep them sterile and prevent any outside contamination of the broth stock. Keep an eye on it and be careful. When not in use, have it set to an orange flame, then when you need to use it for the vial rims have it set to a blue flame.
3. Label the small vials in which the bacteria will be stored. It is standard practice to add your name and the date so you know to whom the sample belongs and how old it is when the time comes to thaw the bacteria.
4. Use one of the pipettes to add 300 μL of the 50 % glycerol solution to the small plastic vials lined up in a stand. How many you make is up to you but be aware of storage space.
5. Give the bacteria in the LB medium a light swirl so that any sediment is dispersed into solution. It should appear cloudy (see Fig. 6.32a).
6. Open the glass vial and briefly pass the rim of the vial through the burner (set to a blue flame; see Fig. 6.32b).
7. Use the other pipette to add 700 μL of the bacteria in the LB medium to the plastic vials containing the glycerol solution. In each plastic vial, you will have 1 mL of bacteria stock and glycerol ready to freeze.
8. Once you have finished filling the plastic vials, pass the rim of the glass vial containing the bacteria stock through the burner briefly once more and seal it. You may now turn off the burner and place the bacteria stock away.
9. Plunge the plastic vials containing the glycerol and bacteria into the shallow liquid nitrogen until they are frozen. Be careful that the liquid nitrogen does not splash (see Fig. 6.32c).
10. Fish your frozen vials out of the liquid nitrogen using a sieve or spoon and place them immediately into the freezer. They will now be kept for a year or two (see Fig. 6.32d).

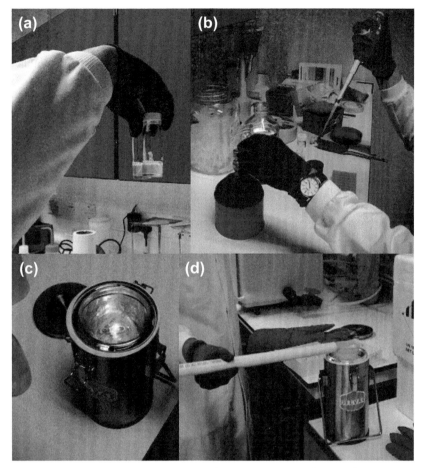

FIGURE 6.32 Photographs of the various steps in the procedure (a)—(d) as described in the text.

To revive the bacteria for use:

1. Turn on the burner and set it to a blue flame.
2. Pass a metal loop or wire briefly through the flame to heat it. It should not be red hot but just hot enough to melt ice on contact.
3. Open a frozen vial and dip the hot metal loop into the ice. This should melt a small amount, and although the heat may kill some of the bacteria there will be many that live and stay on the wire.
4. Use the wire to plate out the bacteria on agar. You should see a colony growing on the plate within 24—48 h.

TESTING THE MINIMUM INHIBITORY CONCENTRATION OF AN ANTIBIOTIC

With thanks to Matthew Avison and Noor Haida Mohd Kaus

There are many biological assays that can be used to determine the health and viability of a cell. The choice of which to use is highly dependent on the nature of the experiment as cells are sensitive to their environments and can be easily influenced. Similarly, some assays are unsuitable as they can make a cell suspension cloudy or the wrong colour for spectroscopic investigation. The importance of studying how nanomaterials interact with both animals and the environment has resulted in the development of assays specifically to look at the interaction of nanomaterials (Bhabra et al., 2009). Herein a starting point for 96-well plate assays is provided that does not rely on overly delicate cells or difficult chemical assays. This is a general procedure for testing the efficacy of an antibiotic drug or nanoparticle on any given bacterial organism. The examples provided in this experiment are from an investigation into potential antibiotic nanoparticles. Using this method, you will be able to determine how effective a drug is in comparison to a control drug for inhibiting the growth of a given microorganism. The lowest concentration at which the bacterial culture ceases to grow is known as the minimum inhibitory concentration (MIC) and is a good indicator of how well the tested molecule behaves as an antibiotic. This test is routinely used for assessing β-lactam-containing antibiotics mostly used in the treatment of Gram-positive bacterial infections. The β-lactam group acts to break down the peptidoglycan cell wall in Gram-positive bacteria which eventually kills them. Sadly, antibiotics lose their efficacy because the bacteria evolve to counter the drug. Although the mechanisms can differ, this is usually by producing a β-lactamase enzyme which breaks up the β-lactam and prevents is from working. β-Lactams have poor efficacy against Gram-negative bacteria like *E. coli*, because although they possess a thinner peptidoglycan layer, they also have an outer membrane which protects them from the drug penetrating into the cell. Food poisoning is mostly caused by Gram-negative bacteria such as campylobacter, *E. coli* and salmonella, and their resistance to antibiotics is one of the reasons why they can be difficult to treat.

Because microorganisms multiply so rapidly, they can quickly evolve to develop resistance. There are a very few 'last line of defence' antibiotics like vancomycin and amikacin (which is not a β-lactam and has some adverse side effects) kept at hospitals only to be used for the very worst cases. The practice of using antibiotics sparingly and making sure that a patient finishes the entire course is vital to prolonging these effective tools. However, from a bacterium point of view, the drugs are another environmental pressure that they will eventually adapt to overcome. Because of this, it is vital to keep searching for the chemical means to stave off infection. Most of the research in this field is oriented towards the chemical modification of existing successful drugs but increasingly sophisticated methods are being employed to negate the natural defences of bacteria. This is where nanotechnology can also help by providing effective delivery methods and even new modes of action that are designed to be non-negotiable with natural bacterial defence mechanisms. Rapid screening of candidate drugs speeds up the process and, in this recipe, a standard assessment test is outlined which will allow you to compare potential drugs with those in existing clinical use.

At a basic level we will expose a number of identical colonies to a range of drug concentrations and see how low the concentration can get before it loses the ability to kill the bacteria or stop it from growing further. Rows in a 96-well plate are each partially filled with aliquots from a single dilute colony of bacteria having a known optical density. It is important that all the bacteria comes from the same colony and is in the same concentration or else you will not be able to directly compare the results. A control drug, such as penicillin G, is used as a positive control and this will occupy one row. Another row without any antibiotics in is run to act as a negative control. All the other rows can have the test drugs in them. The last well of every row is filled with a concentrated solution of each test drug for that row or water in the case of the negative control. A multiheaded pipette is then used to place an aliquot from the concentrated solution into the colony next to it in the row. The well is mixed up to ensure that the medium and drug are homogenised and then a small aliquot of this is added to the next well in the row. This is repeated so that the drug becomes serially diluted along each column in the row. The 96-well plate is placed in an incubator for 16−24 h during which the bacteria will either flourish in the medium or will be killed or stopped from growing further by the presence of the antibiotic. The optical density of each well is measured by a plate well reader to establish which candidate drug has had an effect and what concentration was effective.

In this recipe, we will use the Gram-positive bacteria *Bacillus subtilis* as it is relatively safe to work with and requires biohazard level 1 procedures. Biohazard level 1 precautions should be used in the confines of a laboratory specialised in working with biological materials.

To run an MIC test, you will need:

All materials and solutions should be sterilised by autoclaving before use.

1. A stock solution of 3×10^{-4} mol/L penicillin G sodium salt for use as a positive control. This can be prepared by dissolving 10.6 mg in 100 mL of sterile water. You may need to sonicate the solution to get it to dissolve properly. You do not have to use penicillin G and anything that has a known efficacy against your bacterium can be used for comparison.
2. Stock solutions of the candidate antibiotics to be tested also made up to 3×10^{-4} mol/L.
3. A colony of *Bacillus subtilis* grown in lysogeny broth (LB) medium. This can be prepared by scraping a glycerol stock of the bacteria and inoculating a solution of LB medium and incubating it for a few hours. You are only going to use a very small amount of the stock, so making up about 100 mL will do.
4. At least 100 mL of a stock solution of sterile LB medium. You will not use all of it but the volume will make pipetting less laborious.
5. At least two sterile 96-well plates. You will want to run each row at least three times for statistical validity. You may need more plates if you plan to test more candidate drugs than there are rows for on the one-well plate. An extra spare plate is needed for determining the optical density of the bacterial colony.
6. A 96-well plate reader that will measure the optical density of samples at 650 nm.
7. An 8-channel volumetric pipette in the 50−100 μL range and sterile tips. It is good to use the tips that come provided in stands which you can push the pipette onto without having to handle them. You will also need a single-channel volumetric pipette in the same range.
8. A static incubator set to 37 °C.
9. A Petri dish.

To run an MIC test:

1. Add 200 μL each of the stock positive control and test drug solutions to the last well in the row to be tested. For example, you would add 200 μL of the stock penicillin G solution to well A12.

2. Pour some of the LB medium into the Petri dish and use this as a reservoir for the 8-channel pipette. Add 100 μL of broth to each well from columns 11 down to 1 in each row. *Do not* add medium to the wells containing stock solutions in column 12.

3. Change the pipette tips and take 100 μL from each of the stock candidate drug solutions in column 12 and dispense them into column 11 for each testing row. Flush the pipette up and down to mix the contents of the well.

4. Now take 100 μL of the contents of the wells in column 11 and transfer this into column 10. Repeat steps 3 and 4 moving down the row so that you end up with 100 μL of solution in each well. Every time you move down the row, you dilute the concentration of the drug by half. Do not transfer material from column 2 to column 1 and instead discard the 100-μL aliquot from column 2. The broth in the wells of column 1 will act as the negative controls for each testing row. For every testing row, there should be 100 μL in each well across the plate.

5. Before we can add bacteria to the test plate, we must dilute the concentration from the stock. Use the single-channel pipette and transfer 200 μL of the stock bacterial culture into a well on the spare plate.

6. Change the tip and transfer 200 μL of LB medium into a separate well.

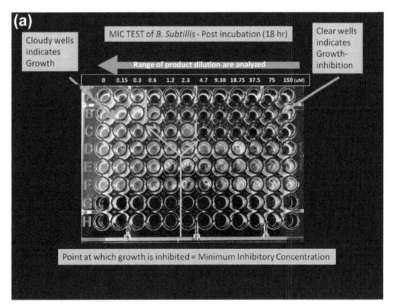

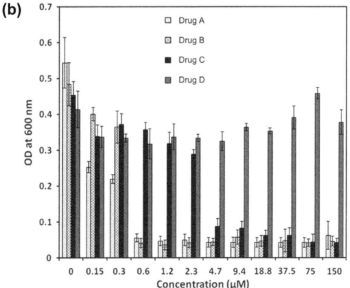

FIGURE 6.33   (a) A picture of a 96-well plate used for running an MIC test. (b) A plot of the optical density versus the drug concentration. Low OD values are observed where the test drug prevents bacterial growth.

7. Use the well plate reader to determine the optical density of the broth and the bacterial culture. Add microlitre amounts of the culture to tens of mL of LB medium until the optical density figure reads 0.002 units for a 200-μL aliquot. Once this is done, proceed immediately to the next step as the bacteria will begin to grow. You will want to finish pipetting the solution into the plate wells within a few minutes so that the bacteria does not get the chance to grow before it is added to the test.

8. Pour the diluted bacterial colony solution into a Petri dish and use the multichannel pipette to dispense 100-microlitre aliquots into the test wells from column 11 to column 1. Do not add bacterial culture to column 12. Take care not to dip the ends of the pipette into the wells as this may transfer drug or bacteria between the columns and result in spurious results. When you have finished, there should be 200 μL of solution in each well.

9. Take a reading with the plate well reader and then place into an incubator at 37 °C. Take further readings every 2 h for 24 h (Fig. 6.33).

In samples where the drug has inhibited growth, the well should remain optically transparent. The drug-free controls should all give similar high optical density readings but it does not matter too much if they are different. Outliers should be immediately obvious. Plot the optical density versus concentration after 24 h to get a picture of the minimum inhibitory concentration for each candidate drug. The figures for each row can be normalised against the drug-free controls in column 1 for comparison across the rows and between candidates. As can be seen from the plot below the MIC for penicillin, G was found to lie between 0.3 and 0.6 μmol/L. Another MIC test can be performed with a narrower concentration range to get a more accurate figure.

REFERENCES

Andersen, E.S., et al., 2009. Self-assembly of a nanoscale DNA box with a controllable lid. Nature 459 (7243), 73–76.

Bhabra, G., et al., 2009. Nanoparticles can cause DNA damage across a cellular barrier. Nature Nanotechnology 4 (12), 876–883.

Castro, C.E., et al., 2011. A primer to scaffolded DNA origami. Nature Methods 8 (3), 221–229.

Crick, F., 1970. Central dogma of molecular biology. Nature 227 (2), 561–563.

Dietz, H., Douglas, S.M., Shih, W.M., 2009. Folding DNA into twisted and curved nanoscale shapes. Science 325 (5941), 725–730 (New York, N.Y.).

Douglas, S.M., et al., 2009. Rapid prototyping of 3D DNA-origami shapes with caDNAno. Nucleic Acids Research 37 (15), 5001–5006.

Gietz, R.D., Woods, R.A., 2002. Transformation of yeast by lithium acetate/single-stranded carrier DNA/polyethylene glycol method. Methods in Enzymology 350 (2001), 87–96.

Lawyer, F.C., et al., 1993. High-level expression, purification, and enzymatic characterization of full-length Thermus aquaticus DNA polymerase and a truncated form deficient in 5′ to 3′ exonuclease activity. Genome Research 2 (4), 275–287.

Madigan, M., Martinko, J., Parker, J., 2000. In: Hall, P. (Ed.), Brock Biology of Microorganisms.

McCarthy, J., 2004. Tackling the challenges of interdisciplinary bioscience. Nature reviews. Molecular Cell Biology 5 (11), 933–937.

Mullis, K., Faloona, F., 1987. Specific synthesis of DNA in-vitro via a polymerase-catalyzed chain reaction. Methods Enzymol. 155, 335–350.

Rothemund, P.W.K., 2006. Folding DNA to create nanoscale shapes and patterns. Nature 440 (7082), 297–302.

Sambrook, J., Russell, D., 2001. Molecular Cloning: A Laboratory Manual. Cold Spring Harbour Laboratory Press.

Schiestl, R.H., Gietz, R.D., 1989. High efficiency transformation of intact yeast cells using single stranded nucleic acids as a carrier. Current Genetics 16 (5), 339–346.

Schreiber, R., et al., 2011. DNA origami-templated growth of arbitrarily shaped metal nanoparticles. Small 7 (13), 1795–1799.

Sherman, F., 1991. Getting started with yeast. Methods Enzymol. 41 (2002), 3–41.

van den Burg, B., 2003. Extremophiles as a source for novel enzymes. Current Opinion in Microbiology 6 (3), 213–218.

Page references followed by "f" indicate figure, "t" indicate table and "b" indicate box.

Printed and bound by CPI Group (UK) Ltd, Croydon, CR0 4YY

03/10/2024

01040728-0001